Ford Escape, Mazda Tribute & Mercury Mariner Automotive Repair Manual

by Mike Stubblefield, Jeff Killingsworth and John H Haynes

Member of the Guild of Motoring Writers

Models covered:

Ford Escape - 2001 through 2017

Mazda Tribute - 2001 through 2011

Mercury Mariner - 2005 through 2011

Does not include information specific to hybrid models

(36022-10AA17)

Haynes Group Limited
Haynes North America, Inc.

www.haynes.com

Acknowledgements

Certain wiring diagrams originated exclusively for Haynes North America, Inc. by Valley Forge Technical Information Services. Technical writers who contributed to this project include Demian Hurst, Scott "Gonzo" Weaver, Rob Maddox, Ralph Rendina and John Wegmann.

© **Haynes North America, Inc. 2003, 2008, 2011, 2013, 2018**

With permission from Haynes Group Limited

A book in the Haynes Automotive Repair Manual Series

ISBN 13: 978-1-62092-288-0
ISBN 10: 1-62092-288-6

Library of Congress Control Number: 2018946025

While every attempt is made to ensure that the information in this manual is correct, no liability can be accepted by the authors or publishers for loss, damage or injury caused by any errors in, or omissions from, the information given.

"Ford" and the Ford logo are registered trademarks of Ford Motor Company. Ford Motor Company is not a sponsor or affiliate of Haynes Group Limited or Haynes North America, Inc. and is not a contributor to the content of this manual.

18-464

Contents

Haynes mechanic and photographer with a 2015 Ford Escape

About this manual

Its purpose

The purpose of this manual is to help you get the best value from your vehicle. It can do so in several ways. It can help you decide what work must be done, even if you choose to have it done by a dealer service department or a repair shop; it provides information and procedures for routine maintenance and servicing; and it offers diagnostic and repair procedures to follow when trouble occurs.

We hope you use the manual to tackle the work yourself. For many simpler jobs, doing it yourself may be quicker than arranging an appointment to get the vehicle into a shop and making the trips to leave it and pick it up. More importantly, a lot of money can be saved by avoiding the expense the shop must pass on to you to cover its labor and overhead costs. An added benefit is the sense of satisfaction and accomplishment that you feel after doing the job yourself.

Using the manual

The manual is divided into Chapters. Each Chapter is divided into numbered Sections, which are headed in bold type between horizontal lines. Each Section consists of consecutively numbered paragraphs.

At the beginning of each numbered Section you will be referred to any illustrations which apply to the procedures in that Section. The reference numbers used in illustration captions pinpoint the pertinent Section and the Step within that Section. That is, illustration 3.2 means the illustration refers to Section 3 and Step (or paragraph) 2 within that Section.

Procedures, once described in the text, are not normally repeated. When it's necessary to refer to another Chapter, the reference will be given as Chapter and Section number. Cross references given without use of the word "Chapter" apply to Sections and/or paragraphs in the same Chapter. For example, "see Section 8" means in the same Chapter.

References to the left or right side of the vehicle assume you are sitting in the driver's seat, facing forward.

Even though we have prepared this manual with extreme care, neither the publisher nor the author can accept responsibility for any errors in, or omissions from, the information given.

NOTE

A **Note** provides information necessary to properly complete a procedure or information which will make the procedure easier to understand.

CAUTION

A **Caution** provides a special procedure or special steps which must be taken while completing the procedure where the Caution is found. Not heeding a Caution can result in damage to the assembly being worked on.

WARNING

A **Warning** provides a special procedure or special steps which must be taken while completing the procedure where the Warning is found. Not heeding a Warning can result in personal injury.

Introduction

1   This manual covers the Ford Escape, Mazda Tribute and Mercury Mariner. The available engines are:
1.5L EcoBoost four-cylinder (2017 models)
1.6L EcoBoost four-cylinder
    (2013 through 2016 models)
2.0L Zetec four-cylinder
    (2004 and earlier models)
2.0L EcoBoost four-cylinder
    (2013 and later models)
2.3L Duratec four-cylinder
    (2005 through 2008 models)
2.5L Duratec four-cylinder
    (2009 and later models)
3.0L Duratec V6 (2008 and earlier models)
3.0L Modular V6 (2009 through 2012 models)
2   The engine drives the front wheels through either a five-speed manual or a four- or six-speed automatic transaxle via independent driveaxles. On 4WD/AWD models the rear wheels are also propelled, via a driveshaft, rear differential, and two rear driveaxles.
3   Suspension is independent at all four wheels, with MacPherson struts being used at the front end. Trailing arms, control arms, coil springs and telescopic shock absorbers are used at the rear. The rack-and-pinion steering unit is mounted on the suspension crossmember.
4   The brakes are disc at the front and either drum or disc at the rear, with power assist standard. An Anti-lock Brake System (ABS) was available as an option on early models, while later models were equipped with ABS as standard equipment.

Vehicle identification numbers

Modifications are a continuing and unpublicized process in vehicle manufacturing. Since spare parts manuals and lists are compiled on a numerical basis, the individual vehicle numbers are essential to correctly identify the component required.

Vehicle Identification Number (VIN)

This very important identification number is stamped on a plate attached to the dashboard inside the windshield on the driver's side of the vehicle (see illustration). The VIN also appears on the Vehicle Certificate of Title and Registration. It contains information such as where and when the vehicle was manufactured, the model year and the body style.

Manufacturer's Certification Regulation label

The Manufacturer's Certification Regulation label is attached to the driver's side door end or post (see illustration). The label contains the name of the manufacturer, the month and year of production, the Gross Vehicle Weight Rating (GVWR), the Gross Axle Weight Rating (GAWR) and the certification statement.

VIN model year code

Counting from the left, the model year code letter designation is the 10th character.

On all models covered by this manual the model year codes are:

1	2001
2	2002
3	2003
4	2004
5	2005
6	2006
7	2007
8	2008
9	2009
A	2010
B	2011
C	2012
D	2013
E	2014
F	2015
G	2016
H	2017

On the models covered by this manual the engine codes are:

B   2.0L - Zetec four-cylinder (2001 through 2004)
1   3.0L - Duretec V6 (2001 through 2008)
G   3.0L - Modular V6 (2009 through 2012)
7   2.5L - Duratec four-cylinder (2009 and later)
9   2.0L - EcoBoost four-cylinder (2013 and later)
X   1.6L - EcoBoost four-cylinder (2013 through 2016)
D   1.5L - EcoBoost four-cylinder (2017)

The Vehicle Identification Number (VIN) is visible through the driver's side of the windshield

Engine number

On four-cylinder models, the engine identification number is stamped into a machined pad on the left-front side of the engine, near the transaxle.

On V6 models, the engine identification number is stamped into a machined pad on the left end (driver's side) of the engine block (see illustration).

Location of the Manufacturer's Certification Regulation label

Location of the engine identification number - V6 engine

Buying parts

Replacement parts are available from many sources, which generally fall into one of two categories - authorized dealer parts departments and independent retail auto parts stores. Our advice concerning these parts is as follows:

Retail auto parts stores: Good auto parts stores will stock frequently needed components which wear out relatively fast, such as clutch components, exhaust systems, brake parts, tune-up parts, etc. These stores often supply new or reconditioned parts on an exchange basis, which can save a considerable amount of money. Discount auto parts stores are often very good places to buy materials and parts needed for general vehicle maintenance such as oil, grease, filters, spark plugs, belts, touch-up paint, bulbs, etc. They also usually sell tools and general accessories, have convenient hours, charge lower prices and can often be found not far from home.

Authorized dealer parts department: This is the best source for parts which are unique to the vehicle and not generally available elsewhere (such as major engine parts, transmission parts, trim pieces, etc.).

Warranty information: If the vehicle is still covered under warranty, be sure that any replacement parts purchased - regardless of the source - do not invalidate the warranty!

To be sure of obtaining the correct parts, have engine and chassis numbers available and, if possible, take the old parts along for positive identification.

Maintenance techniques

Maintenance techniques

There are a number of techniques involved in maintenance and repair that will be referred to throughout this manual. Application of these techniques will enable the home mechanic to be more efficient, better organized and capable of performing the various tasks properly, which will ensure that the repair job is thorough and complete.

Fasteners

Fasteners are nuts, bolts, studs and screws used to hold two or more parts together. There are a few things to keep in mind when working with fasteners. Almost all of them use a locking device of some type, either a lockwasher, locknut, locking tab or thread adhesive. All threaded fasteners should be clean and straight, with undamaged threads and undamaged corners on the hex head where the wrench fits. Develop the habit of replacing all damaged nuts and bolts with new ones. Special locknuts with nylon or fiber inserts can only be used once. If they are removed, they lose their locking ability and must be replaced with new ones.

Rusted nuts and bolts should be treated with a penetrating fluid to ease removal and prevent breakage. Some mechanics use turpentine in a spout-type oil can, which works quite well. After applying the rust penetrant, let it work for a few minutes before trying to loosen the nut or bolt. Badly rusted fasteners may have to be chiseled or sawed off or removed with a special nut breaker, available at tool stores.

If a bolt or stud breaks off in an assembly, it can be drilled and removed with a special tool commonly available for this purpose. Most automotive machine shops can perform this task, as well as other repair procedures, such as the repair of threaded holes that have been stripped out.

Flat washers and lockwashers, when removed from an assembly, should always be replaced exactly as removed. Replace any damaged washers with new ones. Never use a lockwasher on any soft metal surface (such as aluminum), thin sheet metal or plastic.

Fastener sizes

For a number of reasons, automobile manufacturers are making wider and wider use of metric fasteners. Therefore, it is important to be able to tell the difference between standard (sometimes called U.S. or SAE) and metric hardware, since they cannot be interchanged.

All bolts, whether standard or metric, are sized according to diameter, thread pitch and length. For example, a standard 1/2 - 13 x 1 bolt is 1/2 inch in diameter, has 13 threads per inch and is 1 inch long. An M12 - 1.75 x 25 metric bolt is 12 mm in diameter, has a thread pitch of 1.75 mm (the distance between threads) and is 25 mm long. The two bolts are nearly identical, and easily confused, but they are not interchangeable.

In addition to the differences in diameter, thread pitch and length, metric and standard bolts can also be distinguished by examining the bolt heads. To begin with, the distance across the flats on a standard bolt head is measured in inches, while the same dimension on a metric bolt is sized in millimeters (the same is true for nuts). As a result, a standard wrench should not be used on a metric bolt and a metric wrench should not be used on a standard bolt. Also, most standard bolts have slashes radiating out from the center of the head to denote the grade or strength of the bolt, which is an indication of the amount of torque that can be applied to it. The greater the number of slashes, the greater the strength of the bolt. Grades 0 through 5 are commonly used on automobiles. Metric bolts have a property class (grade) number, rather than a slash, molded into their heads to indicate bolt strength. In this case, the higher the number, the stronger the bolt. Property class numbers 8.8, 9.8 and 10.9 are commonly used on automobiles.

Strength markings can also be used to distinguish standard hex nuts from metric hex nuts. Many standard nuts have dots stamped into one side, while metric nuts are marked with a number. The greater the number of dots, or the higher the number, the greater the strength of the nut.

Metric studs are also marked on their ends according to property class (grade). Larger studs are numbered (the same as metric bolts), while smaller studs carry a geometric code to denote grade.

It should be noted that many fasteners, especially Grades 0 through 2, have no distinguishing marks on them. When such is the case, the only way to determine whether it is standard or metric is to measure the thread pitch or compare it to a known fastener of the same size.

Standard fasteners are often referred to as SAE, as opposed to metric. However, it should be noted that SAE technically refers to a non-metric fine thread fastener only. Coarse thread non-metric fasteners are referred to as USS sizes.

Since fasteners of the same size (both standard and metric) may have different strength ratings, be sure to reinstall any bolts, studs or nuts removed from your vehicle in their original locations. Also, when replacing a fastener with a new one, make sure that the new one has a strength rating equal to or greater than the original.

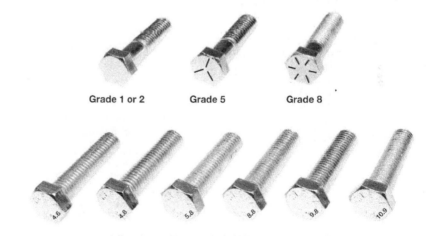

Grade 1 or 2   Grade 5   Grade 8

Bolt strength marking (standard/SAE/USS; bottom - metric)

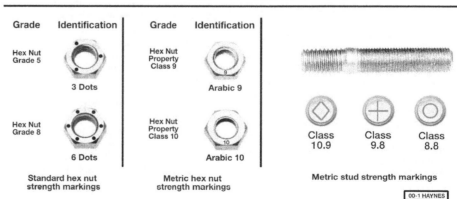

Grade	Identification	Grade	Identification
Hex Nut Grade 5	3 Dots	Hex Nut Property Class 9	Arabic 9
Hex Nut Grade 8	6 Dots	Hex Nut Property Class 10	Arabic 10
Standard hex nut strength markings		Metric hex nut strength markings	

Class 10.9   Class 9.8   Class 8.8

Metric stud strength markings

00-1 HAYNES

Tightening sequences and procedures

Most threaded fasteners should be tightened to a specific torque value (torque is the twisting force applied to a threaded component such as a nut or bolt). Overtightening the fastener can weaken it and cause it to break, while undertightening can cause it to eventually come loose. Bolts, screws and studs, depending on the material they are made of and their thread diameters, have specific torque values, many of which are noted in the Specifications at the beginning of each Chapter. Be sure to follow the torque recommendations closely. For fasteners not assigned a specific torque, a general torque value chart is presented here as a guide. These torque values are for dry (unlubricated) fasteners threaded into steel or cast iron (not aluminum). As was previously mentioned, the size and grade of a fastener determine the amount of torque that can safely be applied to it. The figures listed here are approximate for Grade 2 and Grade 3 fasteners. Higher grades can tolerate higher torque values.

Fasteners laid out in a pattern, such as cylinder head bolts, oil pan bolts, differential cover bolts, etc., must be loosened or tightened in sequence to avoid warping the component. This sequence will normally be shown in the appropriate Chapter. If a specific pattern is not given, the following procedures can be used to prevent warping.

Initially, the bolts or nuts should be assembled finger-tight only. Next, they should be tightened one full turn each, in a criss-cross or diagonal pattern. After each one has been tightened one full turn, return to the first one and tighten them all one-half turn, following the same pattern. Finally, tighten each of them one-quarter turn at a time until each fastener has been tightened to the proper torque. To loosen and remove the fasteners, the procedure would be reversed.

Component disassembly

Component disassembly should be done with care and purpose to help ensure that the parts go back together properly. Always keep track of the sequence in which parts are removed. Make note of special characteristics or marks on parts that can be installed more than one way, such as a grooved thrust washer on a shaft. It is a good idea to lay the disassembled parts out on a clean surface in the order that they were removed. It may also be helpful to make sketches or take instant photos of components before removal.

When removing fasteners from a component, keep track of their locations. Sometimes threading a bolt back in a part, or putting the washers and nut back on a stud, can prevent mix-ups later. If nuts and bolts cannot be returned to their original locations, they should be kept in a compartmented box or a series of small boxes. A cupcake or muffin tin is ideal for this purpose, since each cavity can hold the bolts and nuts from a particular area (i.e. oil pan bolts, valve cover bolts, engine mount bolts, etc.). A pan of this type is especially helpful when working on assemblies with very small parts, such as the

Metric thread sizes	Ft-lbs	Nm
M-6	6 to 9	9 to 12
M-8	14 to 21	19 to 28
M-10	28 to 40	38 to 54
M-12	50 to 71	68 to 96
M-14	80 to 140	109 to 154

Pipe thread sizes	Ft-lbs	Nm
1/8	5 to 8	7 to 10
1/4	12 to 18	17 to 24
3/8	22 to 33	30 to 44
1/2	25 to 35	34 to 47

U.S. thread sizes	Ft-lbs	Nm
1/4 - 20	6 to 9	9 to 12
5/16 - 18	12 to 18	17 to 24
5/16 - 24	14 to 20	19 to 27
3/8 - 16	22 to 32	30 to 43
3/8 - 24	27 to 38	37 to 51
7/16 - 14	40 to 55	55 to 74
7/16 - 20	40 to 60	55 to 81
1/2 - 13	55 to 80	75 to 108

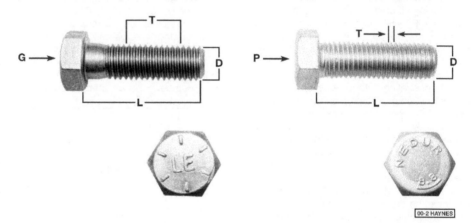

Standard (SAE and USS) bolt dimensions/ grade marks

G   Grade marks (bolt strength)
L   Length (in inches)
T   Thread pitch (number of threads per inch)
D   Nominal diameter (in inches)

Metric bolt dimensions/grade marks

P   Property class (bolt strength)
L   Length (in millimeters)
T   Thread pitch (distance between threads in millimeters)
D   Diameter

carburetor, alternator, valve train or interior dash and trim pieces. The cavities can be marked with paint or tape to identify the contents.

Whenever wiring looms, harnesses or connectors are separated, it is a good idea to identify the two halves with numbered pieces of masking tape so they can be easily reconnected.

Gasket sealing surfaces

Throughout any vehicle, gaskets are used to seal the mating surfaces between two parts and keep lubricants, fluids, vacuum or pressure contained in an assembly.

Many times these gaskets are coated with a liquid or paste-type gasket sealing compound before assembly. Age, heat and pressure can sometimes cause the two

parts to stick together so tightly that they are very difficult to separate. Often, the assembly can be loosened by striking it with a soft-face hammer near the mating surfaces. A regular hammer can be used if a block of wood is placed between the hammer and the part. Do not hammer on cast parts or parts that could be easily damaged. With any particularly stubborn part, always recheck to make sure that every fastener has been removed.

Avoid using a screwdriver or bar to pry apart an assembly, as they can easily mar the gasket sealing surfaces of the parts, which must remain smooth. If prying is absolutely necessary, use an old broom handle, but keep in mind that extra clean up will be necessary if the wood splinters.

After the parts are separated, the old gasket must be carefully removed and the gasket surfaces cleaned. If you're working on cast iron or aluminum parts, stubborn gasket material can be soaked with rust penetrant or treated with a special chemical to soften it so it can be easily scraped off.

Caution: *Never use gasket removal solutions or caustic chemicals on plastic or other composite components.* A scraper can be fashioned from a piece of copper tubing by flattening and sharpening one end. Copper is recommended because it is usually softer than the surfaces to be scraped, which reduces the chance of gouging the part. Some gaskets can be removed with a wire brush, but regardless of the method used, the mating surfaces must be left clean and smooth. If for some reason the gasket surface is gouged, then a gasket sealer thick enough to fill scratches will have to be used during reassembly of the components. For most applications, a non-drying (or semi-drying) gasket sealer should be used.

Hose removal tips

Warning: *If the vehicle is equipped with air conditioning, do not disconnect any of the A/C hoses without first having the system depressurized by a dealer service department or a service station.*

Hose removal precautions closely parallel gasket removal precautions. Avoid scratching or gouging the surface that the hose mates against or the connection may leak. This is especially true for radiator hoses. Because of various chemical reactions, the rubber in hoses can bond itself to the metal spigot that the hose fits over. To remove a hose, first loosen the hose clamps that secure it to the spigot. Then, with slip-joint pliers, grab the hose at the clamp and rotate it around the spigot. Work it back and forth until it is completely free, then pull it off. Silicone or other lubricants will ease removal if they can be applied between the hose and the outside of the spigot. Apply the same lubricant to the inside of the hose and the outside of the spigot to simplify installation.

As a last resort (and if the hose is to be replaced with a new one anyway), the rubber can be slit with a knife and the hose peeled from the spigot. If this must be done, be careful that the metal connection is not damaged.

If a hose clamp is broken or damaged, do not reuse it. Wire-type clamps usually weaken with age, so it is a good idea to replace them with screw-type clamps whenever a hose is removed.

Jacking and towing

Jacking

Warning: *The jack supplied with the vehicle should only be used for changing a tire or placing jackstands under the frame. Never work under the vehicle or start the engine while this jack is being used as the only means of support.*

1   The vehicle should be on level ground. Place the shift lever in Park, if you have an automatic, or Reverse if you have a manual transaxle. Block the wheel diagonally opposite the wheel being changed. Set the parking brake.

2   Remove the spare tire and jack from stowage. Remove the wheel cover and trim ring (if so equipped) with the tapered end of the lug nut wrench by inserting and twisting the handle and then prying against the back of the wheel cover. Loosen, but do not remove, the lug nuts (one-half turn is sufficient).

3   Place the scissors-type jack under the vehicle and adjust the jack height until it engages with the proper jacking point. There is a front and rear jacking point on each side of the vehicle (see illustrations).

4   Turn the jack handle clockwise until the tire clears the ground. Remove the lug nuts and pull the wheel off. Replace it with the spare.

5   Install the lug nuts with the beveled edges facing in. Tighten them snugly. Don't attempt to tighten them completely until the vehicle is lowered or it could slip off the jack. Turn the jack handle counterclockwise to lower the vehicle. Remove the jack and tighten the lug nuts in a diagonal pattern.

6   Install the cover (and trim ring, if used) and be sure it's snapped into place all the way around.

7   Stow the tire, jack and wrench. Unblock the wheels.

Front jacking location (place the jack head under the control arm rear mounting bolt) - 2012 and earlier models

Rear jacking location (place the jack head under the protrusion on the trailing arm) - 2012 and earlier models

Place the jack in the notched area of the rocker panel flange - 2013 and later models

Front and rear jacking points - 2013 and later models

Towing

8   Two-wheel drive models can be towed from the front with the front wheels off the ground, using a wheel lift type tow truck. If towed from the rear, the front wheels must be placed on a dolly. Four-wheel drive models must be towed with all four wheels off the ground. A sling-type tow truck cannot be used, as body damage will result. The best way to tow the vehicle is with a flat-bed car carrier.

9   In an emergency the vehicle can be towed a short distance with a cable or chain attached to one of the towing eyelets located under the front or rear bumpers. The driver must remain in the vehicle to operate the steering and brakes (remember that power steering and power brakes will not work with the engine off). Do not exceed 35 mph or 50 miles.

Booster battery (jump) starting

1   Observe these precautions when using a booster battery to start a vehicle:

   a) *Before connecting the booster battery, make sure the ignition switch is in the Off position.*
   b) *Turn off the lights, heater and other electrical loads.*
   c) *Your eyes should be shielded. Safety goggles are a good idea.*
   d) *Make sure the booster battery is the same voltage as the dead one in the vehicle.*
   e) *The two vehicles MUST NOT TOUCH each other!*
   f) *Make sure the transaxle is in Neutral (manual) or Park (automatic).*
   g) *If the booster battery is not a maintenance-free type, remove the vent caps and lay a cloth over the vent holes.*

2   Connect the red jumper cable to the positive (+) terminals of each battery (see illustration).

3   Connect one end of the black jumper cable to the negative (-) terminal of the booster battery. The other end of this cable should be connected to a good ground on the vehicle to be started, such as a bolt or bracket on the body.

4   Start the engine using the booster battery, then, with the engine running at idle speed, disconnect the jumper cables in the reverse order of connection.

Note: *On vehicles equipped with an automatic transaxle, if the battery has been run down or disconnected, the Powertrain Control Module (PCM) must relearn its idle and fuel mixture trim strategy for optimum drivability and performance (see Chapter 5, Section 1 for this procedure).*

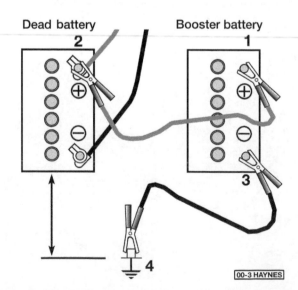

Make the booster battery cable connections in the numerical order shown (note that the negative cable of the booster battery is NOT attached to the negative terminal of the dead battery)

DECIMALS to MILLIMETERS

Decimal	mm	Decimal	mm
0.001	0.0254	0.500	12.7000
0.002	0.0508	0.510	12.9540
0.003	0.0762	0.520	13.2080
0.004	0.1016	0.530	13.4620
0.005	0.1270	0.540	13.7160
0.006	0.1524	0.550	13.9700
0.007	0.1778	0.560	14.2240
0.008	0.2032	0.570	14.4780
0.009	0.2286	0.580	14.7320
		0.590	14.9860
0.010	0.2540		
0.020	0.5080		
0.030	0.7620		
0.040	1.0160	0.600	15.2400
0.050	1.2700	0.610	15.4940
0.060	1.5240	0.620	15.7480
0.070	1.7780	0.630	16.0020
0.080	2.0320	0.640	16.2560
0.090	2.2860	0.650	16.5100
		0.660	16.7640
0.100	2.5400	0.670	17.0180
0.110	2.7940	0.680	17.2720
0.120	3.0480	0.690	17.5260
0.130	3.3020		
0.140	3.5560		
0.150	3.8100		
0.160	4.0640	0.700	17.7800
0.170	4.3180	0.710	18.0340
0.180	4.5720	0.720	18.2880
0.190	4.8260	0.730	18.5420
		0.740	18.7960
0.200	5.0800	0.750	19.0500
0.210	5.3340	0.760	19.3040
0.220	5.5880	0.770	19.5580
0.230	5.8420	0.780	19.8120
0.240	6.0960	0.790	20.0660
0.250	6.3500		
0.260	6.6040		
0.270	6.8580	0.800	20.3200
0.280	7.1120	0.810	20.5740
0.290	7.3660	0.820	21.8280
		0.830	21.0820
0.300	7.6200	0.840	21.3360
0.310	7.8740	0.850	21.5900
0.320	8.1280	0.860	21.8440
0.330	8.3820	0.870	22.0980
0.340	8.6360	0.880	22.3520
0.350	8.8900	0.890	22.6060
0.360	9.1440		
0.370	9.3980		
0.380	9.6520		
0.390	9.9060		
		0.900	22.8600
0.400	10.1600	0.910	23.1140
0.410	10.4140	0.920	23.3680
0.420	10.6680	0.930	23.6220
0.430	10.9220	0.940	23.8760
0.440	11.1760	0.950	24.1300
0.450	11.4300	0.960	24.3840
0.460	11.6840	0.970	24.6380
0.470	11.9380	0.980	24.8920
0.480	12.1920	0.990	25.1460
0.490	12.4460	1.000	25.4000

FRACTIONS to DECIMALS to MILLIMETERS

Fraction	Decimal	mm	Fraction	Decimal	mm
1/64	0.0156	0.3969	33/64	0.5156	13.0969
1/32	0.0312	0.7938	17/32	0.5312	13.4938
3/64	0.0469	1.1906	35/64	0.5469	13.8906
1/16	0.0625	1.5875	9/16	0.5625	14.2875
5/64	0.0781	1.9844	37/64	0.5781	14.6844
3/32	0.0938	2.3812	19/32	0.5938	15.0812
7/64	0.1094	2.7781	39/64	0.6094	15.4781
1/8	0.1250	3.1750	5/8	0.6250	15.8750
9/64	0.1406	3.5719	41/64	0.6406	16.2719
5/32	0.1562	3.9688	21/32	0.6562	16.6688
11/64	0.1719	4.3656	43/64	0.6719	17.0656
3/16	0.1875	4.7625	11/16	0.6875	17.4625
13/64	0.2031	5.1594	45/64	0.7031	17.8594
7/32	0.2188	5.5562	23/32	0.7188	18.2562
15/64	0.2344	5.9531	47/64	0.7344	18.6531
1/4	0.2500	6.3500	3/4	0.7500	19.0500
17/64	0.2656	6.7469	49/64	0.7656	19.4469
9/32	0.2812	7.1438	25/32	0.7812	19.8438
19/64	0.2969	7.5406	51/64	0.7969	20.2406
5/16	0.3125	7.9375	13/16	0.8125	20.6375
21/64	0.3281	8.3344	53/64	0.8281	21.0344
11/32	0.3438	8.7312	27/32	0.8438	21.4312
23/64	0.3594	9.1281	55/64	0.8594	21.8281
3/8	0.3750	9.5250	7/8	0.8750	22.2250
25/64	0.3906	9.9219	57/64	0.8906	22.6219
13/32	0.4062	10.3188	29/32	0.9062	23.0188
27/64	0.4219	10.7156	59/64	0.9219	23.4156
7/16	0.4375	11.1125	15/16	0.9375	23.8125
29/64	0.4531	11.5094	61/64	0.9531	24.2094
15/32	0.4688	11.9062	31/32	0.9688	24.6062
31/64	0.4844	12.3031	63/64	0.9844	25.0031
1/2	0.5000	12.7000	1	1.0000	25.4000

Conversion factors

Length (distance)

Inches (in)	X	25.4	= Millimeters (mm)	X	0.0394	= Inches (in)
Feet (ft)	X	0.305	= Meters (m)	X	3.281	= Feet (ft)
Miles	X	1.609	= Kilometers (km)	X	0.621	= Miles

Volume (capacity)

Cubic inches (cu in; in³)	X	16.387	= Cubic centimeters (cc; cm³)	X	0.061	= Cubic inches (cu in; in³)
Imperial pints (Imp pt)	X	0.568	= Liters (l)	X	1.76	= Imperial pints (Imp pt)
Imperial quarts (Imp qt)	X	1.137	= Liters (l)	X	0.88	= Imperial quarts (Imp qt)
Imperial quarts (Imp qt)	X	1.201	= US quarts (US qt)	X	0.833	= Imperial quarts (Imp qt)
US quarts (US qt)	X	0.946	= Liters (l)	X	1.057	= US quarts (US qt)
Imperial gallons (Imp gal)	X	4.546	= Liters (l)	X	0.22	= Imperial gallons (Imp gal)
Imperial gallons (Imp gal)	X	1.201	= US gallons (US gal)	X	0.833	= Imperial gallons (Imp gal)
US gallons (US gal)	X	3.785	= Liters (l)	X	0.264	= US gallons (US gal)

Mass (weight)

Ounces (oz)	X	28.35	= Grams (g)	X	0.035	= Ounces (oz)
Pounds (lb)	X	0.454	= Kilograms (kg)	X	2.205	= Pounds (lb)

Force

Ounces-force (ozf; oz)	X	0.278	= Newtons (N)	X	3.6	= Ounces-force (ozf; oz)
Pounds-force (lbf; lb)	X	4.448	= Newtons (N)	X	0.225	= Pounds-force (lbf; lb)
Newtons (N)	X	0.1	= Kilograms-force (kgf; kg)	X	9.81	= Newtons (N)

Pressure

Pounds-force per square inch (psi; lbf/in²; lb/in²)	X	0.070	= Kilograms-force per square centimeter (kgf/cm²; kg/cm²)	X	14.223	= Pounds-force per square inch (psi; lbf/in²; lb/in²)
Pounds-force per square inch (psi; lbf/in²; lb/in²)	X	0.068	= Atmospheres (atm)	X	14.696	= Pounds-force per square inch (psi; lbf/in²; lb/in²)
Pounds-force per square inch (psi; lbf/in²; lb/in²)	X	0.069	= Bars	X	14.5	= Pounds-force per square inch (psi; lbf/in²; lb/in²)
Pounds-force per square inch (psi; lbf/in²; lb/in²)	X	6.895	= Kilopascals (kPa)	X	0.145	= Pounds-force per square inch (psi; lbf/in²; lb/in²)
Kilopascals (kPa)	X	0.01	= Kilograms-force per square centimeter (kgf/cm²; kg/cm²)	X	98.1	= Kilopascals (kPa)

Torque (moment of force)

Pounds-force inches (lbf in; lb in)	X	1.152	= Kilograms-force centimeter (kgf cm; kg cm)	X	0.868	= Pounds-force inches (lbf in; lb in)
Pounds-force inches (lbf in; lb in)	X	0.113	= Newton meters (Nm)	X	8.85	= Pounds-force inches (lbf in; lb in)
Pounds-force inches (lbf in; lb in)	X	0.083	= Pounds-force feet (lbf ft; lb ft)	X	12	= Pounds-force inches (lbf in; lb in)
Pounds-force feet (lbf ft; lb ft)	X	0.138	= Kilograms-force meters (kgf m; kg m)	X	7.233	= Pounds-force feet (lbf ft; lb ft)
Pounds-force feet (lbf ft; lb ft)	X	1.356	= Newton meters (Nm)	X	0.738	= Pounds-force feet (lbf ft; lb ft)
Newton meters (Nm)	X	0.102	= Kilograms-force meters (kgf m; kg m)	X	9.804	= Newton meters (Nm)

Vacuum

Inches mercury (in. Hg)	X	3.377	= Kilopascals (kPa)	X	0.2961	= Inches mercury
Inches mercury (in. Hg)	X	25.4	= Millimeters mercury (mm Hg)	X	0.0394	= Inches mercury

Power

Horsepower (hp)	X	745.7	= Watts (W)	X	0.0013	= Horsepower (hp)

Velocity (speed)

Miles per hour (miles/hr; mph)	X	1.609	= Kilometers per hour (km/hr; kph)	X	0.621	= Miles per hour (miles/hr; mph)

Fuel consumption*

Miles per gallon, Imperial (mpg)	X	0.354	= Kilometers per liter (km/l)	X	2.825	= Miles per gallon, Imperial (mpg)
Miles per gallon, US (mpg)	X	0.425	= Kilometers per liter (km/l)	X	2.352	= Miles per gallon, US (mpg)

Temperature

Degrees Fahrenheit = (°C x 1.8) + 32

Degrees Celsius (Degrees Centigrade; °C) = (°F - 32) x 0.56

*It is common practice to convert from miles per gallon (mpg) to liters/100 kilometers (l/100km), where mpg (Imperial) x l/100 km = 282 and mpg (US) x l/100 km = 235

Safety first!

Regardless of how enthusiastic you may be about getting on with the job at hand, take the time to ensure that your safety is not jeopardized. A moment's lack of attention can result in an accident, as can failure to observe certain simple safety precautions. The possibility of an accident will always exist, and the following points should not be considered a comprehensive list of all dangers. Rather, they are intended to make you aware of the risks and to encourage a safety conscious approach to all work you carry out on your vehicle.

Essential DOs and DON'Ts

DON'T rely on a jack when working under the vehicle. Always use approved jackstands to support the weight of the vehicle and place them under the recommended lift or support points.

DON'T attempt to loosen extremely tight fasteners (i.e. wheel lug nuts) while the vehicle is on a jack - it may fall.

DON'T start the engine without first making sure that the transmission is in Neutral (or Park where applicable) and the parking brake is set.

DON'T remove the radiator cap from a hot cooling system - let it cool or cover it with a cloth and release the pressure gradually.

DON'T attempt to drain the engine oil until you are sure it has cooled to the point that it will not burn you.

DON'T touch any part of the engine or exhaust system until it has cooled sufficiently to avoid burns.

DON'T siphon toxic liquids such as gasoline, antifreeze and brake fluid by mouth, or allow them to remain on your skin.

DON'T inhale brake lining dust - it is potentially hazardous (see *Asbestos* below).

DON'T allow spilled oil or grease to remain on the floor - wipe it up before someone slips on it.

DON'T use loose fitting wrenches or other tools which may slip and cause injury.

DON'T push on wrenches when loosening or tightening nuts or bolts. Always try to pull the wrench toward you. If the situation calls for pushing the wrench away, push with an open hand to avoid scraped knuckles if the wrench should slip.

DON'T attempt to lift a heavy component alone - get someone to help you.

DON'T *rush or take unsafe shortcuts to finish a job.*

DON'T allow children or animals in or around the vehicle while you are working on it.

DO wear eye protection when using power tools such as a drill, sander, bench grinder, etc. and when working under a vehicle.

DO keep loose clothing and long hair well out of the way of moving parts.

DO make sure that any hoist used has a safe working load rating adequate for the job.

DO get someone to check on you periodically when working alone on a vehicle.

DO carry out work in a logical sequence and make sure that everything is correctly assembled and tightened.

DO keep chemicals and fluids tightly capped and out of the reach of children and pets.

DO remember that your vehicle's safety affects that of yourself and others. If in doubt on any point, get professional advice.

Steering, suspension and brakes

These systems are essential to driving safety, so make sure you have a qualified shop or individual check your work. Also, compressed suspension springs can cause injury if released suddenly - be sure to use a spring compressor.

Airbags

Airbags are explosive devices that can **CAUSE** injury if they deploy while you're working on the vehicle. Follow the manufacturer's instructions to disable the airbag whenever you're working in the vicinity of airbag components.

Asbestos

Certain friction, insulating, sealing, and other products - such as brake linings, brake bands, clutch linings, torque converters, gaskets, etc. - may contain asbestos or other hazardous friction material. Extreme care must be taken to avoid inhalation of dust from such products, since it is hazardous to health. If in doubt, assume that they do contain asbestos.

Fire

Remember at all times that gasoline is highly flammable. Never smoke or have any kind of open flame around when working on a vehicle. But the risk does not end there. A spark caused by an electrical short circuit, by two metal surfaces contacting each other, or even by static electricity built up in your body under certain conditions, can ignite gasoline vapors, which in a confined space are highly explosive. Do not, under any circumstances, use gasoline for cleaning parts. Use an approved safety solvent.

Always disconnect the battery ground (-) cable at the battery before working on any part of the fuel system or electrical system. Never risk spilling fuel on a hot engine or exhaust component. It is strongly recommended that a fire extinguisher suitable for use on fuel and electrical fires be kept handy in the garage or workshop at all times. Never try to extinguish a fuel or electrical fire with water.

Fumes

Certain fumes are highly toxic and can quickly cause unconsciousness and even death if inhaled to any extent. Gasoline vapor falls into this category, as do the vapors from some cleaning solvents. Any draining or pouring of such volatile fluids should be done in a well ventilated area.

When using cleaning fluids and solvents, read the instructions on the container carefully. Never use materials from unmarked containers.

Never run the engine in an enclosed space, such as a garage. Exhaust fumes contain carbon monoxide, which is extremely poisonous. If you need to run the engine, always do so in the open air, or at least have the rear of the vehicle outside the work area.

The battery

Never create a spark or allow a bare light bulb near a battery. They normally give off a certain amount of hydrogen gas, which is highly explosive.

Always disconnect the battery ground (-) cable at the battery before working on the fuel or electrical systems.

If possible, loosen the filler caps or cover when charging the battery from an external source (this does not apply to sealed or maintenance-free batteries). Do not charge at an excessive rate or the battery may burst.

Take care when adding water to a non maintenance-free battery and when carrying a battery. The electrolyte, even when diluted, is very corrosive and should not be allowed to contact clothing or skin.

Always wear eye protection when cleaning the battery to prevent the caustic deposits from entering your eyes.

Household current

When using an electric power tool, inspection light, etc., which operates on household current, always make sure that the tool is correctly connected to its plug and that, where necessary, it is properly grounded. Do not use such items in damp conditions and, again, do not create a spark or apply excessive heat in the vicinity of fuel or fuel vapor.

Secondary ignition system voltage

A severe electric shock can result from touching certain parts of the ignition system (such as the spark plug wires) when the engine is running or being cranked, particularly if components are damp or the insulation is defective. In the case of an electronic ignition system, the secondary system voltage is much higher and could prove fatal.

Hydrofluoric acid

This extremely corrosive acid is formed when certain types of synthetic rubber, found in some O-rings, oil seals, fuel hoses, etc. are exposed to temperatures above 750-degrees F (400-degrees C). The rubber changes into a charred or sticky substance containing the acid. *Once formed, the acid remains dangerous for years. If it gets onto the skin, it may be necessary to amputate the limb concerned.*

When dealing with a vehicle which has suffered a fire, or with components salvaged from such a vehicle, wear protective gloves and discard them after use.

Troubleshooting

Contents

This section provides an easy reference guide to the more common problems which may occur during the operation of your vehicle. These problems and their possible causes are grouped under headings denoting various components or systems, such as Engine, Cooling system, etc. They also refer you to the chapter and/or section which deals with the problem.

Remember that successful troubleshooting is not a mysterious black art practiced only by professional mechanics. It is simply the result of the right knowledge combined with an intelligent, systematic approach to the problem. Always work by a process of elimination, starting with the simplest solution and working through to the most complex - and never overlook the obvious. Anyone can run the gas tank dry or leave the lights on overnight, so don't assume that you are exempt from such oversights.

Finally, always establish a clear idea of why a problem has occurred and take steps to ensure that it doesn't happen again. If the electrical system fails because of a poor connection, check the other connections in the system to make sure that they don't fail as well. If a particular fuse continues to blow, find out why - don't just replace one fuse after another. Remember, failure of a small component can often be indicative of potential failure or incorrect functioning of a more important component or system.

Engine

1 Engine will not rotate when attempting to start

1 Battery terminal connections loose or corroded (Chapter 5).
2 Battery discharged or faulty (Chapter 1 and 5).
3 Automatic transaxle not completely engaged in Park (Chapter 7B) or clutch pedal not completely depressed (Chapter 7A).
4 Broken, loose or disconnected wiring in the starting circuit (Chapter 5 and 12).
5 Starter motor pinion jammed in flywheel ring gear (Chapter 5).
6 Starter solenoid faulty (Chapter 5).
7 Starter motor faulty (Chapter 5).
8 Ignition switch faulty (Chapter 12).
9 Starter pinion or flywheel teeth worn or broken (Chapter 5).

2 Engine rotates but will not start

1 Fuel tank empty.
2 Battery discharged (engine rotates slowly) (Chapter 5).
3 Battery terminal connections loose or corroded (Chapter 1).
4 Leaking fuel injector(s), faulty fuel pump, pressure regulator, etc. (Chapter 4).
5 Broken or stripped timing belt (Chapter 2A) or broken timing chain (Chapter 2A or 2B).
6 Ignition components damp or damaged (Chapter 5).
7 Worn, faulty or incorrectly gapped spark plugs (Chapter 1).
8 Broken, loose or disconnected wiring in the starting circuit (Chapter 5).
9 Broken, loose or disconnected wires at the ignition coil or faulty coil (Chapter 5).
10 Defective crankshaft or camshaft sensor (Chapter 6).

3 Engine hard to start when cold

1 Battery discharged or low (Chapter 1).
2 Malfunctioning fuel system (Chapter 4).
3 Faulty coolant temperature sensor or intake air temperature sensor (Chapter 6).
4 Faulty ignition system (Chapter 5).

4 Engine hard to start when hot

1 Air filter clogged (Chapter 1).
2 Fuel not reaching the fuel injection system (Chapter 4).
3 Loose or corroded battery connections, especially ground (Chapter 1).
4 Faulty coolant temperature sensor or intake air temperature sensor (Chapter 6).

5 Starter motor noisy or excessively rough in engagement

1 Pinion or flywheel gear teeth worn or broken (Chapter 5).
2 Starter motor mounting bolts loose or missing (Chapter 5).

6 Engine starts but stops immediately

1 Loose or faulty electrical connections at ignition coil or alternator (Chapter 5).
2 Insufficient fuel reaching the fuel injector(s) (Chapters 1 and 4).
3 Vacuum leak at the gasket between the intake manifold/plenum and throttle body (Chapter 4).

7 Oil puddle under engine

1 Oil pan gasket and/or oil pan drain bolt washer leaking (Chapter 2A, 2B).
2 Oil pressure sending unit leaking (Chapter 2C).
3 Valve cover leaking (Chapter 2A, 2B).
4 Engine oil seals leaking (Chapter 2A, 2B).
5 Oil pump housing leaking (Chapter 2A, 2B).

8 Engine lopes while idling or idles erratically

1 Vacuum leakage (Chapters 2A, 2B or 4).
2 Leaking EGR valve (Chapter 6).
3 Air filter clogged (Chapter 1).
4 Malfunction in the fuel injection or engine control system (Chapters 4 and 6).

5 Leaking head gasket (Chapter 2A or 2B).
6 Timing belt or chain and/or sprockets worn (Chapter 2A or 2B).
7 Camshaft lobes worn (Chapter 2A or 2B).

9 Engine misses at idle speed

1 Spark plugs worn or not gapped properly (Chapter 1).
2 Faulty spark plug wires (Chapter 1).
3 Vacuum leaks (Chapter 1).
4 Incorrect ignition timing (Chapters 1 and 5).
5 Uneven or low compression (Chapter 2A or 2B).
6 Problem with the fuel injection system (Chapter 4).

10 Engine misses throughout driving speed range

1 Fuel filter clogged and/or impurities in the fuel system (Chapters 1 and 4).
2 Low fuel pressure (Chapter 4).
3 Faulty or incorrectly gapped spark plugs (Chapter 1).
4 Leaking spark plug wires (Chapter 1 or 5).
5 Faulty emission system components (Chapter 6).
6 Low or uneven cylinder compression pressures (Chapter 2A or 2B).
7 Weak or faulty ignition system (Chapter 5).
8 Vacuum leak in fuel injection system, intake manifold, air control valve or vacuum hoses (Chapters 4 and 6).

11 Engine stumbles on acceleration

1 Spark plugs fouled (Chapter 1).
2 Problem with fuel injection or engine control system (Chapters 4 and 6).
3 Fuel filter clogged (Chapters 1 and 4).
4 Intake manifold air leak (Chapters 2A, 2B and 4).
5 Problem with the emissions control system (Chapter 6).

12 Engine surges while holding accelerator steady

1   Intake air leak (Chapter 4).
2   Fuel pump or fuel pressure regulator faulty (Chapter 4).
3   Problem with the fuel injection system (Chapter 4).
4   Problem with the emissions control system (Chapter 6).

13 Engine stalls

1   Idle speed incorrect (Chapter 1).
2   Fuel filter clogged and/or water and impurities in the fuel system (Chapters 1 and 4).
3   Distributor components damp or damaged (Chapter 5).
4   Faulty emissions system components (Chapter 6).
5   Faulty or incorrectly gapped spark plugs (Chapter 1).
6   Faulty spark plug wires (Chapter 1).
7   Vacuum leak in the fuel injection system, intake manifold or vacuum hoses (Chapter 2A, 2B or 4).

14 Engine lacks power

1   Obstructed exhaust system (Chapter 4).
2   Defective spark plug wires or faulty coil (Chapters 1 and 5).
3   Faulty or incorrectly gapped spark plugs (Chapter 1).
4   Problem with the fuel injection system (Chapter 4).
5   Plugged air filter (Chapter 1).
6   Brakes binding (Chapter 9).
7   Automatic transaxle fluid level incorrect (Chapter 1).
8   Clutch slipping (Chapter 8).
9   Fuel filter clogged and/or impurities in the fuel system (Chapters 1 and 4).
10   Emission control system not functioning properly (Chapter 6).
11   Low or uneven cylinder compression pressures (Chapter 2C).

15 Engine backfires

1   Emission control system not functioning properly (Chapter 6).
2   Problem with the fuel injection system (Chapter 4).
3   Vacuum leak at fuel injector(s), intake manifold, air control valve or vacuum hoses (Chapters 2A, 2B and 4).
4   Valve clearances incorrectly set and/or valves sticking (Chapter 2A or 2B).

16 Pinging or knocking engine sounds during acceleration or uphill

1   Incorrect grade of fuel.
2   Fuel injection system faulty (Chapter 4).

3   Improper or damaged spark plugs or wires (Chapter 1).
4   Knock sensor defective (Chapter 6).
5   EGR valve not functioning (Chapter 6).
6   Vacuum leak (Chapters 2A, 2B and 4).

17 Engine runs with oil pressure light on

1   Low oil level (Chapter 1).
2   Idle rpm below specification (Chapter 1).
3   Short in wiring circuit (Chapter 12).
4   Faulty oil pressure sender (Chapter 2C).
5   Worn engine bearings and/or oil pump (Chapter 2A or 2B).

Engine electrical systems

18 Battery will not hold a charge

1   Alternator drivebelt defective or not adjusted properly (Chapter 1).
2   Battery electrolyte level low (Chapter 1).
3   Battery terminals loose or corroded (Chapter 1).
4   Alternator not charging properly (Chapter 5).
5   Loose, broken or faulty wiring in the charging circuit (Chapter 5).
6   Short in vehicle wiring (Chapter 12).
7   Internally defective battery (Chapters 1 and 5).

19 Alternator light fails to go out

1   Faulty alternator or charging circuit (Chapter 5).
2   Alternator drivebelt defective or out of adjustment (Chapter 1).
3   Alternator voltage regulator inoperative (Chapter 5).

20 Alternator light fails to come on when key is turned on

1   Warning light bulb defective (Chapter 12).
2   Fault in the printed circuit, dash wiring or bulb holder (Chapter 12).

Fuel system

21 Excessive fuel consumption

1   Dirty or clogged air filter element (Chapter 1).
2   Emissions system not functioning properly (Chapter 6).
3   Fuel injection system not functioning properly (Chapter 4).

4   Low tire pressure or incorrect tire size (Chapter 1).

22 Fuel leakage and/or fuel odor

1   Leaking fuel feed or return line (Chapters 1 and 4).
2   Tank overfilled.
3   Evaporative canister filter clogged (Chapters 1 and 6).
4   Problem with the fuel injection system (Chapter 4).

Cooling system

23 Overheating

1   Insufficient coolant in system (Chapter 1).
2   Water pump drivebelt defective or out of adjustment (Chapter 1).
3   Radiator core blocked or grille restricted (Chapter 3).
4   Thermostat faulty (Chapter 3).
5   Electric coolant fan inoperative or blades broken (Chapter 3).
6   Expansion tank cap not maintaining proper pressure (Chapter 3).

24 Overcooling

1   Faulty thermostat (Chapter 3).
2   Inaccurate temperature gauge sending unit (Chapter 3).

25 External coolant leakage

1   Deteriorated/damaged hoses; loose clamps (Chapters 1 and 3).
2   Water pump defective (Chapter 3).
3   Leakage from radiator core or coolant reservoir (Chapter 3).
4   Engine drain or water jacket core plugs leaking (Chapter 2C).

26 Internal coolant leakage

1   Leaking cylinder head gasket (Chapter 2A or 2B).
2   Cracked cylinder bore or cylinder head (Chapter 2A, 2B or 2C).

27 Coolant loss

1   Too much coolant in reservoir (Chapter 1).
2   Coolant boiling away because of overheating (Chapter 3).
3   Internal or external leakage (Chapter 3).
4   Faulty radiator or expansion tank cap (Chapter 3).

28 Poor coolant circulation

1 Inoperative water pump (Chapter 3).
2 Restriction in cooling system (Chapters 1 and 3).
3 Water pump drivebelt defective/out of adjustment (Chapter 1).
4 Thermostat sticking (Chapter 3).

Clutch

29 Pedal travels to floor - no pressure or very little resistance

1 Master or release cylinder faulty (Chapter 8).
2 Hose/pipe burst or leaking.
3 Connections leaking.
4 No fluid in reservoir (Chapter 1).
5 If fluid level in reservoir rises as pedal is depressed, master cylinder center valve seal is faulty (Chapter 8).
6 If there is fluid on dust seal at master cylinder, piston primary seal is leaking (Chapter 8).
7 Broken release bearing or fork (Chapter 8).
8 Faulty pressure plate diaphragm spring (Chapter 8).

30 Fluid in area of master cylinder dust cover and on pedal

Rear seal failure in master cylinder (Chapter 8).

31 Fluid on release cylinder

Release cylinder seal faulty (Chapter 8).

32 Pedal feels spongy when depressed

Air in system (Chapter 8).

33 Unable to select gears

1 Faulty transaxle (Chapter 7A).
2 Faulty clutch disc or pressure plate (Chapter 8).
3 Faulty release lever or release bearing (Chapter 8).
4 Faulty shift lever assembly or linkage/control cables (Chapter 7A).

34 Clutch slips (engine speed increases with no increase in vehicle speed)

1 Clutch plate worn (Chapter 8).

2 Clutch plate is oil soaked by leaking rear main seal (Chapters 2A and 2B).
3 Clutch plate not seated.
4 Warped pressure plate or flywheel (Chapters 2A, 2B and 8).
5 Weak diaphragm springs (Chapter 8).
6 Clutch plate overheated. Allow to cool.

35 Grabbing (chattering) as clutch is engaged

1 Oil on clutch plate lining, burned or glazed facings (Chapter 8).
2 Worn or loose engine or transaxle mounts (Chapters 2A, 2B and 7A).
3 Worn splines on clutch plate hub (Chapter 8).
4 Warped pressure plate or flywheel (Chapters 2A, 2B or 8).
5 Burned or smeared resin on flywheel or pressure plate (Chapter 8).

36 Transaxle rattling (clicking)

1 Release lever loose (Chapter 7A).
2 Clutch plate damper spring failure (Chapter 8).

37 Noise in clutch area

1 Fork shaft improperly installed.
2 Faulty release bearing (Chapter 8).

38 Clutch pedal stays on floor

1 Clutch master cylinder piston binding in bore (Chapter 8).
2 Broken release bearing or fork (Chapter 8).

39 High pedal effort

1 Master or release cylinder piston binding in bore (Chapter 8).
2 Pressure plate faulty (Chapter 8).

Manual transaxle

40 Knocking noise at low speeds

1 Worn driveaxle constant velocity (CV) joints (Chapter 8).
2 Worn side gear shaft counterbore in differential case.*

41 Noise most pronounced when turning

Differential gear noise.*

42 Clunk on acceleration or deceleration

1 Loose engine or transaxle mounts (Chapter 7A).
2 Worn differential pinion shaft in case.*
3 Worn side gear shaft counterbore in differential case.*
4 Worn or damaged driveaxle inboard CV joints (Chapter 8).

43 Clicking noise in turns

Worn or damaged driveaxle outboard CV joint (Chapter 8).

44 Vibration

1 Rough wheel bearing (Chapter 10).
2 Damaged driveaxle (Chapter 8).
3 Out-of-round tires.
4 Tire out of balance.
5 Worn driveaxle CV joint (Chapter 8).

45 Noisy in neutral with engine running

1 Damaged input gear bearing.*
2 Damaged clutch release bearing (Chapter 8).

46 Noisy in one particular gear

1 Damaged or worn constant mesh gears.*
2 Damaged or worn synchronizers.*
3 Bent reverse fork.*
4 Damaged fourth speed gear or output gear.*
5 Worn or damaged reverse idler gear or idler bushing.*

47 Noisy in all gears

1 Insufficient lubricant (Chapter 1).
2 Damaged or worn bearings.*
3 Worn or damaged input gear shaft and/or output gear shaft.*

Note: * *Although the corrective action necessary to remedy the symptoms described is beyond the scope of this manual, the above information should be helpful in isolating the cause of the condition so that the owner can communicate clearly with a professional mechanic.*

48 Slips out of gear

1 Worn or improperly adjusted linkage (Chapter 7A).
2 Transaxle loose on engine (Chapter 7A).
3 Shift linkage does not work freely, binds (Chapter 7A).
4 Input gear bearing retainer broken or loose.*
5 Worn shift fork.*

49 Leaks lubricant

1 Driveaxle (side gear shaft) seals worn (Chapter 7A).
2 Excessive amount of lubricant in transaxle (Chapter 1).
3 Loose or broken input gear shaft bearing retainer.*
4 Input gear bearing retainer O-ring and/or lip seal damaged.*

50 Locked in gear

1 Lock pin or interlock pin missing.*
Note: *Although the corrective action necessary to remedy the symptoms described is beyond the scope of this manual, the above information should be helpful in isolating the cause of the condition so that the owner can communicate clearly with a professional mechanic.*

Automatic transaxle

51 Fluid leakage

1 Automatic transaxle fluid is a deep red color. Fluid leaks should not be confused with engine oil, which can easily be blown onto the transaxle by air flow.
2 To pinpoint a leak, first remove all built-up dirt and grime from the transaxle housing with degreasing agents and/or steam cleaning. Then drive the vehicle at low speeds so air flow will not blow the leak far from its source. Raise the vehicle and determine where the leak is coming from. Common areas of leakage are:

Dipstick tube.
Transaxle oil lines.
Speed sensor (Chapter 6).
Driveaxle oil seals (Chapter 7A).

52 Transaxle fluid brown or has a burned smell

Transaxle fluid overheated (Chapter 1).

53 General shift mechanism problems

1 Chapter 7B deals with checking and adjusting the shift linkage on automatic

transaxles. Common problems which may be attributed to poorly adjusted linkage are:
a) *Engine starting in gears other than Park or Neutral.*
b) *Indicator on shifter pointing to a gear other than the one actually being used.*
c) *Vehicle moves when in Park.*
2 Refer to Chapter 7B for the shift linkage adjustment procedure.

54 Transaxle slips, shifts roughly, is noisy or has no drive in forward or reverse gears

There are many probable causes for the above problems, but the home mechanic should be concerned with only one possibility - fluid level. Before taking the vehicle to a repair shop, check the level and condition of the fluid as described in Chapter 1 . Correct the fluid level as necessary or change the fluid and filter if needed. If the problem persists, have a professional diagnose the cause.

Driveaxles

55 Clicking noise in turns

Worn or damaged outboard CV joint (Chapter 8).

56 Shudder or vibration during acceleration

1 Excessive toe-in. Have the wheels aligned.
2 Worn or damaged inboard or outboard CV joints (Chapter 8).

57 Vibration at highway speeds

1 Out-of-balance front wheels and/or tires.
2 Out-of-round front tires.
3 Worn driveaxle CV joint(s) (Chapter 8).

Brakes

58 Vehicle pulls to one side during braking

1 Incorrect tire pressures (Chapter 1).
2 Front end out of alignment (have the front end aligned).
3 Front, or rear, tire sizes not matched to one another.
4 Restricted brake lines or hoses (Chapter 9).

5 Malfunctioning drum brake or caliper assembly (Chapter 9).
6 Loose suspension parts (Chapter 10).
7 Excessive wear of brake shoe or pad material or disc/drum on one side (Chapter 9).
8 Contamination (grease or brake fluid) of brake shoe or pad material or disc/drum on one side (Chapter 9).

59 Noise (high-pitched squeal when the brakes are applied)

Brake pads or shoes worn out. Replace pads or shoes with new ones immediately (Chapter 9).

60 Brake roughness or chatter (pedal pulsates)

1 Excessive disc lateral runout (Chapter 9).
2 Uneven pad wear (Chapter 9).
3 Worn disc (Chapter 9).

61 Excessive brake pedal effort required to stop vehicle

1 Malfunctioning power brake booster (Chapter 9).
2 Partial system failure (Chapter 9).
3 Excessively worn pads or shoes (Chapter 9).
4 Piston in caliper or wheel cylinder stuck (Chapter 9).
5 Brake pads or shoes contaminated with oil or grease (Chapter 9).
6 Brake disc grooved and/or glazed (Chapter 9).
7 New pads or shoes installed and not yet seated. It will take a while for the new material to seat against the disc or drum.

62 Excessive brake pedal travel

1 Partial brake system failure (Chapter 9).
2 Insufficient fluid in master cylinder (Chapter 1).
3 Air trapped in system (Chapter 9).

63 Dragging brakes

1 Master cylinder pistons not returning correctly (Chapter 9).
2 Restricted brakes lines or hoses (Chapter 9).
3 Incorrect parking brake adjustment (Chapter 9).

64 Grabbing or uneven braking action

1 Binding brake pedal mechanism (Chapter 9).
2 Contaminated brake linings (Chapter 9).

65 Brake pedal feels spongy when depressed

1 Air in hydraulic lines (Chapter 9).
2 Master cylinder mounting bolts loose (Chapter 9).
3 Master cylinder defective (Chapter 9)

66 Brake pedal travels to the floor with little resistance

1 Little or no fluid in the master cylinder reservoir caused by leaking caliper or wheel cylinder piston(s) (Chapter 9).
2 Loose or damaged brake lines (Chapter 9).

67 Parking brake does not hold

Parking brake improperly adjusted (Chapter 9).

Suspension and steering systems

68 Vehicle pulls to one side

1 Mismatched or uneven tires.
2 Broken or sagging springs (Chapter 9).
3 Wheel alignment incorrect. Have the wheels professionally aligned.
4 Front brake dragging (Chapter 9).

69 Abnormal or excessive tire wear

1 Wheel alignment out-of-specification. Have the wheels aligned.
2 Sagging or broken springs (Chapter 10).
3 Tire out-of-balance.
4 Worn strut damper (Chapter 10).
5 Overloaded vehicle.
6 Tires not rotated regularly.

70 Wheel makes a thumping noise

1 Blister or bump on tire (Chapter 1).
2 Improper strut damper action (Chapter 10).

71 Shimmy, shake or vibration

1 Tire or wheel out-of-balance or out-of-round.
2 Worn wheel bearings (Chapter 10).
3 Worn tie-rod ends (Chapter 10).
4 Worn balljoints (Chapter 10).
5 Excessive wheel runout.
6 Blister or bump on tire (Chapter 10).

72 Hard steering

1 Worn balljoints and/or tie-rod ends (Chapter 10).
2 Wheel alignment out-of-specifications. Have the wheels professionally aligned.
3 Low tire pressure(s) (Chapter 1).
4 Worn steering gear (Chapter 10).

73 Poor returnability of steering to center

1 Worn balljoints and/or tie-rod ends (Chapter 10).
2 Binding in steering column (Chapter 10).
3 Lack of power steering fluid (Chapter 10).
4 Wheel alignment out-of-specifications. Have the wheels professionally aligned.

74 Abnormal noise at the front end

1 Worn balljoints and/or tie-rod ends (Chapter 10).
2 Damaged strut mounting (Chapter 10).
3 Worn control arm bushings or tie-rod ends (Chapter 10).
4 Loose stabilizer bar or worn bushings (Chapter 10).
5 Loose wheel lug nuts (Chapter 10).
6 Loose suspension bolts (Chapter 10).

75 Wander or poor steering stability

1 Mismatched or uneven tires.
2 Worn balljoints and/or tie-rod ends (Chapter 10).
3 Worn strut assemblies (Chapter 10).
4 Loose stabilizer bar (Chapter 10).
5 Broken or sagging springs (Chapter 10).
6 Wheels out of alignment. Have the wheels professionally aligned.

76 Erratic steering when braking

1 Wheel bearings worn (Chapter 10).
2 Broken or sagging springs (Chapter 10).
3 Leaking wheel cylinder or caliper (Chapter 9).
4 Warped discs or drums (Chapter 9).

77 Excessive pitching and/or rolling around corners or during braking

1 Loose stabilizer bar (Chapter 10).
2 Worn struts or mountings (Chapter 10).
3 Broken or sagging springs (Chapter 10).
4 Overloaded vehicle.

78 Suspension bottoms

1 Overloaded vehicle.
2 Sagging springs (Chapter 10).

79 Cupped tires

1 Front wheel or rear wheel alignment out-of-specifications. Have the wheels professionally aligned.
2 Worn struts or shock absorbers (Chapter 10).
3 Wheel bearings worn (Chapter 10).
4 Excessive tire or wheel runout.
5 Worn balljoints (Chapter 10).

80 Excessive tire wear on outside edge

1 Inflation pressures incorrect (Chapter 1).
2 Excessive speed in turns.
3 Wheel alignment incorrect (excessive toe-in). Have professionally aligned.
4 Suspension arm bent (Chapter 10).

81 Excessive tire wear on inside edge

1 Inflation pressures incorrect (Chapter 10).
2 Wheel alignment incorrect (toe-out). Have professionally aligned.
3 Loose or damaged steering components (Chapter 10).

82 Tire tread worn in one place

1 Tires out-of-balance.
2 Damaged wheel. Inspect and replace if necessary.
3 Defective tire (Chapter 1).

83 Excessive play or looseness in steering system

1 Wheel bearing(s) worn (Chapter 10).
2 Tie-rod end loose (Chapter 10).
3 Steering gear loose (Chapter 10).
4 Worn or loose steering shaft U-joint (Chapter 10).

84 Rattling or clicking noise in steering gear

1 Steering gear loose (Chapter 10).
2 Steering gear defective (Chapter 10).

Notes

Chapter 1
Tune-up and routine maintenance

Contents

Specifications

Recommended lubricants and fluids

Note: *Listed here are manufacturer recommendations at the time this manual was written. Manufacturers occasionally upgrade their fluid and lubricant specifications, so check with your local auto parts store for current recommendations.*

Engine oil
  Type .... API "certified for gasoline engines"
  Viscosity
    All except 2013 2.0L models .... SAE 5W-20
    2013 2.0L models .... SAE 5W-30
Fuel .... Unleaded gasoline, 87 octane
Automatic transaxle fluid
  2001 through 2006 models .... MERCON® automatic transmission fluid

Caution: *Do not mix Mercon® V or dual usage Mercon®/Mercon® LV automatic transmission fluid with MERCON®.*

  2007 and 2008 models .... MERCON®V automatic transmission fluid
  2009 and later models .... MERCON LV automatic transmission fluid
Manual transaxle lubricant .... SAE 75W-90 gear oil
Transfer case (4WD models)
  With automatic transaxle .... SAE 75W-140
  With manual transaxle
    2001 through 2008 models .... SAE 80W-90 Premium Rear Axle Lubricant
    2009 and later models .... SAE 75W-140 Synthetic Rear Axle Lubricant
Rear differential lubricant (4WD models) .... SAE 80W-90 Premium Rear Axle Lubricant
Brake fluid
  2012 and earlier models .... DOT 3 brake fluid
  2013 and later models .... DOT 4 brake fluid
Power steering fluid .... MERCON automatic transmission fluid (2007 and earlier models only)
Clutch fluid .... DOT 3 brake fluid

Engine coolant*
  2007 and earlier models... 50/50 mixture of Motorcraft Premium gold engine coolant (yellow colored) or Motorcraft Premium engine coolant (green colored) and distilled water
  2008 and through 2011 models.. 50/50 mixture of Motorcraft Premium gold engine coolant (yellow colored) and distilled water
  2012 and later models.. 50/50 mixture of Motorcraft Premium orange engine coolant (orange colored) and distilled water

* Do not mix coolants of different colors. Doing so might damage the cooling system and/or the engine. The manufacturer specifies either a green, yellow or orange colored coolant to be used in these systems, depending on what was originally installed in the vehicle.

Capacities*

Engine oil (including filter)
  Four-cylinder engine
    1.5L engines .... 4.28 qts (4.05 liters)
    1.6L engines .... 4.3 qts (4.01 liters)
    2.0L engines
      2004 and earlier models.. 4.5 qts (4.25 liters)
      2013 and later models... 5.7 qts (5.4 liters)
    2.3L engines .... 4.5 qts (4.25 liters)
    2.5L engines
      2009 through 2012 models.. 5.3 qts (5.0 liters)
      2013 and later models... 5.7 qts (5.4 liters)
  V6 engine
    2004 and earlier models .... 5.5 qts (5.2 liters)
    2005 through 2012 models .... 6.0 qts (5.7 liters)
Coolant
  Four-cylinder engine
    1.5L and 1.6L engines .... Up to 8.5 qts (8.0 liters)
    2.0L engines
      2004 and earlier models.. Up to 7.4 qts (7.0 liters)
      2013 and later models... Up to 9.7 qts (9.2 liters)
    2.3L engines
      With manual transmission .... Up to 7.0 qts (6.6 liters)
      With automatic transmission .... Up to 8.0 qts (7.6 liters)
    2.5L engines
      2009 through 2012 models
        With manual transmission .... Up to 7.0 qts (6.6 liters)
        With automatic transmission .... Up to 8.0 qts (7.6 liters)
      2013 and later models... Up to 9.2 qts (8.7 liters)
  V6 engine .... Up to 10.5 qts (10.0 liters)
Automatic transaxle (dry fill)
  2006 and earlier models.. Up to 10.0 qts (9.5 liters)
  2011 and later models.. Up to 9.5 qts (8.5 liters)

Note: *Since this is a dry-fill specification, the amount required during a routine fluid change will be substantially less. The best way to determine the amount of fluid to add during a routine fluid change is to measure the amount drained. Begin the refill procedure by initially adding 1/3 of the amount drained. Then, with the engine running, add 1/2-pint at a time (cycling the shifter through each gear position between additions) until the level is correct on the dipstick. It is important to not overfill the transaxle. You will, however, need to purchase a few extra quarts, since the fluid replacement procedure involves flushing the torque converter (see Section 25).*

Manual transaxle
  2WD models... Up to 2.85 qts (2.7 liters)
  4WD models... Up to 2.32 qts (2.2 liters)
Transfer case (4WD models)
  With automatic transaxle... Up to 12 ounces (0.35 liters)
  With manual transaxle.. Up to 12 ounces (0.35 liters)
Rear differential (4WD models) .... Up to 1.47 qts (1.4 liters)

Note: **All capacities approximate. Add as necessary to bring up to appropriate level.*

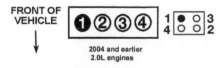

Cylinder location (2004 and earlier 2.0L
engine coil terminal arrangement)
(four-cylinder engines)

Cylinder location (2005 and later
four-cylinder engines)

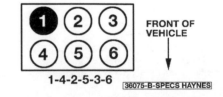

1-4-2-5-3-6

Cylinder location (V6 engine)

Ignition system
Spark plug type and gap
  Type
    Four-cylinder engines
      2007 and earlier
        Cylinders 1 and 3 .... Motorcraft AZFS-32 or equivalent
        Cylinders 2 and 4 .... Motorcraft AZFS-32FE or equivalent
      2008 (2.3L) .... Motorcraft AYSF-32-YPC or equivalent
      2009 and later models .... Motorcraft 12405, or equivalent
    V6 engines
      2007 and earlier .... Motorcraft AWSF-32F or equivalent
      2008 .... Motorcraft AGSF-32N or equivalent
      2009 through 2012 .... Motorcraft CGSF-22N or equivalent
  Gap
    2.0L four-cylinder engine (2004 and earlier models) .... 0.048 to 0.052 inch (1.22 to 1.32 mm)
    2.3L four-cylinder engine .... 0.054 inch (1.37 mm)
    All 2009 and later models, except 1.5L engines .... 0.030 inch (0.75 mm)
    1.5L engines .... 0.028 inch (0.70 mm)
    V6 engine
      2008 and earlier .... 0.052 to 0.056 inch (1.3 to 1.4 mm)
      2009 through 2012 .... 0.045 to 0.049 inch (1.14 to 1.24 mm)
Engine firing order
  Four-cylinder engines
    All models except 1.5L engines 1-3-4-2
    1.5L engines 1-2-4-3
  V6 engine 1-4-2-5-3-6

Clutch pedal
Freeplay .... 0.22 to 0.59 inch (5.58 to 15 mm)
Height .... 8.35 to 8.54 inches (212 to 217 mm)

Brakes
Disc brake pad lining thickness (minimum) .... 1/8 inch (3 mm)
Drum brake shoe lining thickness (minimum)................... 1/16 inch (1.5 mm)
Parking brake adjustment.. 3 to 5 clicks

Torque specifications

Ft-lbs (unless otherwise indicated)   Nm

Note: *One foot pound (ft lb) of torque is equivalent to 12 inch-pounds (in-lbs) of torque. Torque values below approximately 15 ft-lbs are expressed in inch-pounds, since most foot-pound torque wrenches are not accurate at these smaller values.*

	Ft-lbs	Nm
Engine oil drain plug		
Four-cylinder engines		
2008 and earlier models	18	24
2009 and later models	21	28
V6 engine	19	26
Automatic transaxle drain plug		
CD4E (2001 through 2008 models)	18	24
6F35 (2009 and later models)	106 in-lbs	
eCVT (2005 through 2012 models)	30	40
Automatic transaxle check plug		
6F35 (2009 and later models		
M10 plug	71 in-lbs	8
M20 plug	26	35
eCVT	30	40
Manual transaxle drain plug		
2008 and earlier models	35	47
2009 and later models	26	35
Rear differential cover bolts (4WD)	17	23
Rear differential fill plug (4WD)	20	27
Spark plugs		
2008 and earlier models	132 in-lbs	15
2009 through 2012 models	106 in-lbs	12
2013 and later models	115 in-lbs	13
V6 engines	132 in-lbs	15
Drivebelt tensioner bolts		
Four-cylinder engines		
2.0L, 2.3L and 2.5L models	18	25
1.5L and 1.6L models	35	48
V6 engine		
2001 through 2004 models	18	25
2005 through 2008 models	33	45
2009 and later models	18	25
Water pump drivebelt tensioner mounting bolt (V6 engine)	89 in-lbs	10
Wheel lug nuts		
2012 and earlier models	98	133
2013 and later models	100	135

2.2a Typical engine compartment components (2008 and earlier V6 engine)

1	Brake fluid reservoir	5	Automatic transaxle fluid dipstick	9	Power steering fluid reservoir
2	Fuse/relay block	6	Engine oil dipstick	10	Battery
3	Air filter housing	7	Radiator hose	11	Windshield washer fluid reservoir
4	Coolant expansion tank	8	Engine oil filler cap		

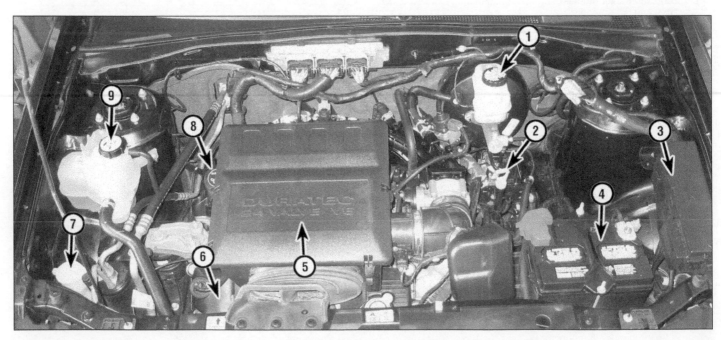

2.2b Typical engine compartment components (2009 and later V6 engine)

1	Brake fluid reservoir	4	Battery	7	Windshield washer fluid reservoir
2	Automatic transaxle fluid dipstick	5	Air filter housing	8	Engine oil filler cap
3	Fuse/relay block	6	Engine oil dipstick (vicinity)	9	Coolant expansion tank

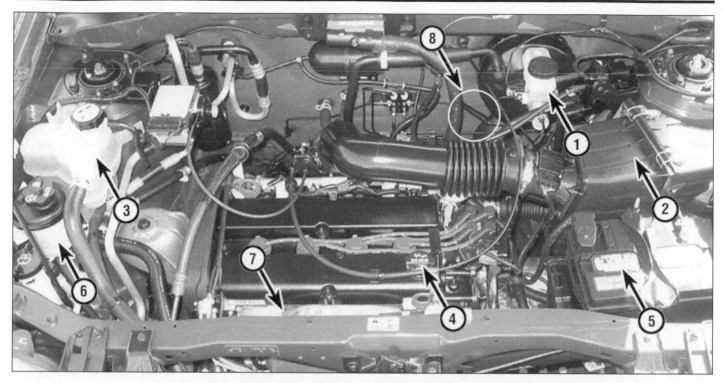

2.2c Typical engine compartment components (early four-cylinder engine)

1	Brake/clutch fluid reservoir	4	Engine oil filler cap	7	Engine oil dipstick (not visible in
2	Air filter housing	5	Battery		this photo)
3	Coolant reservoir	6	Power steering fluid reservoir	8	Automatic transaxle dipstick location

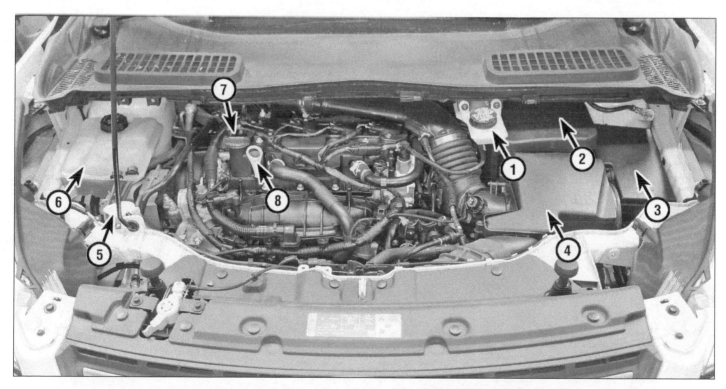

2.2d Typical engine compartment components (1.6L EcoBoost four-cylinder engine)

1	Brake fluid reservoir	4	Air filter housing	7	Engine oil filler cap
2	Battery	5	Windshield washer fluid reservoir	8	Engine oil dipstick
3	Power distribution/relay center	6	Coolant reservoir		

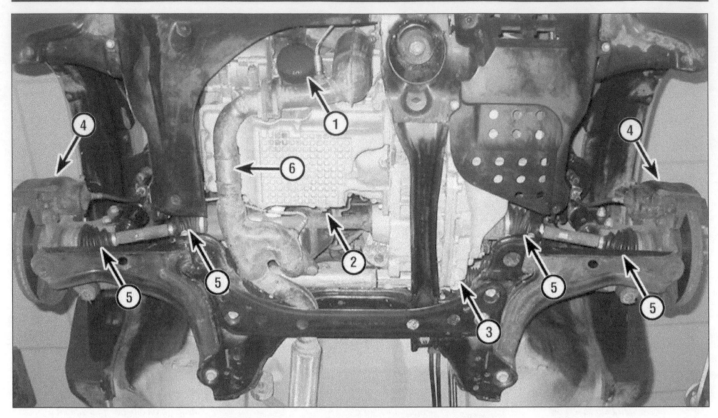

2.2e Typical engine underside components (V6 engine)

1	Engine oil filter	3	Automatic transaxle drain plug	5	Driveaxle boot
2	Engine oil drain plug	4	Front disc brake caliper	6	Exhaust pipe

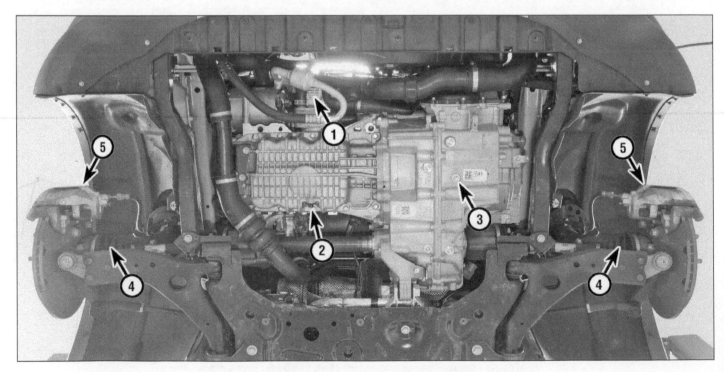

2.2f Typical engine underside components (1.6L EcoBoost engine)

1	Engine oil filter	3	Automatic transaxle drain plug	5	Front disc brake caliper
2	Engine oil drain plug	4	Driveaxle boot		

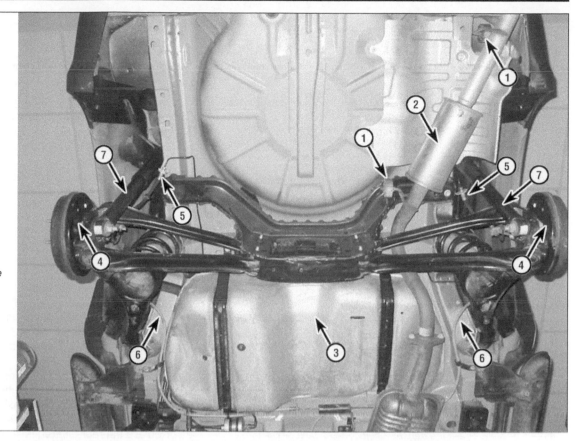

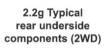

2.2g Typical rear underside components (2WD)

1   Exhaust system hanger
2   Muffler
3   Fuel tank
4   Rear drum brake assembly
5   Brake hose
6   Parking brake cable
7   Shock absorber

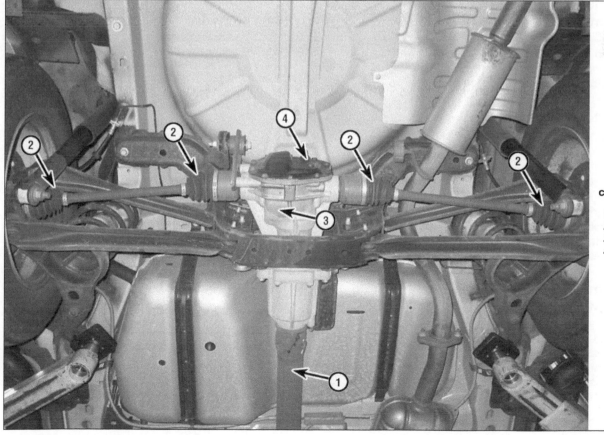

2.2h Typical rear underside components (4WD)

1   Driveshaft
2   Driveaxle boot
3   Differential
4   Differential filler plug

1 Maintenance schedule

1   The maintenance intervals in this manual are provided with the assumption that you, not the dealer, will be doing the work. These are the minimum maintenance intervals recommended by the factory for vehicles that are driven daily. If you wish to keep your vehicle in peak condition at all times, you may wish to perform some of these procedures even more often. Because frequent maintenance enhances the efficiency, performance and resale value of your car, we encourage you to do so. If you drive in dusty areas, tow a trailer, idle or drive at low speeds for extended periods or drive for short distances (less than four miles) in below freezing temperatures, shorter intervals are also recommended.

2   When your vehicle is new, it should be serviced by a factory authorized dealer service department to protect the factory warranty. In many cases, the initial maintenance check is done at no cost to the owner.

Every 250 miles or weekly, whichever comes first

Check the engine oil level (Section 4)
Check the engine coolant level (Section 4)
Check the brake and clutch fluid level (Section 4)
Check the windshield washer fluid level (Section 4)
Check the power steering fluid level (Section 4)
Check the automatic transaxle fluid level (Section 4)
Note: *This check only applies to models with a dipstick. For models without a dipstick, check the transaxle fluid level at each oil change.*
Check the tires and tire pressures (Section 5)

Every 3000 miles or 3 months, whichever comes first

All items listed above plus:
Change the engine oil and oil filter (Section 6)
Check the automatic transaxle fluid level (Section 4)

Every 7500 miles or 6 months, whichever comes first

All items listed above plus:
Inspect (and replace, if necessary) the windshield wiper blades (Section 7)
Check and service the battery (Section 8)
Check the cooling system (Section 9)
Rotate the tires (Section 10)
Check the seat belts (Section 11)

Every 15,000 miles or 12 months, whichever comes first

All items listed above plus:
Check all underhood hoses (Section 12)
Inspect the brake system (Section 13)*
Inspect the suspension and steering components (Section 14)*

Check (and replace, if necessary) the cabin air filter (2005 and later models; Section 19)
Check the fuel system (Section 15)
Check the manual transaxle lubricant level (Section 16)
Check the transfer case lubricant level (4WD models) (Section 28)
Check (and replace, if necessary) the air filter (Section 18)*
Check the driveaxle boots (Section 14)

Every 30,000 miles or 24 months, whichever comes first

All items listed above plus:
Check the exhaust system (Section 20)
Replace the fuel filter (2008 and earlier models) (Section 21)
Service the cooling system (drain, flush and refill) (Section 22)
Change the brake fluid (Section 22)
Check/adjust the engine drivebelts (Section 24)
Replace the automatic transaxle fluid (Section 25)**
Replace the manual transaxle lubricant (Section 26)
Replace the rear differential lubricant (4WD) (Section 27)
Replace the transfer case lubricant (4WD) (Section 28)

Every 60,000 miles or 48 months, whichever comes first

Replace the Positive Crankcase Ventilation (PCV) valve (four-cylinder engines) (Section 29)
Check (and replace, if necessary) the spark plugs (conventional, non-platinum or iridium type) (Section 30)
Check the ignition coils (V6 engines and 2005 and later four-cylinder engines) (Section 31)
Inspect (and replace, if necessary) the spark plug wires (2004 and earlier four-cylinder engines) (Section 32)
Check and adjust the valve clearances (four-cylinder engines) (see Chapter 2A)
Inspect (and replace, if necessary) the engine timing belt (four-cylinder engines) (see Chapter 2A)

Every 90,000 miles

Replace the engine timing belt (four-cylinder engines) (See Chapter 2A)

Every 100,000 miles

Replace the spark plugs (platinum-tipped type) (Section 30)
Positive crankcase ventilation (PCV) replacement (V6 engines) (see Chapter 6)

*This item is affected by "severe" operating conditions as described below. If your vehicle is operated under "severe" conditions, perform all maintenance indicated with an asterisk (*) at 3000 mile/3 month intervals. Severe conditions are indicated if you mainly operate your vehicle under one or more of the following conditions:

a) *Operating in dusty areas*
b) *Towing a trailer*
c) *Idling for extended periods and/or low speed operation*
d) *Operating when outside temperatures remain below freezing and when most trips are less than 4 miles*

** If operated under one or more of the following conditions, change the manual or automatic transaxle fluid and differential lubricant every 15,000 miles:

a) *In heavy city traffic where the outside temperature regularly reaches 90-degrees F (32-degrees C) or higher*
b) *In hilly or mountainous terrain*

2 Introduction

1   This Chapter is designed to help the home mechanic maintain the Ford Escape, Mazda Tribute or Mercury Mariner with the goals of maximum performance, economy, safety and reliability in mind.
2   Included is a master maintenance schedule, followed by procedures dealing specifically with each item on the schedule. Visual checks, adjustments, component replacement and other helpful items are included. Refer to the accompanying illustrations of the engine compartment and the underside of the vehicle for the locations of various components.
3   Servicing the vehicle, in accordance with the mileage/time maintenance schedule and the step-by-step procedures will result in a planned maintenance program that should produce a long and reliable service life. Keep in mind that it is a comprehensive plan, so maintaining some items but not others at the specified intervals will not produce the same results.
4   As you service the vehicle, you will discover that many of the procedures can - and should - be grouped together because of the nature of the particular procedure you're performing or because of the close proximity of two otherwise unrelated components to one another.
5   For example, if the vehicle is raised for chassis lubrication, you should inspect the exhaust, suspension, steering and fuel systems while you're under the vehicle. When you're rotating the tires, it makes good sense to check the brakes since the wheels are already removed. Finally, let's suppose you have to borrow or rent a torque wrench. Even if you only need it to tighten the spark plugs, you might as well check the torque of as many critical fasteners as time allows.
6   The first step in this maintenance program is to prepare yourself before the actual work begins. Read through all the procedures you're planning to do, then gather up all the parts and tools needed. If it looks like you might run into problems during a particular job, seek advice from a mechanic or an experienced do-it-yourselfer.

Owner's manual and VECI label information

7   Your vehicle owner's manual was written for your year and model and contains very specific information on component locations, specifications, fuse ratings, part numbers, etc. The owner's manual is an important resource for the do-it-yourselfer to have; if one was not supplied with your vehicle, it can generally be ordered from a dealer parts department.
8   Among other important information, the Vehicle Emissions Control Information (VECI) label contains specifications and procedures for applicable tune-up adjustments and, in some instances, spark plugs (see Chapter 6 for more information on the VECI label). The information on this label is the exact maintenance data recommended by the manufacturer. This data often varies by intended operating altitude, local emissions regulations, month of manufacture, etc.
9   This Chapter contains procedural details, safety information and more ambitious maintenance intervals than you might find in manufacturer's literature. However, you may also find procedures or specifications in your owner's manual or VECI label that differ with what's printed here. In these cases, the owner's manual or VECI label can be considered correct, since it is specific to your particular vehicle.

3 Tune-up general information

1   The term tune-up is used in this manual to represent a combination of individual operations rather than one specific procedure.
2   If, from the time the vehicle is new, the routine maintenance schedule is followed closely and frequent checks are made of fluid levels and high wear items, as suggested throughout this manual, the engine will be kept in relatively good running condition and the need for additional work will be minimized.
3   More likely than not, however, there will be times when the engine is running poorly due to lack of regular maintenance. This is even more likely if a used vehicle, which has not received regular and frequent maintenance checks, is purchased. In such cases, an engine tune-up will be needed outside of the regular routine maintenance intervals.

4   The first step in any tune-up or diagnostic procedure to help correct a poor running engine is a cylinder compression check. A compression check (see Chapter 2C) will help determine the condition of internal engine components and should be used as a guide for tune-up and repair procedures. If, for instance, a compression check indicates serious internal engine wear, a conventional tune-up will not improve the performance of the engine and would be a waste of time and money. Because of its importance, the compression check should be done by someone with the right equipment and the knowledge to use it properly.
5   The following procedures are those most often needed to bring a generally poor running engine back into a proper state of tune.

Minor tune-up

Check all engine related fluids (Section 4)
Clean, inspect and test the battery (Section 8)
Check the cooling system (Section 9)
Check all underhood hoses (Section 12)
Check the fuel system (Section 15)
Check the air filter (Section 18)

Major tune-up

6   All items listed under Minor tune-up, plus . . .
Replace the air filter (Section 18)
Replace the fuel filter (Section 21)
Check the drivebelt (Section 24)
Replace the PCV valve (Section 29)
Replace the spark plugs (Section 30)

4 Fluid level checks (every 250 miles or weekly)

1   Fluids are an essential part of the lubrication, cooling, brake and windshield washer systems. Because the fluids gradually become depleted and/or contaminated during normal operation of the vehicle, they must be periodically replenished. See *Recommended lubricants and fluids* at the beginning of this Chapter before adding fluid to any of the following components.
Note: *The vehicle must be on level ground when fluid levels are checked.*

4.2a The oil dipstick is located on the forward side of the engine on V6 engines

4.2b On 2012 and earlier four-cylinder engines the oil dipstick is located on the forward side of the engine next to the exhaust manifold

4.2c On 2013 and later four-cylinder engines, the oil dipstick (1) is located next to the oil filler cap (2) on the top of the engine

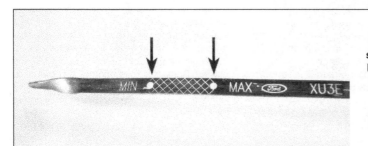

4.4 The oil level should be in the safe range - if it's below the MIN or ADD mark, add enough oil to bring it up to or near the MAX or FULL mark

Engine oil

2   The oil level is checked with a dipstick which extends down into the oil pan (see illustrations).

3   The oil level should be checked before the vehicle has been driven, or about five minutes after the engine has been shut off. If the oil is checked immediately after driving the vehicle, some of the oil will remain in the upper part of the engine, resulting in an inaccurate reading on the dipstick.

4   Pull the dipstick out of the tube and wipe all the oil from the end with a clean rag or paper towel. Insert the clean dipstick all the way back into the tube and pull it out again. Note the oil at the end of the dipstick. At its highest point, the level should be between the MIN and MAX marks or triangle cut outs on the dipstick (see illustration).

5   It takes about one quart of oil to raise the level from the MIN mark to the MAX mark on the dipstick. Do not allow the level to drop below the MIN mark or oil starvation may cause engine damage. Conversely, overfilling the engine (adding oil above the MAX mark) may cause oil fouled spark plugs, oil leaks or oil seal failures. Maintaining the oil level above the MAX mark can cause excessive oil consumption.

6   To add oil, remove the filler cap from the valve cover (see illustration). After adding oil, wait a few minutes to allow the level to stabilize, then pull out the dipstick and check the level again. Add more oil if required. Install the filler cap and tighten it by hand only.

7   Checking the oil level is an important preventive maintenance step. A consistently low oil level indicates oil leakage through damaged seals, defective gaskets or past worn rings or valve guides. If the oil looks milky in color or has water droplets in it, the cylinder head gasket(s) may be blown or the head(s) or block may be cracked. The engine should be checked immediately. The condition of the oil should also be checked. Whenever you check the oil level, slide your thumb and index finger up the dipstick before wiping off the oil.

4.6 The oil filler cap is located on the valve cover - always make sure the area around the opening is clean before unscrewing the cap to prevent dirt from contaminating the engine (V6 engine shown)

If you see small dirt or metal particles clinging to the dipstick, the oil should be changed (see Section 6).

Engine coolant

Warning: *Do not allow antifreeze to come in contact with your skin or painted surfaces of the vehicle. Flush contaminated areas immediately with plenty of water. Don't store new coolant or leave old coolant lying around where it's accessible to children or pets - they're attracted by its sweet smell. Ingestion of even a small amount of coolant can be fatal! Wipe up garage floor and drip pan spills immediately. Keep antifreeze containers covered and repair cooling system leaks as soon as they're noticed.*

8   All vehicles covered by this manual are equipped with a pressurized coolant recovery system. A plastic expansion tank located at the front of the engine compartment is connected by a hose to the radiator (see illustration). As the engine heats up during operation, the expanding coolant fills the tank.

4.8 The cooling system expansion tank is located at the right side of the engine compartment

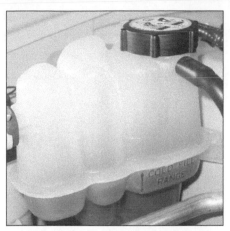

4.9a When the engine is cold, the engine coolant level should be at the COLD FULL mark on 2012 and earlier models

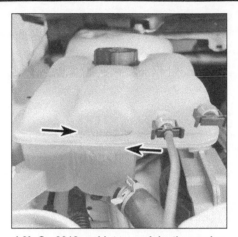

4.9b On 2013 and later models, the engine coolant level should be between the MIN and MAX marks when the engine is cold

4.14 The clutch master cylinder is located next to the brake master cylinder and the two share a common reservoir

9   The coolant level in the tank should be checked regularly. The level in the tank varies with the temperature of the engine. When the engine is cold, the coolant level should be at the COLD FULL mark on the reservoir for 2012 and earlier models or between the MIN and MAX lines on 2013 and later models. If it isn't, remove the cap from the tank and add a 50/50 mixture of ethylene glycol based antifreeze and water (see illustrations).

Warning: *Do not remove the expansion tank cap to check the coolant level when the engine is warm!*

10   Drive the vehicle, let the engine cool completely then recheck the coolant level. Don't use rust inhibitors or additives. If only a small amount of coolant is required to bring the system up to the proper level, water can be used. However, repeated additions of water will dilute the antifreeze and water solution. In order to maintain the proper ratio of antifreeze and water, always top up the coolant level with the correct mixture. An empty plastic milk jug or bleach bottle makes an excellent container for mixing coolant.

11   If the coolant level drops consistently, there may be a leak in the system. Inspect the radiator, hoses, filler cap, drain plugs and water pump (see Section 9). If no leaks are noted, have the expansion tank cap pressure tested by a service station.

12   If you have to remove the expansion tank cap wait until the engine has cooled completely, then wrap a thick cloth around the cap and unscrew it slowly, stopping if you hear a hissing noise. If coolant or steam escapes, let the engine cool down longer, then remove the cap.

13   Check the condition of the coolant as well. It should be relatively clear. If it's brown or rust colored, the system should be drained, flushed and refilled. Even if the coolant appears to be normal, the corrosion inhibitors wear out, so it must be replaced at the specified intervals.

Brake and clutch fluid

14   The brake master cylinder is mounted on the front of the power booster unit in the engine compartment. The hydraulic clutch master cylinder used on manual transaxle vehicles is located next to the brake master cylinder (see illustration).

15   The brake master cylinder and the clutch master cylinder share a common reservoir. To check the fluid level of either system, simply look at the MAX and MIN marks on the brake fluid reservoir (see illustrations).

16   If the level is low, wipe the top of the reservoir cover with a clean rag to prevent contamination of the brake system before lifting the cover.

17   Add only the specified brake fluid to the reservoir (refer to *Recommended lubricants and fluids* at the front of this Chapter or to your owner's manual). Mixing different types of brake fluid can damage the system. Fill the brake master cylinder reservoir only to the MAX line.

Warning: *Use caution when filling the reservoir - brake fluid can harm your eyes and damage painted surfaces. Do not use brake fluid that is more than one year old or has been left open. Brake fluid absorbs moisture from the air. Excess moisture can cause a dangerous loss of braking.*

18   While the reservoir cap is removed, inspect the master cylinder reservoir for contamination. If deposits, dirt particles or water droplets are present, the system should be drained and refilled.

19   After filling the reservoir to the proper level, make sure the lid is properly seated to prevent fluid leakage.

20   The fluid in the brake master cylinder will drop slightly as the brake pads at each wheel wear down during normal operation. If the master cylinder requires repeated replenishing to keep it at the proper level, this is an indication of leakage in the brake or clutch system, which should be corrected immediately. If the brake system shows an indication of leakage check all brake lines and connections, along with the calipers, wheel cylinders and booster (see Section 13 for more information). If the hydraulic clutch system shows an indication of leakage, check all clutch lines

4.15a The brake fluid level should be kept between the MIN and MAX marks on the translucent plastic reservoir; the same reservoir contains the clutch fluid and is connected to the clutch master cylinder by a hose

4.15b On later models, brake fluid reservoir is connected to the master cylinder by a brake hose and mounted to the cowl panel. The brake fluid level should be kept between the MIN and MAX marks on the translucent plastic reservoir

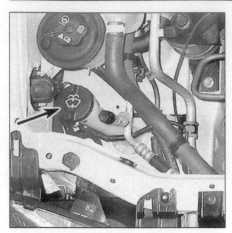

4.22 The windshield/rear window washer fluid reservoir is located in the right front corner of the engine compartment - 2012 and earlier models shown, later models similar

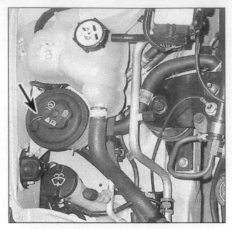

4.25 The power steering fluid reservoir is located at the right side of the engine compartment next to the windshield washer fluid reservoir on 2007 and earlier models

4.28 At normal operating temperature, the power steering fluid level should be between the MAX and MIN marks

and connections along with the clutch release cylinder (see Chapter 9 for more information).

21   If, upon checking the brake or clutch master cylinder fluid level, you discover the reservoir empty or nearly empty, the systems should be bled (see Chapter 8 and Chapter 9).

Windshield washer fluid

22   Fluid for the windshield washer system is stored in a plastic reservoir located at the right front of the engine compartment (see illustration).

23   In milder climates, plain water can be used in the reservoir, but it should be kept no more than 2/3 full to allow for expansion if the water freezes. In colder climates, use windshield washer system antifreeze, available at any auto parts store, to lower the freezing point of the fluid. Mix the antifreeze with water in accordance with the manufacturer's directions on the container.

Caution: *Do not use cooling system antifreeze - it will damage the vehicle's paint.*

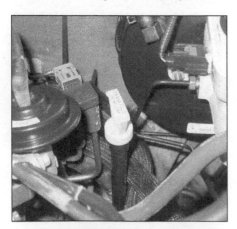

4.34a The automatic transaxle dipstick is located at the rear of the engine compartment or near the left end of the engine

Power steering fluid

Note: *This procedure applies only to 2007 and earlier models. Later models have electric power steering and require no regular maintenance.*

24   Check the power steering fluid level periodically to avoid steering system problems, such as damage to the pump.

Caution: *DO NOT hold the steering wheel against either stop (extreme left or right turn) for more than five seconds. If you do, the power steering pump could be damaged.*

25   The power steering reservoir, located at the right side of the engine compartment (see illustration), has MIN and MAX fluid level marks on the side. The fluid level can be seen without removing the reservoir cap.

26   Park the vehicle on level ground and apply the parking brake.

27   Run the engine until it has reached normal operating temperature. With the engine at idle, turn the steering wheel back and forth about 10 times to get any air out of the steering system. Shut the engine off with the wheels in the straight-ahead position.

28   Note the fluid level on the side of the reservoir. It should be between the two marks (see illustration).

29   Add small amounts of fluid until the level is correct.

Caution: *Do not overfill the reservoir. If too*

much fluid is added, remove the excess with a clean syringe or suction pump.

30   Check the power steering hoses and connections for leaks and wear.

Automatic transaxle fluid
Models with a dipstick (2014 and earlier models, except eCVT transaxle)

31   The level of the automatic transaxle fluid should be carefully maintained. Low fluid level can lead to slipping or loss of drive, while overfilling can cause foaming, loss of fluid and transaxle damage.

32   The transaxle fluid level should only be checked when the transaxle is hot (at its normal operating temperature). If the vehicle has just been driven over 10 miles (15 miles in a frigid climate), and the fluid temperature is 160 to 175-degrees F, the transaxle is hot.

Caution: *If the vehicle has just been driven for a long time at high speed or in city traffic in hot weather, or if it has been pulling a trailer, an accurate fluid level reading cannot be obtained. Allow the fluid to cool down for about 30 minutes.*

33   If the vehicle has not just been driven, park the vehicle on level ground, set the parking brake and start the engine. While the engine is idling, depress the brake pedal and move the selector lever through all the gear ranges, beginning and ending in Park.

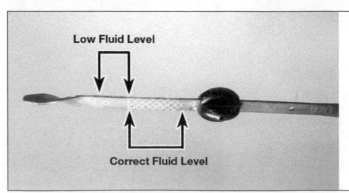

4.34b At operating temperature, the automatic transaxle fluid level should be kept in the cross-hatched area of the dipstick

Low Fluid Level

Correct Fluid Level

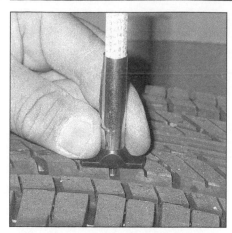

5.2 A tire tread depth indicator should be used to monitor tire wear - they are available at auto parts stores and service stations and cost very little

UNDERINFLATION

CUPPING

OVERINFLATION

Cupping may be caused by:
- Underinflation and/or mechanical irregularities such as out-of-balance condition of wheel and/or tire, and bent or damaged wheel.
- Loose or worn steering tie-rod or steering idler arm.
- Loose, damaged or worn front suspension parts.

INCORRECT TOE-IN OR EXTREME CAMBER

FEATHERING DUE TO MISALIGNMENT

5.3 This chart will help you determine the condition of your tires, the probable cause(s) of abnormal wear and the corrective action necessary

34   With the engine still idling, remove the dipstick from its tube (see illustration). Check the level of the fluid on the dipstick (see illustration) and note its condition.
35   Wipe the fluid from the dipstick with a clean rag and reinsert it back into the filler tube until the cap seats.
36   Pull the dipstick out again and note the fluid level. The fluid level should be in the operating temperature range (the cross-hatched area). If the level is at the low side of either range, add the specified automatic transmission fluid through the dipstick tube with a funnel.
37   Add just enough of the recommended fluid to fill the transaxle to the proper level. It takes about one pint to raise the level from the low mark to the high mark when the fluid is hot, so add the fluid a little at a time and keep checking the level until it is correct.
38   The condition of the fluid should also be checked along with the level. If the fluid at the end of the dipstick is black or a dark reddish brown color, or if it emits a burned smell, the fluid should be changed (see Section 25). If you are in doubt about the condition of the fluid, purchase some new fluid and compare the two for color and smell.

Models without a dipstick (2005 through 2012 eCVT transaxle and all 2015 and later models)
39   On these models it is not necessary to check the automatic transaxle fluid weekly; at every oil change is sufficient. See Section 25 for the fluid level checking procedure.

5   Tire and tire pressure checks (every 250 miles or weekly)

1   Periodic inspection of the tires may spare you the inconvenience of being stranded with a flat tire. It can also provide you with vital information regarding possible problems in the steering and suspension systems before major damage occurs.
2   The original tires on this vehicle are equipped with 1/2-inch wide bands that will appear when tread depth reaches 1/16 inch, at which point they can be considered worn out. Tread wear can be monitored with a simple, inexpensive device known as a tread depth indicator (see illustration).
3   Note any abnormal tread wear (see illustration). Tread pattern irregularities such as cupping, flat spots and more wear on one side than the other are indications of front end alignment and/or balance problems. If any of these conditions are noted, take the vehicle to a tire shop or service station to correct the problem.
4   Look closely for cuts, punctures and embedded nails or tacks. Sometimes a tire will hold air pressure for a short time or leak down very slowly after a nail has embedded itself in the tread. If a slow leak persists, check the valve stem core to make sure it is tight (see illustration). Examine the tread for an object that may have embedded itself in the tire or for a "plug" that may have begun to leak (radial tire punctures are repaired with a plug that is installed in a puncture). If a puncture is suspected, it can be easily verified by spraying a solution of soapy water onto the puncture area (see illustration). The soapy solution will bubble if there is a leak. Unless the puncture is unusually large, a tire shop or service station can usually repair the tire.
5   Carefully inspect the inner sidewall of each tire for evidence of brake fluid leakage. If you see any, inspect the brakes immediately.
6   Correct air pressure adds miles to the life span of the tires, improves mileage and

5.4a If a tire loses air on a steady basis, check the valve core first to make sure it's snug (special inexpensive wrenches are commonly available at auto parts stores)

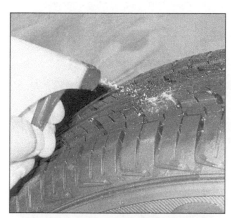

5.4b If the valve core is tight, raise the corner of the vehicle with the low tire and spray a soapy water solution onto the tread as the tire is turned slowly - slow leaks will cause small bubbles to appear

enhances overall ride quality. Tire pressure cannot be accurately estimated by looking at a tire, especially if it's a radial. A tire pressure gauge is essential. Keep an accurate gauge in the glove compartment. The pressure gauges attached to the nozzles of air hoses at gas stations are often inaccurate.

7   Always check tire pressure when the tires are cold. Cold, in this case, means the vehicle has not been driven over a mile in the three hours preceding a tire pressure check. A pressure rise of four to eight pounds is not uncommon once the tires are warm.

8   Unscrew the valve cap protruding from the wheel or hubcap and push the gauge firmly onto the valve stem (see illustration). Note the reading on the gauge and compare the figure to the recommended tire pressure shown on the tire placard on the driver's side door. Be sure to reinstall the valve cap to keep dirt and moisture out of the valve stem mechanism. Check all four tires and, if necessary, add enough air to bring them up to the recommended pressure.

9   Don't forget to keep the spare tire inflated to the specified pressure (refer to the pressure molded into the tire sidewall).

6   Engine oil and filter change (every 3000 miles or 3 months)

Note: *Some models are equipped with an oil life indicator system that illuminates a light or message on the instrument panel when the system deems it necessary to change the oil. A number of factors are taken into consideration to determine when the oil should be considered worn out. Generally, this system will allow the vehicle to accumulate more miles between oil changes than the traditional 3000-mile interval, but we believe that frequent oil changes is cheap insurance and will prolong engine life. If you do decide not to change your oil every 3000 miles and rely on the oil life indicator instead, make sure you don't exceed 10,000 miles before the oil is changed, regardless of what the oil life indicator shows.*

1   Frequent oil changes are the most impor-

5.8 To extend the life of your tires, check the air pressure at least once a week with an accurate gauge (don't forget the spare!)

tant preventive maintenance procedures that can be done by the home mechanic. As engine oil ages, it becomes diluted and contaminated, which leads to premature engine wear.

2   Make sure that you have all the necessary tools before you begin this procedure (see illustration). You should also have plenty of rags or newspapers handy for mopping up oil spills.

3   Access to the oil drain plug and filter will be improved if the vehicle can be lifted on a hoist, driven onto ramps or supported by jackstands.

Warning: *Do not work under a vehicle supported only by a jack - always use jackstands!*

4   If you haven't changed the oil on this vehicle before, get under it and locate the oil drain plug and the oil filter. The exhaust components will be warm as you work, so note how they are routed to avoid touching them when you are under the vehicle.

5   Start the engine and allow it to reach normal operating temperature - oil and sludge will flow out more easily when warm. If new oil, a filter or tools are needed, use the vehicle to go get them and warm up the engine/oil at the same time. Park on a level surface and shut

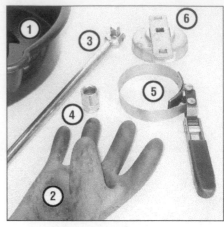

6.2 These tools are required when changing the engine oil and filter

1   **Drain pan** - *It should be fairly shallow in depth, but wide to prevent spills*
2   **Rubber gloves** - *When removing the drain plug and filter, you will get oil on your hands (the gloves will prevent burns)*
3   **Breaker bar** - *Sometimes the oil drain plug is tight, and a long breaker bar is needed to loosen it*
4   **Socket** - *To be used with the breaker bar or a ratchet (must be the correct size to fit the drain plug - six-point preferred)*
5   **Filter wrench** - *This is a metal band-type wrench, which requires clearance around the filter to be effective*
6   **Filter wrench** - *This type fits on the bottom of the filter and can be turned with a ratchet or breaker bar (different-size wrenches are available for different types of filters)*

off the engine when it's warmed up. Remove the oil filler cap from the valve cover.

6   Raise the vehicle and support it on jackstands. Make sure it is safely supported and remove the engine splash shield (see illustration).

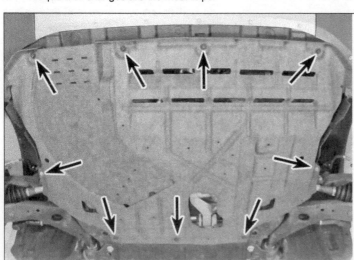

6.6 Typical engine splash shield fastener locations

6.7 Use a proper size box-end wrench or socket to remove the oil drain plug and avoid rounding it off

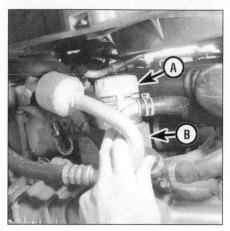

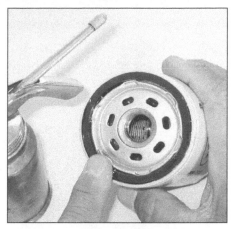

6.12a Use an oil filter wrench to remove the filter (V6 engine shown)

6.12b On 1.5L and 1.6L models, the oil filter (A) is mounted to the engine oil cooler (B)

6.15 Lubricate the oil filter gasket with clean engine oil before installing the filter on the engine

7   Being careful not to touch the hot exhaust components, position a drain pan under the plug in the bottom of the engine, then remove the plug (see illustration). It's a good idea to wear a rubber glove while unscrewing the plug the final few turns to avoid being scalded by hot oil.

8   It may be necessary to move the drain pan slightly as oil flow slows to a trickle. Inspect the old oil for the presence of metal particles.

9   After all the oil has drained, wipe off the drain plug with a clean rag. Any small metal particles clinging to the plug would immediately contaminate the new oil

10   Clean the area around the drain plug opening, reinstall the plug and tighten it securely, but don't strip the threads.

11   Move the drain pan into position under the oil filter.

12   Loosen the oil filter by turning it counterclockwise with a filter wrench (see illustrations). Any standard filter wrench will work.

13   Once the filter is loose, use your hands to unscrew it from the block. Just as the filter is detached from the block, immediately tilt the open end up to prevent the oil inside the filter from spilling out.

14   Using a clean rag, wipe off the mount-ing surface on the block. Also, make sure that none of the old gasket remains stuck to the mounting surface. It can be removed with a scraper if necessary.

15   Compare the old filter with the new one to make sure they are the same type. Smear some engine oil on the rubber gasket of the new filter and screw it into place (see illustration). Overtightening the filter will damage the gasket, so don't use a filter wrench. Most filter manufacturers recommend tightening the filter by hand only. Normally they should be tightened 3/4-turn after the gasket contacts the block, but be sure to follow the directions on the filter or container.

16   Remove all tools and materials from under the vehicle, being careful not to spill the oil in the drain pan, then lower the vehicle.

17   Add new oil to the engine through the oil filler cap. Use a funnel to prevent oil from spilling onto the top of the engine. Pour four quarts of fresh oil into the engine. Wait a few minutes to allow the oil to drain into the pan, then check the level on the dipstick (see Section 4 if necessary). If the oil level is in the OK range, install the filler cap.

18   Start the engine and run it for about a minute. While the engine is running, look under the vehicle and check for leaks at the oil pan drain plug and around the oil filter. If either one is leaking, stop the engine and tighten the plug or filter slightly.

19   Wait a few minutes, then recheck the level on the dipstick. Add oil as necessary to bring the level into the OK range.

20   During the first few trips after an oil change, make it a point to check frequently for leaks and proper oil level.

21   The old oil drained from the engine cannot be reused in its present state and should be disposed of. Check with your local auto parts store, disposal facility or environmental agency to see if they will accept the oil for recycling. After the oil has cooled it can be drained into a container (capped plastic jugs, topped bottles, milk cartons, etc.) for transport to one of these disposal sites. Don't dispose of the oil by pouring it on the ground or down a drain!

7   Windshield/rear window wiper blade inspection and replacement (every 7500 miles or 6 months)

1   The windshield wiper and blade assembly should be inspected periodically for damage, loose components and cracked or worn blade elements.

2   Road film can build up on the wiper blades and affect their efficiency, so they should be washed regularly with a mild detergent solution.

3   If the wiper blade elements are cracked, worn or warped, or no longer clean adequately, they should be replaced with new ones.

Windshield wiper blades

4   On early models, lift the arm assembly away from the glass for clearance, press on the release lever, then slide the wiper blade assembly out of the hook in the end of the arm (see illustrations).

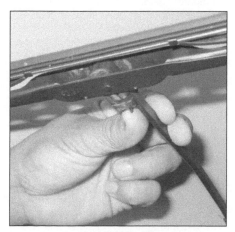

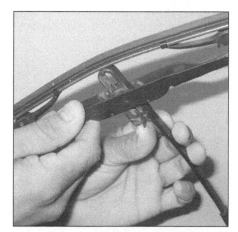

7.4a On early models, depress the release lever . . .

7.4b . . . and pull the wiper blade out of the arm

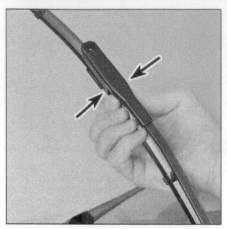

7.5a On later models, press the locking buttons together . . .

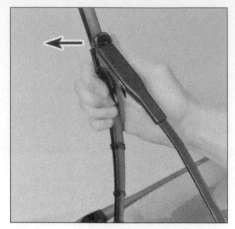

7.5b . . . and separate the wiper blade from the arm

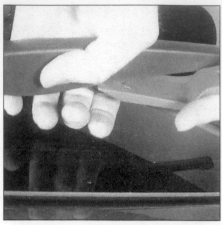

7.8 Pry the wiper blade straight out of the arm

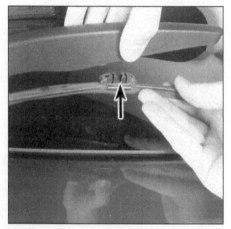

7.9 Align the opening in the blade with the pin in the arm, then push it into place

5   On later models, lift the arm assembly away from the glass for clearance, press the locking buttons together to the arm, then separate the wiper blade assembly from the wiper arm (see illustrations).
6   Attach the new wiper to the arm. Connection can be confirmed by an audible click.

Rear window wiper blade

7   On early models, lift the arm assembly away from the glass for clearance, rotate the blade upwards then pull the blade out from the wiper arm.
8   On later models, lift the wiper arm away from the glass, then use a plastic trim tool to pry the blade from the arm (see illustration).
9   To install, align the slot in the wiper blade with the pin in the wiper arm, then push until it clicks into place (see illustration).

8   Battery check, maintenance and charging (every 7500 miles or 6 months)

Warning: *Certain precautions must be followed when checking and servicing the bat-*

tery. Hydrogen gas, which is highly flammable, is always present in the battery cells, so keep lighted tobacco and all other open flames and sparks away from the battery. The electrolyte inside the battery is actually diluted sulfuric acid, which will cause injury if splashed on your skin or in your eyes. It will also ruin clothes and painted surfaces. When removing the battery cables, always detach the negative cable first and hook it up last!

1   A routine preventive maintenance program for the battery in your vehicle is the only way to ensure quick and reliable starts. But before performing any battery maintenance, make sure that you have the proper equipment necessary to work safely around the battery (see illustration).
2   There are also several precautions that should be taken whenever battery maintenance is performed. Before servicing the battery, always turn the engine and all accessories off and disconnect the cables from the negative terminal of the battery (see Chapter 5).
3   The battery produces hydrogen gas, which is both flammable and explosive. Never create a spark, smoke or light a match around the battery. Always charge the battery in a ventilated area.
4   Electrolyte contains poisonous and corrosive sulfuric acid. Do not allow it to get in your eyes, on your skin or your clothes. Never ingest it. Wear protective safety glasses when working near the battery. Keep children away from the battery.
5   Note the external condition of the battery. If the positive terminal and cable clamp on your vehicle's battery is equipped with a rubber protector, make sure that it's not torn or damaged. It should completely cover the terminal. Look for any corroded or loose connections, cracks in the case or cover or loose hold-down clamps. Also check the entire length of each cable for cracks and frayed conductors.
6   If corrosion, which looks like white, fluffy deposits (see illustration) is evident, particularly around the terminals, the battery should be removed for cleaning. Loosen the cable clamp bolts with a wrench, being careful to

remove the ground cable first, and slide them off the terminals (see illustration). Then disconnect the hold-down clamp bolt and nut, remove the clamp and lift the battery from the engine compartment.
7   Clean the cable clamps thoroughly with a battery brush or a terminal cleaner and a solution of warm water and baking soda (see illustration). Wash the terminals and the top of the battery case with the same solution but make sure that the solution doesn't get into the battery. When cleaning the cables, terminals and battery top, wear safety goggles and rubber gloves to prevent any solution from coming in contact with your eyes or hands. Wear old clothes too - even diluted, sulfuric acid splashed onto clothes will burn holes in them. If the terminals have been extensively corroded, clean them up with a terminal cleaner (see illustration). Thoroughly wash all cleaned areas with plain water.
8   Make sure that the battery tray is in good condition and the hold-down clamp fasteners are tight (see illustration). If the battery is removed from the tray, make sure no parts remain in the bottom of the tray when the battery is reinstalled. When reinstalling the hold-down clamp bolts, do not overtighten them.
9   Information on removing and installing the battery can be found in Chapter 5. If you disconnected the cable(s) from the negative and/or positive battery terminals, the powertrain control module (PCM) must relearn its idle and fuel trim strategy for optimum driveability and performance (see Chapter 5, Section 1 for this procedure). Information on jump starting can be found at the front of this manual. For more detailed battery checking procedures, refer to the Haynes Automotive Electrical Manual.

Cleaning

10   Corrosion on the hold-down components, battery case and surrounding areas can be removed with a solution of water and baking soda. Thoroughly rinse all cleaned areas with plain water.

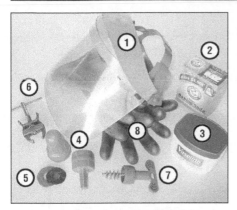

8.1 Tools and materials required for battery maintenance

1 **Face shield/safety goggles** - *When removing corrosion with a brush, the acidic particles can easily fly up into your eyes*

2 **Baking soda** - *A solution of baking soda and water can be used to neutralize corrosion*

3 **Petroleum jelly** - *A layer of this on the battery posts will help prevent corrosion*

4 **Battery post/cable cleaner** - *This wire brush cleaning tool will remove all traces of corrosion from the battery posts and cable clamps*

5 **Treated felt washers** - *Placing one of these on each post, directly under the cable clamps, will help prevent corrosion*

6 **Puller** - *Sometimes the cable clamps are very difficult to pull off the posts, even after the nut/bolt has been completely loosened. This tool pulls the clamp straight up and off the post without damage*

7 **Battery post/cable cleaner** - *Here is another cleaning tool which is a slightly different version of Number 4 above, but it does the same thing*

8 **Rubber gloves** - *Another safety item to consider when servicing the battery; remember that's acid inside the battery!*

11   Any metal parts of the vehicle damaged by corrosion should be covered with a zinc-based primer, then painted.

Charging

Warning: *When batteries are being charged, hydrogen gas, which is very explosive and flammable, is produced. Do not smoke or allow open flames near a charging or a recently charged battery. Wear eye protection when near the battery during charging. Also, make sure the charger is unplugged before connecting or disconnecting the battery from the charger.*

12   Slow-rate charging is the best way to restore a battery that's discharged to the point where it will not start the engine. It's also a

8.6a Battery terminal corrosion usually appears as light, fluffy powder

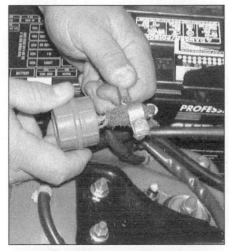

8.7a When cleaning the cable clamps, all corrosion must be removed

good way to maintain the battery charge in a vehicle that's only driven a few miles between starts. Maintaining the battery charge is particularly important in the winter when the battery must work harder to start the engine and electrical accessories that drain the battery are in greater use.

13   It's best to use a one or two-amp battery charger (sometimes called a "trickle" charger). They are the safest and put the least strain on the battery. They are also the least expensive. For a faster charge, you can use a higher amperage charger, but don't use one rated more than 1/10th the amp/hour rating of the battery. Rapid boost charges that claim to restore the power of the battery in one to two hours are hardest on the battery and can damage batteries not in good condition. This type of charging should only be used in emergency situations.

14   The average time necessary to charge a battery should be listed in the instructions that come with the charger. As a general rule, a trickle charger will charge a battery in 12 to 16 hours.

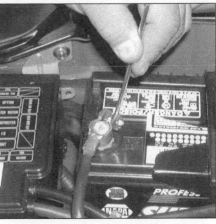

8.6b Removing a cable from the battery post with a wrench - sometimes a pair of special battery pliers are required for this procedure if corrosion has caused deterioration of the nut hex (always remove the ground (-) cable first and hook it up last!)

8.7b Regardless of the type of tool used to clean the battery posts, a clean, shiny surface should be the result

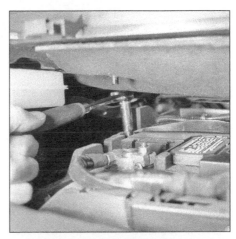

8.8 Make sure the battery hold-down fasteners are tight

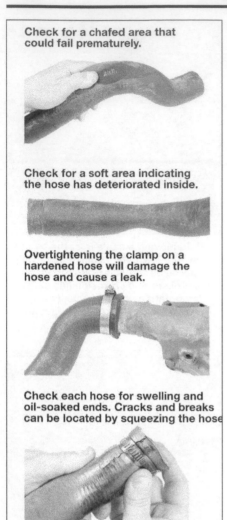

Check for a chafed area that could fail prematurely.

Check for a soft area indicating the hose has deteriorated inside.

Overtightening the clamp on a hardened hose will damage the hose and cause a leak.

Check each hose for swelling and oil-soaked ends. Cracks and breaks can be located by squeezing the hose

9.4 Hoses, like drivebelts, have a habit of failing at the worst possible time - to prevent the inconvenience of a blown radiator or heater hose, inspect them carefully as shown here

9   Cooling system check (every 7,500 miles or 6 months)

1   Many major engine failures can be caused by a faulty cooling system.
2   The engine must be cold for the cooling system check, so perform the following procedure before the vehicle is driven for the day or after it has been shut off for at least three hours.
3   Remove the pressure-relief cap from the expansion tank at the right side of the engine compartment. Clean the cap thoroughly, inside and out, with clean water. The presence of rust or corrosion in the expansion tank means the coolant should be changed (see Section 22). The coolant inside the expansion tank should be relatively clean and transparent. If it's rust colored, drain the system and refill it with new coolant.

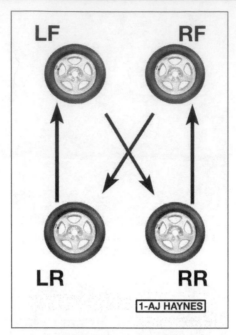

10.2a Four-tire rotation pattern

4   Carefully check the radiator hoses and the smaller diameter heater hoses. Inspect each coolant hose along its entire length, replacing any hose that is cracked, swollen or deteriorated (see illustration). Cracks will show up better if the hose is squeezed. Pay close attention to hose clamps that secure the hoses to cooling system components. Hose clamps can pinch and puncture hoses, resulting in coolant leaks.
5   Make sure that all hose connections are tight. A leak in the cooling system will usually show up as white or rust colored deposits on the area adjoining the leak. If wire-type clamps are used on the hoses, it may be a good idea to replace them with screw-type clamps.
6   Clean the front of the radiator and air conditioning condenser with compressed air, if available, or a soft brush. Remove all bugs, leaves, etc., embedded in the radiator fins. Be extremely careful not to damage the cooling fins or cut your fingers on them.
7   If the coolant level has been dropping consistently and no leaks are detectable, have the expansion tank cap and cooling system pressure checked at a service station.

10   Tire rotation (every 7,500 miles or 6 months)

Caution: *Some models may have high-performance tires, which have a directional tread pattern installed on them. These tires must be installed on the same sides so that they rotate in the correct direction. This is indicated by arrows located on the tire sidewalls.*

1   The tires should be rotated at the specified intervals and whenever uneven wear is noticed. Since the vehicle will be raised and

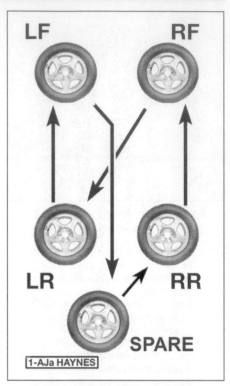

10.2b Five tire rotation pattern

the tires removed anyway, check the brakes also (see Section 13).
2   Radial tires must be rotated in a specific pattern (see illustrations). If your vehicle has a compact spare tire, don't include it in the rotation pattern.
3   Refer to the information in Jacking and towing at the front of this manual for the proper procedure to follow when raising the vehicle and changing a tire. If the brakes must be checked, don't apply the parking brake as stated.
4   The vehicle must be raised on a hoist or supported on jackstands to get all four wheels off the ground. Make sure the vehicle is safely supported!
5   After the rotation procedure is finished, check and adjust the tire pressures as necessary and be sure to check the lug nut tightness.

11   Seat belt check (every 7,500 miles or 6 months)

1   Check seat belts, buckles, latch plates and guide loops for obvious damage and signs of wear.
2   See if the seat belt reminder light comes on when the key is turned to the Run or Start position. A chime should also sound.
3   The seat belts are designed to lock up during a sudden stop or impact, yet allow free movement during normal driving. Make sure the retractors return the belt against your chest while driving and rewind the belt fully when the buckle is unlatched.

4   If any of the above checks reveal problems with the seat belt system, replace parts as necessary.

12   Underhood hose check and replacement (every 15,000 miles or 12 months)

Warning: *Replacement of air conditioning hoses must be left to a dealer service department or air conditioning shop that has the equipment to depressurize the system safely. Never remove air conditioning components or hoses until the system has been depressurized.*

General

1   High temperatures under the hood can cause deterioration of the rubber and plastic hoses used for engine, accessory and emission systems operation. Periodic inspection should be made for cracks, loose clamps, material hardening and leaks.
2   Information specific to the cooling system hoses can be found in Section 9.
3   Most (but not all) hoses are secured to the fittings with clamps. Where clamps are used, check to be sure they haven't lost their tension, allowing the hose to leak. If clamps aren't used, make sure the hose has not expanded and/or hardened where it slips over the fitting, allowing it to leak.

PCV system hose

4   To reduce hydrocarbon emissions, crankcase blow-by gas is vented through the PCV valve in the rocker arm cover to the intake manifold via a rubber hose on most models. The blow-by gases mix with incoming air in the intake manifold before being burned in the combustion chambers.
5   Check the PCV hose for cracks, leaks and other damage. Disconnect it from the valve cover and the intake manifold and check the inside for obstructions. If it's clogged, clean it out with solvent.

Vacuum hoses

6   It's quite common for vacuum hoses, especially those in the emissions system, to be color coded or identified by colored stripes molded into them. Various systems require hoses with different wall thickness, collapse resistance and temperature resistance. When replacing hoses, be sure the new ones are made of the same material.
7   Often the only effective way to check a hose is to remove it completely from the vehicle. If more than one hose is removed, be sure to label the hoses and fittings to ensure correct installation.
8   When checking vacuum hoses, be sure to include any plastic T-fittings in the check. Inspect the fittings for cracks and the hose where it fits over each fitting for distortion, which could cause leakage.
9   A small piece of vacuum hose (1/4-inch inside diameter) can be used as a stethoscope to detect vacuum leaks. Hold one end of the hose to your ear and probe around vacuum hoses and fittings, listening for the "hissing" sound characteristic of a vacuum leak.
Warning: *When probing with the vacuum hose stethoscope, be careful not to come into contact with moving engine components such as drivebelts, the cooling fan, etc.*

Fuel hose

Warning: *Gasoline is flammable, so take extra precautions when you work on any part of the fuel system. Don't smoke or allow open flames or bare light bulbs near the work area, and don't work in a garage where a gas-type appliance (such as a water heater or clothes dryer) is present. Since fuel is carcinogenic, wear fuel-resistant gloves when there's a possibility of being exposed to fuel, and, if you spill any fuel on your skin, rinse it off immediately with soap and water. Mop up any spills immediately and do not store fuel-soaked rags where they could ignite. The fuel system is under constant pressure, so, if any fuel lines are to be disconnected, the fuel pressure in the system must be relieved first (see Chapter 4 for more information). When you perform any kind of work on the fuel system, wear safety glasses and have a Class B type fire extinguisher on hand.*
10   The fuel lines are usually under pressure, so if any fuel lines are to be disconnected be prepared to catch spilled fuel. Refer to Chapter 4 for the fuel system pressure relief procedure.
Warning: *Your vehicle is equipped with fuel injection and you must relieve the fuel system pressure before servicing the fuel lines.*
11   Check all flexible fuel lines for deterioration and chafing. Check especially for cracks in areas where the hose bends and just before fittings, such as where a hose attaches to the fuel pump, fuel filter and fuel injection unit.
12   When replacing a hose, use only hose that is specifically designed for your fuel injection system.
13   Spring-type clamps are sometimes used on fuel return or vapor lines. These clamps often lose their tension over a period of time, and can be "sprung" during removal. Replace all spring-type clamps with screw clamps whenever a hose is replaced. Some fuel lines use spring-lock type couplings, which require a special tool to disconnect. See Chapter 4 for more information on this type of coupling.

Metal lines

14   Sections of metal line are often used for fuel line between the fuel pump and the fuel injection unit. Check carefully to make sure the line isn't bent, crimped or cracked.
15   If a section of metal fuel line must be replaced, use seamless steel tubing only, since copper and aluminum tubing do not have the strength necessary to withstand vibration caused by the engine.
16   Check the metal brake lines where they enter the master cylinder and brake proportioning unit (if used) for cracks in the lines and loose fittings. Any sign of brake fluid leakage calls for an immediate thorough inspection of the brake system.

13   Brake check (every 15,000 miles or 12 months)

Warning: *Dust created by the brake system is harmful to your health. Never blow it out with compressed air and don't inhale any of it. An approved filtering mask should be worn when working on brakes. Do not, under any circumstances, use petroleum-based solvents to clean brake parts. Use brake system cleaner only!*
1   The brakes should be inspected every time the wheels are removed or whenever a defect is suspected. Indications of a potential brake system problem include the vehicle pulling to one side when the brake pedal is depressed, noises coming from the brakes when they are applied, excessive brake pedal travel, a pulsating pedal and leakage of fluid, usually seen on the inside of the tire or wheel.
Note: *It is normal for a vehicle equipped with an Anti-lock Brake System (ABS) to exhibit brake pedal pulsations during severe braking conditions.*

Disc brakes

2   Disc brakes can be visually checked without removing any parts except the wheels. Remove the hub caps (if applicable) and loosen the wheel lug nuts a quarter turn each.
3   Raise the vehicle and place it securely on jackstands.
Warning: *Never work under a vehicle that is supported only by a jack!*
4   Remove the wheels. Now visible is the disc brake caliper which contains the pads. There is an outer brake pad and an inner pad. Both must be checked for wear.
Note: *Usually the inner pad wears faster than the outer pad.*
5   Measure the thickness of the outer pad at each end of the caliper and the inner pad through the inspection hole in the caliper body (see illustration). Compare the measurement with the limit given in this Chapter's Specifications; if any brake pad thickness is less than specified, then all brake pads must be replaced (see Chapter 9).

13.5 You will find an inspection hole like this in each caliper through which you can view the thickness of remaining friction material for the inner pad

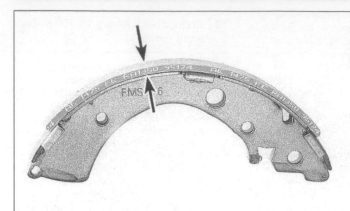

13.14 If the lining is bonded to the brake shoe, measure the lining thickness from the outer surface to the metal shoe, as shown here; if the lining is riveted to the shoe, measure from the lining outer surface to the rivet head

13.17 Carefully peel back the wheel cylinder boot and check for leaking fluid indicating that the cylinder must be replaced or rebuilt

6   If you're in doubt as to the exact pad thickness or quality, remove them for measurement and further inspection (see Chapter 9).

7   Check the disc for score marks, wear and burned spots. If any of these conditions exist, the disc should be removed for servicing or replacement (see Chapter 9).

8   Before installing the wheels, check all the brake lines and hoses for damage, wear, deformation, cracks, corrosion, leakage, bends and twists, particularly in the vicinity of the rubber hoses and calipers.

9   Install the wheels, lower the vehicle and tighten the wheel lug nuts to the torque given in this Chapter's Specifications.

Drum brakes

10   Remove the hub caps (if applicable) and loosen the wheel lug nuts a quarter turn each.

11   Raise the rear of the vehicle and support it securely on jackstands. Block the front wheels to prevent the vehicle from rolling, however, do not apply the parking brake or it will lock the drums in place. Remove the rear wheels.

Warning: *Never work under a vehicle that is supported only by a jack!*

12   Remove the brake drum as described in Chapter 9.

13   With the drum removed, carefully clean off any accumulations of dirt and dust using brake system cleaner.

Warning: *DO NOT blow the dust out with compressed air and don't inhale any of it.*

14   Measure the thickness of the lining material on both leading and trailing brake shoes (see illustration). Compare the measurement with the limit given in this Chapter's Specifications, if any brake shoe thickness is less than specified, then all brake shoes must be replaced (see Chapter 9).

15   Inspect the brake shoes for uneven wear patterns, cracks, glazing and delamination and replace if necessary. If the shoes have been saturated with brake fluid, oil or grease, this also necessitates replacement (see Chapter 9).

16   Make sure all the brake assembly springs are connected and in good condition.

17   Check the brake wheel cylinder for signs of fluid leakage. Carefully pry back the rub-

ber dust boots on the wheel cylinder (see illustration). Any leakage here is an indication that the wheel cylinders must be overhauled immediately (see Chapter 9). Also, check all hoses and connections for signs of leakage.

18   Clean the inside of the drum with brake system cleaner. Again, be careful not to breathe the dust.

19   Inspect the inside of the drum for cracks, score marks, deep scratches and "hard spots" which will appear as small discolored areas. If imperfections cannot be removed with fine emery cloth, the drum must be taken to an automotive machine shop for resurfacing.

20   Repeat the procedure for the remaining wheel.

21   Install the wheels, lower the vehicle and tighten the wheel lug nuts to the torque given in this Chapter's Specifications.

Parking brake

22   Slowly pull up on the parking brake and count the number of clicks you hear until the handle is up as far as it will go. The adjustment is correct if you hear the specified number of clicks (see this Chapter's Specifications). If you hear more or fewer clicks, it's time to adjust the parking brake (see Chapter 9).

23   An alternative method of checking the parking brake is to park the vehicle on a steep hill with the engine running (so you can apply the brakes if necessary) with the parking brake set and the transaxle in Neutral. If the parking brake cannot prevent the vehicle from rolling, it needs adjustment (see Chapter 9).

---

14   Steering, suspension and driveaxle boot check (every 15,000 miles or 12 months)

Note: *For detailed illustrations of the steering and suspension components, refer to Chapter 10.*

With the wheels on the ground

1   With the vehicle stopped and the front wheels pointed straight ahead, rock the steering wheel gently back and forth. If freeplay is excessive, a front wheel bearing, steering shaft universal joint or lower arm balljoint is worn or the steering gear is out of adjustment

or broken. Refer to Chapter 10 for the appropriate repair procedure.

2   Other symptoms, such as excessive vehicle body movement over rough roads, swaying (leaning) around corners and binding as the steering wheel is turned, may indicate faulty steering and/or suspension components.

3   Check the shock absorbers by pushing down and releasing the vehicle several times at each corner. If the vehicle does not come back to a level position within one or two bounces, the shocks/struts are worn and must be replaced. When bouncing the vehicle up and down, listen for squeaks and noises from the suspension components.

4   Check the shock absorbers for evidence of fluid leakage (see illustration). A light film of fluid is no cause for concern. Make sure that any fluid noted is from the shocks and not from some other source. If leakage is noted, replace the shocks as a set.

5   Check the shocks to be sure they are securely mounted and undamaged. Check the upper mounts for damage and wear. If damage or wear is noted, replace the shocks as a set (front and rear).

6   If the shocks must be replaced, refer to Chapter 10 for the procedure.

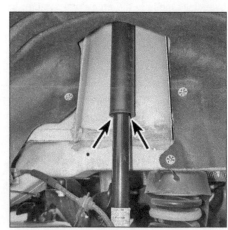

14.4 Check the shocks for leakage at the indicated area

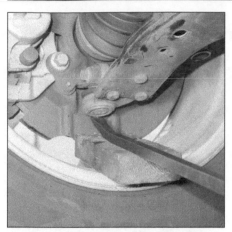

14.10 To check a balljoint for wear, try to pry the control arm up and down to make sure there is no play in the balljoint (if there is, replace it)

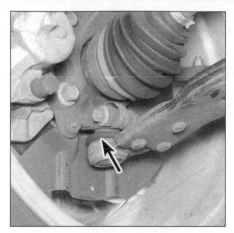

14.11 Check the balljoint boot for damage also

14.14 Flex the driveaxle boots by hand to check for cracks and/or leaking grease

Under the vehicle

7   Raise the vehicle with a floor jack and support it securely on jackstands. See *Jacking and towing* at the front of this book for the proper jacking points.

8   Check the tires for irregular wear patterns and proper inflation. See Section 5 for information regarding tire wear and Chapter 10 for information on wheel bearing replacement.

9   Inspect the universal joint between the steering shaft and the steering gear housing. Check the steering gear housing for lubricant leakage. Make sure that the dust seals and boots are not damaged and that the boot clamps are not loose. Check the steering linkage for looseness or damage. Check the tie-rod ends for excessive play. Look for loose bolts, broken or disconnected parts and deteriorated rubber bushings on all suspension and steering components. While an assistant turns the steering wheel from side to side, check the steering components for free movement, chafing and binding. If the steering components do not seem to be reacting with the movement of the steering wheel, try to determine where the slack is located.

10   Check the balljoints for wear by trying to move each control arm up and down with a pry bar (see illustration) to ensure that its balljoint has no play. If any balljoint does have play, replace it. See Chapter 10 for the balljoint replacement procedure.

11   Inspect the balljoint boots for damage and leaking grease (see illustration). Replace the balljoints with new ones if they are damaged (see Chapter 10).

12   At the rear of the vehicle, inspect the suspension arm bushings for deterioration. Additional information on suspension components can be found in Chapter 10.

Driveaxle boots

13   The driveaxle boots are very important because they prevent dirt, water and foreign material from entering and damaging the constant velocity (CV) joints. Oil and grease can cause the boot material to deteriorate prematurely, so it's a good idea to wash the boots with soap and water. Because it constantly pivots back and forth following the steering action of the front hub, the outer CV boot wears out sooner and should be inspected regularly.

14   Inspect the boots for tears and cracks as well as loose clamps (see illustration). If there is any evidence of cracks or leaking lubricant, they must be replaced as described in Chapter 10.

15 Fuel system check (every 15,000 miles or 12 months)

Warning: *Gasoline is flammable, so take extra precautions when you work on any part of the fuel system. Don't smoke or allow open flames or bare light bulbs near the work area, and don't work in a garage where a gas-type appliance (such as a water heater or clothes dryer) is present. Since fuel is carcinogenic, wear fuel-resistant gloves when there's a possibility of being exposed to fuel, and, if you spill any fuel on your skin, rinse it off immediately with soap and water. Mop up any spills immediately and do not store fuel-soaked rags where they could ignite. When you perform any kind of work on the fuel system, wear safety glasses and have a Class B type fire extinguisher on hand. The fuel system is under constant pressure, so, before any lines are disconnected, the fuel system pressure must be relieved (see Chapter 4).*

1   If you smell gasoline while driving or after the vehicle has been sitting in the sun, inspect the fuel system immediately.

2   Remove the fuel filler cap and inspect it for damage and corrosion. The gasket should have an unbroken sealing imprint. If the gasket is damaged or corroded, install a new cap.

3   Inspect the fuel feed line for cracks. Make sure that the connections between the fuel lines and the fuel injection system and between the fuel lines and the in-line fuel filter are tight.

Warning: *Your vehicle is fuel injected, so you must relieve the fuel system pressure before servicing fuel system components. The fuel system pressure relief procedure is outlined in Chapter 4.*

4   Since some components of the fuel system - the fuel tank and part of the fuel feed line, for example - are underneath the vehicle, they can be inspected more easily with the vehicle raised on a hoist. If that's not possible, raise the vehicle and support it on jackstands.

5   With the vehicle raised and safely supported, inspect the gas tank and filler neck for punctures, cracks and other damage. The connection between the filler neck and the tank is particularly critical. Sometimes a rubber filler neck will leak because of loose clamps or deteriorated rubber. Inspect all fuel tank mounting brackets and straps to be sure that the tank is securely attached to the vehicle.

Warning: *Do not, under any circumstances, try to repair a fuel tank (except rubber components). A welding torch or any open flame can easily cause fuel vapors inside the tank to explode.*

6   Carefully check all rubber hoses and metal lines leading away from the fuel tank. Check for loose connections, deteriorated hoses, crimped lines and other damage. Repair or replace damaged sections as necessary (see Chapter 4).

16 Manual transaxle lubricant level check (every 15,000 miles or 12 months)

1   The manual transaxle does not have a dipstick. To check the fluid level, raise the vehicle and support it securely on jackstands. On the front side of the transaxle housing you will see a plug. Remove it. If the lubricant level is correct, it should be up to the lower edge of the hole.

2   If the transaxle needs more lubricant (if the level is not up to the hole), use a syringe

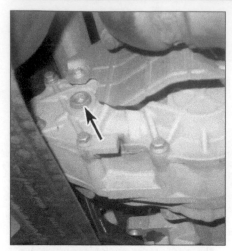

17.2 Remove the check/fill plug and use your finger as a dipstick to check the transfer case lubricant level.

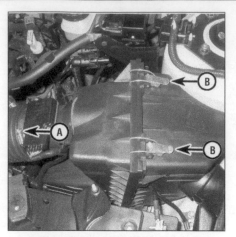

18.1a Loosen the intake hose clamp (A), then unlatch these clips (B) . . .

18.1b . . . pull the cover out of the way and lift the element out

or a gear oil pump to add more. Stop filling the transaxle when the lubricant begins to run out the hole.

3   Install the plug and tighten it securely. Drive the vehicle a short distance, then check for leaks.

17 Transfer case lubricant level check (every 15,000 miles or 12 months)

1   Raise the vehicle and support it securely on jackstands.

2   Using a ratchet or breaker bar, unscrew the check/fill plug from the transfer case (see illustration).

3   Use your little finger to reach inside the housing to feel the lubricant level. The level should be at or near the bottom of the plug hole. If it isn't, add the recommended lubricant through the plug hole with a syringe or squeeze bottle.

4   Install and tighten the plug. Check for leaks after the first few miles of driving.

18 Air filter check and replacement (every 15,000 miles or 12 months)

All 2012 and earlier models, except 2009 and later V6 models

1   The air filter is located inside a housing at the left (driver's) side of the engine compartment. To remove the air filter, loosen the clamp securing the inlet tube to the air filter cover, release the clamps that secure the two halves of the air cleaner housing together, then separate the cover halves and remove the air filter element (see illustrations).

2   Inspect the outer surface of the filter element. If it is dirty, replace it. If it is only moderately dusty, it can be reused by blowing it clean from the back to the front surface with compressed air. Because it is a pleated paper type filter, it cannot be washed or oiled. If it cannot be cleaned satisfactorily with compressed air, discard and replace it. While the cover is off, be careful not to drop anything down into the housing.

Caution: *Never drive the vehicle with the air cleaner removed. Excessive engine wear*

could result and backfiring could even cause a fire under the hood.

3   Wipe out the inside of the air cleaner housing.

4   Place the new filter into the air cleaner housing, making sure it seats properly.

5   Installation of the housing is the reverse of removal.

2009 and later V6 models

6   Release the clips retaining the air filter cover, then remove the cover (see illustrations).

7   Remove the filter element from the housing (see illustration).

8   Perform Steps 2 through 4.

9   Install the filter housing cover, making sure the retaining clips snap into place.

2013 and later models
1.5L, 1.6L and 2.0L engines

10   The air filter is located inside a housing at the left (driver's) side of the engine compartment. Remove the air cleaner cover fasteners and cover (see illustrations) then remove the cover.

11   Remove the air filter element from the air filter housing (see illustration).

18.6a Flip up the air filter housing cover clips . . .

18.6b . . . remove the cover . . .

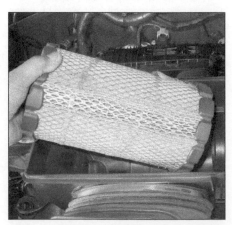

18.7 . . . then remove the air filter element from the housing

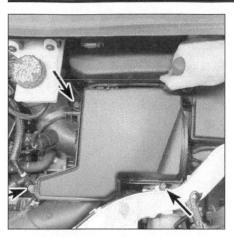

18.10a Remove the air cleaner cover fasteners . . .

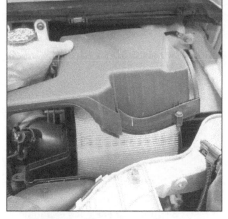

18.10b . . . then remove the cover

18.11 Remove the filter from the housing

12   Wipe out the inside of the air cleaner housing.

13   Install the new air filter element. Slide the open end of the air filter element in first. The tab of the closed end of the air filter should be facing down (see illustration).

14   Install the cover and tighten the cover fasteners securely.

2.5L engine

15   Release the clips retaining the air filter cover, then lift the cover up.

16   Remove the filter element from the housing, noting which side is facing up.

17   Wipe out the inside of the air cleaner housing.

18   Install the filter housing cover, making sure the retaining clips snap into place.

19   Cabin air filter replacement (2005 and later models)

Note: *Not all 2005 through 2012 models were equipped with a cabin air filter from the factory, but one can be installed by following this procedure. 2004 and earlier models do not have the provision for installing a cabin air filter.*

18.13 Make sure the tab on the filter is facing down and is aligned with the fork in the bottom of the housing

2005 through 2012 models

1   Remove the right-side cowl cover and vent tray (see Chapter 11).

2   Remove the cabin air filter element.

3   Installation is the reverse of removal.

2013 and later models

4   Remove the instrument panel insulator

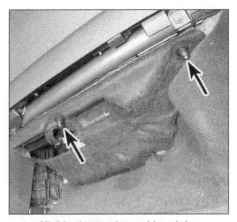

19.4 Instrument panel insulator retainer locations

retainers (see illustration) and pull the instrument panel insulator down.

5   Remove the screw securing the cabin air filter cover (see illustration).

6   Press the filter cover, release the tabs together (see illustration) and remove the cabin air filter cover.

7   Remove the cabin filter, noting the direction of the air flow arrows on the filter (see illustration).

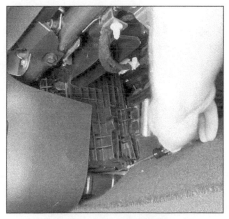

19.5 Remove the cabin filter cover, reverse Torx fastener

19.6 Press the filter cover, release the tabs together to release it

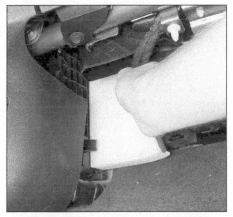

19.7 Remove the filter and note the direction of air flow arrows on the filter

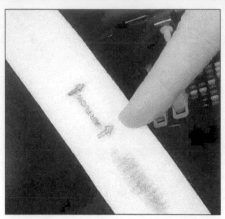

19.8 Install the filter with the air flow arrows pointing towards the front of the vehicle

20.2 Be sure to check each exhaust system rubber hanger for damage

21.1 The fuel filter is located in front of the fuel tank

8   Installation is the reverse of removal, making sure the filter is oriented correctly (see illustration).

20   Exhaust system check (every 30,000 miles or 24 months)

1   With the engine cold (at least three hours after the vehicle has been driven), check the complete exhaust system from the engine to the end of the tailpipe. Ideally, the inspection should be done with the vehicle on a hoist to permit unrestricted access. If a hoist isn't available, raise the vehicle and support it securely on jackstands.
2   Check the exhaust pipes and connections for evidence of leaks, severe corrosion and damage. Make sure that all brackets and hangers are in good condition and tight (see illustration).
3   At the same time, inspect the underside of the body for holes, corrosion, open seams, etc., which may allow exhaust gases to enter the passenger compartment. Seal all body openings with silicone or body putty.
4   Rattles and other noises can often be traced to the exhaust system, especially the mounts and hangers. Try to move the pipes, muffler and catalytic converter. If the components can come in contact with the body or suspension parts, secure the exhaust system with new mounts.
5   Check the running condition of the engine by inspecting inside the end of the tailpipe. The exhaust deposits here are an indication of engine state-of-tune. If the pipe is black and sooty or coated with white deposits, the engine may need a tune-up, including a thorough fuel system inspection and adjustment.

21   Fuel filter replacement (every 30,000 miles or 24 months)

Warning: *Gasoline is extremely flammable, so take extra precautions when you work on any*

part of the fuel system. Don't smoke or allow open flames or bare light bulbs near the work area, and don't work in a garage where a gas-type appliance (such as a water heater or clothes dryer) is present. Since fuel is carcinogenic, wear fuel-resistant gloves when there's a possibility of being exposed to fuel, and, if you spill any fuel on your skin, rinse it off immediately with soap and water. Mop up any spills immediately and do not store fuel-soaked rags where they could ignite. When you perform any kind of work on the fuel system, wear safety glasses and have a Class B type fire extinguisher on hand.
Note: *This procedure applies only to 2008 and earlier models. Later vehicles have a filter that is part of the fuel pump module. See Chapter 4 for more information.*
1   The fuel filter is mounted under the vehicle on the right side, in front of the gas tank (see illustration).
2   Relieve the fuel system pressure (see Chapter 4).
3   If necessary, raise the vehicle and support it securely on jackstands. Inspect the fittings at both ends of the filter to see if they're clean. If more than a light coating of dust is present, clean the fittings before proceeding.
4   Release the clips holding the fuel lines to the filter (see illustration).
Note: *On some models, spring-lock couplings are used instead of retaining clips. See Chapter 4 for information on how to disconnect spring-lock couplings.*
5   Detach the fuel hoses, one at a time, from the filter. Be prepared for fuel spillage.
6   After the lines are detached, check the fittings for damage and distortion. If they were damaged in any way during removal, new ones must be used when the lines are reattached to the new filter (if new clips are packaged with the filter, be sure to use them in place of the originals).
7   Remove the fuel filter from the mounting clamp, while noting the direction the fuel filter is installed.
8   Install the new filter in the same direction. Carefully push each hose onto the filter until it's seated against the collar on the fitting, then install the clips. Make sure the clips are

securely attached to the hose fittings - if they come off, the hoses could back off the filter and a fire could result! On models with spring-lock couplings, push the fittings onto the filter until they click into place.
9   Start the engine and check for fuel leaks.

22   Cooling system servicing (draining, flushing and refilling) (every 30,000 miles or 24 months)

Warning: *Wait until the engine is completely cool before beginning this procedure.*
Warning: *Do not allow antifreeze to come in contact with your skin or painted surfaces of the vehicle. Rinse off spills immediately with plenty of water. Antifreeze is highly toxic if ingested. Never leave antifreeze lying around in an open container or in puddles on the floor; children and pets are attracted by its sweet smell and may drink it. Check with local authorities about disposing of used antifreeze. Many communities have collection centers which will see that antifreeze is disposed of safely. Never dump used antifreeze on the ground or pour it into drains.*
Caution: *Do not mix coolants of different colors. Doing so might damage the cooling system and/or the engine. The manufacturer specifies either a green colored coolant or a yellow colored coolant to be used in these systems. Read the warning label in the engine compartment for additional information.*
Note: *Non-toxic antifreeze is now manufactured and available at local auto parts stores, but even this type must be disposed of properly.*
1   Periodically, the cooling system should be drained, flushed and refilled to replenish the antifreeze mixture and prevent formation of rust and corrosion, which can impair the performance of the cooling system and cause engine damage. When the cooling system is serviced, all hoses and the expansion tank cap should be checked and replaced if necessary.

21.4 Use a small screwdriver to pry off the fuel line fitting retaining clips at both ends of the filter - 2008 and earlier models

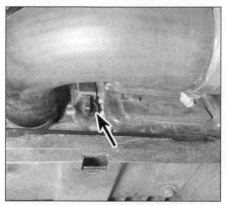

22.4a The radiator drain fitting is located at the bottom of the radiator - 2013 and later models

22.4b On 2012 and earlier models, the radiator drain fitting is located at the bottom of the radiator - before opening the valve, push a short length of rubber hose onto the plastic fitting to prevent the coolant from splashing

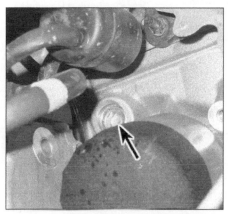

22.5 On the V6 engine, remove the engine block drain plugs from both sides of the engine

Draining

2   Apply the parking brake and block the wheels. If the vehicle has just been driven, wait several hours to allow the engine to cool down before beginning this procedure.

3   Once the engine is completely cool, remove the expansion tank cap.

4   Move a large container under the radiator drain to catch the coolant. Attach a length of hose to the drain fitting to direct the coolant into the container, then open the drain fitting (a pair of pliers may be required to turn it) (see illustration).

5   After the coolant stops flowing out of the radiator, move the container under the engine block drain plug(s) and allow the coolant in the block to drain (see illustration).

6   While the coolant is draining, check the condition of the radiator hoses, heater hoses and clamps (see Section 9 if necessary). Replace any damaged clamps or hoses.

7   Reinstall the block drain plugs and tighten them securely.

Flushing

8   Fill the cooling system with clean water, following the *Refilling* procedure (see Step 15).

9   Start the engine and allow it to reach

normal operating temperature, then rev up the engine a few times.

10   Turn the engine off and allow it to cool completely, then drain the system as described earlier.

11   Repeat Steps 8 through 10 until the water being drained is free of contaminants.

12   In severe cases of contamination or clogging of the radiator, remove the radiator (see Chapter 3) and have a radiator repair facility clean and repair it if necessary.

13   Many deposits can be removed by the chemical action of a cleaner available at auto parts stores. Follow the procedure outlined in the manufacturer's instructions.

14   When the coolant is regularly drained and the system refilled with the correct antifreeze/water mixture, there should be no need to use chemical cleaners or descalers.

Refilling

15   Close and tighten the radiator drain.

16   Place the heater temperature control in the maximum heat position. If you're working on a 2.3L or 2.5L four-cylinder engine, open the bleed valve on the back of the water outlet.

17   Slowly add new coolant (a 50/50 mixture of water and antifreeze) to the expansion tank until the level is at the MAX fill mark on the expansion tank. If you're working on a 2.3L or 2.5L four-cylinder engine, add fluid until it flows from the bleed valve, then close the valve. Continue filling the expansion tank until the level is up to the MAX fill line.

18   Install the expansion tank cap and run the engine in a well-ventilated area until the thermostat opens (coolant will begin flowing through the radiator and the upper radiator hose will become hot).

19   Turn the engine off and let it cool. Add more coolant mixture to bring the level to the MAX fill mark on the expansion tank.

20   Squeeze the upper radiator hose to expel air, then add more coolant mixture if necessary. Replace the expansion tank cap.

21   Start the engine, allow it to reach normal operating temperature and check for leaks. Also, set the heater and blower controls to

the maximum setting and check to see that the heater output from the air ducts is warm. This is a good indication that all air has been purged from the cooling system.

23   Brake fluid change (every 30,000 miles or 24 months)

Warning: *Brake fluid can harm your eyes and damage painted surfaces, so use extreme caution when handling or pouring it. Do not use brake fluid that has been standing open or is more than one year old. Brake fluid absorbs moisture from the air. Excess moisture can cause a dangerous loss of braking effectiveness.*

1   At the specified intervals, the brake fluid should be drained and replaced. Since the brake fluid may drip or splash when pouring it, place plenty of rags around the master cylinder to protect any surrounding painted surfaces.

2   Before beginning work, purchase the specified brake fluid (see *Recommended lubricants and fluids* at the beginning of this Chapter).

3   Remove the cap from the master cylinder reservoir.

4   Using a hand suction pump or similar device, withdraw the fluid from the master cylinder reservoir.

5   Add new fluid to the master cylinder until it rises to the base of the filler neck.

6   Bleed the brake system as described in Chapter 9 at all four brakes until new and uncontaminated fluid is expelled from the bleeder screw. Be sure to maintain the fluid level in the master cylinder as you perform the bleeding process. If you allow the master cylinder to run dry, air will enter the system.

7   Refill the master cylinder with fluid and check the operation of the brakes. The pedal should feel solid when depressed, with no sponginess.

Warning: *Do not operate the vehicle if you are in doubt about the effectiveness of the brake system.*

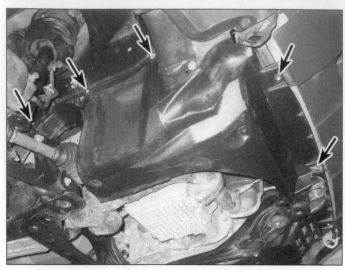

24.2 Remove the lower splash shield to gain
access to the drivebelt - 2012 and earlier models shown
(if you're working on a 2013 or later model, see illustration 6.6)

ACCEPTABLE

Cracks Running Across
"V" Portions of Belt

1/2"

Missing Two or More Adjacent
Ribs 1/2" or longer

UNACCEPTABLE

24.4 Small cracks in the
underside of a V-ribbed belt
are acceptable - lengthwise
cracks, or missing pieces that
cause the belt to make noise,
are cause for replacement

Cracks Running Parallel
to "V" Portions of Belt

24   Drivebelt check and replacement (every 30,000 miles or 24 months)

1   The drivebelt(s) is/are located at the front of the engine and plays an important role in the overall operation of the engine and its components. V6 models use a separate drivebelt to drive the water pump. Due to its function and material make up, the belt is prone to wear and should be periodically inspected. The belt drives the alternator, power steering pump (2007 and earlier models), water pump (four-cylinder models) and air conditioning compressor (2012 and earlier four-cylinder models). On 2013 and later 2.0L and 2.5L engines, two belts are used; the outer belt drives the air conditioning compressor and the inner belt drives the alternator and water pump. Although the belt(s) should be inspected at the recommended intervals, replacement may not be necessary for more than 100,000 miles.

Check

2   Since the drivebelt is located very close to the right-hand side of the engine compartment, it is possible to gain better access by raising the front of the vehicle and removing the right-hand wheel, then unbolting the lower splash shield from the underbody (see illustration). Be sure to support the front of the vehicle securely on jackstands.

3   With the engine stopped, inspect the full length of the drivebelt for cracks and separation of the belt plies. It will be necessary to turn the engine (using a wrench or socket and bar on the crankshaft pulley bolt) in order to move the belt from the pulleys so that the belt can be inspected thoroughly. Twist the belt between the pulleys so that both sides can be viewed. Also check for fraying, and glazing which gives the belt a shiny appearance. Check the pulleys for nicks, cracks, distortion and corrosion.

4   Note that it is not unusual for a ribbed

belt to exhibit small cracks in the edges of the belt ribs, and unless these are extensive or very deep, belt replacement is not essential (see illustration).

Replacement
2012 and earlier models - accessory drivebelt

5   To remove the drivebelt, loosen the right front wheel lug nuts, then raise the front of the vehicle and support it on jackstands. Remove the right front wheel and remove the lower splash shield from the underbody.

6   Note how the drivebelt is routed, then remove the belt from the pulleys. If you're working on a four-cylinder engine, use a wrench on the tensioner center bolt and turn the tensioner clockwise to release the drivebelt tension. If you're working on a V6 engine, insert a 3/8-inch drive ratchet or breaker bar into the tensioner hole and pull the handle clockwise to release the drivebelt tension (see illustration).

24.6 Rotate the tensioner arm to
relieve belt tension

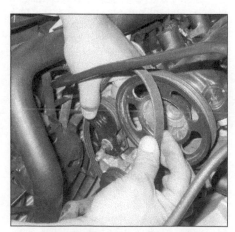

24.10 Rotate the tensioner and
remove the belt

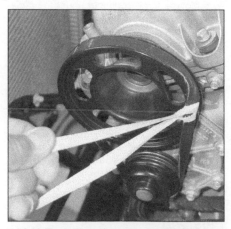

24.13 Removing the water pump belt on
a 2005 or later V6 engine. Warning: *While
doing this, rotate the engine by hand only
(do not use the starter)*

7   Fit the new drivebelt onto the crankshaft, alternator, power steering pump, and air conditioning compressor pulleys, as applicable, then turn the tensioner counterclockwise and locate the drivebelt on the pulley. Make sure that the drivebelt is correctly seated in all of the pulley grooves, then release the tensioner.

8   Install the lower splash shield and wheel, then lower the car to the ground. Tighten the lug nuts to the torque listed in this Chapter's Specifications.

2012 and earlier water pump drivebelt - V6 engine

Replacement

9   The water pump drivebelt is located at the left end of the engine and is driven by a pulley attached to the end of the front cylinder bank intake camshaft. The belt and pulley are protected by a cover.

2004 and earlier models

10   To replace the belt, remove the cover (see illustration 4.1 in Chapter 2B), rotate the tensioner clockwise and remove the belt (see illustration). Slowly release the tensioner.

11   Route the new belt over the pulleys, again rotating the tensioner to allow the belt to be installed, then release the belt tensioner. Make sure the belt is positioned properly on the pulleys. Reinstall the cover and tighten the fasteners securely.

2005 through 2012 models

12   The water pump belt on these models is of a unique design, called a "stretchy belt," which provides tension without the use of a mechanical tensioner.

13   Disconnect the cable from the negative terminal of the battery (see Chapter 5). Remove the cover from over the pulley. Insert a length of flexible material such as a leather or plastic strap under the belt at the pulley at the end of the camshaft. Have an assistant rotate the engine (clockwise) by using a socket and breaker bar on the crankshaft pulley bolt, while you feed the remover strap between the belt and the pulley (see illustra-

tion). Pull the strap quickly to force the belt from the pulley on the camshaft.

Caution: *Do not use hard plastic or metal tools to pry the belt off; it can be easily damaged.*

Note: *If the belt is not going to be re-used, you can simply cut it off.*

14   Route the new belt under the water pump pulley, then over the pulley on the camshaft, and have your assistant rotate the engine again; the belt should pop over the pulley on the camshaft. Make sure the belt is positioned properly on both pulleys.

15   Reinstall the cover and tighten the fasteners securely.

2013 and later models

2.0L and 2.5L engine air conditioning drivebelt and late production 1.5L engine accessory drivebelt

16   The 2.0L and 2.5L engine air conditioning belt and late build 1.5L engine accessory drivebelt are of a unique design, called a "stretchy belt," which provides tension without the use of a mechanical tensioner.

17   To remove the drivebelt, loosen the right front wheel lug nuts, then raise the front of the vehicle and support it on jackstands. Remove the right front wheel and remove the inner fender splash shield fasteners, (see Chapter 11, Section 10, if necessary).

18   Raise the vehicle and remove the engine splash shield (see illustration 6.6).

19   Disconnect the cable from the negative terminal of the battery (see Chapter 5).

20   Insert a length of flexible material such as a leather or plastic strap under the belt at the pulley at the end of the crankshaft. Have an assistant rotate the engine (clockwise) by using a socket and breaker bar on the crankshaft pulley bolt, while you feed the remover strap between the belt and the pulley. Pull the strap quickly to force the belt from the pulley on the crankshaft.

Caution: *Do not use hard plastic or metal tools to pry the belt off; it can be easily damaged.*

Note: *If the belt is not going to be re-used, you can simply cut it off.*

21   Route the new belt over the air conditioning pulley or all pulleys on the late build 1.5L, then partially over the crankshaft pulley, and have your assistant rotate the engine again; the belt should pop over the pulley on the crankshaft. Make sure the belt is positioned properly on all pulleys.

Accessory drivebelt

22   To remove the drivebelt, loosen the right front wheel lug nuts, then raise the front of the vehicle and support it on jackstands. Remove the right front wheel and remove the inner fender splash shield fasteners, (see Chapter 11, Section 10, if necessary).

23   On 2.0L and 2.5L models, remove the air conditioning belt (see Steps 17 through 20).

24   Note how the drivebelt is routed. If you're working on a 2.0L or 2.5L engines, from the top of the vehicle, use a wrench on the tensioner body and turn the tensioner counterclockwise to release the drivebelt tension. If you're working on a early build 1.5L or 1.6L engine, from under the vehicle, use a wrench on the center of the tensioner and turn the tensioner counterclockwise to release the drivebelt tension (see illustration).

25   On 2.0L and 2.5L engines, remove the drivebelt. Fit the new drivebelt onto the crankshaft, idler pulleys and alternator, as applicable, then turn the tensioner counterclockwise and locate the drivebelt on the pulley. Make sure that the drivebelt is correctly seated in all of the pulley grooves, then release the tensioner.

26   On early build 1.5L engines and 1.6L engines, the tensioner must be removed to replace the belt. Remove the tensioner (see Steps 34 and 35) and remove the tensioner with the belt. Install the tensioner with the belt in place (see Steps 36), then fit the new drivebelt onto the crankshaft, idler pulleys and alternator, as applicable. Rotate the tensioner counterclockwise and locate the drivebelt on the pulley. Make sure that the drivebelt is correctly seated in all of the pulley grooves, then release the tensioner.

27   Remaining installation is the reverse of removal.

Tensioner replacement

2012 and earlier models

28   Remove the drivebelt as described previously.

Accessory drivebelt tensioner

29   On four-cylinder models, remove the two bolts securing the tensioner to the engine block. On 2008 and earlier V6 models, remove the bolt in the center of the tensioner, then detach the tensioner from the engine (see illustration). On 2009 and later V6 engines, remove the three tensioner mounting bolts and detach the tensioner.

30   Installation is the reverse of removal. Be sure to tighten the tensioners bolt(s) to the torque listed in this Chapter's Specifications.

24.24 Rotate the tensioner arm counterclockwise to relieve belt tension - 1.6L engine shown

24.29 On V6 models, remove the bolt securing the tensioner to the block

24.31 Remove the bolt securing the water pump drivebelt tensioner

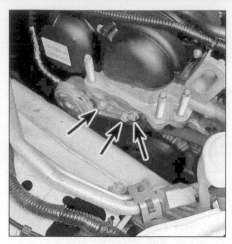

24.35 Tensioner mounting bolt locations - early build 1.5L engines and 1.6L engines

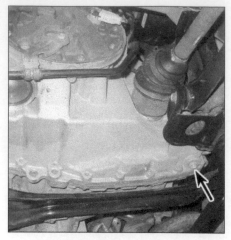

25.6a Transaxle drain plug (2002 model with a CD4E transaxle)

Water pump drivebelt tensioner (2004 and earlier V6 engines)

31   Unscrew the bolt securing the tensioner, then remove the tensioner (see illustration).
32   Installation is the reverse of removal. Be sure to tighten the tensioner bolts to the torque listed in this Chapter's Specifications.

2013 and later models

33   Remove the drivebelt as described previously.
34   On early build 1.5L engines and 1.6L engines, reposition the coolant expansion tank (see Chapter 3), then support the engine from below and remove the engine mount (see Chapter 2A, Section 18).
Caution: *Place a wood block on the jack head to protect the oil pan.*
35   Remove the tensioner mounting bolts and tensioner (see illustration).
36   Installation is the reverse of removal. Be sure to tighten the tensioner bolts to the torque listed in this Chapter's Specifications.

25   Automatic transaxle fluid change (every 30,000 miles or 24 months)

1   The automatic transaxle fluid should be changed at the recommended intervals.
2   Before beginning work, purchase the specified transmission fluid (see *Recommended lubricants and fluids* at the front of this Chapter).
3   Other tools necessary for this job include jackstands to support the vehicle in a raised position, wrenches, drain pan capable of holding at least four quarts, newspapers and clean rags.
4   The fluid should be drained immediately after the vehicle has been driven. Hot fluid is more effective than cold fluid at removing built up sediment.
Warning: *Fluid temperature can exceed 350-degrees F in a hot transaxle. Wear protective gloves.*
5   After the vehicle has been driven to warm up the fluid, raise the front of the vehicle and

support it securely on jackstands. Remove the transaxle splash shield for access to the transaxle drain plug.
Warning: *Never work under a vehicle that is supported only by a jack!*
6   Place the drain pan under the drain plug in the transaxle pan and remove the drain plug (see illustrations). Be sure the drain pan is in position, as fluid will come out with some force. Once the fluid is drained, reinstall the drain plug and tighten it to the torque listed in this Chapter's Specifications. Measure the amount of fluid drained and write down this figure for reference when refilling.
Note: *Before installing the drain plug, wrap the threads with Teflon tape.*

2014 and earlier models (except eCVT transaxle)

Note: *This portion of the procedure applies to models with a transaxle fluid dipstick.*
7   Lower the vehicle
8   With the engine off, add new fluid to the transaxle through the dipstick tube (see *Rec-*

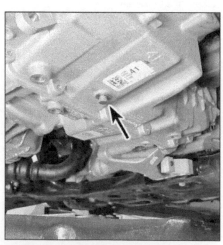

25.6b Transaxle drain plug (2015 model with a 6F35 transaxle)

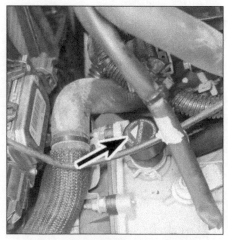

25.10 Transaxle fill tube vent cap (6F35 transaxle, air filter housing removed for clarity)

25.11 Automatic transaxle fluid level check plug (2015 and later 6F35 transaxle)

ommended lubricants and fluids for the recommended fluid type). Begin the refill procedure by initially adding 1/3 of the amount drained. Then, with the engine running, add 1/2-pint at a time (cycling the shifter through each gear position between additions) until the level is correct on the dipstick.

9   If desired, repeat Steps 5 through 8 once to flush any contaminated fluid from the torque converter.

2012 and earlier eCVT transaxle and all 2015 and later models - fluid fill and level check

Note: *The vehicle must be level to perform an accurate fluid level check. This will require raising the rear of the vehicle and supporting it securely on jackstands.*

10   6F35 (six-speed) transaxle only: Working at the top of the transaxle, squeeze the clamp and remove the vent cap from the transaxle fill tube (see illustration).

11   Remove the check plug (6F35 transaxle) (see illustration) or check/fill plug (eCVT transaxle) from the side of the transaxle case.

12   Fill the transaxle with the recommended fluid until the fluid is even with the check plug hole. On eCVT transaxles a pump will be necessary to pump the fluid into the hole.

13   Install the plug and tighten it securely.

14   Start the engine and, while depressing the brake pedal, cycle the shifter through each gear position and return it to Park. Remove the check plug and note whether or not fluid flows from the hole. If it does, allow it to run out until it does not. If no fluid flows from the hole, add fluid until it does.

15   If desired, repeat the fluid drain and fill procedure one or two more times to flush any contaminated fluid from the torque converter.

16   Reinstall the check plug and tighten it to the torque listed in this Chapter's Specifications. On 6F35 transaxles, reinstall the fill tube vent cap.

17   Drive the vehicle a few miles until the fluid is up to normal operating temperature, then recheck the fluid level, adding fluid (or allowing the excess to flow out) as necessary.

Fluid disposal

18   The old fluid drained from the transaxle cannot be reused in its present state and should be disposed of. Check with your local auto parts store, disposal facility or environmental agency to see if they will accept the fluid for recycling. After the fluid has cooled it can be drained into a container (capped plastic jugs, topped bottles, milk cartons, etc.) for transport to one of these disposal sites. Don't dispose of the fluid by pouring it on the ground or down a drain!

---

26   Manual transaxle lubricant change (every 30,000 miles or 24 months)

1   Raise the vehicle and support it securely on jackstands.

2   Move a drain pan, rags, newspapers and wrenches under the transaxle.

3   Remove the transaxle fill plug on the front of the case and the drain plug at the bottom of the case, then allow the lubricant to drain into the pan.

4   After the lubricant has drained completely, reinstall the drain plug and tighten it securely.

5   Using a hand pump, syringe or funnel, fill the transaxle with the specified lubricant until it is level with the lower edge of the filler hole. Reinstall the fill plug and tighten it securely.

6   Lower the vehicle.

7   Drive the vehicle for a short distance, then check the drain and fill plugs for leakage.

Lubricant disposal

8   The old lubricant drained from the transaxle cannot be reused in its present state and should be disposed of. Check with your local auto parts store, disposal facility or environmental agency to see if they will accept the lubricant for recycling. After the lubricant has cooled it can be drained into a container (capped plastic jugs, topped bottles, milk cartons, etc.) for transport to one of these disposal sites. Don't dispose of the lubricant by pouring it on the ground or down a drain!

---

27   Differential lubricant change (4WD models) (every 30,000 miles or 24 months)

Drain

1   This procedure should be performed after the vehicle has been driven so the lubricant will be warm and therefore flow out of the differential more easily.

2   Raise the vehicle and support it securely on jackstands.

3   The easiest way to drain the differential(s) is to remove the lubricant through the filler plug hole with a suction pump. If the differential cover gasket is leaking, it will be necessary to remove the cover to drain the lubricant (which will also allow you to inspect the differential).

Changing the lubricant with a suction pump (all models)

4   Remove the filler plug from the differential (see illustration).

5   Insert the flexible hose.

6   Work the hose down to the bottom of the differential housing and pump the lubricant out.

Changing the lubricant by removing the cover (2012 and earlier models only)

7   Move a drain pan, rags, newspapers and wrenches under the vehicle.

8   Remove the bolts on the lower half of the cover. Loosen the bolts on the upper half and use them to loosely retain the cover. Allow the oil to drain into the pan, then completely remove the cover.

9   Using a lint-free rag, clean the inside of the cover and the accessible areas of the differential housing. As this is done, check for chipped gears and metal particles in the lubricant, indicating that the differential should be more thoroughly inspected and/or repaired.

10   Thoroughly clean the gasket mating surfaces of the differential housing and the cover plate. Use a gasket scraper or putty knife to remove all traces of the old gasket.

11   Apply a thin bead of RTV sealant to the cover flange (see illustration). Make sure the bolt holes align properly, then install the cover and tighten the fasteners to the torque listed in this Chapter's Specifications.

Refill

12   Use a hand pump, syringe or funnel to fill the differential housing with the specified lubricant until it's level with the bottom of the filler plug hole.

13   Install the fill plug and tighten it securely.

27.4 Rear differential check/fill plug - early model shown, later models similar

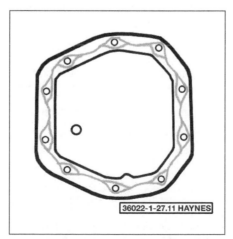

27.11 Apply a continuous thin bead of RTV sealant to the cover

36022-1-27.11 HAYNES

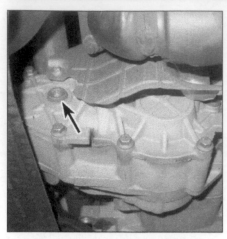

28.3a First remove the transfer case check/fill plug . . .

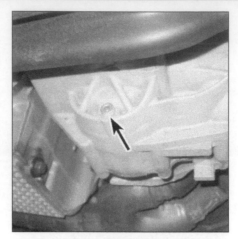

28.3b . . . then remove the transfer case drain plug - early models shown, later models similar

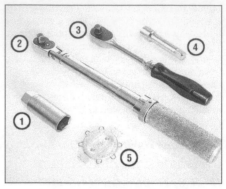

30.2 Tools required for changing spark plugs

1   **Spark plug socket** - *This will have special padding inside to protect the spark plug's porcelain insulator*
2   **Torque wrench** - *Although not mandatory, using this tool is the best way to ensure the plugs are tightened properly*
3   **Ratchet** - *Standard hand tool to fit the spark plug socket*
4   **Extension** - *Depending on model and accessories, you may need special extensions and universal joints to reach one or more of the plugs*
5   **Spark plug gap gauge** - *This gauge for checking the gap comes in a variety of styles. Make sure the gap for your engine is included*

Lubricant disposal

14   The old lubricant drained from the differential cannot be reused in its present state and should be disposed of. Check with your local auto parts store, disposal facility or environmental agency to see if they will accept the lubricant for recycling. After the lubricant has cooled it can be drained into a container (capped plastic jugs, topped bottles, milk cartons, etc.) for transport to one of these disposal sites. Don't dispose of the lubricant by pouring it on the ground or down a drain!

28   Transfer case lubricant change (4WD models) (every 30,000 miles or 24 months)

1   Drive the vehicle for at least 15 minutes to warm the lubricant in the case. Perform this warm-up procedure with 4WD engaged, if possible.
2   Raise the vehicle and support it securely on jackstands.
3   Remove the check/fill plug, then the drain plug and allow the old lubricant to drain completely (see illustrations).
4   After the lubricant has drained completely, reinstall the plug and tighten it securely.
5   Fill the case with the specified lubricant until it is level with the lower edge of the filler hole.
6   Install the check/fill plug and tighten it securely.
7   Drive the vehicle for a short distance, then check the drain and fill plugs for leakage.

Lubricant disposal

8   The old lubricant drained from the transfer case cannot be reused in its present state and should be disposed of. Check with your local auto parts store, disposal facility or environmental agency to see if they will accept the lubricant for recycling. After the lubricant

has cooled it can be drained into a container (capped plastic jugs, topped bottles, milk cartons, etc.) for transport to one of these disposal sites. Don't dispose of the lubricant by pouring it on the ground or down a drain!

29   Positive Crankcase Ventilation (PCV) valve replacement (four-cylinder models) (every 60,000 miles or 48 months)

Note: *For additional information on the PCV system, the PCV valve on V6 models and the oil separators on both engines, refer to Chapter 6.*
Note: *This Section applies to models with four-cylinder engines only.*

2004 and earlier models
1   The PCV valve is located in the valve cover.
2   Start the engine and allow it to idle, then disconnect the PCV valve from the valve cover and feel for vacuum at the end of the valve. If vacuum is felt, the PCV valve/system is working properly (see Chapter 6 for additional PCV system information).
3   If no vacuum is felt, remove the valve and check for vacuum at the hose. If vacuum is present at the hose but not at the valve, replace the valve. If no vacuum is felt at the hose, check for a plugged or cracked hose between the PCV valve and the intake plenum.
4   Check the rubber grommet in the valve cover for cracks and distortion. If it's damaged, replace it.
5   If the valve is clogged, the hose might also be plugged. Remove the hose between the valve and the intake manifold and clean it with solvent.
6   After cleaning the hose, inspect it for damage, wear and deterioration. Make sure it fits snugly on the fittings.
7   If necessary, install a new PCV valve.

2005 through 2012 models
8   Refer to Chapter 2A and remove the intake manifold.
9   Detach the hose from the PCV valve.
10   Push in the two tabs, then remove the PCV valve from the crankcase oil separator.
11   Installation is the reverse of removal.

2013 and later models
Note: *Similar to some 2012 and earlier 4-cylinder models, 2013 and later models utilize an oil separator as opposed to individual components. The oil separator is attached to the front of the engine block with a hose going to the intake manifold.*
12   Refer to Chapter 6. The PCV valve is an integral part of the crankcase oil separator.

30   Spark plug check and replacement (see Maintenance schedule for service intervals)

1   On these vehicles the spark plugs are located at the top of the engine.
2   In most cases, the tools necessary for spark plug replacement include a spark plug socket which fits onto a ratchet (spark plug sockets are padded inside to prevent damage to the porcelain insulators on the new plugs), various extensions and a gap gauge to check and adjust the gaps on the new plugs (see

30.5 Spark plug manufacturers recommend using a wire-type gauge when checking the gap on platinum or iridium spark plugs

30.7a V6 models and 2005 and later four-cylinder models are equipped with individual coils which must be removed to access the spark plugs

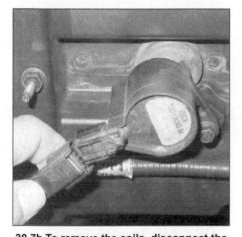

30.7b To remove the coils, disconnect the electrical connector . . .

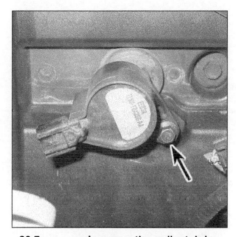

30.7c . . . and remove the coil retaining screw . . .

30.7d . . . then pull straight up and out to remove the coil

illustration). A torque wrench should be used to tighten the new plugs.

3   The best approach when replacing the spark plugs is to purchase the new ones in advance, adjust them to the proper gap and replace the plugs one at a time. When buying the new spark plugs, be sure to obtain the correct plug type for your particular engine. This information can be found in this Chapter's Specifications at the beginning of this Chapter or in the factory owner's manual.

4   Allow the engine to cool completely before attempting to remove any of the plugs. These engines are equipped with aluminum cylinder heads, which can be damaged if the spark plugs are removed when the engine is hot. While you are waiting for the engine to cool, check the new plugs for defects and adjust the gaps.

5   The gap is checked by inserting the proper-thickness gauge between the electrodes at the tip of the plug (see illustration). The gap between the electrodes should be the same as the one specified in this Chapter's Specifications. The gauge should just slide between the electrodes with a slight amount of drag. If the gap is incorrect, use the adjuster on the

gauge body to bend the curved side electrode slightly until the proper gap is obtained. If the side electrode is not exactly over the center electrode, bend it with the adjuster until it is. Check for cracks in the porcelain insulator (if any are found, the plug should not be used).
Note: *We recommend using a wire-type gauge when checking platinum- or iridium-type spark plugs. Other types of gauges may scrape the thin coating from the electrodes, thus dramatically shortening the life of the plugs.*

6   Remove the engine cover on models so equipped. On V6 models remove the intake manifold (see Chapter 2B).

7   Some engines are equipped with individual ignition coils which must be removed first to access the spark plugs (see illustrations). On other engines, remove the spark plug wire from one spark plug. Pull only on the boot at the end of the wire - do not pull on the wire. A plug wire removal tool should be used if available.

8   If compressed air is available, use it to blow any dirt or foreign material away from the spark plug hole. The idea here is to eliminate the possibility of debris falling into the cylinder as the spark plug is removed.

9   Place the spark plug socket over the plug

30.9 Use a ratchet and extension to remove the spark plugs

and remove it from the engine by turning it in a counterclockwise direction (see illustration).

10   Compare the spark plug to those shown in the inside back cover to get an indication of the general running condition of the engine (see illustration).

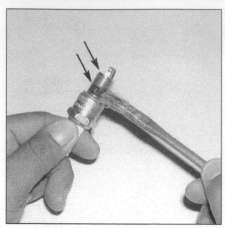

30.11a Apply a thin coat of anti-seize compound to the spark plug threads

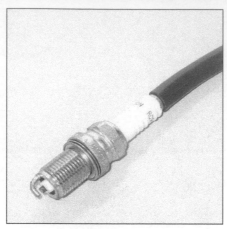

30.11b A length of snug-fitting rubber hose will save time and prevent damaged threads when installing the spark plugs

11   Apply a small amount of anti-seize compound to the spark plug threads (see illustration). Install one of the new plugs into the hole until you can no longer turn it with your fingers, then tighten it with a torque wrench (if available) or the ratchet. It is a good idea to slip a short length of rubber hose over the end of the plug to use as a tool to thread it into place (see illustration). The hose will grip the plug well enough to turn it, but will start to slip if the plug begins to cross-thread in the hole - this will prevent damaged threads and the accompanying repair costs.

12   On models equipped with coil-over plugs, before pushing the ignition coil onto the end of the plug, inspect the ignition coil following the procedures outlined in Section 31. On 2004 and earlier four-cylinder engines, inspect the plug wire following the procedures outlined in Section 32.

13   Repeat the procedure for the remaining spark plugs.

31   Ignition coil check (V6 engines and 2005 and later four-cylinder engines) (every 60,000 miles or 48 months)

1   Remove the ignition coils (see illustrations 30.7b, 30.7c and 30.7d). Clean the coil(s) with a dampened cloth and dry them thoroughly.

2   Inspect each coil for cracks, damage and carbon tracking. If damage exists, replace the coil.

32   Spark plug wire check and replacement (2004 and earlier four-cylinder engines) (every 60,000 miles or 48 months)

1   The spark plug wires should be checked at the recommended intervals or whenever new spark plugs are installed.

2   Begin this procedure by making a visual check of the spark plug wires while the engine is running. In a darkened garage (make sure there is adequate ventilation) or at night while using a flashlight, start the engine and observe each plug wire. Be careful not to come into contact with any moving engine parts. If possible, use an insulated or non-conductive object to wiggle each wire. If there is a break in the wire, you will see arcing or a small blue spark coming from the damaged area. Secondary ignition voltage increases with engine speed and sometimes a damaged wire will not produce an arc at idle speed. Have an assistant press the accelerator pedal to raise the engine speed to approximately 2000 rpm. Check the spark plug wires for arcing as stated previously. If arcing is noticed, replace all spark plug wires.

3   Perform the following checks with the engine Off. The wires should be inspected one at a time to prevent mixing up the order that is essential for proper engine operation. **Note:** *Due to the short length of the spark plug wire, always disconnect the spark plug wire from the ignition coil pack first.*

4   With the engine cool, disconnect the spark plug wire from the ignition coil pack. Pull only on the boot at the end of the wire; don't pull on the wire itself. Use a twisting motion to free the boot/wire from the coil. Disconnect the same spark plug wire from the spark plug, using the same twisting method while pulling on the boot. Disconnect the spark plug wire from any retaining clips as necessary and remove it from the engine.

5   Check inside the boot for corrosion, which will look like a white, crusty powder (don't mistake the white dielectric grease used on some plug wire boots for corrosion protection).

6   Now push the wire and boot back onto the end of the spark plug. It should be a tight fit on the plug end. If not, remove the wire and use a pair of pliers to carefully crimp the metal connector inside the wire boot until the fit is snug.

7   Now push the wire and boot back into the end of the ignition coil terminal. It should be a tight fit in the terminal. If not, remove the wire and use a pair of pliers to carefully crimp the metal connector inside the wire boot until the fit is snug.

8   Now, using a cloth, clean each wire along its entire length. Remove all built-up dirt and grease. As this is done, inspect for burned areas, cracks and any other form of damage. Bend the wires in several places to ensure that the conductive material inside hasn't hardened. Repeat the procedure for the remaining wires.

9   If new spark plug wires are required, purchase a complete set for your particular engine. The terminals and rubber boots should already be installed on the wires. Replace the wires one at a time to avoid mixing up the firing order and make sure the terminals are securely seated on the coil pack and the spark plugs.

10   Attach the plug wire to the new spark plug and to the ignition coil pack using a twisting motion on the boot until it is firmly seated. Attach the spark plug wire to any retaining clips to keep the wires in their proper location on the valve cover.

Chapter 2 Part A
Four-cylinder engines

Contents

Specifications

General

1.5L engine

Engine type	Four-cylinder, in-line, 16 valve, turbocharged direct injection, DOHC
Displacement	91.53 cubic inches (1500 cc)
Engine VIN code	D
Bore	3.1 inches (79.015 mm)
Stroke	3.0 inches (76.4 mm)
Compression ratio	10: 1
Compression	See Chapter 2C
Firing order	1-2-4-3
Oil pressure	See Chapter 2C

1.6L engine

Engine type	Four-cylinder, In-line, 16 valve, turbocharged direct injection, DOHC
Displacement	97.63 cubic inches (1600 cc)
Engine VIN code	X
Bore	3.1 inches (79 mm)
Stroke	3.2 inches (81.4 mm)
Compression ratio	10: 1
Compression	See Chapter 2C
Firing order	1-3-4-2
Oil pressure	See Chapter 2C

General (Continued)

2.0L engine

Engine type	Four-cylinder, in-line, 16 valve, DOHC
Displacement	122 cubic inches (1989 cc)
Engine VIN code	
2004 and earlier models	B
2013 and later models	9
Bore	
2004 and earlier models	3.34 inches (84.8 mm)
2013 and later models	3.4449 inches (87.5 mm)
Stroke	
2004 and earlier models	3.46 inches (88.0 mm)
2013 and later models	3.2717 inches (83.1 mm)
Compression ratio	
2004 and earlier models	9.6: 1
2013 and later models	9.3: 1
Firing order..	1-3-4-2
Oil pressure...	See Chapter 2C

**Cylinder location
(2004 and earlier 2.0L engine coil
terminal arrangement)**

2.3L engine

Engine type	Four cylinder, in-line, 16 valve, DOHC
Displacement	140 cubic inches (2261 cc)
Engine VIN code	Z
Bore	3.44 inches (87.5 mm)
Stroke	3.70 (94.0 mm)
Compression ratio...	9.7: 1
Compression...	See Chapter 2C
Firing order...	1-3-4-2
Oil pressure..	See Chapter 2C

2.5L engine

Engine type	Four-cylinder, in-line, 16 valve, DOHC
Displacement	153 cubic inches (2507 cc)
Engine VIN code	7
Bore	3.504 inches (89.0 mm)
Stroke	3.937 inches (100.0 mm)
Compression ratio...	9.7: 1
Compression...	See Chapter 2C
Firing order...	1-3-4-2
Oil pressure..	See Chapter 2C

Camshafts

1.5L and 1.6L engines

Lobe height	
Intake	0.310 inch (7.9 mm)
Exhaust...	0.260 inch (6.6 mm)
Thrust bearing clearance	
1.5L engine	0.8689 to 0.8705 inch (22.07 to 22.11 mm)
1.6L engine	1.218 to 1.219 inch (30.96 to 30.98 mm)
Journal diameter	0.9827 to 0.9835 inch (24.96 to 24.98 mm)
Journal-to-bore clearance	0.0008 to 0.0028 inch (0.02 to 0.07 mm)
End play	0.003 to 0.008 inch (0.07 to 0.2 mm)

2.0L engine

Lobe height	
Intake	
2004 and earlier models....................................	0.350 inch (8.9 mm)
2013 and later models......................................	0.327 inch (8.3 mm)
Exhaust	
2004 and earlier models....................................	0.343 inch (8.7 mm)
2013 and later models......................................	0.291 inch (7.4 mm)
Bearing journal diameter	
2004 and earlier models	1.022 to 1.023 inches (25.96 to 25.98 mm)
2013 and later models	0.9827 to 0.9835 inch (24.96 to 24.98 mm)
End play	0.003 to 0.009 inch (0.08 to 0.23 mm)

2.3L and 2.5L engines

Lobe height	
Intake	0.324 inch (8.23 mm)
Exhaust...	0.307 inch (7.8 mm)
Bearing journal diameter	0.982 to 0.983 inch (24.96 to 24.98 mm)
End play	0.003 to 0.009 inches (0.08 to 0.23 mm)

Valve clearances (cold)

1.5L engine
Intake .... 0.009 to 0.013 inch (0.24 to 0.33 mm)
Exhaust .... 0.015 to 0.019 inch (0.38 to 0.47 mm)
1.6L engine
Intake .... 0.010 to 0.014 inch (0.255 to 0.345 mm)
Exhaust .... 0.014 to 0.018 inch (0.355 to 0.445 mm)
2.0L engine
Intake
2004 and earlier models .... 0.004 to 0.007 inch (0.11 to 0.18 mm)
2013 and later models .... 0.007 to 0.012 inch (0.19 to 0.31 mm)
Exhaust
2004 and earlier models .... 0.010 to 0.013 inch (0.27 to 0.33 mm)
2013 and later models .... 0.012 to 0.017 inch (0.30 to 0.42 mm)
2.3L engine
Intake .... 0.008 to 0.011 inches (0.22 to 0.28 mm)
Exhaust .... 0.010 to 0.013 inches (0.27 to 0.33 mm)
2.5L engine
Intake .... 0.009 to 0.011 inch (0.23 to 0.28 mm)
Exhaust .... 0.010 to 0.013 inch (0.27 to 0.33 mm)

Warpage limits

Cylinder head gasket surfaces (head and block) .... 0.002 inch (0.05 mm)
Intake and exhaust manifolds .... 0.008 inch (0.20 mm)

Torque specifications

Note: *One foot-pound (ft-lb) of torque is equivalent to 12 inch-pounds (in-lbs) of torque. Torque values below approximately 15 ft-lbs are expressed in inch-pounds, since most foot-pound torque wrenches are not accurate at these smaller values.*

	Ft-lbs (unless otherwise indicated)	Nm
Camshaft bearing cap bolts		
1.5L and 1.6L engines		
Step 1, bolts 1-16	62 in-lbs	7
Step 2, bolts 17-20	89 in-lbs	10
Step 3, bolts 1-16	Tighten an additional 45 degrees	
Step 4, bolts 17 and 18	Tighten an additional 70 degrees	
Step 5, bolts 19 and 20	Tighten an additional 63 degrees	
2.0L engine		
2004 and earlier models		
Step 1	89 in-lbs	10
Step 2	168 in-lbs	19
2013 and later models		
Step 1	62 in-lbs	7
Step 2	142 in-lbs	16
2.3L engine		
Step 1	62 in-lbs	7
Step 2	144 in-lbs	16
2.5L engine		
Step 1	Finger-tight	
Step 2	62 in-lbs	7
Step 3	142 in-lbs	16
Camshaft sprocket bolts		
1.5L and 1.6L engines (VCT sprockets)	18	25
2.0L engine		
2004 and earlier models		
Intake sprocket	50	68
Exhaust sprocket		
Stage 1	36	50
Stage 2	Remove special tools: TDC peg and cam alignment timing tool	
Stage 3	85 to 92	115 to 125
2013 and later models		
Stage 1	62 inch-lbs	7
Stage 2	142 inch-lbs	16
2.3L engine	53	72
2.5L engine		
Stage 1	30	40
Stage 2	60-degrees	
Brake vacuum pump adapter-to-intake camshaft		
(2013 and later 2.0L models)	46	63

Torque specifications (continued)   Ft-lbs (unless otherwise indicated)   Nm

Note: *One foot-pound (ft-lb) of torque is equivalent to 12 inch-pounds (in-lbs) of torque. Torque values below approximately 15 ft-lbs are expressed in inch-pounds, since most foot-pound torque wrenches are not accurate at these smaller values.*

	Ft-lbs (unless otherwise indicated)	Nm
Crankshaft pulley bolt		
1.5 and 1.6L engines*		
Stage 1	74	100
Stage 2	Tighten an additional 90-degrees	
Stage 3	Wait two seconds	
Stage 4	Tighten an additional 15-degrees	
2.0L engine		
2004 and earlier	85	115
2013 and later		
Stage 1	74	100
Stage 2	Tighten an additional 90-degrees	
2.3L and 2.5L engines*		
Stage 1	74	100
Stage 2	Tighten an additional 90-degrees	
Cylinder head bolts - 1.5L and 1.6L engines (in sequence - see illustration 13.51a)*		
Step 1	44 in-lbs	5
Step 2	133 in-lbs	15
Step 3	26	35
Step 4	Tighten an additional 90 degrees	
Step 5	Tighten an additional 90 degrees	
Cylinder head bolts - 2004 and earlier 2.0L engine (in sequence - see illustration 13.51b)*		
Step 1	15	20
Step 2	30	40
Step 3	Tighten an additional 90-degrees	
Cylinder head bolts - 2013 and later 2.0L engine (in sequence - see illustration 13.51c)*		
Step 1	62 in-lbs	7
Step 2	133 in-lbs	15
Step 3	41	55
Step 4	Tighten an additional 90 degrees	
Step 5	Tighten an additional 90 degrees	
Cylinder head bolts - 2.3L and 2.5L engines (in sequence - see illustration 13.51c)*		
Step 1	44 in-lbs	5
Step 2	132 in-lbs	15
Step 3	33	44
Step 4	Tighten an additional 90-degrees	
Step 5	Tighten an additional 90-degrees	
Drivebelt idler pulley bolts		
2.0L, 1.5L and 1.6L engines	35	47
2.3L and 2.5L engines	18	24
Valve cover bolts		
2004 and earlier 2.0L models	80 in-lbs	9
2005 and later 2.0L, 2.3L and 2.5L models	89 in-lbs	10
1.5L and 1.6L models		
Step 1	44 in-lbs	5
Step 2	80 in-lbs	9
Step 3	89 in-lbs	10
High-pressure fuel pump drive housing bolts		
1.6L models (in sequence - see illustration 19.11a)		
Step 1, bolts 1 through 10	27 in-lbs	3
Step 2, bolts 1 through 10	80 in-lbs	9
Step 3, bolts 1 through 10	97 in-lbs	11
Step 4, bolts 1, 2, 5, 6, 7, 8, 9 and 10	124 in-lbs	14
1.5L models (in sequence - see illustration 19.11b)		
Step 1, bolts 1 through 8	35 in-lbs	4
Step 2, bolts 1 through 8	124 in-lbs	14
Step 3, bolts 9 and 10	97 in-lbs	11

* Use NEW bolt(s)

Torque specifications (continued)

	Ft-lbs (unless otherwise indicated)	Nm
Engine mount (RH side)		
1.5L engine		
Mount-to-engine nuts..	85	115
Mount-to-engine studs...	106 in-lbs	12
Mount-to-body bolts..	66	90
1.6L engine		
Mount-to-engine nuts and mount-to-body bolts.............................	66	90
Mount-to-engine studs...	89 in-lbs	10
2004 and earlier 2.0L engine		
Mount upper bracket		
Outer 2 nuts..	72	98
Center bolt...	57	77
Engine mount fasteners..	41	55
2013 and later 2.0L engine		
Mount-to-engine nuts..	59	80
Mount-to-body bolts..	66	90
2.3L engine		
Mount bracket-to-engine bolt/nuts	66	90
Mount-to-body bolts..	41	55
2.5L engine		
Mount bracket-to-engine bolt/nuts	85	115
Mount-to-body bolts..	35	48
Exhaust manifold fasteners		
2.0L engines		
2004 and earlier models (in sequence)	144 in-lbs	16
2013 and later models	Turbocharger fasteners - see Chapter 4	
2.3L and 2.5L engines		
Manifold studs...	156 in-lbs	17
Manifold/catalytic converter fasteners (in sequence)		
Stage 1	35	47
Stage 2	35	47
1.5L engines...	Turbocharger fasteners - see Chapter 4	
1.6L engines		
Manifold studs...	89 in-lbs	10
Manifold nuts		
Stage 1	15	21
Stage 2	15	21
Flywheel/driveplate bolts		
2004 and earlier 2.0L engines..	83	112
All other engines (except 1.5L and 1.6L)		
Step 1	37	50
Step 2	59	80
Step 3	83	112
1.5L and 1.6L engines		
Step 1	133 in-lbs	15
Step 2	18	25
Step 3	22	30
Step 4	Tighten an additional 90 degrees	
Intake manifold nuts and bolts (in sequence)		
1.5L models		
Step 1, bolts 1, 2 and 7..	18 in-lbs	2
Step 2, bolts 3 and 4..	18 in-lbs	2
Step 3, bolts 1 and 2..	159 in-lbs	18
Step 4, bolts 3, 4, 5 and 6..	89 in-lbs	10
Step 5, retighten bolts 1 and 2...	159 in-lbs	18
Step 6, bolt 7..	159 in-lbs	18
Step 7, retighten bolts 3, 4, 5 and 6...	89 in-lbs	10
1.6L models		
Bolts 1...	150 in-lbs	17
Bolts 2...	89 in-lbs	10
2.0L (2004 and earlier), 2.3L and 2.5L models.............................	156 in-lbs	18
2.0L (2013 and later models)	177 in-lbs	20
Lower crankcase-to-engine block (in sequence - see illustration 15.23)......	22	30

Torque specifications (continued)	Ft-lbs (unless otherwise indicated)	Nm

Note: *One foot-pound (ft-lb) of torque is equivalent to 12 inch-pounds (in-lbs) of torque. Torque values below approximately 15 ft-lbs are expressed in inch-pounds, since most foot-pound torque wrenches are not accurate at these smaller values.*

	Ft-lbs	Nm
Oil pick-up pipe bolts		
1.5L and 1.6L models	84 in-lbs	9.5
2.0L, 2.3L and 2.5L models	89 in-lbs	10
Oil pump-to-cylinder block bolts (in sequence)		
1.5L engine		
Step 1, bolts 1, 2, 3 and 6	18 in-lbs	2
Step 2, bolts 1 through 8	97 in-lbs	11
1.6L engine		
Step 1	49 in-lbs	5.5
Step 2	84 in-lbs	9.5
2.0L engine (2004 and earlier models)		
Step 1	53 in-lbs	6
Step 2	Tighten an additional 45-degrees	
2.0L (2013 and later), 2.3L and 2.5L engines		
Step 1	89 in-lbs	10
Step 2	177 in-lbs	20
Oil pump drive gear bolt	18	25
Oil pump chain guide/tensioner bolts	89 in-lbs	10
Oil pan-to-lower crankcase (2004 and earlier 2.0L engine) (in sequence)		
Step 1	53 in-lbs	6
Step 2	108 in-lbs	12
Oil pan-to-engine block bolts (2013 and later 2.0L engine) (in sequence)	177 in-lbs	20
Oil pan-to-bellhousing bolts (1.5L, 1.6L, 2.0L (2013 and later), 2.3L and 2.5L engines)	35	48
Oil pan-to-timing chain cover bolts (2.0L (2013 and later), 2.3L and 2.5L engines)	89 in-lbs	10
Oil pan-to-engine block bolts (1.5L engine) (in sequence)	168 in-lbs	19
Oil pan-to-engine block bolts (1.6L engine) (in sequence)		
Step 1	89 in-lbs	10
Step 2	168 in-lbs	19
Oil pan-to-engine block bolts (2.3L and 2.5L engines) (in sequence)	18	25
Rear crankshaft oil seal retainer bolts - All engines except 1.5L and 1.6L (in sequence)	89 in-lbs	10
Rear crankshaft oil seal retainer bolts - 1.5L and 1.6L engines (in sequence)		
Stage 1	35 in-lbs	4
Stage 2	89 in-lbs	10
Timing belt covers		
1.5L and 1.6L engines	89 in-lbs	10
2.0L engine (2004 and earlier models)		
Upper cover bolts	71 in-lbs	8
Lower cover bolts	62 in-lbs	7
Timing belt guide pulley bolt - 2.0L engine (2004 and earlier)	18	25
Timing belt tensioner pulley bolt - 2.0L engine (2004 and earlier)	18	25
Timing chain guide bolts (2.0L (2013 and later), 2.3L and 2.5L engines)	89 in-lbs	10
Timing chain tensioner bolts (2.0L (2013 and later), 2.3L and 2.5L engines)	89 in-lbs	10
Timing chain cover bolts - in sequence (2013 and later 2.0L engines)		
Bolts, 1 through 3	35	48
Bolts, 4 through 22	89 in-lbs	10
Timing chain cover bolts - in sequence (2.3L and 2.5L engines)		
8 mm bolts	89 in-lbs	10
13 mm bolts	35	48

1   General information

How to use this Chapter

1   This Part of Chapter 2 is devoted to repair procedures possible while the engine is still installed in the vehicle. Since these procedures are based on the assumption that the engine is installed in the vehicle, if the engine has been removed from the vehicle and mounted on a stand, some of the preliminary dismantling steps outlined will not apply.

2   Information concerning engine/transaxle removal and replacement and engine overhaul, can be found in Part C of this Chapter.

Engine description

3   These engines are sixteen-valve, double overhead camshaft (DOHC), four-cylinder, in-line type, mounted transversely at the front of the vehicle, with the transmission on its left-hand end. The 2004 and earlier 2.0L engine is equipped with plastic timing belt covers, a fiber-glass-reinforced plastic intake manifold, a cast-iron cylinder block/crankcase and aluminum cylinder head. The 2.0L engine also includes a cast aluminum alloy lower crankcase that is bolted to the underside of the engine block, with a pressed-steel oil pan bolted under that. This arrangement offers greater rigidity than the normal oil pan arrangement, and helps to reduce engine vibration.

4   The 1.5L, 1.6L, 2013 and later 2.0L, 2.3L and 2.5L engines incorporate an aluminum cylinder head and an aluminum cylinder block.

5   The two camshafts are driven by a timing belt on the 1.5L, 1.6L, (2004 and earlier 2.0L engine), or timing chain (2013 and later 2.0L, 2.3L and 2.5L engines), each operating eight valves via conventional cam followers. Each camshaft rotates in five bearings that are line-bored directly in the cylinder head and the (bolted-on) bearing caps. This means that the bearing caps are not available separately from the cylinder head, and must not be interchanged with caps from another engine.

6   In addition to the features of the 2004 and earlier 2.0L Zetec engine, the 2013 and later 2.0L EcoBoost, 2.3L and 2.5L engines incorporate an all aluminum timing chain cover, engine block and oil pan.

7   When working on this engine, note that Torx-type (both male and female heads) and hexagon socket (Allen head) fasteners are widely used. A good selection of sockets, with the necessary adapters, will be required, so that these can be unscrewed without damage and, on reassembly, tightened to the torque wrench settings specified.

Lubrication system

8   Lubrication is by means of an eccentric-rotor trochoidal pump, which is mounted on the crankshaft right-hand end, and draws oil through a strainer located in the oil pan. The pump forces oil through an externally-mounted full-flow cartridge-type filter. From the filter, the oil is pumped into a main gallery in the cylinder block/crankcase, from where it is distributed to the crankshaft (main bearings) and cylinder head.

9   The connecting rod bearings are supplied with oil via internal drillings in the crankshaft. Each piston crown is cooled by a spray of oil directed at its underside by a jet. These jets are fed by passages off the crankshaft oil supply galleries, with spring-loaded valves to ensure that the jets open only when there is sufficient pressure to guarantee a good oil supply to the rest of the engine components. Where the jets are not installed, separate blanking plugs are provided so that the passages are sealed, but can be cleaned at overhaul.

10   The cylinder head is provided with two oil galleries, one on the intake side and one on the exhaust, to ensure constant oil supply to the camshaft bearings and cam followers. A retaining valve (inserted into the cylinder head's top surface, in the middle, on the intake side) prevents these galleries from being drained when the engine is switched off. The valve incorporates a ventilation hole in its upper end, to allow air bubbles to escape from the system when the engine is restarted.

11   While the crankshaft and camshaft bearings receive a pressurized supply, the camshaft lobes and valves are lubricated by splash, as are all other engine components.

2   Repair operations possible with the engine in the vehicle

1   Many major repair operations can be accomplished without removing the engine from the vehicle.

2   Clean the engine compartment and the exterior of the engine with some type of degreaser before any work is done. It will make the job easier and help keep dirt out of the internal areas of the engine.

3   Depending on the components involved, it may be helpful to remove the hood to improve access to the engine as repairs are performed (refer to Chapter 11 if necessary). Cover the fenders to prevent damage to the paint. Special pads are available, but an old bedspread or blanket will also work.

4   If vacuum, exhaust, oil or coolant leaks develop, indicating a need for gasket or seal replacement, the repairs can generally be made with the engine in the vehicle. The intake and exhaust manifold gaskets, oil pan gasket, camshaft and crankshaft oil seals and cylinder head gasket are all accessible with the engine in place.

5   Exterior engine components, such as the intake and exhaust manifolds, the oil pan, the oil pump, the water pump, the starter motor, the alternator and the fuel system components can be removed for repair with the engine in place.

6   Since the camshaft(s) and cylinder head can be removed without pulling the engine, valve component servicing can also be accomplished with the engine in the vehicle. Replacement of the timing belt or chain and sprockets is also possible with the engine in the vehicle.

7   In extreme cases caused by a lack of necessary equipment, repair or replacement of piston rings, pistons, connecting rods and rod bearings is possible with the engine in the vehicle. However, this practice is not recommended because of the cleaning and preparation work that must be done to the components involved.

3   Top Dead Center (TDC) for number one piston - locating

1   Top Dead Center (TDC) is the highest point of the cylinder that each piston reaches as the crankshaft turns. Each piston reaches its TDC position at the end of its compression stroke, and then again at the end of its exhaust stroke. For the purpose of engine timing, TDC on the compression stroke for the Number 1 piston is used. The Number 1 cylinder is at the timing belt end of the engine. Proceed as follows.

2012 and earlier (and all 2.5L) models

2   Disconnect the cable from the battery negative terminal (see Chapter 5).

3   Remove all the spark plugs as described in Chapter 1, then remove the valve cover as described in Section 4.

Note: On models with 2.3L and 2.5L engines, the small hole in the crankshaft pulley should be in the 6 o'clock position (see illustration 8.16).

4   Loosen the right front wheel lug nuts, raise the front of the vehicle and support it securely on jackstands, then remove the wheel. Remove the fender splash shield. Using a wrench or socket on the crankshaft pulley bolt, rotate the crankshaft clockwise until the intake valves for the Number 1 cylinder have opened and just closed again. The TDC notch on the crankshaft pulley should also meet up with the alignment mark on the lower crankcase (above oil pan) (see illustration).

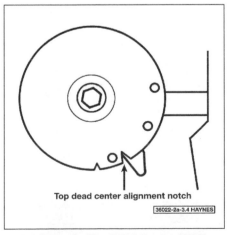

Top dead center alignment notch

36022-2a-3.4 HAYNES

3.4 Align the Top Dead Center (TDC) notch with the casting on the lower crankcase - 2.0L engine only

3.5a Turn the engine so that the camshaft end slots are aligned . . .

3.5b . . . then insert the metal strip into the slots to locate and set the camshafts to TDC

3.6 Remove the timing hole blanking plug - catalytic converter removed for clarity (2.0L engine shown)

5   The camshafts each have a machined slot at the transaxle end of the engine. Both slots will be completely horizontal, and at the same height as the cylinder head machined surface, when the engine is at TDC on the Number 1 cylinder. Manufacturer service tool 303-465 is used to check this position, and to positively locate the camshafts in position. Fortunately, a substitute tool can be made from a strip of metal 5 mm thick. While the strip's thickness is critical, its length and width are not, but should be approximately 180 to 230 mm long by 20 to 30 mm wide (see illustrations).

6   A TDC timing hole is provided on the front of the cylinder block to permit the crankshaft to be located more accurately at TDC. The blanking plug is located behind the catalytic converter, and access is not easy - also, take care against burning if the engine is still warm (see illustration).

Note: *On 2.3L and 2.5L engines, the blanking plug hole is located on the rear of the cylinder block, near the crankshaft pulley (see illustrations 3.23 and 3.24)*

7   Unscrew the timing pin blanking plug and screw in the timing pin (manufacturer service tool - for 2.0L engines: 303-574, and

for 2.3L and 2.5L engines: 303-507). This tool is obtainable from dealers or a tool supplier. An alternative pin can be made from an M10 diameter bolt, cut down so that the length from the underside of the bolt head to the tip is exactly 63.4 mm (see illustrations). It may be necessary to slightly turn the crankshaft either way (remove the tool from the camshafts first) to be able to fully insert the timing pin.

Caution: *Never use the timing pin as a means to stop the engine from rotating - tool breakage and/or engine damage can result.*

8   Turn the engine forward slowly until the crankshaft comes into contact with the timing pin - in this position, the engine is set to TDC on the Number 1 cylinder.

9   Before rotating the crankshaft again, make sure that the timing pin is removed. When operations are complete, do not forget to install the blanking plug.

10   If the timing pin is not available, insert a length of wooden dowel (about 150 mm/ 6 in long) or similar into the Number 1 spark plug hole until it rests on the piston crown. Turn the engine back from its TDC position, then forward (taking care not to allow the dowel to be trapped in the cylinder) until the dowel stops rising - the piston is now at the top of its compres-

sion stroke and the dowel can be removed.

11   There is a dead area around TDC (as the piston stops rising, pauses and then begins to descend) which makes it difficult to find the exact location of TDC by this method. If accuracy is required, either carefully establish the exact mid-point of the dead area (perhaps by using a dial gauge and probe), or refer to Step 5.

12   Once the Number 1 cylinder has been positioned at TDC on the compression stroke, TDC for any of the other cylinders can then be located by rotating the crankshaft clockwise 180° at a time and following the firing order (see this Chapter's Specifications).

13   Reconnect the battery. After you're done, the Powertrain Control Module (PCM) must relearn its idle and fuel trim strategy for optimum driveability and performance (see Chapter 5, Section 1).

2013 and later (except 2.5L) models

Note: *You will need special tools for this procedure: the camshaft positioning tool (303-1565 for 2.0L engines, or equivalent) and the crankshaft timing pin (303-748 for 1.5L and 1.6L engines and 303-507 for 2.0L engines).*

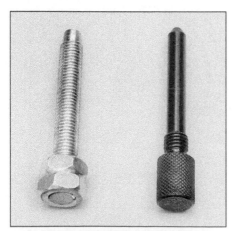

3.7a Homemade (left) and factory (right) timing pins

3.7b Insert the timing pin in the hole . . .

3.7c . . . and screw it fully into position

3.22 Remove the rear intake camshaft cap

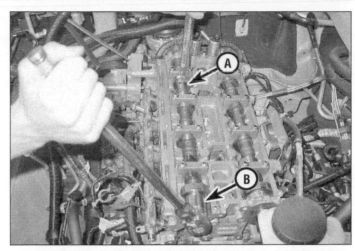

3.23 Prevent the camshaft from turning by holding it with a wrench on the hex surface (A), then unscrew the vacuum pump adapter (B)

All models

14   Disconnect the cable from the negative battery terminal (see Chapter 5).
15   Remove the passenger's side inner fender splash shield (see Chapter 11).
16   Remove the under engine splash shield.
17   Remove the drivebelt (see Chapter 1).

1.5L and 1.6L models

Note: *1.5L and 1.6L models do not have markings etched into the crankshaft pulley - TDC is obtained by viewing the markings on the camshaft (VCT) sprockets*

18   Remove the timing belt cover (see Section 9)
19   With the engine supported, rotate the crankshaft pulley clockwise until the camshaft (VCT) sprocket marks are both in the 11 o'clock position.

2.0L models

20   Remove the brake vacuum pump (see Chapter 9), the high-pressure fuel pump (see Chapter 4) and pump housing. Be sure to properly relieve the fuel system pressures from both the high and low-side fuel system beforehand.

21   Remove the valve cover (see Section 4).
22   Remove the rear intake camshaft bearing cap (see illustration).
23   Prevent the camshaft from turning by using a wrench on the flats of the intake camshaft, then loosen and remove the brake vacuum pump adapter from the end of the camshaft (see illustration).
24   Remove the the crankshaft position sensor (see Chapter 6), then rotate the crankshaft clockwise, until the No.1 piston is 45-degrees BTDC (Before Top Dead Center) using the guide holes on the engine front cover and the crankshaft pulley (see illustration).

All models

Note: *On 1.5L and 1.6L models, the RH driveaxle support bearing bracket will need to be unbolted and removed (Chapter 8) to allow access to the TDC timing hole.*

25   The TDC timing hole is located near the lower right corner of the engine block (on the firewall side), hidden behind the driveaxle support bracket; it provides a means of accurately positioning the no. 1 cylinder at TDC. When you locate this hole, remove the timing pin plug bolt (see illustration).

26   Screw in the timing pin (see illustration). **Caution:** *We don't recommend trying to fabricate a timing pin with a bolt because while you would be able to determine the correct bolt diameter and thread pitch, it is impossible to determine what the length of the bolt should be. These pins come in several lengths, depending on the engine family. There is no way to determine the correct pin length without comparing it to a factory or aftermarket tool designed to be used with this engine. Using a bolt of the wrong length could damage the engine. Also, never use the timing pin as a means to stop the engine from rotating - tool breakage and/or engine damage can result.*
27   Turn the crankshaft slowly clockwise until the crankshaft counterweight comes into contact with the timing pin and the hole on the crankshaft pulley (2.0L engine only) is in a center line with the hole on the timing cover (6 o'clock position) - in this position, the engine is set to TDC on no. 1 cylinder.

2.0L models

28   The camshafts each have a machined slot at the transaxle end of the engine. Both slots will be completely horizontal, and at the

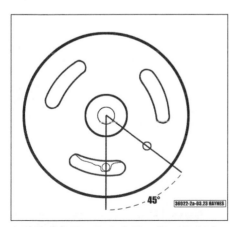

3.24 Rotate the crankshaft pulley clockwise until the pulley guide hole is 45-degrees before the guide mark on the engine cover

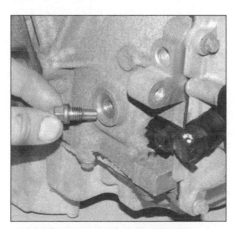

3.25 Remove the timing hole plug . . .

3.26 . . . and insert the timing pin tool

3.28 When the engine is at TDC for cylinder no. 1, the slots in the ends of the camshafts will be horizontal (and the offset portions will be even with the cylinder head surface). Manufacturer tool no. 303-1565 can then be inserted into the slots

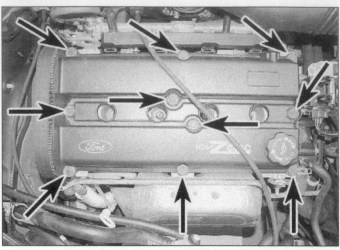

4.8 Remove the retaining bolts

same height as the cylinder head machined surface, when the engine is at TDC on the Number 1 cylinder (see illustration). Manufacturer service tool 303-1565 (2.0L engines) is used to check this position, and to positively locate the camshafts in position. As previously mentioned, the rear intake camshaft cap and the vacuum pump drive adapter (2.0L models only) will have to be removed from the intake camshaft, to allow this tool to be installed.

Note: *The timing slots in the camshafts are offset. If the service tool cannot be installed, remove the timing pin and carefully rotate the crankshaft three-quarters of a turn clockwise and repeat Steps 24 through 27 of this procedure.*

Caution: *Never use the camshaft alignment tool as a means to stop the engine from rotating - engine damage can result.*

29   Before rotating the crankshaft again, make sure that the special tools are removed. Do not forget to install the brake vacuum pump adapter and tighten it to the torque listed in this Chapter's Specifications.

30   Once Number 1 cylinder has been positioned at TDC on the compression stroke, TDC for any of the other cylinders can then be located by rotating the crankshaft clockwise 180-degrees at a time and following the firing order (see this Chapter's Specifications).

---

4   Valve cover - removal and installation

2008 and earlier models (2.0L and 2.3L)

Removal

1   Disconnect the battery cable from the negative battery terminal (see Chapter 5).

2   Remove the air intake duct from the air filter housing (see Chapter 4).

3   Remove the spark plug wires from the spark plugs . On the 2.3L engine, remove the ignition coil assemblies from the spark plugs.

4   Disconnect the accelerator cable and the

cruise control actuator cable from the throttle valve (see Chapter 4).

5   Disconnect the catalytic converter monitor connector and the heated oxygen sensor connector (see Chapter 6). Remove the connectors from the bracket.

6   Remove the wiring harness support nuts from the studs on the valve cover and position the wiring harness off to the side.

7   Remove the ignition coil (see Chapter 5) and the ignition coil bracket from the valve cover (2001 through 2004 only).

8   Working progressively, unscrew the valve cover retaining bolts, noting the (captive) spacer sleeve and rubber seal, then withdraw the cover (see illustration).

9   Discard the cover gasket. This must be replaced whenever it is disturbed. Check that the sealing faces are undamaged and that the rubber seal at each bolt hole is serviceable. Replace any worn or damaged seals.

Installation

10   On installation, clean the cover and cylinder head gasket faces carefully, then install a new gasket onto the valve cover, ensuring that it is located correctly by the rubber seals and spacer sleeves.

11   Install the cover to the cylinder head, ensuring as the cover is tightened that the gasket remains seated.

12   Working in a diagonal sequence from the center outwards, first tighten the cover bolts by hand only. Once all the bolts are hand-tight, go around once more in sequence, and tighten the bolts to the torque listed in this Chapter's Specifications.

13   On 2004 and earlier models, install the spark plug wires, clipping them into place so that they are correctly routed; each is numbered and can also be identified by the numbering on its respective coil terminal. On other models, install the ignition coils.

14   Reconnect the battery. After you're done, the Powertrain Control Module (PCM) must relearn its idle and fuel trim strategy for opti-

mum driveability and performance (see Chapter 5, Section 1).

15   On completion, run the engine and check for signs of oil leakage.

2009 and later models

Removal

16   Disconnect the cable from the negative battery terminal (see Chapter 5).

Note: *On models equipped with "captive" valve cover bolts, the bolts should stay attached to the cover once they are loosened, do not try and remove the bolts from the valve cover.*

17   On 2013 and later models with either a 1.6L, 2.0L or 2.5L engine, remove the cowl panel (see Chapter 11).

1.5L and 1.6L engines

Warning: *The fuel system pressure MUST be relieved (both high and low side pressures) before the valve cover can be removed on 1.5L and 1.6L engines*

18   Relieve the fuel pressure (see Chapter 4).

19   Carefully lift the engine cover up to disengage the cover from the ballstuds and remove the cover.

20   Disconnect the crankcase quick-connect vent tube(s) from the valve cover.

21   Slide back the hose clamps and disconnect the hoses from the sides of the valve cover.

22   Remove the fuel rail (see Chapter 4).

23   Remove the variable camshaft timing (VCT) solenoids (see Chapter 6).

24   Disconnect the camshaft position sensor electrical connector (see Chapter 6).

25   Disconnect the high-pressure fuel pump electrical connector (see Chapter 4), then disconnect the wiring harness(es) from the retainers and position the wiring harnesses out of the way.

26   Loosen the valve cover "captive" mounting bolts, in the reverse order of the tightening sequence (see illustration 4.45a), and remove the valve cover with the bolts.

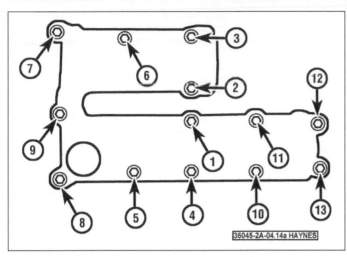

**4.45a Valve cover bolt tightening sequence -
1.5L and 1.6L engines**

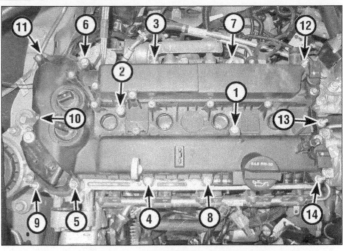

**4.45b Valve cover bolt tightening sequence -
typical later 2.0L and 2.5L engines**

2.0L engines

27   Raise the front of the vehicle and support it securely on jackstands. Remove the engine under cover.

28   Carefully lift the engine cover up to disengage the cover from the ballstuds and remove the cover.

29   Remove the air cleaner housing outlet pipe-to-outlet tube and the connecting turbocharger inlet tubes from the top and rear of the engine (see Chapter 4).

30   Remove the oil dipstick tube from the valve cover.

31   Disconnect the crankcase pressure sensor electrical connector on the crankcase vent tube.

32   Disconnect the vacuum tubes from the brake vacuum pump.

33   Remove the ignition coils (see Chapter 5).

34   Labeling them as necessary, disconnect the electrical connectors to the cylinder head temperature (CHT) sensor, the manifold air pressure and temperature (MAPT) sensor, the variable camshaft timing (VCT) oil control solenoids and camshaft position sensor (and any other connectors that interfere with positioning the harness away from the valve cover) (see Chapter 6) then move the harness out of the way.

35   Disconnect the vacuum tubes (metal or nylon) from the retainers at the front and rear sides of the valve cover and move the vacuum tubes out of the way.

36   Remove the bolt and position aside the interfering wiring harness bracket. The bracket retaining bolt is located near the bottom of the bracket.

37   Loosen the valve cover "captive" mounting bolts and remove the valve cover with the bolts.

2.5L engines

38   Refer to Chapter 5 and remove the ignition coils.

39   Disconnect the PCV system vent hose, then disconnect the interfering wiring harness connectors. Detach the wiring harnesses from

their retainers and set them aside.

40   Remove the metal electrical connector bracket fasteners and position the bracket aside.

41   Remove the valve cover mounting bolts, then lift the valve cover off along with the gaskets.

Installation

42   Carefully clean the sealing surfaces of the valve cover and the cylinder head.

43   Discard the old inner and outer valve cover gasket(s) and replace them with new ones. Also replace the VCT seals on the cover if they look to be damaged.

44   Place small dabs of RTV sealant at the joints at the front of the cylinder head where the timing cover meets it.

Caution: *On 1.5L and 1.6L models, the dabs of RTV sealant must ALSO be applied to the 2 joints where the cylinder head meets the HP injection pump and the 2 joints where the head meets the brake vacuum pump*

Caution: *On 2.0L models, the dabs of RTV sealant must ALSO be applied to the 4 joints on the opposite side (rear) of the cylinder head (at the seems of the "half-domes")*

**5.3 Check the valve clearances with a
feeler gauge of the specified thickness**

Caution: *The valve cover must be installed within five minutes of applying the RTV sealant*

45   Install the valve cover and tighten the bolts, in sequence (see illustrations) to the torque listed in this Chapter's Specifications.

46   The remainder of installation is the reverse of removal. On 1.5L and 1.6L engines, pressurize the fuel system and check for any fuel leaks, which must be immediately repaired (see Chapter 4).

47   Reconnect the battery. After you're done, the Powertrain Control Module (PCM) must relearn its idle and fuel trim strategy for optimum driveability and performance (see Chapter 5, Section 1).

---

5   Valve clearances - checking and adjustment

Note: *While checking the valve clearances is an easy operation, changing the shims requires the removal of the camshaft(s) - owners may prefer to have this work carried out by a dealer.*

Checking

Note: *On 1.5L engines, remove the following items: high-pressure fuel pump drive housing (see Section 19) and brake vacuum pump and pump cap (see Chapter 9).*

1   Remove the valve cover as described in Section 4. Remove the spark plugs as described in Chapter 1.

2   Set the Number 1 cylinder to TDC on the compression stroke as described in Section 3. The intake and exhaust cam lobes of the Number 1 cylinder will be pointing upwards (though not vertical) and the valve clearances can be checked.

3   Working on each valve in turn, measure the clearance between the base of the cam lobe and the shim using feeler gauges (see illustration). Record the thickness of the blade required to give a firm sliding fit on all four valves of the Number 1 cylinder.

4   Now turn the crankshaft clockwise through 180 degrees so that the valves of cylinder Number 3 are pointing upwards. Measure and record the valve clearances for cylinder Number 3. The clearances for cylinders 4 and 2 can be measured after turning the crankshaft through 180° each time. Any measured valve clearances that do not fall in the specified range listed in this Chapter's Specifications must be adjusted.

Adjustment

5   If adjustment is required, the shims will need to be replaced with new shim(s) of the proper thicknesses.
6   To gain access to the shims, remove the camshafts (see Section 12). Be sure to record accurate valve clearance measurements before the camshafts are removed.
7   Remove the shim with a magnetic tool.
8   Record the thickness of the shim (in mm) which is engraved on the side facing away from the camshaft. If the marking is missing or illegible, a micrometer will be needed to establish shim thickness.
9   If the valve clearance is too small, a thinner shim must be installed. If the clearance is too large, a thicker shim must be installed. When the shim thickness and the valve clearance are known, the required thickness of the new shim can be calculated as follows:
10   Sample calculation - clearance too small
Desired clearance (A) = 0.15 mm
Measured clearance (B) = 0.09 mm
Shim thickness found (C) = 2.55 mm
Thickness required (D) = C + B - A = 2.49 mm
11   Sample calculation - clearance too large
Desired clearance (A) = 0.30 mm
Measured clearance (B) = 0.36 mm
Shim thickness found (C) = 2.19 mm
Thickness required (D) = C + B - A = 2.25 mm
12   In reference to the previous clearance measurements taken, obtain the required shim(s) to be replaced.
13   Smear some clean engine oil on the replacement shim(s), then install the correctly sized shim(s) into the valve recesses.
14   Install the camshaft(s) (see Section 12).
15   Repeat the valve clearance checking procedure for all cylinders (see Steps 1 through 4), ensuring that the new measurements fall within the correct specifications.
16   It will be helpful for future adjustment if a record is kept of the thickness of the shims fitted at each position. The shims required can be purchased in advance once the clearances and the existing shim thicknesses are known. It is permissible to interchange shims between valve recess locations to achieve the correct clearances. Be sure to keep an exact record of which shims have been interchanged or replaced.
17   When all the clearances are correct, install the valve cover as described in Section 4. On 1.5L engines, install the high-pressure fuel pump drive housing (see Section 19) and the brake vacuum pump cap and pump (see Chapter 9).

6   Intake manifold - removal and installation

2008 and earlier models (2.0L and 2.3L)

Removal

1   Relieve the fuel system pressure (see Chapter 4)
2   Disconnect the cable from the negative battery terminal (see Chapter 5).
3   Remove the air intake hose (see Chapter 4).
4   Disconnect the accelerator cable and, if equipped, the cruise control cable from the throttle lever arm (see Chapter 4).
5   Remove the fuel rail, fuel pressure regulator and fuel injectors as a single assembly (see Chapter 4).
6   On 2004 and earlier models, remove the breather hose and the PCV valve from the intake manifold (see Chapter 6).
7   Remove the ignition wires and disconnect the ignition coil harness connector (see Chapter 5).
8   Disconnect the vacuum lines from the intake manifold and label them for correct reassembly.
9   Disconnect the PCM harness connector (see Chapter 6).
10   Disconnect the cylinder head temperature (CHT) sensor connector (see Chapter 6). Detach the harness from the clip and separate the harness from the intake manifold.
11   Remove the alternator (see Chapter 5).
12   Remove the EGR valve (see Chapter 6).
13   If you're planning to replace or service the intake manifold, remove the throttle body (see Chapter 4).
Note: *If you're simply removing the intake manifold plenum to remove or service the cylinder head, it's not necessary to remove the throttle body from the intake manifold plenum.*
14   Remove the intake manifold support bracket.
15   Remove the intake manifold fasteners and washers and then remove the intake manifold and the manifold gasket.

Inspection

16   Using a straightedge and feeler gauge, check the intake manifold mating surface for warpage. Check the intake manifold surface on the cylinder head also. If the warpage on either surface exceeds the limit listed in this Chapter's Specifications, the intake manifold and/or the cylinder head must be resurfaced at an automotive machine shop or, if the warpage is too excessive for resurfacing, replaced.

Installation

17   Using a new manifold gasket, install the intake manifold. Tighten the bolts and nuts in several stages, working from the center out, to the torque listed in this Chapter's Specifications.

18   Installation is otherwise the reverse of removal.
19   Reconnect the battery. After you're done, the Powertrain Control Module (PCM) must relearn its idle and fuel trim strategy for optimum driveability and performance (see Chapter 5, Section 1).

2009 and later models

Removal

1.5L and 1.6L engines

20   Disconnect the cable from the negative battery terminal (see Chapter 5).
21   Carefully lift the engine cover up to disengage the cover from the ballstuds and remove the cover.
22   Raise the front of the vehicle and support it securely on jackstands. Remove the engine under cover.
23   On 1.5L engines, remove the air filter housing (see Chapter 4), the evaporative emissions canister purge valve (see Chapter 6), and the intercooler (see Chapter 4).
24   Remove the alternator (see Chapter 5).
25   Remove the throttle body intake pipe (intercooler outlet pipe) by disconnecting all hoses or connectors (if equipped), loosening the hose clamps at the throttle body and opposite end, removing any pipe bracket nuts, then disconnecting and removing the pipe (see Chapter 4). On 1.5L engines, position the throttle body end of the pipe aside.
26   Disconnect the electrical connectors to the knock sensors (see Chapter 6).
27   Disconnect any crankcase vent tubes from the intake manifold and valve cover. Label the hoses to their correct locations as necessary.
28   Disconnect all remaining interfering sensor electrical connectors and wiring harness retainers, labeling them if necessary. Move the harnesses aside.
29   On 1.5L engines, remove the intake manifold support bracket bolt near the throttle body, then remove the coolant tube bracket bolt below the throttle body and move the coolant tubes out of the way.
30   Loosen the intake manifold "captive" bolts and carefully remove the intake manifold.
Note: *The intake manifold is secured using "captive" bolts, the bolts should stay attached to the intake manifold once they are loosened, do not try and remove the bolts from the intake manifold.*

2.0L engines

31   Remove the engine cover by pulling straight up on the cover to disengage the grommets from the cover studs.
Caution: *The engine cover must be pulled straight up, do not pull the cover forward or sideways to remove it. The cover mounting points and studs are easily damaged if the cover is not pulled straight upward from the underside of the cover.*
32   Disconnect the cable from the negative battery terminal (see Chapter 5).

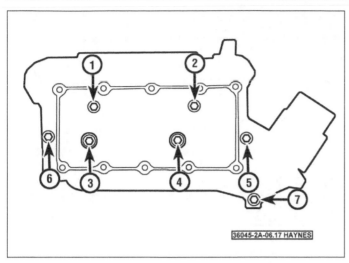

6.53a 1.5L engine - intake manifold bolt tightening sequence

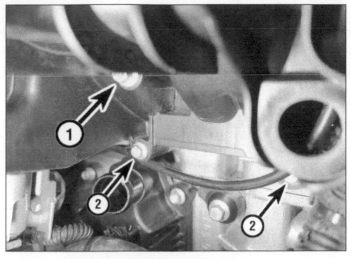

6.53b 1.6L engine - intake manifold bolt tightening sequence (LH side)

33   Raise the vehicle and support it securely on jackstands.

34   Remove the engine under splash shield fasteners and shield.

35   Remove the throttle body (see Chapter 4).

36   Disconnect the electrical connectors from the Fuel Rail Pressure (FRP) sensor and the Manifold Absolute Pressure (MAP) sensor. In the same area, disconnect the quick-connect tube from the manifold and free the tube from the bracket clip.

37   Release the fuel line clips at the top of the intake manifold and set it aside.

38   Disconnect any remaining electrical connectors/clips or breather hoses that interfere with the removal of the manifold. Be sure to label the connections to ensure correct reconnection.

Note: *Some early models may have a plug at the LH side of the manifold in place of the brake vacuum hose. If this plug is missing and there was no brake hose attached, install a new factory obtained plug and securing strap.*

39   Disconnect the brake vacuum hose from the left-hand side of the intake manifold by pushing the red locking ring inward and pulling the hose out.

40   Remove the mounting bolts and pull the intake manifold away from the engine enough to access and disconnect the crankcase vent hose from the oil separator.

2.5L engine

41   Disconnect the cable from the negative battery terminal (see Chapter 5).

42   On 2013 and later models (at this stage), disconnect the air intake duct from the throttle body and air filter housing, and remove the duct.

43   Raise the front of the vehicle and support it securely on jackstands.

44   On 2012 and earlier models, relieve the fuel system pressure and remove the fuel rail (see Chapter 4).

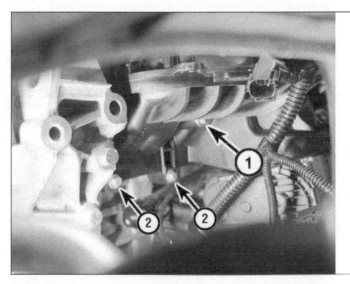

6.53c 1.6L engine - intake manifold bolt tightening sequence (RH side)

45   Disconnect the brake vacuum hose from the left-hand side of the intake manifold by pushing the locking ring inward and pulling the hose out.

46   Disconnect the vapor return line quick-connect fitting from the intake manifold.

Note: *There may be some wiring harness retaining clips that cannot be easily accessed from above - in this case, remove the engine under shield and detach the harness clip(s) by working from beneath the engine*

47   Disconnect all interfering electrical connectors (labeling them as necessary) and wiring harness or hose retaining clips, then move the harnesses aside from the intake manifold.

48   On 2012 and earlier models (at this stage), disconnect the air intake duct from the throttle body and air filter housing, and remove the duct.

49   Remove the bottom support bolt from the intake manifold, then remove the main intake manifold bolts.

50   Move the manifold aside to reach the

PCV vent tube and on 2012 and earlier models, the EGR tube. Detach the tube(s).

51   Remove the intake manifold and its gaskets.

Installation

1.5L and 1.6L engines

52   Inspect the intake manifold gaskets and replace them if they're damaged.

53   Install the intake manifold and tighten the "captive" bolts, in sequence (see illustrations 6.53a, b and c) to the torque values listed in this Chapter's Specifications.

54   Installation is the reverse of removal. Tighten all fasteners to the torque values listed in this Chapter's Specifications.

2.0L and 2.5L models

55   Inspect the intake manifold gaskets and replace them if they're damaged.

56   Installation is the reverse of removal. Tighten all fasteners to the torque values listed in this Chapter's Specifications.

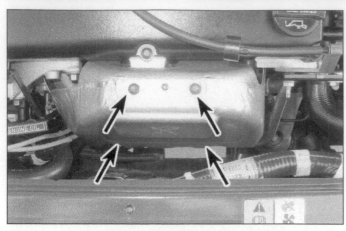

7.5 Remove the heat shield mounting bolts

**7.16 Remove the exhaust manifold mounting nuts -
1.6L model shown, early 2.5L model similar**

7   Exhaust manifold - removal and installation

Warning: *Allow the engine to cool completely before beginning this procedure.*

2008 and earlier models (2.0L and 2.3L)

Removal

1   Raise the vehicle and place it securely on jackstands.

2   Remove the exhaust pipe and the catalytic converter (see Chapter 4). Remove and discard the old flange gasket.

3   Remove the dipstick tube bracket bolt and lift the assembly from the engine compartment.

4   Unplug the electrical connector for the oxygen sensor and then remove the oxygen sensor from the exhaust manifold (see Chapter 6).

5   Remove the exhaust manifold heat shield (see illustration).

6   Remove the exhaust manifold mounting bolts and then remove the exhaust manifold and the old manifold gasket.

Inspection

7   Inspect the exhaust manifold for cracks and any other obvious damage. If the manifold is cracked or damaged in any way, replace it.

8   Using a wire brush, clean up the threads of the exhaust manifold bolts and then inspect the threads for damage. Replace any bolts that have thread damage.

9   Using a scraper, remove all traces of gasket material from the mating surfaces and inspect them for wear and cracks.

Caution: *When removing gasket material from any surface, especially aluminum, be very careful not to scratch or gouge the gasket surface. Any damage to the surface may leak after reassembly. Gasket removal solvents are available from auto parts stores and may prove helpful.*

10   Using a straightedge and feeler gauge, inspect the exhaust manifold mating surface for warpage. Check the exhaust manifold sur-

face on the cylinder head also. If the warpage on any surface exceeds the limits listed in this Chapter's Specifications, the exhaust manifold and/or cylinder head must be replaced or resurfaced at an automotive machine shop.

Installation

11   Coat the threads of the exhaust manifold bolts and studs with an anti-seize compound. Install a new gasket, install the manifold and install the fasteners. Tighten the bolts and nuts in several stages, working from the center out, to the torque listed in this Chapter's Specifications.

12   The remainder of installation is the reverse of removal. Run the engine and check for exhaust leaks.

2009 and later models

Note: *On 1.5L, 2.0L and 2013 and later 2.5L engines, the exhaust manifold has been integrated into the cylinder - therefore, this procedure is applicable only to 2009 through 2012 2.5L engines and 1.6L engines*

Removal

13   Raise the vehicle and support it securely on jackstands.

14   On 1.6L engines, remove the turbocharger (see Chapter 4).

15   On 2.5L engines: Remove the front exhaust pipe (downpipe) from the exhaust manifold, then remove the intermediate exhaust pipe. Disconnect the upstream oxygen sensor electrical connector. Remove the exhaust manifold heat shield.

16   Remove the exhaust manifold mounting nuts. Discard them - they must be replaced with new ones (see illustration).

17   Remove the exhaust manifold and its gasket. Discard the gasket.

18   Remove the exhaust manifold mounting studs and discard them.

19   Refer to Steps 7 through 10 and inspect the exhaust manifold. Replace it or have it machined flat if necessary.

Installation

20   Install the new exhaust manifold mounting studs. Tighten them to the torque listed in this Chapter's Specifications.

21   Place a new gasket over the studs, then install the manifold and the new nuts.

22   On 1.6L models, tighten the nuts in several steps to the torque listed in this Chapter's specifications, working from the center out. Be sure to perform the tightening sequence at least twice.

23   On 2.5L models, tighten the nuts in the sequence shown (see illustration) to the torque listed in this Chapter's Specifications.

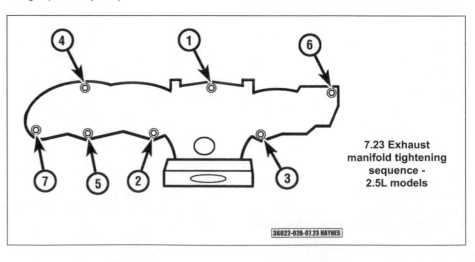

7.23 Exhaust manifold tightening sequence - 2.5L models

36022-02A-07.23 HAYNES

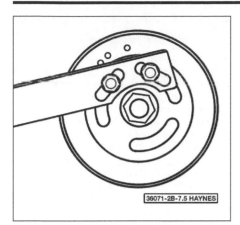

8.10 The crankshaft pulley holding tool installed

8.16 Installing the M6 bolt to verify TDC

8.26 The camshaft alignment tool installed

24   The remainder of installation is the reverse of removal.

Note: *Tighten the catalytic converter mounting nuts to the proper torque before installing the converter bracket bolts.*

---

8   Crankshaft pulley - removal and installation

2004 and earlier 2.0L engines

Removal
1   Remove the drivebelt (see Chapter 1).
2   If the pulley is being removed as part of another procedure such as timing belt replacement, it will be easier to set the engine to TDC now (see Section 3) before removing the crankshaft pulley bolt.
Note: *NEVER use the timing pin as a means of locking the crankshaft - it is not strong enough for this, and will shear off. Always ensure that the timing pin is removed before the crankshaft pulley bolt (or similar fasteners) is loosened or tightened.*
3   The crankshaft must now be held or locked to prevent its rotation while the pulley bolt is unscrewed. There are two recommended procedures. One method is to use a chain wrench around the crankshaft pulley with a piece of rubber or old drivebelt material under the chain to prevent any damage to the pulley. Another method to prevent crankshaft rotation is to lock the flywheel by removing the starter (see Chapter 5) and wedging a screwdriver in the flywheel/driveplate ring gear teeth.
4   Unscrew the pulley bolt and remove the pulley.

Installation
5   Installation is the reverse of the removal procedure. Ensure that the pulley's keyway is aligned with the crankshaft's locating key. Tighten the pulley bolt to the torque setting listed in this Chapter's Specifications. Lock the crankshaft using the same method as for loosening.

2.3L and 2.5L engines
Removal
Caution: *Once the crankshaft pulley is loosened, the crankshaft (timing) sprocket will be*

loosened as well. The engine is considered out-of-time at this point. The installation procedure in this Section must be followed exactly to re-time the engine properly. Severe engine damage may occur otherwise.
6   Loosen the right-front wheel lug nuts, then raise the front of the vehicle and support it securely on jackstands. Remove the right-front wheel and fenderwell splash shield.
7   Remove the auxiliary drivebelt (see Chapter 1).
8   Remove the valve cover (see Section 4)
9   Set the engine to TDC (see Section 3).
Caution: *Make sure that the camshaft alignment plate and the threaded crankshaft alignment bolt are installed correctly before proceeding.*
Caution: *Use of a pry bar or similar tool (to hold the crankshaft from turning) can damage the crankshaft pulley.*
10   The crankshaft must be held to prevent its rotation while the pulley bolt is unscrewed. A special tool to hold the crankshaft pulley is needed for this (see illustration), and is available at most auto supply stores or equipment rental locations. A suitable equivalent can be fabricated from a length of strap steel.
11   Insert the service tool into the spaces in the front face of the pulley to hold it in place while removing the pulley bolt with a 1/2 inch drive impact gun.
Caution: *The crankshaft pulley bolt is recommended to be removed with a 1/2-inch (or greater) drive impact gun to ensure the engine remains at constant TDC.*
Caution: *Failure to hold the crankshaft pulley securely while removing the pulley bolt could result in engine damage. Moreover, NEVER rely on the timing pin or the camshaft service tool as a means of locking the crankshaft - they are designed for calibration only. Engine damage could occur by not using these tools for their intended purpose.*
12   Unscrew the pulley bolt the rest of the way and remove the holding tool.
Note: *The diamond washer must be discarded and replaced upon installing*
13   Remove the pulley and diamond washer, making sure to label or mark which side is installed towards the engine. Discard diamond washer and obtain a new one for installation.

Installation
14   Lightly coat the crankshaft front seal with clean engine oil, then install the diamond washer and crankshaft pulley.
Note: *If the seal shows signs of leakage, you may want to replace it before installing the crankshaft pulley (see Section 16).*
15   Install a new crankshaft pulley bolt and hand-tighten only. Attempt to closely align the hole in the pulley with the threaded hole in the timing chain cover.
Note: *This aligns the pulley correctly with the crankshaft.*
16   Install a crankshaft pulley alignment bolt (M6 x 18 mm) through the pulley and into the front engine cover. Rotate the pulley as necessary to do this (see illustration).
17   DO NOT use an impact gun to tighten the crankshaft pulley bolt
Caution: *DO NOT use an impact gun to tighten the crankshaft pulley bolt*
18   Insert the large holding tool into the pulley and tighten the crankshaft pulley bolt to the torque listed in this Chapter's Specifications.
19   Remove the crankshaft pulley holding tool and the threaded crankshaft alignment bolt from the pulley and front engine cover.
20   Remove the timing pin from the cylinder block.
21   Remove the camshaft alignment tool.
22   Rotate the engine clockwise, one and three-quarters (not quite two revolutions) by turning the crankshaft pulley bolt with a wrench or large socket.
23   Bring the engine to TDC (see Section 3).
Note: *Rotate the engine in the clockwise direction only.*
24   Install the timing pin into the cylinder block.
25   Install the crankshaft pulley alignment bolt (M6 x 18 mm). If it cannot be installed, the crankshaft pulley must be removed and aligned so that it can. Repeat Steps 14 through 23 until crankshaft pulley alignment is correct.
26   With the crankshaft pulley alignment bolt (M6 x 18 mm) installed, install the camshaft alignment tool and check the position of the camshafts (see illustration). If the tool cannot be installed, the engine timing must be corrected by repeating Steps 14 through 24.

27   The correct engine timing is achieved when the camshaft alignment tool, timing pin and the crankshaft pulley alignment bolt tool can be placed simultaneously.

28   Once correct engine timing is achieved, remove all the alignment tools and bolts and install the timing pin plug.

29   The remainder of installation is the reverse of removal.

1.5L, 1.6L and 2013 and later 2.0L engines

Removal

Caution: *Once the crankshaft pulley is loosened, the crankshaft (timing) sprocket will be loosened as well. The engine is considered out-of-time at this point. The installation procedure in this Section must be followed exactly to re-time the engine properly. Severe engine damage may occur otherwise.*

Note: *You will need several special tools for the timing procedure (see Section 9 for 1.5L and 1.6L models and Section 10 for 2.0L models): On 1.5L and 1.6L models use the camshaft positioning tool 303-1552, the variable camshaft sprocket timing (VCT) locking tool 303-1097, crankshaft pulley alignment tool 303-1550, flywheel locking tools 303-393A, 303-393-02 and the timing pin 303-748. On 2.0L models use the camshaft positioning tool 303-1565 and the timing pin (303-507).*

Caution: *Only rotate the crankshaft in the clockwise direction.*

30   Raise the vehicle and support it on jackstands. Make sure it is safely supported, then remove the right inner fender splash shield, (see Chapter 11).

31   Remove the drivebelt and drivebelt tensioner (see Chapter 1).

1.5L and 1.6L engines

Warning: *When loosening the crankshaft pulley bolt, great force is exerted on the rotation of the engine. The manufacturer recommends that a special engine holding tool and brackets (Part numbers 303-1502, 303-050 [support brackets], 303-F072 and 303-290B-18 [engine holding device]) should be used on 1.6L models instead of just the floor jack placed under the oil pan throughout the procedure. If you opt not to use this method, great care must be taken to prevent the jack/wood from dislodging and the engine potentially falling and causing serious injury or damage. In this case, we recommend reinstalling the engine mount temporarily when loosening the pulley bolt.*

32   Remove the alternator and starter motor (see Chapter 5).

33   While carefully supporting the engine with a jack and block of wood placed underneath the engine oil pan, remove the timing belt cover (see Section 9).

34   The TDC timing hole is covered by the passenger's side driveaxle intermediate shaft support bearing bracket. Remove the driveaxle intermediate shaft and support bearing bracket (see Chapter 8).

35   Rotate the crankshaft until the camshaft phaser timing marks are pointing at the 11 o'clock position.

36   Remove the TDC timing hole plug (see Section 3) and insert the timing pin (303-748).

37   Slowly rotate the crankshaft clockwise until the crankshaft balance weight contacts the end of the timing pin. Both camshaft phaser timing marks should now be at the 12 o'clock position, the engine is now at TDC.

38   Assemble the flywheel locking tools (303-393A, 303-393-02). Using the starter mounting bolts, install the tools to the starter opening, making sure the tool engages the teeth on the flywheel/driveplate. Tighten the tool securely in place.

Caution: *Failure to hold the crankshaft pulley securely (by means of the flywheel locking tool) while removing the pulley bolt could result in severe engine damage. NEVER use the timing pin tool as a means of locking the crankshaft - it is designed only for calibration and achieving TDC. Engine damage could occur by using this tool for anything other than its intended purpose.*

39   Remove the crankshaft pulley bolt and pulley.

Caution: *Always replace the crankshaft pulley bolt with a new one during installation.*

Note: *Do not remove the timing belt. If the timing belt is removed, see Section 9 for the installation procedure.*

2013 and later 2.0L engines

Note: *This procedure requires the use of a factory scan tool (or equivalent) to perform the Misfire Monitor Neutral Profile Correction Procedure at the end of installation to initialize spark timing. If this tool cannot be obtained, the procedure is better left to a dealer service department or qualified repair shop.*

40   Remove the brake vacuum pump (see Chapter 9).

41   Remove the valve cover (see Section 4).

42   Remove the rear intake camshaft cap bolts and cap (see Section 12).

43   Place a wrench onto the flats of the intake camshaft to prevent the intake camshaft from turning, then remove the brake vacuum pump adapter from the end of the camshaft (see Section 3).

44   Remove the crankshaft position sensor (see Chapter 6).

45   Rotate the crankshaft clockwise until the Number1 piston is 45-degrees BTDC. The hole in the crankshaft pulley should be 45-degrees (4 o'clock position) from the guide holes on the engine front cover (6 o'clock position) (see Section 3).

46   Remove the bolt from the side of the engine block and install the timing pin (303-507) into the cylinder block.

Caution: *The crankshaft TDC timing pin will contact the crankshaft and prevent it from turning past TDC. The crankshaft can still be rotated in the counterclockwise direction. The crankshaft must remain at the TDC position during the crankshaft pulley removal and installation procedure.*

47   Rotate the crankshaft pulley clockwise until the crankshaft contacts the timing pin (303-507).

48   Install the camshaft alignment tool (303-1565) into the slot in the end of the intake camshaft and bolt the tool in place.

49   Place a strap wrench around the crankshaft pulley to prevent it from moving, then remove the crankshaft pulley bolt. It may be necessary to use an impact gun (with at least a 1/2-inch drive) to remove the crankshaft pulley bolt.

Caution: *Failure to hold the crankshaft pulley securely while removing the pulley bolt could result in severe engine damage. NEVER use the timing pin tool as a means of locking the crankshaft - it is designed only for calibration. Engine damage could occur by using this tool for anything other than its intended purpose.*

50   Remove and discard the crankshaft pulley bolt, then remove the pulley.

Note: *The crankshaft sprocket diamond washer may come off with the crankshaft pulley, make sure to locate the washer.*

Installation

1.5L and 1.6L engines

51   Remove the crankshaft sensor (see Chapter 6).

52   Install the crankshaft pulley alignment tool (part number 303-1550) where the CKP sensor was mounted.

53   Lightly coat the crankshaft front seal with clean engine oil.

Note: *If the seal shows signs of leakage, you may want to replace it before installing the crankshaft pulley (see Section 11).*

54   Install the crankshaft pulley over the tool, align the pulley with the tool, then install the new crankshaft pulley bolt hand-tight.

55   Install the Variable Camshaft Timing (VCT) locking tool (part number 303-1097) between the two camshaft sprockets. The three ears on each side of the tool should lock into each VCT sprocket (see Section 9 if necessary).

Note: *It may be necessary to rotate the camshafts only slightly to install the special tool.*

56   With the flywheel locking tools (part numbers 303-393A, 303-393-02) still installed (see Step 37), tighten the pulley bolt to the torque listed in this Chapter's Specifications.

Caution: *Do not use an impact gun to tighten the crankshaft pulley bolt.*

57   Remove the crankshaft pulley alignment tool, the the VCT locking tool from the camshafts, and the flywheel locking tools and the timing pin from the cylinder block.

58   The VCT sprocket timing marks should be at the 12 o'clock position. If the variable camshaft sprocket timing (VCT) locking tool (303-1097) cannot be installed, repeat Steps 34 through 38, then reinstall the pulley (see Steps 50 through 55).

59   Install the crankshaft position sensor (see Chapter 6).

60   The remainder of installation is the reverse of removal. Tighten all related fasteners to the torque settings listed in this Chapter's Specifications.

2013 and later 2.0L engines

61   Install a new diamond washer onto the nose of the crankshaft.

62   Lightly coat the crankshaft front seal with clean engine oil, then install the crankshaft pulley.
Note: *If the seal shows signs of leakage, you may want to replace it before installing the crankshaft pulley (see Section 11).*
63   Install the crankshaft pulley with the small hole in the pulley at the 6 o'clock position.
64   Insert an M6 bolt through the hole in the crankshaft pulley. Hand-tighten the bolt ONLY. This will correctly align the crankshaft pulley.
65   Install the crankshaft pulley bolt, then, using the strap wrench to hold the pulley, tighten the crankshaft pulley bolt to the torque listed in this Chapter's Specifications.
Caution: *Do not use an impact gun to tighten the crankshaft pulley bolt.*
66   Remove the M6 bolt from the pulley.
67   Install the Crankshaft Position Sensor (see Chapter 6), then install the Crankshaft Position (CKP) sensor alignment tool (#303-1521) onto/over the sensor and the pulley sprocket teeth. Tighten the sensor bolts with the tool installed. Once the bolts are tightened, remove the sensor alignment tool.
68   Remove the timing pin from the cylinder block and remove the camshaft alignment tool. Install the timing pin plug.
69   Double check that the engine timing is correct at TDC (see Section 3).
70   Install the vacuum pump adapter to the end of the intake camshaft and tighten the adapter to the torque listed in this Chapter's Specifications.
71   Install the intake camshaft rear bearing cap (see Section 12).
72   Installation is the reverse of removal. Tighten all related fasteners to the torque settings listed in this Chapter's Specifications or their respective Chapter's Specifications.
73   Have a dealer service department or correct scan tool capable of performing the *Misfire Monitor Neutral Profile Correction Procedure* (see the Note at beginning of *Removal*).

9   Timing belt - removal, inspection and installation

2004 and earlier 2.0L engines
Removal
Caution: *The timing system is complex. Severe engine damage will occur if you make any mistakes. Do not attempt this procedure unless you are highly experienced with this type of repair. If you are at all unsure of your abilities, consult an expert. Double-check all your work and be sure everything is correct before you attempt to start the engine.*
1   Disconnect the battery cable from the negative battery terminal (see Chapter 5).
2   Remove the air intake duct from the air filter housing (see Chapter 4).
3   Remove the spark plugs (see Chapter 1).

9.14 Unscrew the bolts and remove the timing belt lower cover

4   Disconnect the accelerator cable and the cruise control actuator cable from the throttle valve (see Chapter 6).
5   Disconnect the catalytic converter monitor connector and the heated oxygen sensor connector (see Chapter 6). Remove the connectors from the bracket.
6   Remove the wiring harness support nuts from the studs on the valve cover and position the wiring harness off to the side.
7   Remove the valve cover (see Section 4).
8   Drain the coolant (see Chapter 3). Remove the coolant pipe mounting bolts and position the coolant pipe off to the side.
9   Loosen the right front wheel lug nuts. Raise the vehicle and secure it on jackstands.
10   Remove the front wheel and front fender splash shield on the timing belt side of the engine compartment.
11   Position the engine at TDC for Number 1 (see Section 3).
Note: *NEVER use the timing pin as a means of locking the crankshaft - it is not strong enough for this, and will shear off. Always ensure that*

9.17 Unscrew and remove the engine mount studs

the timing pin is removed before the crankshaft pulley bolt (or similar fasteners) is loosened or tightened. Follow the recommended procedure in Section 8 for locking the flywheel.
12   Remove the three water pump pulley bolts (see Chapter 3). Separate the water pump pulley from the timing belt cover.
13   Remove the crankshaft pulley (see Section 8).
14   Remove the bolts from the lower timing belt cover (see illustration).
15   Connect a hoist or an engine support fixture to the engine (see Chapter 2C). Alternatively, a floor jack and block of wood can be positioned under the oil pan to support the engine.
16   Remove the upper engine mount next to the timing belt cover (see Section 18).
17   Remove the studs from the engine mount bracket (see illustration).
18   Remove the knock sensor electrical connector from the upper timing belt cover (see Chapter 6).
19   Remove the upper timing belt cover (see illustrations).

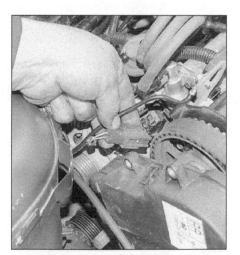
9.19a Disconnect the harness connector next to the upper timing belt cover

9.19b Remove the bolts and lift off the upper timing belt cover

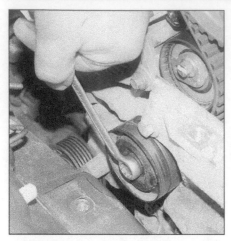

9.20 Remove the bolt for the idler pulley

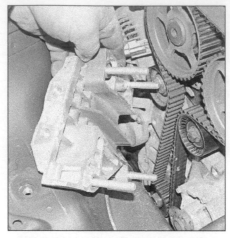

9.21 Remove the engine mount bracket

9.24 Unscrew the bolt, then unhook and remove the timing belt tensioner

20   Remove the accessory drivebelt idler pulley bolt and pulley from the timing belt cover (see illustration).

21   Remove the engine mount bracket (see illustration).

22   Before removing the timing belt, note that manufacturer service tool 303-376 will be needed (see Section 3) to set the camshafts to their TDC position. Fortunately, a substitute tool can be made from a strip of metal 5 mm thick (while the strip's thickness is critical, its length and width are not, but should be approximately 180 to 230 mm long by 20 to 30 mm wide) (see illustration 3.5b).

23   If the timing belt is to be re-used (and this is not recommended), mark it with paint to indicate its direction of rotation, clockwise, viewed from the timing belt end of the engine.

24   Loosen the timing belt tensioner bolt and turn the tensioner clockwise, using an Allen key in the hole provided. Now unscrew the bolt and unhook the tensioner from the inner shield (see illustration).

25   Ensuring that the sprockets are turned as little as possible, slide the timing belt off the sprockets and pulleys, and remove it.

26   If the timing belt is not being installed right now or if the belt is being removed as part of another procedure, such as cylinder head removal, temporarily install the engine right-hand mount and tighten the bolts securely.

Inspection

27   If the old belt is likely to be reinstalled, check it carefully for any signs of uneven wear, splitting, or cracks, especially at the roots of the belt teeth (see illustration 9.78).

28   Even if a new belt is to be installed, check the old one for signs of oil or coolant. If evident, trace the source of the leak and rectify it, then clean the engine timing belt area and related components to remove all traces of oil or coolant. Do not simply install a new belt in this instance, or its service life will be greatly reduced, increasing the risk of engine damage if it fails in service.

29   Always replace the belt if there is the slightest doubt about its condition. As a pre-

caution against engine damage, the belt must be replaced as a matter of course at the intervals given in Chapter 1. If the vehicle's history is unknown, the belt should be replaced irrespective of its apparent condition whenever the engine is overhauled.

30   Similarly check the belt tensioner, replacing it if there is any doubt about its condition. Check also the sprockets and pulleys for signs of wear or damage (and particularly cracking), and ensure that the tensioner and guide pulleys rotate smoothly on their bearings. Replace any worn or damaged components.

Note: *It is considered good practice by many professional mechanics to replace tensioner and guide pulley assemblies as a matter of course, whenever the timing belt is replaced.*

Installation

Caution: *Before starting the engine, carefully rotate the crankshaft by hand through at least two full revolutions (use a socket and breaker bar on the crankshaft pulley center bolt). If you feel any resistance, STOP! There is something wrong - most likely, valves are contacting the pistons. You must find the problem before*

proceeding. Check your work and see if any updated repair information is available.

31   Without turning the crankshaft more than a few degrees, check that the engine is still set to TDC on the Number 1 cylinder (see Section 3).

32   The TDC position of the camshafts must now be set, and for this, the camshaft sprocket bolts must be loosened. A holding tool will be required to prevent the camshaft sprockets from rotating while their bolts are slackened and retightened; either obtain manufacturer service tool 303-465, or fabricate a substitute as follows. Find two lengths of steel strip, one approximately two feet long and the other about eight inches long, and three bolts with nuts and washers; one nut and bolt forming the pivot of a forked tool, with the remaining nuts and bolts at the tips of the forks, to engage with the sprocket spokes (see illustration).

Note: *Do not use the camshaft TDC alignment tool (whether genuine manufacturer or not) to prevent rotation while the camshaft sprocket bolts are loosened or tightened; the risk of damage to the camshaft concerned and to the cylinder head is far too great. Use only a holding tool applied directly to the sprockets, as described.*

9.32 Use a sprocket holding tool while loosening the camshaft sprocket bolts

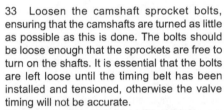

9.38a Use an Allen key to set the tensioner pointer in the center of the square window . . .

9.38b . . . then hold it while the tensioner bolt is tightened to the specified torque

33 Loosen the camshaft sprocket bolts, ensuring that the camshafts are turned as little as possible as this is done. The bolts should be loose enough that the sprockets are free to turn on the shafts. It is essential that the bolts are left loose until the timing belt has been installed and tensioned, otherwise the valve timing will not be accurate.

34 The tool described in Step 22 is now required. Rest the tool on the cylinder head mating surface, and slide it into the slots in the opposite end (transaxle side) of both camshafts. The tool should slip snugly into both slots while resting on the cylinder head mating surface. If one camshaft is only slightly out of alignment, rotate the camshaft gently and carefully until the tool will fit. Once the camshafts are set at TDC, the positions of the sprockets are less important - they can move independently of the camshafts, to allow the timing belt to be accurately aligned with the teeth on the sprockets.

35 When installing the belt, note that the tensioner must not be installed until after the belt is fully positioned around the sprockets and pulley(s).

36 Install the timing belt; if the original is being reinstalled, ensure that the marks and notes made on removal are followed, so that the belt is installed to run in the same direction. Starting at the crankshaft sprocket, work counterclockwise around the guide pulley(s) and camshaft sprockets. Ensure that the belt teeth engage correctly with those on the sprockets.

Note: *The crankshaft gear does not have alignment marks to verify TDC position. Instead it will be necessary to install the crankshaft pulley and check the alignment mark with the relief mark on the case (see Section 3).*

37 Any slack in the belt should be kept on the tensioner side - the front run must be kept tensioned, without altering the position of the crankshaft or the camshafts. If necessary, the camshaft sprockets can be turned slightly in relation to the camshafts (which remain fixed by the aligning tool).

38 When the timing belt is correctly installed on the sprockets and guide pulley(s), the belt tensioner can be installed. Hook the tensioner into the timing belt inner shield, and insert the bolt loosely. Using the Allen key, turn the tensioner counterclockwise until the arrow (Type A) (see illustrations) or fork (Type B) is aligned with the mark or square hole on the bracket, then tighten the tensioner bolt to its specified torque.

Note: *There are two types of indicators on the tensioner. Type A uses an arrow that aligns with the dot in the square box. Type B uses a fork that aligns with a mark directly in the middle of the two prongs on the fork.*

Note: *The timing belt tensioner automatically adjusts the tension by internal spring forces - checking the tension once set, for instance, by depressing or twisting the timing belt, will not give a meaningful result.*

39 Tighten both camshaft sprocket bolts to the torque listed in this Chapter's Specifications, holding the sprockets stationary using the tool described in Step 32.

40 Remove the camshaft aligning tool and the timing pin (see Section 3). Temporarily install the crankshaft pulley, and rotate the crankshaft through two full turns clockwise to settle and tension the timing belt, returning it to TDC. Check that the timing pin can be fully installed, then install the camshaft aligning tool; it should slip into place. If the checks are not correct, go back to Step 31 and perform the timing procedures again.

41 If one camshaft is slightly out of line, attach the forked holding tool to its sprocket (see Step 32), adjust its position as required, and check that any slack created in the belt has been taken up by the tensioner. Rotate the crankshaft two turns clockwise and install the camshaft aligning tool to check that it now fits as it should. If all is well, proceed to Step 43.

42 If either camshaft is significantly out of line, use the holding tool described in Step 32 to prevent its sprocket from rotating while its retaining bolt is loosened - the camshaft can then be rotated carefully until the camshaft aligning tool will slip into place. Take care not to disturb the relationship of the sprocket to the timing belt. Without disturbing the sprocket's new position on the camshaft, tighten the sprocket bolt to the torque listed in this Chapter's Specifica-

tions. Remove the camshaft aligning tool and timing pin, and rotate the crankshaft two more turns clockwise, and install the tool to check that it now fits as it should.

43 The remainder of the reassembly procedure is the reverse of removal, noting the following points:

a) *Make sure that the timing pin and camshaft aligning tool are removed.*

b) *When replacing the engine right-hand mount, remember to install the timing belt upper cover.*

c) *Tighten all fasteners to the specified torque settings.*

44 After you reconnect the battery, the Powertrain Control Module (PCM) must relearn its idle and fuel trim strategy for optimum driveability and performance (see Chapter 5, Section 1).

1.5L and 1.6L engines
Removal

Caution: *The timing system is complex. Severe engine damage will occur if you make any mistakes. Do not attempt this procedure unless you are highly experienced with this type of repair. If you are at all unsure of your abilities, consult an expert. Double-check all your work and be sure everything is correct before you attempt to start the engine.*

Caution: *The crankshaft pulley and crankshaft (timing) sprocket are not keyed to the crankshaft. Once the crankshaft pulley bolt is loosened, the engine is considered out-of-time. The installation procedure in this Section must be followed exactly to re-time the engine properly, or severe engine damage will occur.*

Note: *Only rotate the crankshaft in the clockwise direction.*

45 Disconnect the negative battery cable (see Chapter 5).

46 Loosen the right-side wheel lug nuts, then raise the vehicle and support it securely on jackstands. Remove the wheel.

47 Remove the right-side inner fender splash shield (see Chapter 11).

48 Remove the under-vehicle splash shield.

49 Lift the engine cover up and off of the ballstuds, then remove the cover.

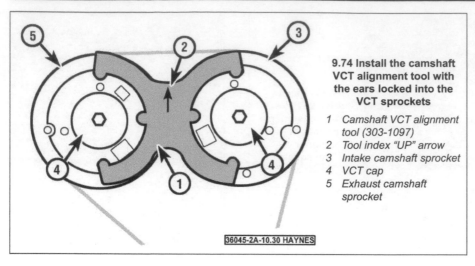

9.74 Install the camshaft VCT alignment tool with the ears locked into the VCT sprockets

1 *Camshaft VCT alignment tool (303-1097)*
2 *Tool index "UP" arrow*
3 *Intake camshaft sprocket*
4 *VCT cap*
5 *Exhaust camshaft sprocket*

36045-2A-10.30 HAYNES

50   Remove the cowl panel (see Chapter 11).
51   Remove the air filter housing outlet pipe and turbocharger inlet pipe (see Chapter 4).
52   Remove the charge air cooler (CAC) inlet and outlet tubes (see Chapter 4), then remove the CAC bracket.
53   Remove the coolant expansion tank (see Chapter 3).
54   Loosen the water pump pulley mounting bolts.
55   Remove the drivebelt (see Chapter 1).
56   Remove the alternator and the starter motor (see Chapter 5).
57   Remove the right-side driveaxle and intermediate shaft bearing support bracket (see Chapter 8).
58   Position a jack and block of wood under the engine oil pan, then carefully raise the engine just enough to take the weight off the right-side mount.
59   Remove the right-side engine mount (see Section 18).
60   With the mount removed, raise the right side of the engine slightly, then remove the drivebelt tensioner (see Chapter 1).
61   Remove the vacuum line retainer bracket from the rear side of the cylinder head, then remove the cover stud bolt.
62   Slide the side inspection cover down to disengage the clip and remove the side timing belt cover at the rear.
63   Remove the water pump pulley mounting bolts and pulley.
64   Disengage the harness clips from the cover, then remove the timing belt cover fasteners. Using a reverse Torx socket, remove the rearmost engine mount bracket stud from the bracket.
Note: *The timing belt cover bolts may be different lengths - make note of their locations while removing them*
65   Tilt the top of the cover down, then pull the cover out and away from the engine.
66   Remove the valve cover (see Section 4).
67   Remove the right-side engine mount bracket bolts and bracket.
68   Attach an engine support fixture to the cylinder head. If no hooks are provided, use a bolt of the proper size and thread pitch to

attach the support fixture chain to a hole in the cylinder head.
Note: *Engine support fixtures can be obtained at most equipment rental yards and some auto parts stores.*
69   Rotate the crankshaft until the camshaft phaser timing marks are pointing at the 11 o'clock position.
70   Remove the TDC timing hole plug (see Section 3) and insert the timing pin (303-748) into the hole.
71   Slowly rotate the crankshaft clockwise until the crankshaft balance weight contacts the end of the timing pin. The engine is now at TDC. The camshaft phaser VTC timing marks should now be at the 12 o'clock position.
72   Assemble the flywheel locking tools (303-393A, 303-393-02). Using the starter mounting bolts, install the tools to the starter opening, making sure the tool engages the teeth on the flywheel/driveplate, then tighten the tool securely in place.
Caution: *Failure to hold the crankshaft pulley securely while removing the pulley bolt could result in severe engine damage. NEVER use the timing pin tool as a means of locking the crankshaft - it is designed only for calibration. Engine damage could occur by using this tool for anything other than its intended purpose.*

73   Remove the crankshaft pulley bolt and pulley (see Section 8).
Note: *Always replace the crankshaft pulley bolt with a new one during installation.*
Note: *It may necessary to rotate the camshafts slightly to install the special tool.*
74   The VCT sprocket timing marks should be at the 12 o'clock position. Install the variable camshaft sprocket timing (VCT) locking tool (303-1097) between the two camshafts. The three ears on each side of the tool should lock into each VCT sprockets (see illustration).
75   Rotate the timing belt tensioner clockwise until the tab on the tensioner aligns with the tab on the tensioner body, then install a drill bit or Allen wrench through both tabs.
Warning: *The timing belt tensioner spring is under an extreme load. If the drill bit or Allen wrench is removed, the tensioner can spring back and cause personal injury.*
76   Check to see that the timing belt is marked with an arrow to show which side faces out. If there isn't a mark, paint one on (only if the same belt will be reinstalled). Slide the timing belt off the sprockets and check the condition of the tensioner.
77   To remove the tensioner, remove the tensioner mounting bolt, but make sure the drill bit or Allen wrench is secure.

Inspection
78   Inspect the timing belt (see illustration). Look at the backside (the side without the teeth): If it's cracked or peeling, or if it's hard, glossy and inflexible, and leaves no indentation when pressed with your fingernail, replace the belt. Look at the drive side; if teeth are missing, cracked or excessively worn, replace the belt.

Installation
Caution: *Before starting the engine, carefully rotate the crankshaft by hand through at least two full revolutions (use a socket and breaker bar on the crankshaft pulley center bolt). If you feel any resistance, STOP! There is something wrong - most likely, valves are contacting the pistons. You must find the problem before proceeding. Check your work and see if any updated repair information is available.*
79   Install the belt on the crankshaft sprocket

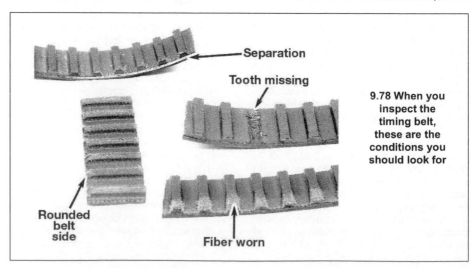

Separation

Tooth missing

9.78 When you inspect the timing belt, these are the conditions you should look for

Rounded belt side

Fiber worn

first, then around the intake camshaft sprocket, exhaust camshaft sprocket and then the tensioner so that the belt is tight and any slack is on the tensioner side.

80   Remove the drill bit or Allen key from the tensioner and allow the tensioner to expand and take up any slack on the tensioner side.

81   Remove the crankshaft position sensor from the oil pump body (see Chapter 6).

82   Install the crankshaft pulley alignment tool (303-1550) where the CKP sensor was mounted.

83   Install the crankshaft pulley over the tool, align the pulley with the tool, then install the old crankshaft pulley bolt hand-tight.

84   Remove the old bolt, install a new bolt and tighten the crankshaft pulley bolt to the torque listed in this Chapter's Specifications.

Caution: *Do not use an impact gun to tighten the crankshaft pulley bolt.*

85   Unbolt and remove the pulley alignment tool.

86   Remove the VCT locking tool (303-1097) from in between the camshafts, the flywheel locking tools (303-393A, 303-393-02), and the timing pin (303-748) from the cylinder block.

87   Slowly rotate the crankshaft clockwise for 1-3/4 turns, then reinstall the timing pin (303-748). Continue to slowly rotate the crankshaft clockwise until the crankshaft balance weight contacts the end of the timing pin.

88   The VCT sprocket timing marks should be at the 12 o'clock position. Install the VCT locking tool (303-1097) between the two camshafts. The three ears on each side of the tool should lock into each VCT sprocket (see illustration 9.74).

89   If the variable camshaft sprocket timing (VCT) locking tool (303-1097) can be installed and also the crank timing pin (simultaneously), the timing belt is installed correctly. If the tools cannot be installed, see Section 8.

90   Install the crankshaft position sensor (see Chapter 6).

91   Install the engine mount bracket nuts and bolts then tighten the fasteners to the torque listed in this Chapter's Specifications.

92   Install the timing belt cover and tighten the fasteners securely.

93   The remaining installation is the reverse of removal. Tighten all other fasteners to the torque settings listed in this Chapter's Specifications or their respective Chapter's specifications.

10   Timing chain cover, timing chain and tensioner - removal and installation

Timing chain cover

Removal

2008 and earlier models

1   Disconnect the negative battery cable (see Chapter 5).

2   Remove the drivebelt and the tensioner (see Chapter 1).

3   Disconnect the crankshaft position (CKP) sensor electrical connector (see Chapter 6). Move the wiring harness away from the engine cover by removing the harness retainers.

4   Remove the (CKP) sensor (see Chapter 6).

Note: *Unfortunately with this engine, anytime this sensor is removed, a new sensor must be installed. The new sensor is packaged with a special sensor alignment tool that is critical for installation.*

5   Remove the crankshaft pulley (see Section 8).

6   Remove the water pump pulley (see Chapter 3).

7   Disconnect the electrical connector to the power steering pressure switch. The switch is located on a high-pressure hose within the power steering system.

8   Remove the single bolt on the lower left corner of the power steering pump and move the pump aside.

9   Remove the bolts and the engine cover.

2009 and later models

10   Raise the vehicle and support it securely on jackstands.

11   Disconnect the cable from the negative battery terminal (see Chapter 5).

12   Remove the engine mount (Section 18).

13   Refer to Chapter 1 and remove the drivebelt, the drivebelt idler pulley and drivebelt tensioner.

14   Remove the crankshaft pulley (see Section 8).

15   Refer to Section 11 and remove the front engine oil seal.

16   Remove the water pump pulley.

17   Remove the crankshaft position (CKP) sensor (see Chapter 6).

18   If equipped, unbolt and detach any brackets or harnesses from the cover.

19   Loosen the front cover bolts evenly, a quarter turn at a time, until they can be removed. Tap the cover sideways with a plastic hammer to break it loose if necessary, then remove it.

Installation

20   Installation is the reverse of removal noting the following:

a) *Clean the mating surfaces of all material.*

Note: *Be careful not to gouge or use any abrasives on the mating surfaces.*

b) *Install the engine cover within four minutes of applying a 2 to 2.5 mm bead of RTV sealant.*

c) *Tighten all bolts to the specified torque settings (using a three-stage sequence - see this Chapter's Specifications) and also using the sequence pattern provided (see illustrations).*

d) *Be sure to re-time the engine as described in Section 8 and also use a new crankshaft position sensor which includes the necessary alignment tool.*

Timing chain and tensioner

Removal

21   Apply the parking brake, then carefully jack up the front of the vehicle and support the lifted portion on jackstands.

22   On 2013 and later 2.0L models, relieve the fuel pressure, then remove the high pressure fuel pump (see Chapter 4). Remove high pressure fuel pump drive unit and tappet (see Section 19).

23   Remove the timing chain cover.

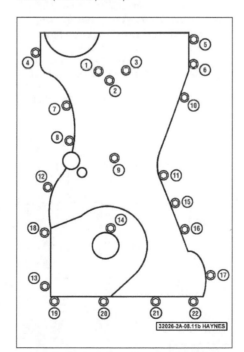

10.20a Timing chain cover bolt tightening sequence - 2013 and later 2.0L engine

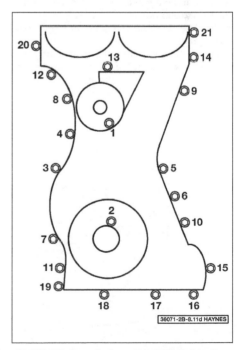

10.20b Timing chain cover bolt tightening sequence - 2.3L engine

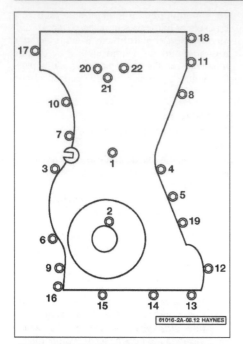

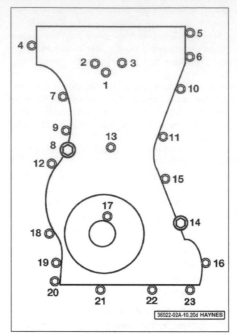

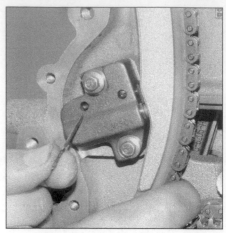

10.25a Compressing the timing chain tensioner and placing the lock pin

10.20c Timing chain cover bolt tightening sequence - 2012 and earlier 2.5L engine

10.20d Timing chain cover bolt tightening sequence - 2013 and later 2.5L engine

Installation

Caution: *DO NOT rely on the camshaft alignment tool to hold the camshaft in place when loosening the sprocket bolts - Use the hexagonal area on the camshaft instead, as a means for counter-rotation*

29   Loosen both camshaft sprocket bolt or VCT sprocket bolts but don't remove them. Use a wrench on the hexagonal area of the camshaft to hold it while turning the camshaft sprocket bolt.

Note: *The hexagonal portion of the camshaft is located just after the second lobe from the front.*

Caution: *Damage to the valves or pistons may occur if the camshafts are rotated during this procedure.*

30   Install the left chain guide (if removed).

31   Install the timing chain.

32   Install the right chain guide.

33   Install the timing chain tensioner and tighten the fasteners to the torque listed in this Chapter's Specifications.

34   If the timing chain tensioner plunger is not pinned in the compressed position, place the tensioner in a soft jaw vice. Spread the ratchet wire apart on the plunger while com-

24   On 2013 and later 2.0L models, install special tool (303-1565) into the slots on the back ends of the camshafts and secure the tool to the cylinder head.

25   Place a pick-type tool in the hole closest to the ratchet to relieve the tension on the ratchet mechanism then carefully compress the timing chain tensioner and place a pin (a drill bit or paper clip will work) into the hole to hold it in the compressed position (see illustrations).

Note: *The tensioner contacts the right-hand chain guide.*

Caution: *Compress only the round plunger on the tensioner and not the ratchet mechanism. The ratchet is next to the plunger and has square sides. If the ratchet needs to be reset, do the following:*

a) *Remove the tensioner and place it lightly in a vise using the plunger and tensioner housing.*

b) *Place a pick-type tool in the hole closest to the ratchet to relieve the tension on the ratchet mechanism.*

c) *While holding the pick tool in place, move the ratchet back into the tensioner, then install a pin into the other hole to keep the plunger and ratchet compressed.*

d   *Remove the tensioner from the vise.*

26   Remove the two tensioner mounting bolts and then the tensioner itself.

27   Remove the loose timing chain guide (right-hand side). Remove the timing chain.

28   The left-hand chain guide can now be removed, and also the camshaft sprockets, if necessary.

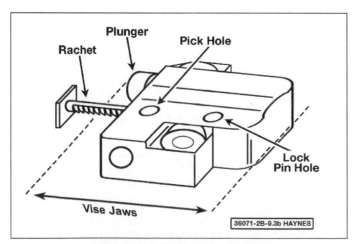

10.25b Timing chain tensioner details

10.36 Compressing the tensioner to release the lock pin - 2.3L engine shown, other engines similar

pressing the plunger into the locking position, then install the lock pin to hold the plunger into the compressed position.

35   See illustration 10.25b for more detailed information on resetting the tensioner.

36   Then remove the pin to release the tensioner to engage the chain guide (see illustration).

37   With the camshaft alignment tool still in place, tighten the camshaft sprocket bolts or VCT sprocket bolts to the torque listed in this Chapter's Specifications, while holding the camshafts in place with a wrench.

Caution: *Do not rely on the camshaft alignment tool to hold the camshafts while tightening the camshaft sprocket bolts. Tool and engine damage may occur.*

38   The timing chain cover can now be installed (see Section 10). Don't forget to remove the special alignment tools before rotating the engine by hand.

---

11   Engine front oil seals - replacement

Crankshaft front oil seal

2004 and earlier 2.0L engine, 1.5L and 1.6L engines

Caution: *Do not rotate the camshaft(s) or crankshaft when the timing belt is removed or damage to the engine may occur.*

1   Remove the timing belt (see Section 9).

2   Pull the crankshaft sprocket from the crankshaft. Remove the Woodruff key (2.0L engines only).

3   On 2004 and earlier 2.0L models, remove the timing belt guide washer from the crankshaft.

4   Wrap the tip of a small screwdriver with tape. Working from below the right inner fender, use the screwdriver to pry the seal out of its bore. Take care to prevent damaging the oil pump assembly, the crankshaft and the seal bore.

Note: *Another procedure for removing the seal is to drill a small hole on each side of the seal and place a self-tapping screw in each hole. Use these screws as a means of pulling the seal out without having to pry on it.*

5   Thoroughly clean and inspect the seal bore and sealing surface on the crankshaft. Minor imperfections can be removed with emery cloth. If there is a groove worn in the crankshaft sealing surface from contact with the seal, installing a new seal will probably not stop the leak.

6   Lubricate the new seal with engine oil and drive the seal into place with a hammer and an appropriate size socket.

7   The remaining steps are the reverse of removal. Position the crankshaft sprocket with the word "FRONT" facing out (see illustration).

8   Reinstall the timing belt and covers (see Section 9).

9   Run the engine and check for oil leaks.

2.3L, 2.5L, and 2013 and later 2.0L engines

10   Remove the crankshaft pulley (see Section 8).

11.7 Make sure the FRONT mark on the crankshaft sprocket is facing out

11   Use a screwdriver or hook tool to carefully pry out the seal.

Note: *Be careful not to damage the timing chain cover bore where the seal is seated or the nose and sealing surface of the crankshaft.*

12   Another procedure for removing the seal is to drill a small hole on each side of the seal and place a self-tapping screw in each hole. Use these screws as a means of pulling the seal out without having to pry on it.

13   Wipe the sealing surfaces in the engine cover and on the crankshaft. Clean and coat them with clean engine oil.

14   Start installing the new seal by pressing it into the timing chain cover.

15   Once started, use a seal driver or a suitable socket of the correct size to carefully drive the seal squarely into place (see illustration).

16   The seal should be flush with the engine cover and remain square when installed.

17   Coat the lip of the seal (where it contacts the crankshaft) with clean engine oil.

18   Install the crankshaft pulley (see Section 8).

Camshaft oil seals

Caution: *Do not rotate the camshaft(s) or crankshaft when the timing belt is removed or damage to the engine may occur.*

19   Position the engine at TDC for Number 1

11.24a Install the seal with the lip toward the journal - 2004 and earlier 2.0L engine shown, other models similar

11.15 A socket of the correct size can be used to install the new seal

piston (see Section 3).

20   Remove the timing belt (see Section 9).

21   On 2004 and earlier models, remove the camshaft sprocket bolts (see illustration 9.32), then remove the sprockets from the camshafts. Make alignment marks on each sprocket so it can be installed in the same position on its camshaft. Also mark each camshaft so you won't get them mixed up, then remove the sprockets. On 1.5L and 1.6L models, remove the VCT sprockets.

22   Note how far the seal is seated in the bore, then carefully pry it out with a small screwdriver. Don't scratch the bore or damage the camshaft in the process. If the camshaft is damaged, the new seal will end up leaking.

Note: *Another procedure for removing the seal is to drill a small hole on each side of the seal and place a self-tapping screw in each hole. Use these screws as a means of pulling the seal out without having to pry on it.*

23   Clean the bore and coat the outer edge of the new seal with engine oil or multi-purpose grease. Also lubricate the seal lip.

24   Using a socket with an outside diameter slightly smaller than the outside diameter of the seal, carefully drive the new seal into place with a hammer (see illustrations). Make sure it's installed squarely and driven in to the same depth as the original.

11.24b Tap the seal into the housing using a large socket

12.6 Remove the VCT oil control solenoid mounting bolts - 2013 and later 2.0L engine shown, 2.5L similar

12.9 Location of the camshaft bearing designations

25   Make sure the camshafts are in their TDC positions, with the tool shown in illustration 3.5b into the slots in the ends of the camshafts.

26   Install the camshaft sprockets (see Section 12). Tighten the bolts to the torque listed in this Chapter's Specifications. On 2004 and earlier 2.0L models, use the tool shown in illustration 9.32 to prevent the camshaft sprockets from turning.

27   Install the timing belt (see Section 9).

28   Run the engine and check for oil leaks near the camshaft seals.

12   Camshafts - removal, inspection and installation

Caution: *These engines are difficult to work on and require some special tools. On any procedure involving the timing chains or belts, the Steps must be read carefully and disassembly must proceed using the special tools, otherwise damage to the engine will result.*

Caution: *The timing system is complex. Severe engine damage will occur if you make any*

mistakes. Do not attempt this procedure unless you are highly experienced with this type of repair. If you are at all unsure of your abilities, consult an expert. Double-check all your work and be sure everything is correct before you attempt to start the engine.

Warning: *Turbocharged models are equipped with Direct Injection (DI) and a high-pressure fuel pump. Fuel pressure in the high-pressure system on Direct Injection (DI) systems is under extremely high pressure. Be sure to correctly perform the fuel pressure relief procedure prior to servicing any of the high-pressure fuel system components to prevent injury (see Chapter 4, Section 2).*

Removal

1   Before removing the camshafts, check the valve clearances (see Section 5).

2   Remove the intake manifold (see Section 6).

3   On turbocharged models, relieve the fuel system pressure and remove the high-pressure fuel pump and drive housing (see Section 19), then remove the brake vacuum pump (see Chapter 9).

4   Remove the timing belt or chain (see Section 9 or Section 10).

5   Remove the camshaft sprockets or VCT units as described in Section 11. Mark them so that each sprocket is reinstalled to its original location and alignment.

Note: *The camshaft phasers and sprockets should be marked with indelible ink so that they can be reinstalled in the same positions. When loosening the camshaft phaser or sprocket bolts, place a wrench on the hexagonal area of the camshaft to prevent it from turning.*

6   On 2013 and later 2.0L and 2.5L models, remove the variable camshaft timing (VCT) solenoid fastener(s) and pull the solenoid(s) out of the front camshaft bearing cap (see illustration).

Note: *On 1.5L and 1.6L models, the variable camshaft timing (VCT) solenoid(s) must be removed before the valve cover can be removed (see Chapter 6).*

7   On 1.5L and 1.6L models, use a screwdriver or hook to carefully pry out the camshaft seals from the cylinder head.

Note: *Be careful not to damage the cylinder head bore where the seals are seated or the nose and sealing surfaces of the camshafts.*

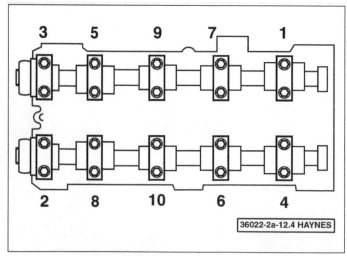

12.10a Camshaft bearing cap loosening sequence (2004 and earlier 2.0L engine) - loosen each pair of bolts on the designated bearing cap in sequence

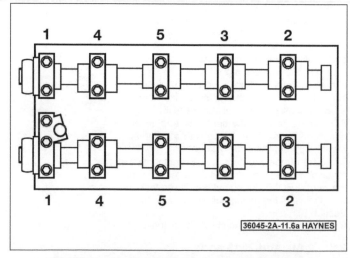

12.10b Camshaft bearing cap loosening sequence (2012 and earlier 2.3L and 2.5L engines) - loosen each pair of bolts on the designated bearing cap in sequence

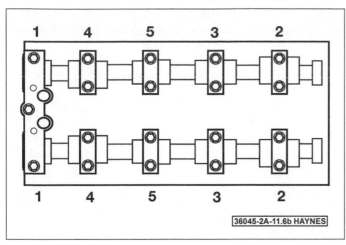

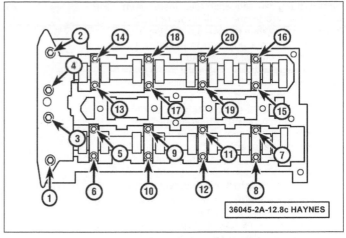

12.10c Camshaft bearing cap loosening sequence (2013 and later 2.0L and 2.5L engines) - loosen each pair of bolts on the designated bearing cap in sequence

12.10d Camshaft bearing cap loosening sequence (1.5L and 1.6L engines)

8   On 1.5L, 1.6L, 2013 and later 2.0L and 2013 and later 2.5L models, remove the bolts from the front one-piece camshaft bearing cap and lift the cap from both camshafts (see illustrations 12.9c and 12.9d). Replace the O-ring located underneath the cap upon installation. **Caution:** *On 1.5L and 1.6L engines, use a heat gun to soften the sealant used to seal the one-piece camshaft cap. The sealant used is very strong and the cap can be severely damaged if the sealant is not softened.*

9   On early models, the camshaft bearing caps have a single-digit identifying number etched on them. The exhaust camshaft's bearing caps are numbered in sequence from 0 (right-hand cap) to 4 (left-hand cap); the intake's are numbered from 5 (right-hand cap) to 9 (left-hand cap). Each cap is to be fitted so that its numbered side faces outward, to the front (exhaust) or to the rear (intake) (see illustration). If no marks are present, or they are hard to see, make your own - the bearing caps must be reinstalled in their original positions.

10   Working in the sequence shown, loosen the camshaft bearing cap bolts progressively by half a turn at a time (see illustrations). Work only as described, to release gradually and evenly the pressure of the valve springs on the caps.

11   Withdraw the caps, noting their markings and the presence of the locating dowels, then remove the camshafts and withdraw their oil seals. The intake camshaft can be identified by the reference lobe for the camshaft position sensor.

12   Obtain sixteen small, clean containers, and number them 1 to 16. Using a rubber suction tool (such as a valve-lapping tool) or magnetic retrieval tool, withdraw each valve shim (see Section 5) and place them in the labeled containers. Do not interchange the shims, or the rate of wear will be greatly increased. Make sure the shims remain with their corresponding valve locations, to ensure correct installation.

Inspection

13   With the camshafts and valve shims removed, check each for signs of obvious wear (scoring, pitting, etc.) and for round-ness and replace if necessary.

14   Measure the outside diameter of each cam follower (valve shim) - take measurements at the top and bottom of each cam follower, then a second set at right-angles to the first; if any measurement is significantly different from the others, the cam follower is tapered or oval (as applicable) and must be replaced (see illustration). If the necessary equipment is available, measure the inside diameter of the corresponding cylinder head bore. No manufacturer's specifications were available at time of writing. If the cam followers or the cylinder head bores are excessively worn, new cam followers and/or a new cylinder head may be required.

15   If the engine's valve components have sounded noisy, it may be just that the valve clearances need adjusting. Although this is part of the routine maintenance schedule in Chapter 1, the extended checking interval and the need for dismantling or special tools may result in the task being overlooked.

16   Visually examine the camshaft lobes for score marks, pitting, galling (wear due to rubbing) and evidence of overheating (blue, discolored areas). Look for flaking away of the hardened surface layer of each lobe. If any such signs are evident, replace the component concerned.

17   Examine the camshaft bearing journals and the cylinder head bearing surfaces for signs of obvious wear or pitting. If any such signs are evident, consult an automotive machine shop for advice. Also check that the bearing oilways in the cylinder head are clear (see illustration).

18   Using a micrometer, measure the diameter of each journal at several points. If the diameter of any one journal is less than the specified value, replace the camshaft. To check the bearing journal running clearance, remove the cam followers, use a suitable solvent and a clean lint-free rag to carefully clean all bearing surfaces, then install the camshafts and bearing caps with a strand of Plastigage across each journal. Tighten the bearing cap

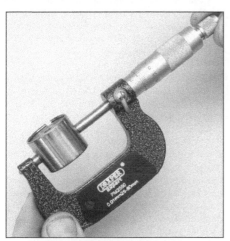

12.14 Measure the cam follower (valve shim) outside diameter at several points

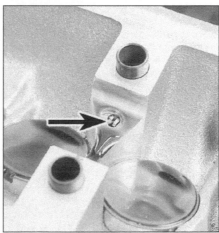

12.17 Check that the camshaft bearing oilways are not blocked with debris

bolts in the given sequence (see *Installation*) to the specified torque setting (do not rotate the camshafts), then remove the bearing caps and use the scale provided to measure the width of the compressed strands. Scrape off the Plastigage with your fingernail or the edge of a credit card - don't scratch or nick the journals or bearing caps.

19   If the running clearance of any bearing is found to be worn to beyond the specified service limits, install a new camshaft and repeat the check; if the clearance is still excessive, the cylinder head must be replaced.

20   To check camshaft endplay, remove the cam followers, clean the bearing surfaces carefully and install the camshafts and bearing caps. Tighten the bearing cap bolts to the specified torque wrench setting, then measure the endplay using a dial indicator mounted on the cylinder head so that its tip bears on the camshaft right-hand end.

21   Tap the camshaft fully towards the gauge, zero the gauge, then tap the camshaft fully away from the gauge and note the gauge reading. If the endplay measured is found to be at or beyond the specified service limit, install a new camshaft and repeat the check. If the clearance is still excessive, the cylinder head must be replaced.

Installation

22   On 1.5L and 1.6L engines, if new parts were installed (camshafts, lifters, valves or a cylinder head) it will be necessary to check the valve adjustment at this point. Follow these steps exactly to prevent severe engine damage. If no original parts were replaced, it is not necessary to check the valve clearance; proceed to Step 24.

a) *Use a permanent marker or paint and make a line on the crankshaft at the 12 o'clock position.*

b) *Remove any special tools - 303-1552 (VCT) locking tool, flywheel locking tools 303-393A, 303-393-02 and the timing pin 303-748.*

c) *Rotate the crankshaft 270-degrees clockwise until the mark is at the 9 o'clock position.*

Note: *By rotating the crankshaft to this position all of the pistons will be below the top of the cylinder block. The camshafts can be installed and rotated so that the valve clearance can be checked without damaging the valves or pistons.*

d) *Place the cam followers (valve shims) into the cylinder head bores, then the camshafts onto the cylinder head. Install the camshaft bearing caps, tightening the bolts a turn at a time (in sequence) until each cap touches the cylinder head, then tighten the bolts, in sequence (see illustration 12.29d) to the torque listed in this Chapter's Specifications.*

e) *Check the valve clearance (see Section 5), rotating only the camshafts. Do NOT rotate the crankshaft.*

12.26 Lubricate the camshaft bearing surface then install the camshafts ensuring that the slot at the transmission end is approximately horizontal

f) *Loosen the camshaft bearing caps in sequence (see illustration 12.9d) two turns at a time until all tension is released from the camshaft bearing caps, following the loosening sequence.*

g) *Remove the camshaft bearing caps and make any adjustments as necessary.*

h) *Install the timing pin 303-748 and rotate the crankshaft, clockwise 90-degrees until the crankshaft touches the timing pin.*

23   As a precaution against the valves hitting the pistons when the camshafts are replaced, remove the timing pin and turn the engine approximately 90-degrees (1/4-turn) back from the TDC position (or 270-degrees clockwise) - this will move all the pistons an equal distance down the bores. This is only a precaution - if the engine is known to be at TDC on the Number 1 cylinder, and the camshafts are replaced as described below, valve-to-piston contact should not occur.

24   On reassembly, liberally oil the cylinder head cam follower bores and the cam followers. Carefully install the cam followers to the cylinder head, ensuring that each cam follower is placed into its original bore, and is the correct way up. Some care will be required to guide the cam followers squarely into their bores.

25   It is highly recommended that new camshaft oil seals are installed. The new seals are installed after the caps are tightened - see Section 11.

26   Liberally oil the camshaft bearings (not the caps) and lobes. Ensuring that each camshaft is in its original location, install the camshafts, locating each so that the slot in its left-hand end is approximately parallel to, and just above, the cylinder head mating surface (see illustration). Check that, as each camshaft is laid in position, the metal strip TDC setting tool will fit into the slot.

27   Ensure that the locating dowels are pressed firmly into their recesses and check that all mating surfaces are completely clean,

12.27 On 2004 and earlier 2.0L models, apply sealant to each of the bearing caps at the timing belt end

unmarked and free from oil. On early 2.0L models, apply a thin film of suitable sealant (the manufacturer recommends a sealant to specification WSK-M2G348-A5) to the cylinder head mating surfaces of each camshaft's front bearing caps (closest to timing belt cover) (see illustration).

28   On 2013 and later models, clean the one-piece camshaft bearing cap surface with brake system cleaner and a lint-free rag. Apply a 1/16-inch (1.5 mm) thick bead of silicone sealant (Flange Sealant/CU7Z-19B508-A) or equivalent to the sealing edges of the one-piece bearing cap.

Note: *On 1.5L and 1.6L models, install a new O-ring to the cylinder head for the front camshaft cap.*

29   Apply a little oil to the camshaft as shown, then install each of the camshaft bearing caps to their previously noted positions, so that its numbered side faces outward, to the front (exhaust) or to the rear (intake).

30   Ensuring that each cap is kept square to the cylinder head as it is tightened down and working in the sequence shown, tighten the camshaft bearing cap bolts slowly and by one turn at a time, until each cap touches the cylinder head (see illustrations).

31   Next, using the same sequence, tightening the bolts to the Step 1 torque listed in this Chapter's Specifications.

32   With all the bolts tightened to the Step 1 torque, tighten them to the Stage 2 torque listed in this Chapter's Specifications.

33   Wipe off all surplus sealant, so that none is left to find its way into any oilways. Follow the sealant manufacturer's instructions as to the time needed for curing; usually, at least an hour must be allowed between application of the sealant and starting the engine (including turning the engine for further reassembly work).

34   Once the caps are fully tightened, it makes sense to check the valve clearances before proceeding - noting that the camshafts have to be removed to allow any of the shims

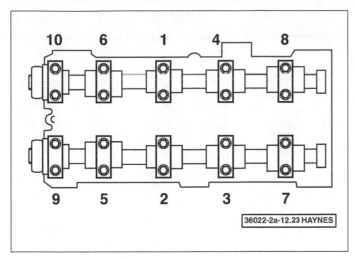

12.30a Camshaft bearing cap tightening sequence
(2004 and earlier 2.0L engine)

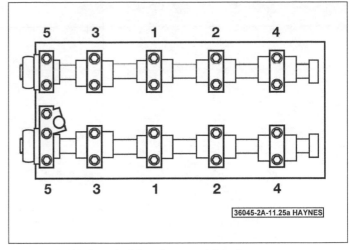

12.30b Camshaft bearing cap tightening sequence
(2012 and earlier 2.3L and 2.5L engines)

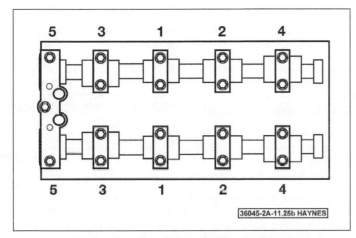

12.30c Camshaft bearing cap tightening sequence
(2013 and later 2.0L and 2.5L engines)

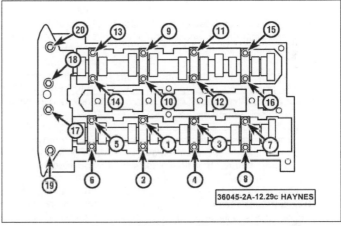

12.30d Camshaft bearing cap tightening sequence
(1.5L and 1.6L engines)

to be changed. Turning the camshafts with the timing belt or chain removed carries a risk of the valves hitting the pistons, so (if not already done) remove the timing pin and turn the crankshaft 90 degrees counterclockwise (or 270 degrees clockwise) first. Set the camshafts to TDC on the Number 1 cylinder using the horizontal setting tool (see Section 3) (see illustration 3.5b)to establish a starting point, then proceed as described in Section 5. When all the clearances have been checked, bring the camshafts, then the crankshaft back to TDC, installing the crank timing pin.

35   Install new camshaft oil seals as described in Section 11.

36   Using the marks and notes made on dismantling to ensure that each is replaced to its original camshaft, install the sprockets (or VCT units) and phaser(s) to the camshafts, tightening the retaining bolts loosely by hand.

37   The remainder of the reassembly procedure for the camshaft sprockets and timing belt for 2004 and earlier 2.0L, 1.5L and 1.6L models is described in Section 9. The tim-

ing chain installation procedure and setting the valve timing for 2013 and later 2.0L, 2.3L and 2.5L models, is described in Section 10.

38   On 2.0L, 2.3L and 2.5L models, before installing the valve cover, check the valve clearance (see Section 5).

39   After you reconnect the battery, the Powertrain Control Module (PCM) must relearn its idle and fuel trim strategy for optimum driveability and performance (see Chapter 5, Section 1).

13   Cylinder head - removal and installation

Warning: *Wait until the engine is completely cool before beginning this procedure.*
Caution: *For all engines mentioned in this chapter, the head bolts are torque-to-yield type and cannot be reused; obtain a set of NEW head bolts.*
Note: *Label all electrical and hose connections that are to be disconnected.*

Removal

2008 and earlier
2.0L and 2.3L models

1   Relieve the fuel pressure (see Chapter 4).

2   Disconnect the battery cable from the negative battery terminal (see Chapter 5).

3   Remove the intake duct from the air filter housing (see Chapter 4).

4   Disconnect the accelerator cable (see Chapter 4) and cruise control cable, if equipped.

5   Drain the cooling system (see Chapter 1).

6   Remove the timing belt (see Section 9) or timing chain (see Section 10).

7   Remove the camshafts (see Section 12).

8   Temporarily install the valve cover, to protect the top of the engine until the cylinder head bolts are removed.

9   Disconnect the camshaft position sensor (see Chapter 6).

10   Remove the fuel rail and injectors as a complete assembly (see Chapter 4).

11   On 2004 and earlier models, remove the alternator and its mounting bracket (see Chapter 5).

13.12 Remove the thermostat housing from the cylinder head

13.14 Remove the nut that secures the dipstick tube

13.16 Unbolt this power steering fluid pipe bracket

12 Remove the thermostat housing from the cylinder head (see illustration).
13 Remove the intake manifold (see Section 6).
14 Remove the nut and detach the dipstick tube from the cylinder head (see illustration).
15 Remove the exhaust manifold (see Section 7).
16 Remove the bolt from the power steering hose support bracket, if so equipped (see illustration) and position the hose off to the side.
17 If an engine support fixture or hoist is being used to hold the engine up (for timing belt removal), reinstall the engine mount bracket and engine mount, then remove the hoist or support fixture. If a floor jack and block of wood is being used to support the engine, it can remain in place.
18 Remove the power steering pump mounting bracket (see illustration).
19 Remove the valve shims (cam followers). Make sure they are marked and stored so they can be returned to their original positions later (see Section 5).

1.5L, 1.6L, 2013 and later 2.0L and 2009 and later 2.5L models
20 Refer to Chapter 4 and relieve the fuel system pressure, then disconnect the cable from

the negative battery terminal (see Chapter 5).
21 Raise the vehicle and support it securely on jackstands. Drain the cooling system (see Chapter 1).
22 Position a jack under the engine oil pan. Place a large wood block between the jack head and the oil pan to prevent oil pan damage, then carefully raise the engine just enough to take the weight off the front mount.
Warning: *DO NOT place any part of your body under the engine when it's supported only by a jack!*
23 Remove the front mount (see Section 18).
24 Remove the variable camshaft timing (VCT) solenoid(s) (see Chapter 6).
25 Remove the timing belt (see Section 9) or the timing chain (see Section 10).
26 Remove the camshaft alignment plate, then mark the camshafts so they can be installed in the same positions and orientations. Also mark each camshaft bearing cap for position and orientation.
27 Remove the camshafts (see Section 12).
28 Remove the valve shims (cam followers). Make sure they are marked and stored so they can be returned to their original positions later (see Section 5).
29 On turbocharged models, remove the fuel rail and fuel injectors (see Chapter 4).

30 Remove the intake manifold (see Section 6).
31 Remove the alternator (see Chapter 5).
32 On 2.5L and 1.6L models, remove the exhaust manifold (see Section 7) or on 2.0L and 1.5L models, remove the turbocharger (see Chapter 4).
33 On 2.5L models, disconnect the wiring and the coolant hose from the EGR valve.
34 Disconnect the various coolant hoses from the cylinder head and the coolant outlet.

All models
Caution: *The head bolts are torque-to-yield bolts that must be replaced with new ones on installation.*
35 Loosen the ten cylinder head bolts progressively and by half a turn at a time, working in the reverse order of the tightening sequence (see illustrations 13.51a, 13.51b and 13.51c).
36 On 1.5L and 1.6L engines, make sure that the cylinder head is at ambient air temperature before removing the cylinder head bolts
37 Lift the cylinder head from the engine compartment (see illustration).
38 If the head is stuck, be careful how you choose to free it. Remember that the cylinder head is made of aluminum alloy, which is

13.18 Unbolt the power steering pump bracket

13.44 Location of the cylinder head locating dowels

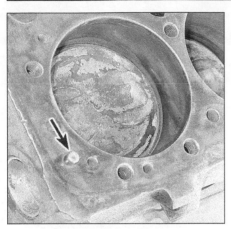

13.47a Position the head gasket over the dowels

13.47b The head gasket teeth must be positioned at the front of the block

easily damaged. Striking the head with tools carries the risk of damage, and the head is located on two dowels, so its movement will be limited. Do not, under any circumstances, pry the head between the mating surfaces, as this will certainly damage the sealing surfaces for the gasket, leading to leaks. Try rocking the head free, to break the seal, taking care not to damage any of the surrounding components.

39 Once the head has been removed, recover the gasket from the two dowels, and discard it.

Inspection

40 The mating faces of the cylinder head and cylinder block must be perfectly clean before replacing the head. Use a hard plastic or wood scraper to remove all traces of gasket and carbon.

41 Take particular care during the cleaning operations, as aluminum alloy is easily damaged. Also, make sure that the carbon is not allowed to enter the oil and water passages - this is particularly important for the lubrication system, as carbon could block the oil supply to the engine's components.

42 To prevent carbon entering the gap between the pistons and bores, smear a little grease in the gap. After cleaning each piston, use a small brush to remove all traces of grease and carbon from the gap, then wipe away the remainder with a clean rag and brake cleaner.

43 Check the mating surfaces of the cylinder block and the cylinder head for nicks, deep scratches and other damage. If slight, they may be removed carefully with a file. Also check the cylinder head gasket surface and the cylinder block gasket surface with a precision straight-edge and feeler gauges. If either surface exceeds the warpage limit listed in this Chapter's Specifications, the manufacturer states that the component out of specification must be replaced. If the gasket mating surface of your cylinder head or block is out of specification or is severely nicked or scratched, you may want to consult with an automotive machine shop for advice.

Installation

44 Wipe clean the mating surfaces of the cylinder head and cylinder block. Check that the two locating dowels are in position in the cylinder block (see illustration).

45 The cylinder head bolt holes must be free from oil or water. This is most important, because a hydraulic lock in a cylinder head bolt hole can cause a fracture of the block casting when the bolt is tightened.

46 The new head gasket is selected according to a number cast on the front face of the cylinder block, in front of the Number 1 cylinder. Consult a dealer or your parts supplier for details.

47 Position a new gasket over the dowels on the cylinder block surface, so that the TOP mark is uppermost; where applicable, the teeth must protrude towards the front of the vehicle (see illustrations).

48 Temporarily install the crankshaft pulley and rotate the crankshaft counterclockwise 90-degrees (or 270-degrees clockwise via the pulley bolt, on models without a keyed crankshaft) so that the Number 1 cylinder's piston is lowered to approximately 20 mm before TDC, thus avoiding any risk of valve/piston contact and damage during reassembly. The other pistons should be at approximately the same level.

49 As the cylinder head is such a heavy and awkward assembly to install, it is helpful to make up a pair of guide studs from two 10 mm (thread size) studs approximately 90 mm long, with a screwdriver slot cut in one end - you can use two of the old cylinder head bolts with their heads cut off. Screw these guide studs, screwdriver slot upwards to permit removal, into the bolt holes at diagonally-opposite corners of the cylinder block surface; ensure that approximately 70 mm of stud protrudes above the gasket.

50 Install the cylinder head, sliding it down the guide studs (if used) and locating it on the dowels. Unscrew the guide studs (if used) when the head is in place.

51 On all models, install NEW head bolts. Coat the threads with a light oil - do not apply more than a light film of oil. Install the cylinder head bolts carefully, and screw them in by hand only until finger-tight.

52 Working progressively and in the sequence shown, first tighten all the bolts to the specified Step 1 torque setting listed in this Chapter's Specifications (see illustrations).

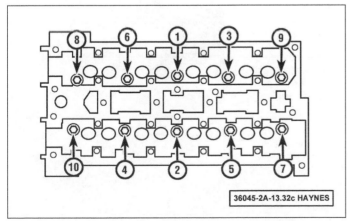

13.52a Cylinder head bolt tightening sequence (1.5L and 1.6L engines)

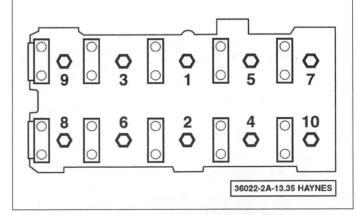

13.52b Cylinder head bolt tightening sequence (2004 and earlier 2.0L engine)

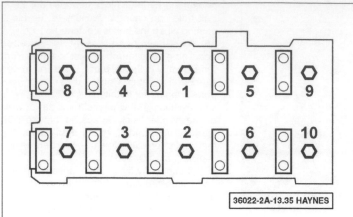

13.52c Cylinder head bolt tightening sequence (2013 and later 2.0L, 2.3L and 2.5L engines)

36022-2A-13.35 HAYNES

13.54 You can use a torque angle gauge, or you can carefully note the starting and stopping points of the wrench handle

2004 and earlier 2.0L engines

53   With all the bolts tightened to Step 1, tighten them in sequence to the Step 2 torque setting.

54   Step 3 involves tightening the bolts through an angle, rather than to a torque. Each bolt in sequence must be rotated through the specified angle - special angle gauges are available from tool outlets (see illustration), but a 90 degree angle is equivalent to a quarter-turn, and this is easily judged by assessing the start and end positions of the socket handle or torque wrench.

55   Once all the bolts have been tightened to Step 3, no subsequent tightening is necessary.

1.5L, 1.6L, 2013 and later 2.0L, 2.3L and 2.5L engines

56   On these engines there are five tightening stages, the final two using the angle torque method (see illustration 13.53).

57   Before proceeding with installation, turn the crankshaft forwards to TDC, and check that the setting tools described in Section 3 can be inserted. Do not turn the engine more than necessary while the timing belt is removed, or the valves may hit the pistons - for instance, if the engine is accidentally turned past the TDC position, turn it back slightly and try again - do not rotate the crankshaft a full turn.

58   Replacement of the other components removed is a reversal of removal, noting the following points:

a) *Install the exhaust manifold/catalytic converter with reference to Chapter 4 if necessary, using new gaskets.*

b) *Install the camshafts as described in Section 12, and the timing belt or timing chain as described in Section 9 or Section 10.*

c) *On turbocharged models, install the turbocharger and high pressure fuel pump (see Chapter 4).*

d) *Tighten all fasteners to the specified torque, where given, and use new gaskets.*

e) *Ensure that all hoses and wiring are correctly routed, and that hose clips and wiring connectors are securely installed.*

f) *Refill the cooling system as described in Chapter 1.*

g) *Check all disturbed joints for signs of oil or coolant leakage once the engine has been restarted and warmed-up to normal operating temperature.*

59   After you reconnect the battery, the Powertrain Control Module (PCM) must relearn its idle and fuel trim strategy for optimum driveability and performance (see Chapter 5, Section 1).

---

14   Oil pan - removal and installation

2008 and earlier 2.0L and 2.3L models

Removal

1   Raise the vehicle and support it securely on jackstands (see *Jacking and Towing*).

2   Drain the engine oil (see Chapter 1), then clean and install the engine oil drain plug, tightening it to the torque listed in the Chapter 1 Specifications. Remove and discard the oil filter, so that it can be replaced with the oil.

3   Remove the catalytic converter from the exhaust system (see Chapter 4).

4   Progressively unscrew the oil pan retaining bolts. Use a rubber mallet to loosen the oil pan seal, then lower the oil pan, turning it as necessary to clear the exhaust system.

5   Unfortunately, the use of sealant can make removal of the oil pan more difficult. Be careful when prying between the mating surfaces, otherwise they will be damaged, resulting in leaks when finished. With care, a putty knife can be used to cut through the sealant.

Installation

6   Thoroughly clean and degrease the mating surfaces of the lower engine block/crankcase and oil pan, removing all traces of sealant, then use a clean rag and brake cleaner to wipe out the oil pan.

Note: *The oil pan must be installed within seven minutes of applying the sealant.*

7   Apply a 1/8-inch wide bead of sealant to the oil pan flange so that the bead is approximately 3/16-inch from the outside edge of the flange. Make sure the bead is around the inside edge of the bolt holes (see illustration).

8   Install the oil pan bolts, tightening them to the Step 1 torque listed in this Chapter's Specifications, in a criss-cross pattern.

9   When all the oil pan bolts have been tightened to Step 1, tighten them again in sequence to the Step 2 setting.

10   Lower the vehicle to the ground. Wait at least one hour for the sealant to cure, or whatever time is indicated by the sealant manufacturer, before refilling the engine with oil. Trim off the excess sealant with a sharp knife. Install a new oil filter with reference to Chapter 1.

2009 and later 2.0L and 2.5L models

Removal

11   Loosen the left front wheel lug nuts. Raise the vehicle and support it securely on jackstands. Remove the left front wheel.

12   Refer to Chapter 4 and remove the throttle body air intake duct. Drain the engine oil (see Chapter 1).

13   Loosen the top two transaxle-to-engine mounting bolts 3/16-inch.

14   On 2.0L engine models, remove the charge air cooler (CAC) pipes (see Chap-

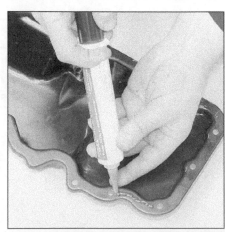

14.7 Apply a bead of sealant to the oil pan flange, inboard of the bolt holes

14.32 Remove the charge air cooler (CAC) pipe center section mounting nuts

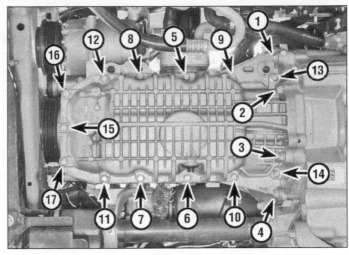

14.37 Oil pan tightening sequence - 1.5L and 1.6L engines

ter 4) and unbolt the charge air cooler pipe center section from the rear of the oil pan.

15 On 4WD models, loosen the two right transaxle-to-engine mounting bolts 3/16-inch then, work under the vehicle and loosen the two engine bracket-to-transfer case bolts 3/16-inch.

16 On 4WD models with a 2.0L engine, remove the transfer case (see Chapter 7C).

17 On 2.0L models, remove the roll restrictor mounting bolts and remove the mount from the frame and transmission bracket.

18 Remove the splash shield from the left front wheel well, then loosen the two left-side transaxle-to-engine bolts 3/16-inch.

19 Remove the four bolts that secure the oil pan to the transaxle, then slide the transaxle away from the engine 3/16-inch.

Note: *Keep the various length oil pan bolts separated so they can be re-installed in their original positions.*

20 Remove the four bolts that secure the front cover to the oil pan then remove the bolts that secure the oil pan to the engine block. Bump the oil pan using a block of wood if necessary to break it loose, then remove it.

Installation

21 Thoroughly clean and degrease the mating surfaces of the engine block, the front cover and the oil pan. Remove all traces of sealant, then use a clean rag to wipe out the oil pan.

22 Apply a 1/8-inch bead of RTV sealant to the oil pan sealing surfaces, inboard of the bolt holes.

Note: *The oil pan must be installed within 10 minutes of applying the sealant.*

23 Install the oil pan and put the front cover and engine block bolts in place finger-tight.

24 Tighten the front cover-to-oil pan bolts first to the torque listed in this Chapter's Specifications.

Note: *The oil pan must have these bolts tightened first to force it as far forward as possible.*

25 Tighten the engine-to-oil pan bolts a little at a time, working from the center outwards

in a criss-cross pattern, to the torque listed in this Chapter's Specifications.

26 Alternate tightening one of the left transaxle mounting bolts and one of the right transaxle mounting bolts a quarter turn at a time until the transaxle is pulled tight to the engine, then tighten all four side bolts to the torque listed in this Chapter's Specifications.

27 Install the four transaxle-to-oil pan bolts and tighten them to the torque listed in this Chapter's Specifications.

28 Install the splash shield in the wheel well, then tighten the remaining transaxle-to-engine bolts and tighten them to the torque listed in this Chapter's Specifications.

29 Install the air intake duct, then fill the engine with new oil when you're sure the RTV sealant has cured according to the manufacturer's directions.

1.5L and 1.6L engines

30 Raise the vehicle and support it securely on jackstands. Remove the engine splash shield fasteners, then remove the shield from under the vehicle.

31 Drain the engine oil (see Chapter 1), then clean and install the engine oil drain plug, tightening it to the torque listed in the Chapter 1 Specifications. Remove and discard the oil filter, so that it can be replaced along with new oil.

32 Remove the charge air cooler (CAC) pipes (see Chapter 4) and unbolt the charge air cooler pipe center section from the oil pan (see illustration).

33 On 1.5L models, unbolt and position the charge air cooler (CAC) pump aside. It is located to the lower-right of the oil filter.

34 Progressively unscrew the oil pan retaining bolts. Use a rubber mallet to loosen the oil pan seal or a small pry bar to pry against the pry pads on the sides of the oil pan, then lower the oil pan.

Note: *Unfortunately, the use of sealant can make removal of the oil pan more difficult. Be careful when prying between the mating surfaces, otherwise they will be damaged, result-*

ing in leaks when finished. With care, a putty knife can be used to cut through the sealant.

35 Thoroughly clean and degrease the mating surfaces of the engine block, the front cover and the oil pan. Remove all traces of sealant, then use a clean rag to wipe out the oil pan.

36 Apply a 0.137 inch (3.5 mm) bead of RTV sealant to the oil pan sealing surfaces, inboard of the bolt holes.

Note: *The oil pan must be installed within five minutes of applying the sealant.*

37 Install the oil pan bolts, tightening them, in sequence (see illustration) to the torque listed in this Chapter's Specifications.

38 Remaining installation is the reverse of removal, noting the following points:

a) *After installing the oil pan, install a new oil filter and refill the crankcase with oil (see Chapter 1).*

b) *Be certain to check for any oil warning lights in the instrument panel after the vehicle has been started and idling.*

15 Oil pump - removal, inspection and installation

2004 and earlier 2.0L engine
Removal

Warning: *The air conditioning system is under high pressure. DO NOT loosen any fittings or remove any components until after the system has been discharged. Air conditioning refrigerant must be properly discharged into an EPA-approved container at a dealership service department or an automotive air conditioning repair facility. Always wear eye protection when disconnecting air conditioning system fittings.*

Note: *While this task is theoretically possible when the engine is in place in the vehicle, in practice, it requires so much preliminary dismantling and is so difficult to carry out due to the restricted access, that owners are advised to remove the engine from the vehicle first (see Chapter 2C). Note, however, that the oil*

15.9a Remove the oil pump cover . . .

15.9b . . . and withdraw the rotors, noting which way they are facing

15.12 Oil pump pressure relief valve components (seen with oil pump removed)

pump pressure relief valve can be removed with the engine in place (see Step 11).

1   Have the air conditioning system discharged (see the Warning above).

2   Remove the timing belt (see Section 9), then withdraw the crankshaft sprocket and the guide washer behind it, noting which way the guide washer is installed (see Section 11).

3   Remove the oil pan (see Section 14).

4   Unscrew the two bolts securing the oil pump pick-up tube to the base of the lower crankcase, and withdraw it. Discard the gasket.

5   Remove the air conditioning compressor (see Chapter 3). Remove the air conditioning bracket bolts and remove it from the lower section of the engine block.

6   Remove the timing belt tensioner (see Section 9).

7   Progressively unscrew and remove the bolts securing the lower crankcase to the base of the engine block and the transaxle. Remove the lower crankcase, and recover any spacer washers, noting where they are installed, as these must be used on replacement. Remove the gasket, and remove any traces of sealant at the joint with the oil pump.

8   Unbolt the pump from the engine block. Withdraw and discard the gasket, then remove the crankshaft right-hand oil seal. Thoroughly clean and degrease all components, particularly the mating surfaces of the pump, the oil pan and the engine block.

Inspection

9   Unscrew the Torx screws and remove the pump cover plate. Noting any identification marks on the rotors, withdraw the rotors (see illustrations).

10   Inspect the rotors for obvious signs of wear or damage, and replace if necessary. If either rotor, the pump body, or its cover plate are scored or damaged, the complete oil pump assembly must be replaced.

11   The oil pressure relief valve can be disassembled, if required, without disturbing the pump. With the vehicle parked on firm level ground, apply the parking brake securely and raise its front end, supporting it securely on jackstands. Remove the front right-hand wheel and fender splash shield to provide

15.17 Check the alignment of the oil pump with the crankcase

access to the oil pressure relief valve.

12   Unscrew the threaded plug and recover the valve spring and plunger (see illustration).

13   Reassembly is the reverse of the disassembly procedure; ensure the spring and valve are reinstalled the correct way and tighten the threaded plug securely.

14   When reassembling the oil pump, oil the housing and components as they are installed; tighten the oil pump cover bolts securely.

15.22 Check the alignment of the lower crankcase and engine block

Installation

15   The oil pump must be primed on installation, by pouring clean engine oil into it and rotating its inner rotor a few turns.

16   Rotate the pump's inner rotor to align with the flats on the crankshaft, then install the pump (and new gasket) and insert the bolts, tightening them lightly at first.

17   Using a suitable straight-edge and feeler

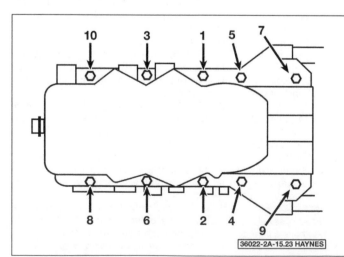

15.23 Lower crankcase bolt tightening sequence (2004 and earlier 2.0L engine)

36022-2A-15.23 HAYNES

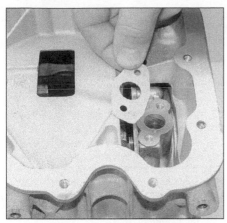

15.24a Using a new gasket . . .

15.24b . . . install the oil pump pickup/strainer pipe and tighten the bolts to the specified torque

15.28 The oil pump pick-up tube is held by two mounting bolts

15.29 The tensioner is secured by one mounting bolt while the guide is held by two

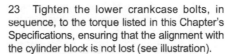

15.31 Using a holding tool on the oil pump drive sprocket to remove the retaining bolt

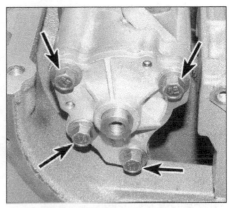

15.33 Remove the four mounting bolts for the oil pump

gauges, check that the pump is both centered exactly around the crankshaft and aligned squarely so that its (oil pan) mating surface is exactly the same amount between 0.012 and 0.031-inch (0.30 to 0.80 mm) below that of the cylinder block/crankcase on each side of the crankshaft (see illustration). Being careful not to disturb the gasket, move the pump into the correct position and tighten its bolts to the torque listed in this Chapter's Specifications.

18   Check that the pump is correctly located. If necessary, unbolt it again and repeat the full procedure to ensure that the pump is correctly aligned.

19   Install a new oil pump housing seal (see Section 10).

20   Apply a little RTV sealant to the joints between the oil pump and cylinder block.

Note: *Once this sealant has been applied, the lower crankcase must be installed and the bolts fully tightened within seven minutes.*

21   Install the new gasket in place, then raise the lower crankcase into position and loosely secure with the bolts (see illustrations).

22   The lower crankcase must now be aligned with the cylinder block before the bolts are tightened. Using a straight-edge and feeler gauges, check the alignment of the end faces (see illustration).

23   Tighten the lower crankcase bolts, in sequence, to the torque listed in this Chapter's Specifications, ensuring that the alignment with the cylinder block is not lost (see illustration).

24   Using grease to stick the gasket in place on the pump, install the pickup/strainer pipe, tightening the retaining bolts to the specified torque (see illustrations).

25   The remainder of reassembly is the reverse of the removal procedure; refer to the relevant text for details where required.

2.3L engine

Note: *Removing the oil pump on this engine requires engine removal. Refer to Chapter 2C for the engine removal procedure. Also, the oil pump is serviced as a complete unit without any sub-assembly or internal inspection.*

26   With the engine removed, remove the timing chain (see Section 10).

27   Remove the oil pan.

28   Remove the oil pick-up mounting bolts and then remove the pick-up tube (see illustration).

29   Remove the chain tensioner mounting bolt and the tensioner, noting the spring location (see illustration).

30   Remove the two mounting bolts for the chain guide and remove the guide.

31   While holding the oil pump drive sprocket with a suitable tool, remove the sprocket bolt from the oil pump, then remove the sprocket (see illustration).

32   Slide the oil pump drive sprocket (and timing chain sprocket) and chain from the end of the crankshaft (if necessary).

Caution: *The special washers on each side of the sprocket must be placed back in their original positions or engine timing will not be maintained after assembly. This may result in engine damage or failure.*

33   Remove the oil pump mounting bolts, then remove the pump (see illustration).

34   Installation is the reverse of removal noting the following points:

a) *Replace all gaskets with new ones.*

b) *Tighten the oil pump mounting bolts to the torque listed in this Chapter's Specifications in a criss-cross pattern.*

c) *After installing the engine, refill the crankcase with oil and the cooling system with the proper type and mixture of antifreeze (see Chapter 1).*

d) *Be certain to check for any oil warning lights in the instrument panel after the vehicle has been started and idling.*

2.5L engine

35   Raise the vehicle and support it securely on jackstands.

36   Refer to Section 10 and remove the timing chain cover.

37   Remove the oil pan (see Section 14).

38   Remove the oil pump pick-up tube and the strainer.

39   Release some tension from the oil pump drive chain tensioner, then remove its two mounting bolts and remove the tensioner.

40   Lift the chain from the oil pump sprocket. Remove the oil pump sprocket mounting bolt and the sprocket.

41   Remove the oil pump mounting bolts and remove the oil pump.

42   Installation is the reverse of removal with special attention to the following.

43   Install the chain tensioner spring shoulder bolt first, then install the tensioner assembly, hooking the tensioner spring around the shoulder bolt.

44   When installing the oil pan, place a straight-edge against the front cover surface, then slide the oil pan forward until it touches the straight-edge. The front sealing surface of the oil pan must be flush with the front cover sealing surface. Refer to Section 14 for additional installation information.

1.5L and 1.6L engines

45   Remove the timing belt (see Section 9), then withdraw the crankshaft sprocket and the guide washer behind it, noting which way the guide washer is installed (see Section 11).

46   Remove the charge air cooler (CAC) pipes (see Chapter 4) and unbolt the charge air cooler pipe center section from the oil pan.

47   Remove the oil pan (see Section 14).

48   Unscrew the three bolts securing the oil pump pick-up tube to the base of the lower crankcase, and withdraw it. Discard the gasket.

49   Remove the crankshaft front seal from the oil pump (see Section 11).

50   Unbolt the pump from the engine block, noting the location of the different length bolts. Withdraw and discard the gasket. Thoroughly clean and degrease all components, particu-larly the mating surfaces of the pump, the oil pan and the engine block.

51   Install a new gasket on to the cylinder block.

52   Before installing the oil pump, pour a couple of tablespoons of engine oil into the pump the rotate the pump sprocket by hand to prime the pump.

53   Install the pump with the mounting bolts in their original locations. After bolting the oil pump in place tighten the bolts, in sequence (see illustration) to the torque listed in this Chapter's Specifications, reinstall the oil pump pickup tube and screen with a new O-ring.

54   Remaining installation is the reverse of removal, noting the following points:

a) *After installing the oil pan, install a new oil filter and refill the crankcase with oil (see Chapter 1).*

b) *Be certain to check for any oil warning lights in the instrument panel after the vehicle has been started and idling.*

2013 and later 2.0L engines

Note: *The oil pump is serviced as a complete unit without any sub-assembly or internal inspection.*

55   Drain the engine oil and remove the oil filter (see Chapter 1).

56   Support the engine from above with a support fixture (see Chapter 2C).

57   Remove the timing chain cover (see Section 10).

58   Remove the charge air cooler (CAC) pipes (see Chapter 4).

59   Remove the air conditioning compressor fasteners (see Chapter 3) and tie the compressor out of the way without disconnecting the compressor lines.

60   Remove the oil pan (see Section 14).

61   Remove the oil pick-up tube (see illustration 15.28) and remove the O-ring from the sealing surface of the tube.

62   Use a screwdriver to pry the end of the oil pump drive chain tensioner's spring from under the shouldered bolt. Remove the two bolts and the tensioner. This step is similar to what is shown in illustration 15.29.

63   Remove the chain from the oil pump sprocket. While holding the oil pump drive sprocket with a suitable tool, remove the sprocket bolt from the oil pump, then remove the sprocket (see illustration 15.31).

64   Remove the oil pump mounting bolts, then remove the pump (see illustration 15.33).

65   Installation is the reverse of removal, noting the following points:

a) *Replace all gaskets with new ones.*

b) *Tighten the oil pump mounting bolts to the torque listed in this Chapter's Specifications in a criss-cross pattern.*

c) *After installing the oil pan, install a new oil filter and refill the crankcase with oil (see Chapter 1).*

d) *Be certain to check for any oil warning lights in the instrument panel after the vehicle has been started and idling.*

16   Flywheel/driveplate - removal, inspection and installation

Removal

1   Remove the transaxle as described in Chapter 7A or Chapter 7B. Now is a good time to check components such as oil seals and replace them if necessary.

2   If you're working on a manual transaxle vehicle, remove the clutch (see Chapter 8). Check the clutch components and replace them now if necessary.

3   Use a center-punch or paint to make alignment marks on the flywheel/driveplate and crankshaft to make replacement easier - the bolt holes are slightly offset, and will only line up one way, but making a mark eliminates the guesswork (and the flywheel is heavy).

4   Hold the flywheel/driveplate stationary using one of the following methods:

a) *If an assistant is available, insert one of the transaxle mounting bolts into the cylinder block, and have the assistant engage a wide-bladed screwdriver with the starter ring gear teeth while the bolts are loosened.*

b) *Alternatively, a piece of angle-iron can be engaged with the ring gear, and located against the transaxle mounting bolt.*

5   Loosen and remove each bolt in turn and ensure that new replacements are obtained for reassembly. These bolts are subjected to severe stresses and so must be replaced, regardless of their apparent condition, whenever they are disturbed.

6   Remove the flywheel/driveplate; do not drop it.

Inspection

Driveplate

7   Driveplates are used on automatic transaxle vehicles. Make sure that the driveplate is not cracked. Inspect the bolt holes for elongation. Also inspect the ring gear teeth for damage from the starter motor. Replace the driveplate if there's any doubt as to its condition.

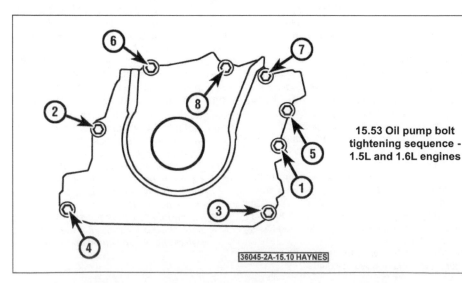

15.53 Oil pump bolt tightening sequence - 1.5L and 1.6L engines

36045-2A-15.10 HAYNES

Flywheel

8 Flywheels are used on manual transaxle vehicles. Clean the flywheel to remove grease and oil. Inspect the surface for cracks, rivet grooves, burned areas and score marks. Light scoring can be removed with emery cloth. Check for cracked and broken ring gear teeth. Lay the flywheel on a flat surface and use a straight-edge to check for warpage.

9 Clean and inspect the mating surfaces of the flywheel and the crankshaft. If the oil seal is leaking, replace it (see Section 17) before replacing the flywheel. If the engine has covered a high mileage, it may be worth installing a new seal as a matter of course, given the amount of work needed to access it.

10 While the flywheel is removed, carefully clean its inboard face, particularly the recesses that serve as the reference points for the crankshaft speed/position sensor. Clean the sensor's tip and check that the sensor is securely fastened.

Installation

Caution: *The flywheel/driveplate bolts must be replaced once the bolts are removed. The flywheel/drivepate bolts are a special type of bolt. Do NOT use standard bolts to install the flywheel/drivepate.*

11 On installation, ensure that the engine/transaxle adapter plate is in place (where necessary), then install the flywheel/driveplate on the crankshaft so that all bolt holes align - it should fit only one way - check this using the marks made on removal. Install the new bolts, tightening them by hand.

12 Lock the flywheel/driveplate by the method used on removal. Working in a diagonal sequence to tighten them evenly and increasing to the final amount, tighten the new bolts to the torque listed in this Chapter's Specifications.

13 The remainder of installation is the reverse of removal.

17 Rear main oil seal - replacement

Note: *The one-piece rear main oil seal is pressed into the rear main oil seal carrier (all engines except 2004 and earlier 2.0L engine)*

1 Remove the automatic transaxle (see Chapter 7B) or manual transaxle (see Chapter 7A) and clutch (see Chapter 8) and the flywheel/driveplate (see Section 16).

2004 and earlier 2.0L engines

2 Pry out the old seal with a flat-blade screwdriver.

Caution: *To prevent an oil leak after the new seal is installed, be very careful not to scratch or otherwise damage the crankshaft sealing surface or the bore in the engine block.*

3 Clean the crankshaft and seal bore in the block thoroughly and de-grease these areas by wiping them with a rag soaked in lacquer thinner or acetone. Lubricate the lip of the new seal and the outer diameter of the crankshaft with engine oil. Make sure the edges of the new oil seal are not rolled over.

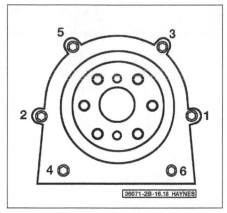

17.15a Rear oil seal carrier tightening sequence (2.3L, 2.5L, and 2013 and later 2.0L engines)

4 Position the new seal onto the crankshaft. Use a special rear main oil seal installation tool or a socket with the exact diameter of the seal to drive the seal in place. Make sure the seal is not off-set; it must be flush along the entire circumference of the seal carrier.

Note: *When installing the new seal, if so marked, the words THIS SIDE OUT on the seal must face out, away from the engine.*

5 The remainder of installation is the reverse of removal.

2005 and later engines

Note: *A special tool (Part #303-328) for installing/seating the rear seal must be obtained for 2.3L, 2.5L and 2013 and later 2.0L models*

6 Remove the oil pan (see Section 14).

7 Unbolt the oil seal and carrier assembly.

8 Clean the mating surface for the oil seal carrier on the cylinder block and the crankshaft. Carefully remove and polish any burrs or raised edges on the crankshaft that may have caused the seal to fail. DO NOT use any metal brushes or power wire wheels to clean the surface - use a plastic scraping tool and rag saturated in brake cleaner to clean the mating surfaces.

9 On 1.5L and 1.6L models, if there was no liner provided with the new seal upon purchase, use a thin (but durable) two-inch wide plastic strip (or a two-liter plastic beverage bottle cut to size) around the inside circumference of the seal to act as a liner for installation. Otherwise, perform the following with the included liner.

10 On 1.5L and 1.6L models, with the plastic seal liner in place, carefully move the new oil seal and carrier into position by sliding it onto the contact surface of the crankshaft without stopping until the seal carrier meets the engine block.

11 On 2.3L, 2.5L and 2013 and later 2.0L models, apply a thin bead of silicone gasket sealant along the bottom mating surface, continuing slightly above to the protruding edges at each end.

12 On 2.3L, 2.5L and 2013 and later 2.0L models, install the special tool (303-328) onto the seal housing.

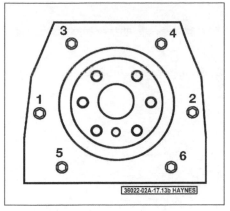

17.15b Rear oil seal carrier tightening sequence (1.5L and 1.6L engines)

13 Install the oil seal carrier bolts and finger-tighten them while holding the carrier in place. On 2.3L, 2.5L and 2013 and later 2.0L models, carefully install the oil seal carrier at an angle to avoid smearing the sealant.

14 On 2.3L, 2.5L and 2013 and later 2.0L models, press the oil seal squarely in place with the special tool (303-328) and a mallet.

15 Tighten the oil seal carrier to the torque listed in this Chapter's Specifications using the proper sequence (see illustrations).

16 On 1.5L and 1.6L models, remove the alignment liner. On 2.3L, 2.5L and 2013 and later 2.0L models, remove the special tool (303-328).

17 The seal is now fully installed. Be sure to check for signs of oil leakage when the engine is operable.

18 Engine mount - check and replacement

1 Engine mounts seldom require attention, but broken or deteriorated mounts should be replaced immediately or the added strain placed on the driveline components may cause damage or wear.

Check

2 During the check, the engine must be raised slightly to remove the weight from the mounts.

3 Raise the vehicle and support it securely on jackstands, then position a jack under the engine oil pan. Place a large wood block between the jack head and the oil pan to prevent oil pan damage, then carefully raise the engine just enough to take the weight off the mounts.

Warning: *DO NOT place any part of your body under the engine when it's supported only by a jack!*

4 Check the mounts to see if the rubber is cracked, hardened or separated from the metal backing. Sometimes the rubber will split right down the center.

5 Check for relative movement between the mount plates and the engine or frame.

18.11 Place a floor jack under the engine with a wood block between the jack head and oil pan

18.12a Remove the bolts and lift the engine mount upper bracket off the engine mount - early models shown

Use a large screwdriver or pry bar to attempt to move the mounts. If movement is noted, lower the engine and tighten the mount fasteners.

6   Rubber preservative may be applied to the mounts to slow deterioration.

Replacement

Note: *This procedure shows the replacement of the main engine mount (RH side). For the replacement of the LH transaxle mount or engine roll restrictor, see Chapter 7B.*

7   Disconnect the battery cable from the battery negative terminal (see Chapter 5), then raise the vehicle and support it securely on jackstands (if not already done).

8   Carefully lift the engine cover up to disengage the cover from the ballstuds and remove the cover, if equipped.

9   Remove the engine splash shield fasteners and remove the splash from under the vehicle, if equipped.

10   On later models, unclip the coolant reservoir and move the reservoir out of the way without disconnecting the coolant lines (see Chapter 3).

11   Place a floor jack under the engine with a wood block between the jack head and oil pan (see illustration) and raise the engine slightly to relieve the weight from the mounts.

12   Remove the fasteners and detach the mount from the frame and engine (see illustrations).

Caution: *Do not disconnect more than one mount at a time, except during engine removal.*

13   Installation is the reverse of removal. Use thread locking compound on the mount bolts and be sure to tighten them securely.

14   After you reconnect the battery, the Powertrain Control Module (PCM) must relearn its idle and fuel trim strategy for optimum driveability and performance (see Chapter 5, Section 1).

19   High-pressure fuel pump drive housing - removal and installation

Warning: *Gasoline is extremely flammable. See* Fuel system warnings *(see Chapter 4, Section 2).*

Note: *The manufacturer recommends replacing the fuel lines anytime they are removed.*

Note: *On 1.5L and 1.6L engines, the high-pressure fuel pump is mounted on a separate drive housing on the top of the cylinder head. On 2.0L engines, the pump and drive housing are mounted on the end of the cylinder head.*

1.5L and 1.6L models

Removal

1   Relieve the fuel system pressure (see Chapter 4).

2   Disconnect the cable from the negative battery terminal (see Chapter 5).

3   Lift the engine cover up and off of the ballstuds, then remove the cover.

4   Remove the high-pressure fuel pump (see Chapter 4, Section 6) and the brake booster vacuum pump (see Chapter 9).

5   Remove the valve cover (see Section 4).

6   Remove the camshaft position (CMP) sensor (see Chapter 6).

Caution: *There is sealant on the drive housing flange mating surface - the edges of the housing flange should be heated with a heat gun prior to removal or damage may occur*

7   Loosen the drive housing mounting bolts, a little at a time using a criss-cross pattern, then remove the pump drive housing and the O-rings.

Installation

8   Clean the drive housing and cylinder head mounting surfaces with a rag saturated in brake system cleaner.

9   Install a new O-ring to the cylinder head at the oil feed hole.

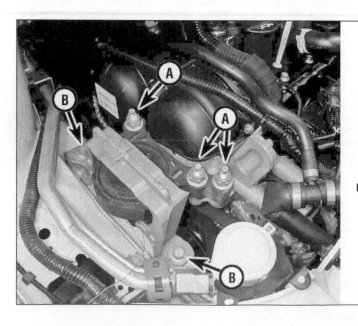

18.12b Remove the nuts (A) and bolts (B) then lift the engine mount from the vehicle - later models shown

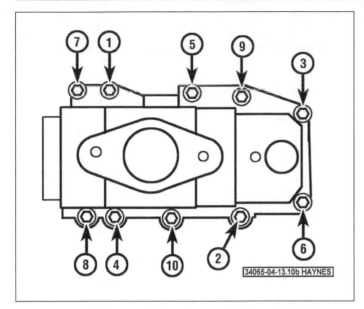

**19.11a High-pressure fuel pump drive housing bolt
tightening sequence - 1.6L models**

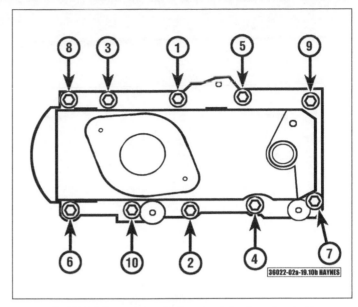

**19.11b High-pressure fuel pump drive housing bolt
tightening sequence - 1.5L models**

10   Apply a continuous 1/16 inch (1.5 mm) thick bead of RTV sealant to the cylinder head, staying on the inside of the bolt holes (and the outside of the oil feed hole) where the drive housing contacts the head.

11   Install the drive housing and tighten the bolts in sequence (see illustrations) to the torque listed in this Chapter's Specifications. **Note:** *The drive housing must be installed and tightened within five minutes of applying the sealant.*

12   The remainder of installation is the reverse of removal. Tighten all fasteners to the correct torque settings listed at the beginning of their respective Chapter(s).

2.0L engines
Removal

13   Relieve the fuel system pressure (see Chapter 4).

14   Disconnect the cable from the negative battery terminal (see Chapter 5).

15   Lift the engine cover up and off of the ballstuds, then remove the cover.

16   Remove the high-pressure fuel pump (see Chapter 4, Section 6).

17   Using a magnet pull the high pressure fuel pump lifter out of the drive housing.

18   Remove the drive housing mounting bolts and separate the housing from the cylinder head.

Installation

19   Clean the drive housing and cylinder head mounting surfaces with a rag saturated in brake system cleaner.

20   Replace the drive housing O-ring gasket if it looks damaged.

21   Install the drive housing and tighten the bolts in a criss-cross pattern to the torque listed in this Chapter's Specifications.

22   The remainder of installation is the reverse of removal, making sure the camshaft is rotated until the camshaft lobe for the fuel pump lifter is at the base or zero lift before installing the fuel pump lifter into the housing. Tighten all fasteners to the correct torque settings listed at the beginning of their respective Chapter(s).

Notes

Chapter 2 Part B
V6 engine

Contents

Specifications

General

Engine type	Double overhead cam (DOHC) V6
Displacement	182 cubic inches (3.0 liters)
Engine VIN code	
2008 and earlier models	1
2009 through 2012 models	G
Bore	3.50 inches (89.0 mm)
Stroke	3.13 inches (79.5 mm)
Compression ratio	10: 1
Compression	See Chapter 2C
Firing order	1-4-2-5-3-6
Oil pressure	See Chapter 2C

Cylinder location diagram - V6 engine

1-4-2-5-3-6

36075-B-SPECS HAYNES

Camshafts

Lobe lift (intake and exhaust)	
2008 and earlier models	0.188 inch (4.77 mm)
2009 through 2012 models	0.200 inch (5.08 mm)
Journal diameter	1.060 to 1.061 inches (26.924 to 26.949 mm)
Bearing inside diameter	1.062 to 1.063 inches (26.975 to 27.000 mm)
Journal-to-bearing oil clearance	
Standard	0.001 to 0.003 inch (0.025 to 0.076 mm)
Service limit	0.0047 (0.119 mm)
Endplay	
Standard	0.001 to 0.0065 inch (0.025 to 0.165 mm)
Service limit	
2008 and earlier models	0.0075 inch (0.191 mm)
2009 through 2012 models	0.009 inch (0.229 mm)

Hydraulic lash adjuster

Diameter	0.6290 to 0.6294 inch (15.977 to 15.987 mm)
Lash adjuster-to-bore clearance	
Standard	0.0007 to 0.0027 inch (0.018 to 0.069 mm)
Minimum	0.0006 inch (0.015 mm)

Warpage limits

Head gasket surface warpage limit	0.005 inch (0.127 mm)

Torque specifications

	Ft-lbs (unless otherwise indicated)	Nm

Note: *One foot-pound (ft-lb) of torque is equivalent to 12 inch-pounds (in-lbs) of torque. Torque values below approximately 15 ft-lbs are expressed in inch-pounds, since most foot-pound torque wrenches are not accurate at these smaller values.*

	Ft-lbs (unless otherwise indicated)	Nm
Air conditioning compressor bolts	18	25
Air conditioning compressor bracket nuts		
2008 and earlier models		
Step 1	18	25
Step 2	Tighten an additional 90-degrees	
2009 through 2012 models	18	25
Camshaft bearing cap bolts		
Step 1	89 in-lbs	10
Step 2	89 in-lbs	10
Camshaft oil seal retainer bolts	89 in-lbs	10
Crankshaft pulley bolt		
Step 1	86	120
Step 2	Loosen 360-degrees	
Step 3	37	50
Step 4	Tighten an additional 90-degrees	
Cylinder head bolts (in sequence - see illustration 11.21a)		
Step 1	30	40
Step 2	Tighten an additional 90-degrees	
Step 3	Loosen 360-degrees	
Step 4	30	40
Step 5	Tighten an additional 90-degrees	
Step 6	Tighten an additional 90-degrees	
Drivebelt tensioner bolt	See Chapter 1	
Driveplate-to-crankshaft bolts	59	80
Engine access cover	53 in-lbs	6
Engine front cover bolts (in sequence - see illustration 8.23)	18	25
Exhaust manifold nuts		
Step 1	15	20
Step 2	15	20
Exhaust manifold studs	106 in-lbs	12
Exhaust pipe-to-exhaust manifold nuts	30	40
Intake manifold assembly (in sequence - see illustrations 5.23a, 5.23b and 5.38)		
Upper intake manifold bolts		
2008 and earlier models	89 in-lbs	10
2009 and later models		
Step 1	89 in-lbs	10
Step 2	Tighten an additional 45-degrees	
Lower intake manifold bolts	89 in-lbs	10
Oil pan bolts	18	25
Oil pan-to-transaxle bolts	30	40
Oil pump screen cover and tube bolts	89 in-lbs	10
Oil pump screen cover and tube nut(s)		
2008 and earlier models		
Step 1	44 in-lbs	5
Step 2	Tighten an additional 45-degrees	
2009 through 2012 models	89 in-lbs	10
Oil pump-to-engine block bolts	89 in-lbs	10
Oil pan baffle bolts		
2008 and earlier models		
Step 1	44 in-lbs	5
Step 2	Tighten an additional 45-degrees	
2009 and later models		
6 mm bolts		
Step 1	44 in-lbs	5
Step 2	Tighten an additional 45-degrees	
8 mm bolts		
Step 1	133 in-lbs	15
Step 2	Tighten an additional 45-degrees	
Phaser (variable camshaft sprocket) bolts	159 in-lbs	18
Timing chain guide bolts	18	25
Timing chain tensioner bolts	18	25
Valve cover bolts (in sequence - see illustrations 4.19a, 4.19b, 4.19c, 4.30a, 4.30b and 4.30c)	89 in-lbs	10

* Note: *Refer to Chapter 2C for additional torque specifications.*

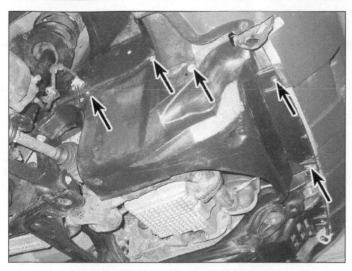

3.4 Inner fender splash shield fasteners

3.6 To position the number 1 piston at TDC, align the mark on the pulley with the mark on the timing chain cover

1 General Information

1   This part of Chapter 2 covers in-vehicle repairs for the 3.0 liter Overhead Camshaft (OHC) V6 Duratec engine. This version of the 3.0L V6 engine features an aluminum engine block and aluminum cylinder heads with dual overhead camshafts and four valves per cylinder.
2   Information on engine removal and installation, as well as general overhaul procedures, is in Chapter 2C.
3   The following repair procedures are based on the assumption that the engine is installed in the vehicle. If the engine has been removed and mounted on a stand, many of the Steps in this part of Chapter 2 will not apply.
4   In this Chapter, "left" and "right" are used to describe locations on the engine. These directions are in relation to the vehicle overall from the position of sitting in the driver's seat. So, "left" means the driver's side, and "right" means the passenger's side of the vehicle.

2 Repair operations possible with the engine in the vehicle

1   Many major repairs can be done without removing the engine from the vehicle. Clean the engine compartment and the exterior of the engine with a pressure washer or degreaser solvent before doing any work. Cleaning the engine and engine compartment will make repairs easier and help to keep dirt out of the engine.
2   It may help to remove the hood for better access to the engine. Refer to Chapter 11, if necessary.
3   If the engine has vacuum, exhaust, oil, or coolant leaks that indicate the need for gasket replacement, repairs can usually be done with the engine in the vehicle. The intake and

exhaust manifold gaskets, the timing chain cover gasket, the oil pan gasket, crankshaft oil seals, and cylinder head gaskets are all accessible with the engine in the vehicle.
4   Exterior engine components, such as the intake and exhaust manifolds, the oil pan, the water pump, the starter motor, the alternator, and many fuel system components also can be serviced with the engine installed.
5   Because the cylinder heads can be removed without pulling the engine, valve train components can be serviced with the engine in the vehicle. The timing chain and sprockets also can be replaced without removing the engine, although clearance is very limited.
6   In some cases - caused by a lack of equipment - pistons, piston rings, connecting rods, and rod bearings can be replaced with the engine in the vehicle. This is not recommended, however, because of the cleaning and preparation that must be done to the parts involved.

3 Top Dead Center (TDC) for number one piston - locating

Note: *Although this engine does not have a distributor, timing marks on the crankshaft pulley and a stationary pointer on the engine front cover will help you to locate Top Dead Center (TDC) for cylinders 1 and 5.*
1   Top Dead Center (TDC) is the highest point in the cylinder that each piston reaches as it travels upward when the crankshaft turns. Each piston reaches TDC on the compression stroke and on the exhaust stroke, but TDC usually refers to piston position on the compression stroke.
2   Positioning one or more pistons at TDC is an essential part of many procedures such as rocker arm removal, valve adjustment, and timing chain replacement.
3   Remove the spark plug from number 1

cylinder and install a compression gauge in its place (see Chapter 2C). Disconnect the cable from the negative terminal of the battery (see Chapter 5).
4   Loosen the right front wheel lug nuts, then raise the front of the vehicle and support it securely on jackstands. Remove the wheel and the inner fender splash shield (see illustration).
5   Rotate the crankshaft by using a socket and breaker bar on the crankshaft pulley bolt (turning the pulley clockwise), until a compression reading is indicated on the gauge. This indicates that the piston is on the compression stroke and is approaching TDC.
6   Continue turning the crankshaft with a socket and breaker bar until the notch in the crankshaft pulley is aligned with the TDC mark on the timing chain cover (see illustration). At this point, the number one cylinder is at TDC on the compression stroke. If the marks aligned but there was no compression, the piston was on the exhaust stroke. Continue rotating the engine until compression is indicated.
7   If you go past TDC, rotate the engine backwards (counterclockwise) until the timing marks indicate that the piston is before TDC. Then rotate the crankshaft clockwise until the marks align. Final rotation should always be clockwise to remove slack from the timing chains.
8   After the number one piston is at TDC on the compression stroke, TDC for any of the remaining cylinders can be located by turning the crankshaft and following the firing order (refer to the Specifications). Divide the crankshaft pulley into three equal sections with chalk marks at three points, each indicating 120-degrees of crankshaft rotation. For example, rotating the engine 120-degrees past TDC for number 1 piston will place the engine at TDC for cylinder number 4. Refer to the firing order for the remaining cylinder numbers.

4.1 Engine cover mounting fasteners

4.2 Disconnect the crankcase ventilation tube from the front valve cover

4   Valve covers - removal and installation

1   Remove the engine cover (see illustration).

Front (left-side) valve cover

2008 and earlier models

2   Disconnect the crankcase ventilation tube (see illustration).
3   Remove the ignition coils (see Chapter 5) from the top of the valve cover.
4   Disconnect the wiring harness from the valve cover and position it off to the side.
5   Remove the engine lifting brackets.
6   Remove the bolt from the power steering pump hose bracket and position the hose off to the side.

2009 through 2012 models

7   Remove the air intake duct from the throttle body (see Chapter 4).
8   Refer to Chapter 5 and remove the ignition coils.

9   Pull the upper radiator hose from its retainers and push it aside.
10   Disconnect the VCT and heated O2 sensor electrical connectors (see Chapter 6), then disconnect the wiring harnesses and retainers and set them out of the way.

All models

11   Loosen the valve cover bolts gradually and evenly until all are loose. Follow the reverse of the tightening sequence (see illustration 4.16a, 4.16b or 4.16c). Remove the valve cover fasteners and lift the valve cover off the engine.
12   Remove and discard the valve cover gaskets and spark plug tube seals (see illustrations). Install new gaskets and seals during reassembly.
13   Inspect the valve cover and cylinder head sealing surfaces for nicks or other damage. Clean the sealing surfaces with clean solvent and a shop towel.
14   Install a new valve cover gasket, making sure the gasket is properly seated in the groove (see illustration). Also install new spark plug tube seals. On 2009 and later models,

install the new gasket to the valve cover by attaching it at the waffle areas, then the corners. Install the gasket around the bolt holes, then the rest of the gasket.
15   Apply a 5/16-inch bead of RTV sealant at the locations shown (see illustrations).
16   Lower the valve cover into position, making sure that the gaskets stay in place. Install the cover fasteners and tighten them, in sequence (see illustrations), to the torque listed in this Chapter's Specifications.
17   The remaining installation is the reverse of removal.

Rear (right-side) valve cover

18   Remove the upper intake manifold (see Section 5).
19   Remove the ignition coils (see Chapter 5) from the top of the valve cover.
20   Disconnect all of the wiring that interferes with valve cover removal and set the harnesses out of the way.
21   On 2008 and earlier models, remove the capacitor from the base of the alternator. Disconnect any interfering hoses or ventilation tubes.

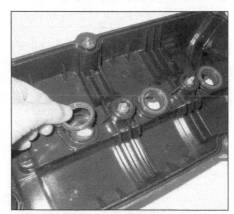

4.12a Remove and discard the old valve cover gasket - it must be replaced during assembly to prevent leakage

4.12b Remove and discard the spark plug tube seals from the valve cover

4.14 Press the new gasket into the groove in the valve cover, making sure it isn't twisted

4.15a Apply a 5/16-inch bead of RTV sealant to the seams of the engine front cover/cylinder head . . .

4.15b . . . and the camshaft seal retainer/cylinder head

22   Loosen the valve cover fasteners gradually and evenly, until all fasteners are loose. Follow the reverse of the tightening sequence (see illustration 4.27a, 4.27b or 4.27c). Then, unscrew and remove the fasteners. Lift the valve cover off the cylinder head.

23   Remove and discard the valve cover gaskets and spark plug tube seals (see illustrations 4.12a and 4.12b). Install new gaskets and seals during reassembly.

24   Inspect the valve cover and cylinder head sealing surfaces for nicks or other damage. Clean the sealing surfaces with clean solvent and a shop towel.

25   Install a new valve cover gasket, making sure the gasket is properly seated in the groove (see illustration 4.14). Also install new spark plug tube seals.

26   Apply a 5/16-inch bead of RTV sealant to the two locations where the front cover contacts the cylinder head (see illustration 4.15a).

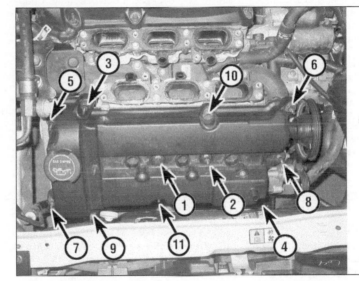

4.16a Valve cover bolt tightening sequence for the left (front) valve cover - 2001 through early production 2006 models

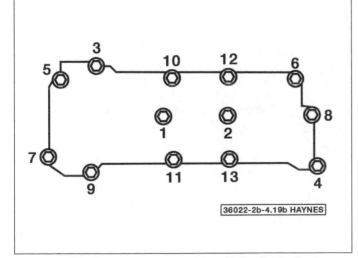

4.16b Left (front) valve cover bolt tightening sequence - late-production 2006 through 2008 models

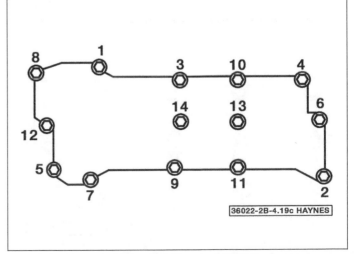

4.16c Left (front) valve cover bolt tightening sequence - 2009 through 2012 models

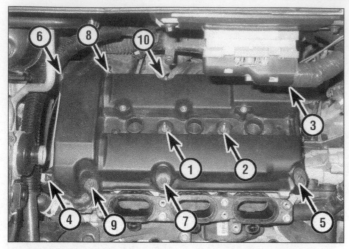

4.27a Valve cover bolt tightening sequence for the right (rear) valve cover - 2001 through early production 2006 models

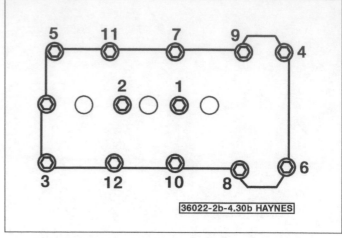

4.27b Right (rear) valve cover bolt tightening sequence - late-production 2006 through 2008 models

27   Install the valve cover fasteners and tighten them gradually and evenly to the torque listed in this Chapter's Specifications. Follow the correct torque sequence (see illustrations).
28   Reinstall all wiring harnesses and brackets.
29   Reinstall the ignition coils (see Chapter 5).
30   Install the upper intake manifold (see Section 5).
31   The remaining installation is the reverse of removal.

5   Intake manifold - removal and installation

Upper intake manifold

Removal

2008 and earlier models

1   Disconnect the cable from the negative battery terminal (see Chapter 5).
2   Remove the air filter housing intake duct (see Chapter 4).
3   Remove the engine access cover (see illustration 4.1).

4   Disconnect the accelerator cable and the cruise control cable, if equipped (see Chapter 4 and Chapter 6).
5   Disconnect the throttle position sensor (TPS) and the idle air control (IAC) valve connectors (see Chapter 6).
6   Disconnect the EGR valve vacuum hose and the EGR pipe (see Chapter 6).
7   Disconnect the EGR valve vacuum regulator solenoid connector (see Chapter 6).
8   Disconnect the vapor management valve vacuum hose.
9   Disconnect the vacuum lines and electrical connectors from the upper intake manifold. Be sure to mark them with tape to insure correct reassembly.
10   Disconnect the power steering pressure sensor from the power steering pump.
11   Remove the transaxle vent hose bracket and position the hose off to the side.
12   Remove the wire harness bracket nut and position the wiring harness off to the side.
13   Loosen the upper intake manifold bolts. Follow the opposite of the tightening sequence (see Illustration 5.23a).

14   Remove and discard the upper intake manifold gaskets (see illustration).

2009 through 2012 models

15   Disconnect the cable from the negative battery terminal (see Chapter 5).
16   Refer to Chapter 4 and remove the air filter housing and the intake air duct.
17   Disconnect all interfering wiring, then lay the wiring harnesses aside.
18   Disconnect the EVAP canister line and the brake vacuum line from the intake manifold. Disconnect the PCV tube from the intake manifold.
19   Remove the bolts from the intake manifold brackets. The brackets may be left in place.
20   Remove the intake manifold mounting bolts, then lift the manifold off. Discard the gaskets.

Installation

Caution: *Be very careful when scraping on aluminum engine parts. Aluminum is soft and gouges easily. Severely gouged parts may require replacement.*

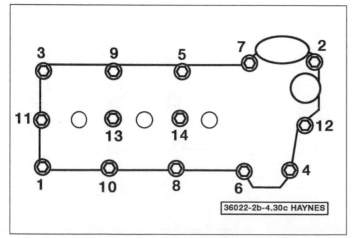

4.27c Right (rear) valve cover bolt tightening sequence - 2009 through 2012 models

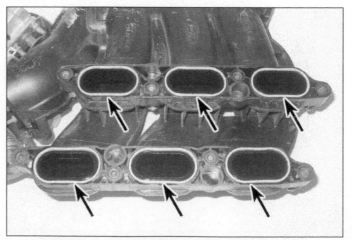

5.14 Remove the gaskets from the upper intake manifold - 2008 and earlier models shown

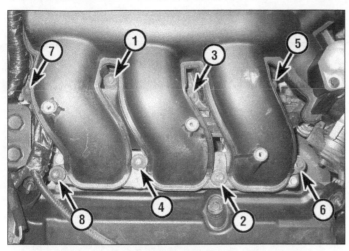

5.23a Upper intake manifold bolt tightening sequence -
2008 and earlier models

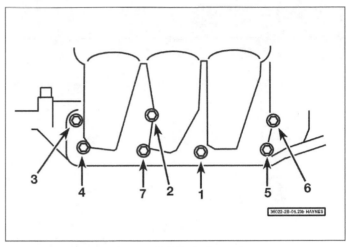

5.23b Upper intake manifold bolt tightening sequence -
2009 through 2012 models

21   If the gasket was leaking, check the mating surfaces for warpage. Check carefully around the mounting points of components such as the IAC valve and the EGR pipe. Replace the manifold if it is cracked or badly warped.

22   Install new gaskets. If the mating surfaces are clean and flat, a new gasket will ensure the joint is sealed. Don't use any kind of sealant on any part of the fuel system or intake manifold.

23   Locate the upper manifold on the lower manifold and install the fasteners. Tighten the fasteners in three or four steps in the sequence shown (see illustrations) to the torque listed in this Chapter's Specifications.

24   Install the remaining parts in the reverse order of removal.

25   Before starting the engine, check the accelerator cable for correct adjustment and the throttle linkage for smooth operation.

26   Reconnect the battery. After you're done, the Powertrain Control Module (PCM) must relearn its idle and fuel trim strategy for optimum driveability and performance (see Chapter 5, Section 1).

27   When the engine is fully warm, check

for fuel and vacuum leaks. Road test the vehicle and check for proper operation of all components.

Lower intake manifold
Removal

28   Relieve the fuel pressure (see Chapter 4).

29   Disconnect the cable from the battery negative terminal (see Chapter 5).

30   Remove the upper intake manifold (see Steps 1 through 14).

31   On 2008 and earlier models, disconnect the fuel supply line from the fuel rail (see Chapter 4). On 2009 through 2012 models, remove the fuel rail (see Chapter 4)

32   On 2008 and earlier models, disconnect the wiring from the fuel injectors, move the wiring harness aside, then disconnect the vacuum hose from the fuel pressure regulator.

33   To prevent warpage, loosen the eight lower manifold bolts gradually and evenly in the reverse of the tightening sequence until all are loose (see illustration 5.38). Then, remove the bolts.

34   Lift the lower intake manifold off the engine. Remove and discard the manifold

gaskets (see illustration).

35   Carefully clean all gasket material from the manifold and cylinder head mating surfaces. Don't nick, scratch or gouge the sealing surfaces. Inspect all parts for cracks or other damage. If the manifold gaskets were leaking, check the mating surfaces for warpage.

Caution: *Be very careful when scraping on aluminum engine parts. Aluminum is soft and gouges easily. Severely gouged parts may require replacement.*

Installation

36   Install new lower intake manifold gaskets on the cylinder heads

37   Place the lower manifold into position on the cylinder heads. Make sure the gaskets are not dislodged.

38   Install the lower manifold bolts. Tighten the bolts gradually and evenly, in the sequence shown (see illustration), to the torque listed in this Chapter's Specifications.

39   The remainder of installation is the reverse of the removal procedure. Tighten the fasteners to the torque listed in this Chapter's Specifications.

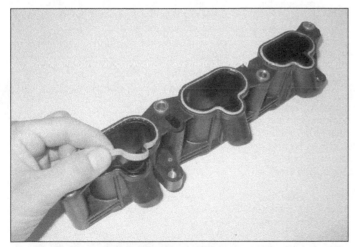

5.34 Remove the gaskets from the lower intake manifold -
2008 and earlier models shown

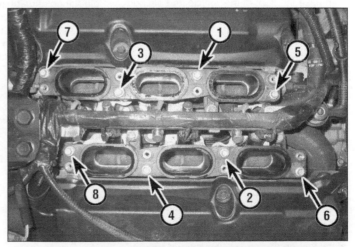

5.38 Lower intake manifold bolt tightening sequence -
2008 and earlier models shown, later models similar

6.7 Remove the front exhaust manifold mounting nuts - three nuts hidden from view

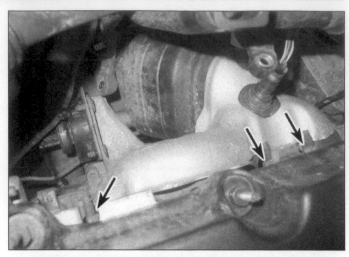

6.18 The rear exhaust manifold mounting nuts are difficult to access (not all are visible here)

40   Reconnect the battery. After you're done, the Powertrain Control Module (PCM) must relearn its idle and fuel trim strategy for optimum driveability and performance (see Chapter 5, Section 1).

---

6   Exhaust manifolds - removal and installation

Warning: *Allow the engine to cool completely before beginning this procedure.*

2008 and earlier models

Front (left-side) exhaust manifold

Warning: *The engine and exhaust system must be completely cool before performing this procedure.*

1   Disconnect the cable from the negative battery terminal (see Chapter 5).
2   Disconnect the electrical connector for the oxygen sensor and the catalytic converter monitor (see Chapter 6).
3   Loosen the right front wheel lug nuts, then raise the front of the vehicle and support it securely on jackstands. Remove the wheel, then remove the fender splash shield.
4   Remove the exhaust crossover pipe (see Chapter 4).
5   Remove the drivebelt (see Chapter 1).
6   Remove the air conditioning compressor bolts and position the compressor off to the side. Do not disconnect the refrigerant lines from the compressor.
7   Remove the six nuts securing the exhaust manifold (see illustration). Remove and discard the manifold gasket.
8   Using a scraper, remove all old gasket material and carbon deposits from the manifold and cylinder head mating surfaces.If the gasket was leaking, check the manifold for warpage and have it resurfaced if necessary.
Caution: *Be very careful when scraping on aluminum engine parts. Aluminum is soft and gouges easily.*

9   Install a new manifold gasket over the cylinder head studs. Lightly oil the stud threads with clean engine oil before installation. Install the manifold and the nuts.
10   Tighten the nuts in three or four steps, in a spiral pattern starting with the center fasteners, to the torque listed in this Chapter's Specifications.
11   The remainder of installation is the reverse of the removal procedure.
12   Start and run the engine and check for exhaust leaks. The Powertrain Control Module (PCM) must relearn its idle and fuel trim strategy for optimum driveability and performance (see Chapter 5, Section 1).

Rear (right-side) exhaust manifold

13   Disconnect the cable from the negative battery terminal (see Chapter 5).
14   Remove the EGR pipe from the exhaust manifold (see Chapter 6).
15   Remove the alternator (see Chapter 5).
16   Remove the oxygen sensor (see Chapter 6).
17   Remove the nuts from the exhaust manifold flange and separate the crossover pipe from the manifold (see Chapter 4).
18   Remove the six exhaust manifold mounting nuts and remove the manifold from the engine (see illustration). Remove and discard the manifold gasket.
19   Using a scraper, remove all gasket material and carbon deposits from the exhaust manifold and cylinder head mating surfaces. If the gasket was leaking, check the manifold for warpage and have it resurfaced if necessary.
Caution: *Be very careful when scraping on aluminum engine parts. Aluminum is soft and gouges easily. Severely gouged parts may require replacement.*
20   Install a new manifold gasket over the cylinder head studs. Lightly oil the stud threads with clean engine oil before installation. Install the manifold and the nuts.
21   Tighten the nuts in three or four steps, in a spiral pattern starting with the center fas-

teners, to the torque listed in this Chapter's Specifications.
22   Complete the installation by reversing the removal procedure.
23   Start and run the engine and check for exhaust leaks. The Powertrain Control Module (PCM) must relearn its idle and fuel trim strategy for optimum driveability and performance (see Chapter 5, Section 1).

2009 through 2012 models
Removal
24   Disconnect the cable from the negative battery terminal (see Chapter 5). Loosen the right-front wheel lug nuts, then raise the vehicle and support it securely on jackstands and remove the wheel.

Left (front) manifold
25   Remove the passenger's side inner fender splash shield (see Chapter 11, Section 10).
26   Refer to Chapter 1 and remove the engine drivebelt.
27   Disconnect the wiring from the air conditioning compressor. Remove the three compressor mounting bolts and move the compressor out of the way.
28   Remove the oil dipstick tube and its bracket.

Right (rear) manifold
29   Detach the EGR fitting nut.

Both manifolds
30   Disconnect the exhaust pipe from the exhaust manifold and discard the gasket.
Note: *The left (front) exhaust manifold has the catalytic converter built into it.*
31   Disconnect the wiring from the oxygen sensor.
32   Remove the heat shield.
33   Remove the six exhaust manifold nuts and discard them.
34   Remove the exhaust manifold and the gasket, then remove and discard the mounting studs.

7.5a To keep the crankshaft from turning, remove the driveplate access cover . . .

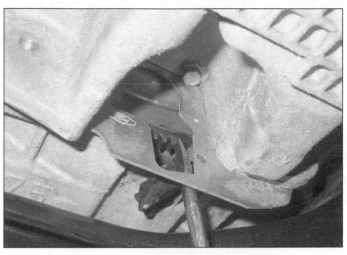

7.5b . . . engage the ring gear teeth with a large screwdriver or prybar

Installation

35   Using a scraper, remove all old gasket material from the manifold and cylinder head mating surfaces. Check the manifold surface for warpage and have it surfaced if necessary.
Caution: *Be very careful when scraping aluminum engine parts. Aluminum is soft and gouges easily.*
36   Install six new studs. Tighten them to the torque listed in this Chapter's Specifications.
37   Install a new gasket over the studs, then install the exhaust manifold. Tighten the new nuts to the torque listed in this Chapter's Specifications, working from the center outward.
38   Install the exhaust pipe to the exhaust manifold using a new gasket. Tighten the nuts to the torque listed in this Chapter's Specifications. Tighten the nuts a second time, then install the exhaust pipe bracket bolts.
39   The remainder of installation is the reverse of removal.

7   Crankshaft pulley and front oil seal - removal and installation

Removal

1   Disconnect the cable from the negative battery terminal (see Chapter 5).
2   Loosen the right front wheel lug nuts. Raise the front of the vehicle and support it securely on jackstands, then remove the wheel.
3   Remove the fender splash shield.
4   Remove the drivebelt (see Chapter 1).
5   Remove the driveplate access cover and hold the ring gear teeth with a large screwdriver or prybar to prevent the crankshaft from turning (see illustrations).
6   Remove the crankshaft pulley retaining bolt.
7   On 2008 and earlier models, remove the

pulley using a puller that uses bolts that screw into the holes of the pulley (see illustration). On 2009 and later models, use a puller that has three hooks that grab the inner hub of the pulley.
Caution: *Never use a puller that attaches to the outer rim of the crankshaft pulley.*
8   Using a seal puller, remove the seal from the engine front cover. Note the orientation of the sealing lip so that the new seal will be installed in the same direction.

Installation

9   Inspect the front cover and the pulley seal surface for nicks, burrs, or other roughness that could damage the new seal. Correct as necessary.
10   Lubricate the new seal with clean engine oil and install it in the engine front cover with a suitable seal driver. Be sure that the lip of the seal faces inward. If a seal driver is unavailable, carefully tap the seal into place with a large socket and hammer until it's flush with the front cover surface.
11   Apply RTV sealant to the keyway and inner bore of the pulley and lubricate the outer sealing surface of the pulley with clean engine oil. Then align the pulley keyway with the crankshaft key and install the pulley with a suitable pulley installation tool (available at auto parts stores). If such a tool is unavailable, start the pulley onto the crankshaft with a soft-faced mallet and finish the installation by tightening the retaining bolt. Tighten the bolt to the torque listed in this Chapter's Specifications.
12   Reinstall the remaining parts in the reverse order of removal. Tighten the wheel lug nuts to the torque listed in the Chapter 1, Specifications.
13   Reconnect the battery, then start the engine and check for oil leaks.
14   The Powertrain Control Module (PCM) must relearn its idle and fuel trim strategy for optimum driveability and performance (see Chapter 5, Section 1).

7.7 Remove the crankshaft pulley with a bolt-type puller on 2008 and earlier models

8   Timing chain cover - removal and installation

Warning: *Wait until the engine is completely cool before beginning this procedure.*
Note: *Because of a lack of clearance, this procedure is very difficult with the engine installed in the vehicle. If major engine work is being performed, it may be easier to remove the engine from the vehicle (see Chapter 2C).*

2008 and earlier models
Removal

1   Disconnect the cable from the negative battery terminal (see Chapter 5).
2   Loosen the right front wheel lug nuts. Raise the front of the vehicle and support it securely on jackstands, then remove the wheel.
3   Drain the engine oil (see Chapter 1).
4   Remove the drivebelt (see Chapter 1).
5   Remove the drivebelt tensioner (see Chapter 1).

8.11 Disconnect the camshaft position sensor (A)
and the crankshaft position sensor (B)

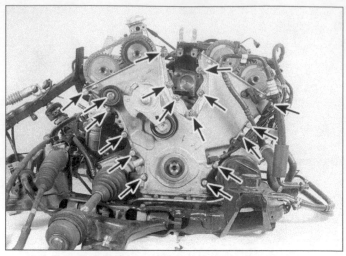

8.15 Remove the cover fasteners - note the locations of all studs
and bolts for installation reference

6   Remove the valve covers (see Section 4).
7   Remove the alternator and alternator
mounting bracket (see Chapter 5).
8   Remove the crankshaft pulley (see Section 7).
9   Drain the cooling system (see Chapter 1)
and remove the coolant expansion tank (see
Chapter 3).
10   Remove the power steering pump (see
Chapter 10).
11   Disconnect the crankshaft position sensor (see illustration).
12   Disconnect the camshaft position sensor
(see Chapter 6).
13   Remove the oil pan (see Section 12).
14   If necessary, loosen the air conditioning compressor mounting bolts and move the
compressor aside as required for access to the
timing chain cover. Remove or disconnect any
remaining wires, hoses, clamps, or brackets
that will interfere with engine cover removal.
Warning: *The air conditioning system is under
high pressure. Do not disconnect the refriger-*
ant lines from the compressor.
15   Loosen the timing chain cover fasteners
gradually and evenly; then remove the fasteners (see illustration).
Note: *Draw a sketch of the timing chain cover
and fasteners. Identify the location of all stud
bolts for installation in their original locations.*
16   Remove the front cover (see illustration).
17   Remove and discard the cover-to-cylinder block gaskets.

Installation

18   Inspect and clean all sealing surfaces of
the timing chain cover and the block.
Caution: *Be very careful when scraping on
aluminum engine parts. Aluminum is soft and
gouges easily. Severely gouged parts may require replacement.*
19   If necessary, replace the crankshaft seal
in the cover (see Section 7).
20   Apply a bead of RTV sealant approximately 1/8-inch wide at the locations shown
(see illustration).

21   Install a new timing chain cover gasket
on the engine block. Make sure the gasket fits
over the alignment dowels.
22   Install the cover and cover fasteners.
Make sure the fasteners are in their original
locations. Tighten the fasteners by hand until
the cover is contacting the block and cylinder
heads around its entire periphery.
23   Following the correct sequence (see
illustration), tighten the bolts to the torque
listed in this Chapter's Specifications.
24   Install the drivebelt tensioner (see Chapter 1).
25   Install the oil pan (see Section 12).
26   Install the crankshaft pulley (see Section 7).
27   Connect the wiring harness connectors
to the camshaft and crankshaft position sensors.
28   Install the power steering pump and
hoses (see Chapter 10).
29   Reinstall the remaining parts in the
reverse order of removal.

8.16 Lift off the timing chain cover

8.20 Apply a bead of RTV sealant at the locations shown
(2008 and earlier models)

2009 through 2012 models

Removal

30   Disconnect the cable from the negative battery terminal (see Chapter 5).

31   Loosen the right front wheel lug nuts. Raise the vehicle and support it securely on jackstands, then remove the wheel.

32   Refer to Section 7 and remove the crankshaft pulley.

33   Remove the alternator mounting nuts and studs, then move the alternator out of the way.

34   Remove the drivebelt tensioner (see Chapter 1).

35   Disconnect the wiring harnesses from the camshaft position (CMP) sensor and the crankshaft position (CKP) sensor.

36   Refer to Chapter 6 and remove the variable camshaft timing (VCT) solenoids.

37   Remove the oil pan-to-front cover bolts.

38   Support the oil pan with a floor jack and a block of wood. Remove the interfering engine mount.

39   Remove the mounting bolts and studs from the timing chain cover. Mark the locations of the studs and bolts so they can be installed in their correct locations. Remove the cover and discard the gaskets.

Installation

40   Inspect and clean all sealing surfaces.

Caution: *Be very careful when scraping aluminum parts. Aluminum is soft and gouges easily. Damaged parts may require replacement. Also, be careful not to damage the front of the oil pan gasket.*

41   Replace the crankshaft seal (see Section 7).

42   Apply 1/4-inch dab of RTV sealant to the joints indicated (see illustration 8.20).

43   Install a new set of gaskets, then install

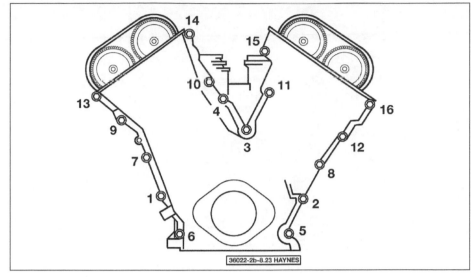

8.23 Timing chain cover bolt tightening sequence

the timing chain cover. Install the bolts and studs to their original locations, then tighten them to the torque listed in this Chapter's Specifications and in the proper sequence (see illustration 8.23).

44   The remainder of installation is the reverse of removal.

45   Fill the crankcase with the recommended oil (see Chapter 1). Fill the power steering reservoir with the correct fluid and refill the cooling system.

46   Reconnect the battery, then start the engine and check for leaks. The Powertrain Control Module (PCM) must relearn its idle and fuel trim strategy for optimum driveability and performance (see Chapter 5, Section 1).

9   Timing chains, tensioners, and chain guides - removal, inspection, and installation

Note: *Because of a lack of clearance, this procedure is very difficult with the engine installed in the vehicle. If major engine work is being performed, it may be easier to remove the engine from the vehicle (see Chapter 2C).*

Removal

Caution: *The timing system is complex. Severe engine damage will occur if you make any mistakes. Do not attempt this procedure unless you are highly experienced with this type of repair. If you are at all unsure of your abilities, consult an expert. Double-check all your work and be sure everything is correct before you attempt to start the engine.*

1   Remove the timing chain cover (see Section 8), then slide the crankshaft position sensor trigger wheel off the crankshaft. Remove the spark plugs (see Chapter 1).

Caution: *Note the exact position of the pulse wheel to insure correct reassembly.*

2   Install the crankshaft pulley retaining bolt, then use a wrench on the bolt to turn the crankshaft clockwise and place the crankshaft keyway at the 11 o'clock position. TDC number 1 is the starting position and the ending position for this procedure. Verify TDC by observing the index marks on the front of the camshaft sprockets. If the number 1 cylinder is at TDC, the index mark on the front (left) bank exhaust camshaft will be directly UP (12 o'clock) in relation to the cylinder head surface and the intake camshaft index mark will be horizontal (9 o'clock) to the cylinder head (see illustration). If not, turn the crankshaft exactly one full turn and again position the crankshaft keyway at 11 o'clock.

Caution: *Turning the crankshaft counterclockwise can cause the timing chains to bind and damage the chains, sprockets and tensioners. Turn the crankshaft clockwise only.*

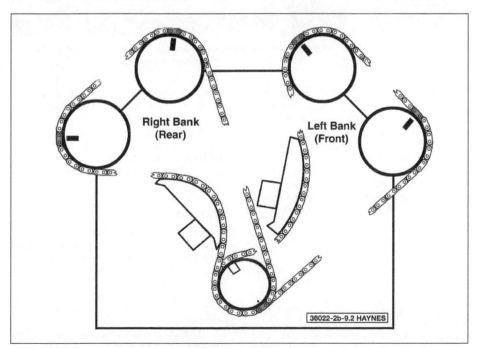

9.2 Positions of the camshaft sprocket timing marks and the crankshaft keyway with the engine set at TDC for the number one piston

9.4a Remove the bolts securing the rear timing chain tensioner and remove the tensioner . . .

9.4b . . . then slide the tensioner arm off its pivot

9.5 Lift the rear timing chain off the sprockets and remove the chain

9.6 Remove the bolts securing the rear timing chain guide, then remove the guide

3   Recheck the marks on the sprockets. If they are aligned and the keyway is at 11 o'clock, the number 1 cylinder is at TDC on the compression stroke. Continue to turn the crankshaft clockwise until the keyway is at the 3 o'clock position, which will set the camshafts on the rear cylinder head in their neutral position.

4   Remove the two bolts securing the timing chain tensioner for the rear chain. Remove the tensioner, then remove the tensioner arm (see illustrations). Mark all parts that will be reused so they can be reinstalled in their original locations.

5   Lift the rear timing chain from the sprockets and remove the chain (see illustration).

6   Remove the bolts securing the rear chain guide and remove the guide (see illustration).

7   Slide the sprocket for the rear timing chain off the crankshaft (see illustration).

8   Rotate the crankshaft 600-degrees (1-2/3 turns) clockwise, until the keyway is in the 11 o'clock position, setting the camshafts

in the front cylinder head in their neutral position. Remove the front timing chain tensioner mounting bolts. Remove the tensioner and tensioner arm (see illustrations).

9   Lift the front timing chain off the sprockets and remove the chain.

10   Remove the chain guide mounting bolts and remove the front chain guide (see illustration). If necessary, slide the crankshaft sprocket for the front timing chain off the crankshaft.

Inspection

Note: *Do not mix parts from the front and rear timing chains and tensioners. Keep the parts separate.*

11   Clean all parts with clean solvent. Dry with compressed air, if available.

12   Inspect the chain tensioners and tensioner arms for excessive wear or other damage.

13   Inspect the timing chain guides for deep

grooves, excessive wear, or other damage.

14   Inspect the timing chain for excessive wear or damage.

15   Inspect the camshaft and crankshaft sprockets for chipped or broken teeth, excessive wear, or damage.

16   Replace any component that is in questionable condition.

Installation

Caution: *Before starting the engine, carefully rotate the crankshaft by hand through at least two full revolutions (use a socket and breaker bar on the crankshaft pulley center bolt). If you feel any resistance, STOP! There is something wrong - most likely, valves are contacting the pistons. You must find the problem before proceeding.*

17   The timing chain tensioners must be fully compressed and locked in place before chain installation. To prepare the chain tensioners for installation:

9.7 Remove the rear timing chain sprocket from the crankshaft

9.8a Remove the bolts that secure the front timing chain tensioner, then remove the tensioner . . .

9.8b . . . and remove the tensioner arm

9.10 Remove the bolts securing the front timing chain guide and remove the guide

a) *Insert a small screwdriver into the access hole in the tensioner and release the pawl mechanism.*

b) *Compress the plunger into the tensioner housing until the plunger tip is below the plate on the pawl.*

c) *Hold the plunger in the compressed position and rotate the plunger one-half turn so the plunger can be removed. Remove the plunger and plunger spring.*

d) *Drain the oil from the tensioner housing and plunger.*

e) *Lubricate the tensioner housing, spring and plunger with clean engine oil. Insert the plunger spring into the tensioner housing. Install the plunger into the housing and push it in until fully compressed. Then, turn the plunger 180-degrees and use a small screwdriver to push back the pawl mechanism into contact with the plunger.*

f) *With the plunger compressed, insert a 1/16-inch drill bit or a straightened paper clip into the small hole above the pawl mechanism to hold the plunger in place.*

g) *Repeat this procedure for the other tensioner.*

18   If removed, install the crankshaft sprocket for the front timing chain. Make sure the crankshaft keyway is still at 11 o'clock.

19   Look at the index marks on the sprockets of the front bank intake and exhaust camshafts. The marks should be at 9 o'clock (intake) and at 12 o'clock (exhaust) (see illustration 9.2). If not, reposition the camshafts.

Caution: *The timing chains have three links that are a different color than the rest of the links. When installed, the colored links on the chain must be aligned with the index marks on the camshaft and crankshaft sprockets (see illustration 9.2).*

20   Install the front timing chain guide. Tighten the mounting bolts to the torque listed in this Chapter's Specifications.

21   Install the front timing chain around the camshaft and crankshaft sprockets. Make

sure the index marks on the sprockets are aligned with the colored links of the chain.

22   Install the front tensioner arm over its pivot dowel. Seat the tensioner arm firmly on the cylinder head and block.

23   Install the front timing chain tensioner assembly (see illustrations 9.8a and 9.8b). Be sure the tensioner plunger is fully compressed and locked in place. Tighten the tensioner mounting bolts to the torque listed in this Chapter's Specifications. Verify that the colored links of the timing chain are aligned with the index marks on the camshaft and crankshaft sprockets, and the crankshaft keyway is at 11 o'clock. If not, remove the timing chain and repeat the installation procedure.

24   Install the crankshaft sprocket for the rear timing chain on the crankshaft (see illustration 9.7).

25   Rotate the crankshaft clockwise until the crankshaft keyway is in the 3 o'clock position. This will correctly position the pistons for installation of the rear timing chain.

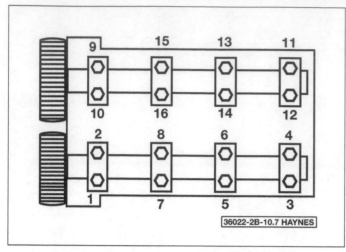

10.3 Rear bank camshaft bearing cap loosening sequence - 2008 and earlier models

10.5 Remove the rocker arms

26 Install the rear timing chain guide. Tighten the mounting bolts to the torque listed in this Chapter's Specifications.

27 Double-check the position of the rear camshafts so the index marks on the sprockets are at (approximately) 3 o'clock (intake) and at 12 o'clock (exhaust).

Caution: *The timing chains have three links that are a different color than the rest of the links. When installed, the colored links on the chain must be aligned with the index marks on the camshaft and crankshaft sprockets(see illustration 9.2).*

28 Install the rear timing chain around the camshaft and crankshaft sprockets. Make sure the colored links of the chain are aligned with the index marks on the front of the camshaft and crankshaft sprockets.

29 Install the rear tensioner arm over its pivot dowel. Seat the tensioner arm firmly on the cylinder head and block.

30 Install the rear timing chain tensioner (see illustrations 9.4a and 9.4b). Be sure the tensioner plunger is fully compressed and locked in place. Tighten the tensioner mounting bolts to specification. Verify that the colored links of the timing chain are aligned with the index marks on the camshaft and crankshaft sprockets, and the crankshaft keyway is at 3 o'clock. If not, remove the timing chain and repeat the installation procedure.

31 Remove the drill bits or wires (locking pins) from the timing chain tensioners.

32 Rotate the crankshaft counterclockwise back to the 11 o'clock position (the Number 1 TDC position). Verify the timing marks on the camshaft sprockets and the crankshaft sprocket are aligned with the colored links (see illustration 9.2).

33 Install the crankshaft position sensor pulse (trigger) wheel on the crankshaft.

Caution: *Make sure the pulse wheel is installed with the crankshaft key in the correct slot; "20-25-34Y-30M" on 2004 and earlier engines, "30" or "30RFF" for 2005 through 2012 engines.*

34 Rotate the engine by hand and verify that there is no binding. If any resistance is felt, STOP; double-check your work and find out why.

35 Install the timing chain cover (see Section 8).

36 Reinstall the remaining parts in the reverse order of removal.

37 Fill the crankcase with the recommended oil (see Chapter 1). Fill the power steering reservoir with the correct fluid and refill the cooling system.

38 Reconnect the battery, then start the engine and check for leaks.The Powertrain Control Module (PCM) must relearn its idle and fuel trim strategy for optimum driveability and performance (see Chapter 5, Section 1).

---

10 Camshafts, hydraulic lash adjusters and rocker arms - removal, inspection and installation

---

Removal

Note: *If you're only removing the camshafts from the rear bank cylinder head, you only have to remove the rear timing chain. If you're removing the camshafts from the front bank cylinder head, both timing chains will have to be removed.*

2008 and earlier models

Rear bank cylinder head

1 Remove the timing chain (see Section 9).

2 Note the location of each camshaft cap. Use a marker on each cap or notes on paper or cardboard if they're not marked. Do not mix any of the camshaft caps.

3 Following the sequence shown (see illustration), gradually loosen the bolts that secure the camshaft bearing caps to the cylinder head, then remove the camshaft caps. It may be necessary to tap the caps lightly with

a soft-faced mallet to loosen them from the locating dowels.

Caution: *The caps must be installed in their original locations. Keep all parts from each camshaft together; never mix parts from one camshaft with those for another.*

4 Mark the intake and exhaust camshafts to prevent reinstalling them in the wrong locations, then lift the camshafts straight up and out of the cylinder head.

5 Mark the positions of the rocker arms so they can be reinstalled in their original locations, then remove the rocker arms (see illustration).

6 Place the rocker arms in a suitable container so they can be separated and identified (see illustration).

7 Lift the hydraulic lash adjusters from their bores in the cylinder head. Identify and separate the adjusters so they can be reinstalled in their original locations.

Front bank cylinder head

8 Remove the water pump drivebelt (see Chapter 1). Also remove the oil seal retainer from the rear of the intake camshaft.

10.6 Place all the parts in a container so they can be separated and identified for installation in their original locations

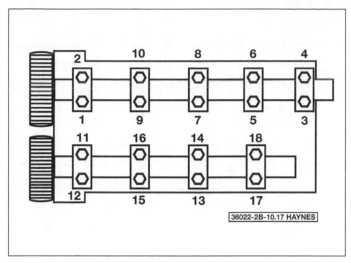

**10.9 Front bank camshaft bearing cap loosening sequence -
2008 and earlier models**

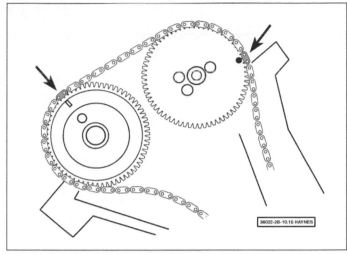

**10.14 The marks on the camshaft sprockets of the rear cylinder
head must be positioned in neutral as shown; if they aren't
aligned as shown, damage may result - 2009 through 2012 models**

9   Repeat Steps 3 to 7 to remove the front (left side) camshafts, rocker arms and lash adjusters. Be sure to loosen the bearing caps on the front camshafts in the sequence shown (see illustration).

2009 through 2012 models

Rear bank cylinder head
10   Refer to Chapter 6 and remove the variable camshaft timing (VCT) solenoid and the camshaft position (CMP) sensor.
11   Loosen the lug nuts of the right front wheel. Raise the vehicle and support it securely on jackstands. Remove the wheel.
12   Remove the inner fender splash shield (see Chapter 11, Section 10).
13   Remove the rear valve cover (see Section 4).
14   Use a ratchet and socket attached to the crankshaft pulley bolt to rotate the engine clockwise until the rear camshafts are in the neutral position; with the mark on the exhaust sprocket (phaser) in the 12 o'clock position

and the mark on the intake sprocket in the 3 o'clock position in relation to the top surface of the cylinder head (see illustration).
Note: *Mark the chain links to the marks on the phaser and sprocket for installation purposes.*
15   Install a timing chain locking tool into the front cover. This type of tool wedges between the opposite sides of the chain a few inches above the crankshaft sprocket. Make sure that the tool is firmly wedged in place. If it comes loose, you'll have to remove the timing chain cover to fix the problem.
Note: *The manufacturer tool is part #303-1175. Equivalent tools are also available at specialty automotive tool dealers.*
16   Use paint to put match marks on the timing chain and the sprocket and phaser. The chain and the sprocket and phaser must be aligned exactly when reassembled.
17   Mark every camshaft bearing cap so they can be kept in order and in the proper orientation.
Caution: *Each cap must be reinstalled in its*

original position and facing the correct direction or engine damage may occur.
18   Loosen the intake camshaft bearing cap bolts evenly and in the proper sequence (see illustration). Remove the bolts from the thrust cap (the one closest to the phaser) and remove the thrust cap.
19   Hold the intake camshaft by inserting a 3/8-inch drive ratchet into the D slot in the rear of the camshaft, then remove the three phaser and three sprocket mounting bolts.
20   Push the phaser and camshaft to the rear of the engine, then remove the phaser from the camshaft and slip the timing chain off. Do the same for the sprocket.
21   Loosen the exhaust camshaft bolts evenly and in the correct sequence (see illustration).
22   Remove the exhaust camshaft thrust cap (the one closest to the sprocket) first to avoid damaging it. Remove the rest of the bearing caps from the intake and exhaust camshafts, then remove the camshafts.

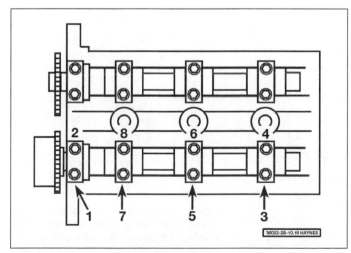

**10.18 Loosen the rear intake camshaft bearing caps in this order -
2009 through 2012 models**

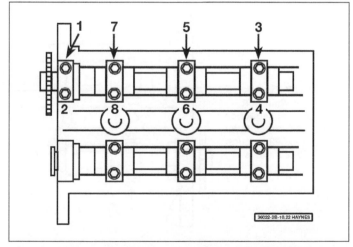

**10.21 Loosening sequence for the exhaust camshaft bearing caps
on the rear cylinder head - 2009 through 2012 models**

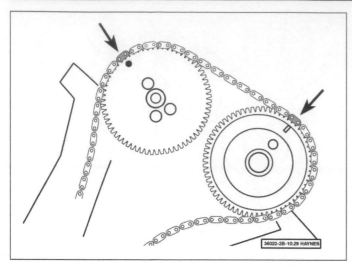

10.28 Front bank camshaft sprocket and phaser marks in the neutral position - 2009 through 2012 models

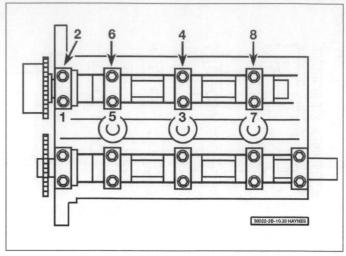

10.32 Intake camshaft bolt loosening sequence for the front cylinder head - 2009 through 2012 models

Front bank cylinder head

Caution: *Do not rotate the engine (Step 28) if the timing chain for the rear cylinder bank has been removed but the rear cylinder bank camshafts have not.*

23   Remove the water pump drivebelt (see Chapter 1), then use the special tools to remove the water pump pulley.

24   Loosen the lug nuts of the right front wheel. Raise the vehicle and support it securely on jackstands. Remove the wheel.

25   Refer to Chapter 6 and remove the Variable Camshaft Timing (VCT) solenoid and the Camshaft Position (CMP) sensor.

26   Remove the inner fender splash shield (see Chapter 11, Section 10).

27   Remove the front valve cover (see Section 4).

28   Use a ratchet and socket on the crankshaft pulley bolt to rotate the engine clockwise until the camshaft sprocket and phaser marks are in the neutral position (see illustration).

29   Install a timing chain locking tool into the front cover. This type of tool wedges between the opposite sides of the chain a few inches above the crankshaft sprocket. Make sure that the tool is firmly wedged in place. If it comes loose, you'll have to remove the timing chain cover to correctly align the timing chain with the sprockets.

Note: *The manufacturer tool is part # 303-1175. Equivalent tools are also available at specialty automotive tool dealers.*

30   Make match marks on the camshaft sprocket and phaser and the timing chain to aid in alignment later.

31   Mark every camshaft bearing cap so they can be kept in order and in the proper orientation.

32   Loosen the bearing cap bolts of the intake camshaft evenly and in the proper sequence (see illustration). Remove the intake camshaft thrust bearing cap (the one closest to the phaser).

33   Hold the intake camshaft by inserting a 3/8-inch drive ratchet into the D slot in the rear of the camshaft, then remove the three phaser and three sprocket mounting bolts.

34   Push the phaser and camshaft to the rear of the engine, then remove the phaser from the camshaft and slip the timing chain off. Do the same for the sprocket.

35   Remove the camshaft oil seal. This can be done with manufacturer special tool #303-409 or an equivalent tool. These are available from specialty automotive tool dealers.

36   Unbolt and remove the camshaft oil seal retainer. Discard the gasket.

37   Loosen the exhaust camshaft cap bolts evenly and in the proper sequence (see illustration).

38   Remove the thrust bearing cap (the one closest to the sprocket). Remove the remaining bearing caps, then remove the camshafts.

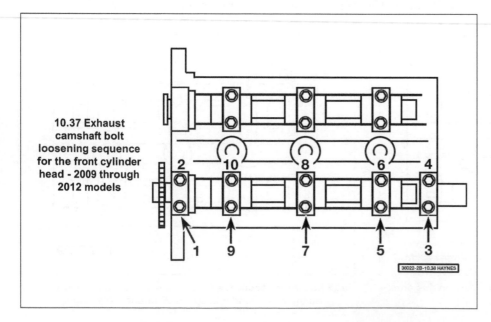

10.37 Exhaust camshaft bolt loosening sequence for the front cylinder head - 2009 through 2012 models

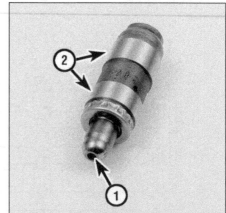

10.39 Inspect the lash adjusters for signs of excessive wear or damage, such as pitting, scoring or signs of overheating (bluing or discoloration), where the tip contacts the rocker arm (1) and the side surfaces that contact the bore in the cylinder head (2)

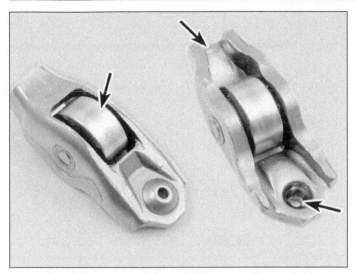

10.41 Check the roller surface (left arrow) of the rocker arm and the areas where the valve stem and lash adjuster contact the rocker arm (right arrows)

10.42 Check the cam lobes for pitting, excessive wear, and scoring. If scoring is excessive, as shown here, replace the camshaft

Inspection

39   Check each hydraulic lash adjuster for excessive wear, scoring, pitting, or an out-of-round condition (see illustration). Replace as necessary.

40   Measure the outside diameter of each adjuster at the top and bottom of the adjuster. Then take a second set of measurements at a right angle to the first. If any measurement is significantly different from the others, the adjuster is tapered or out of round and must be replaced. If the necessary equipment is available, measure the diameter of the lash adjuster and the inside diameter of the corresponding cylinder head bore. Subtract the diameter of the lash adjuster from the bore diameter to obtain the oil clearance. Compare the measurements obtained to those given in this Chapter's Specifications. If the adjusters

or the cylinder head bores are excessively worn, new adjusters or a new cylinder head, or both, may be required. If the valve train is noisy, particularly if the noise persists after a cold start, you can suspect a faulty hydraulic adjuster.

41   Inspect the rocker arms for signs of wear or damage. The areas of wear are the tip that contacts the valve stem, the socket that contacts the lash adjuster and the roller that contacts the camshaft (see illustration).

42   Examine the camshaft lobes for scoring, pitting, galling (wear due to rubbing), and evidence of overheating (blue, discolored areas). Look for flaking of the hardened surface layer of each lobe (see illustration). If any such wear is evident, replace the camshaft.

43   Calculate the camshaft lobe lift by measuring the lobe height and the diameter of the base circle of the lobe (see illustrations). Subtract the base circle measurement from the lobe height to determine the lobe lift. If the lobe lift is less than that listed in this Chapter's Specifications the camshaft lobe is worn and should be replaced.

44   Inspect the camshaft bearing journals and the cylinder head bearing surfaces for pitting or excessive wear. If any such wear is evident, replace the component concerned. Using a micrometer, measure the diameter of each camshaft bearing journal at several points (see illustration). If the diameter of any journal is less than specified, replace the camshaft.

45   To check the bearing journal oil clearance, remove the rocker arms and hydraulic

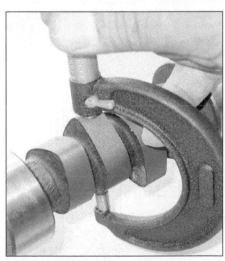

10.43a Measure the camshaft lobe height (greatest dimension) . . .

10.43b . . . and subtract the camshaft lobe base circle diameter (smallest dimension) to obtain the lobe lift specification

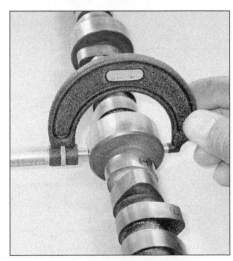

10.44 Measure each journal diameter with a micrometer. If any journal is less than the specified minimum, replace the camshaft

10.45 Lay a strip of Plastigage on each camshaft journal, in line with the camshaft

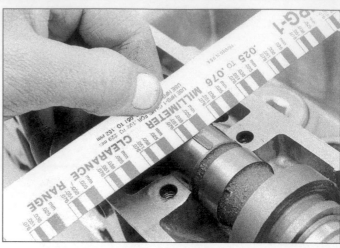

10.46 Compare the width of the crushed Plastigage to the scale on the package to determine the journal oil clearance

lash adjusters (if not already done), use a suitable solvent and a clean lint-free rag to clean all bearing surfaces, then install the camshafts and bearing caps with a piece of Plastigage across each journal (see illustration). Tighten the bearing cap bolts to the specified torque. Don't rotate the camshafts.

46   Remove the bearing caps and measure the width of the flattened Plastigage with the Plastigage scale (see illustration). Scrape off the Plastigage with your fingernail or the edge of a credit card. Don't scratch or nick the journals or bearing caps.

47   If the oil clearance of any bearing is worn beyond the specified service limit, install a new camshaft and repeat the check. If the clearance is still excessive, replace the cylinder head.

48   To check camshaft endplay, remove the hydraulic lash adjusters, clean the bearing surfaces carefully, and install the camshafts and bearing caps. Tighten the bearing cap bolts to the specified torque, then measure the endplay using a dial indicator mounted on the cylinder head so that its tip bears on the camshaft end.

49   Lightly but firmly tap the camshaft fully toward the gauge, zero the gauge, then tap the

camshaft fully away from the gauge and note the gauge reading. If the measured endplay is at or beyond the specified service limit, install a new camshaft thrust cap and repeat the check. If the clearance is still excessive, the camshaft or the cylinder head must be replaced.

Installation

2008 and earlier models

Front bank cylinder head

50   Make sure the crankshaft keyway is at the 11 o'clock position (see illustration 9.2).

51   Lubricate the rocker arms and hydraulic lash adjusters with engine assembly lubricant or fresh engine oil. Install the adjusters into their original bores, then install the rocker arms in their correct locations.

52   Similarly lubricate the camshafts and install them in their correct locations.

53   Install the camshaft bearing caps in their correct locations. Install the cap bolts and tighten by hand until snug. Then, install the camshaft thrust caps and bolts. Tighten the bolts in four to five steps, following the sequence shown (see illustration) to the

torque listed in this Chapter's Specifications.

54   Install the seal retainer and a new seal over the left end of the intake camshaft. Tighten the bolts to the torque listed in this Chapter's Specifications.

55   Install the front timing chain sprocket on the crankshaft. Then install the front timing chain guide, the chain, and the tensioner (see Section 9).

56   Turn the crankshaft clockwise and position the crankshaft keyway at 3 o'clock.

Rear bank cylinder head

57   Lubricate the rocker arms and hydraulic lash adjusters with engine assembly lubricant or fresh engine oil. Install the adjusters into their original bores, then install the rocker arms in their correct locations.

58   Similarly lubricate the camshafts and install them in their correct locations.

59   Install the camshaft bearing caps in their correct locations. Install the cap bolts and tighten by hand until snug. Then, install the camshaft thrust caps and bolts. Tighten the bolts in four to five steps, following the sequence shown (see illustration) to the

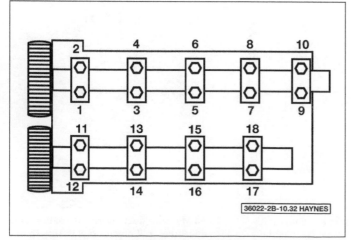

10.53 Front bank camshaft bearing cap tightening sequence - 2008 and earlier models

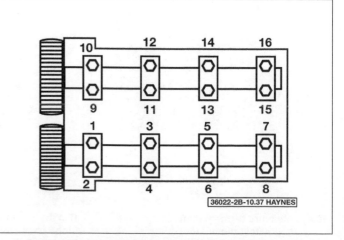

10.59 Rear bank camshaft bearing cap tightening sequence - 2008 and earlier models

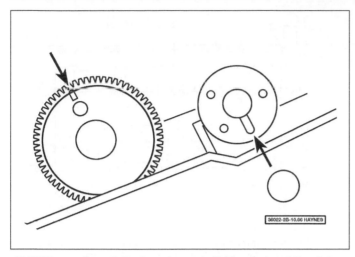

10.65 The mark on the exhaust camshaft sprocket and the slot on the intake camshaft must be aligned as shown before installing the phaser - 2009 through 2012 models

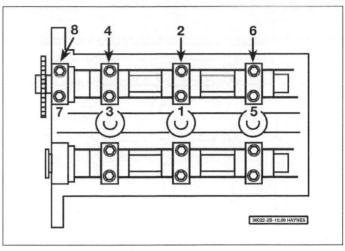

10.68 Exhaust camshaft bolt tightening sequence for the rear cylinder head - 2009 through 2012 models

torque listed in this Chapter's Specifications.

60   Install the rear timing chain, chain guide and tensioner (see Section 9).

61   Install the engine front cover (see Section 8).

62   Reinstall the remaining parts in the reverse order of removal.

63   Refer to Chapter 1 and fill the crankcase with the recommended oil; fill the power steering reservoir with the correct fluid and refill the cooling system.

64   Reconnect the battery, then start the engine and check for leaks. The Powertrain Control Module (PCM) must relearn its idle and fuel trim strategy for optimum driveability and performance (see Chapter 5, Section 1).

2009 through 2012 models

Rear bank cylinder head

65   Lubricate the camshaft bearings with clean oil and carefully install them onto the cylinder head. The dot on the sprocket must be at 12 o'clock and the slot on the intake camshaft must be at 6 o'clock (see illustration).

66   Install the phaser and the sprocket onto the camshafts, aligning the match marks made previously. Tighten the sprocket and phaser bolts to the torque listed in this Chapter's Specifications. Be sure to hold the camshafts by the D slots in the ends as you tighten the bolts.

67   Lubricate the bearing caps and install them in their original positions and facing the original directions. Install the bolts finger tight. Don't install the intake camshaft thrust bearing cap at this time.

68   Tighten the exhaust camshaft bearing cap bolts to the torque listed in this Chapter's Specifications and in the correct sequence (see illustration).

69   Lubricate and install the intake camshaft thrust bearing cap. Tighten all intake camshaft bolts to the torque listed in this Chapter's Specifications and in the proper sequence (see illustration).

70   Again verify that the marks you made earlier on the timing chain and the sprocket and phaser are aligned and that the locking tool hasn't slipped out of place. If the align-

ment isn't correct, refer to Section 9, remove the front cover and set the alignments of all the timing components.

71   When you're sure that all components are properly aligned and all parts are tightened, remove the locking tool from inside the front cover.

72   The remainder of installation is the reverse of removal.

Front bank cylinder head

73   Lubricate the camshaft bearings with clean oil and carefully install them onto the cylinder head. The dot on the sprocket must be at 12 o'clock and the slot on the intake camshaft must also be at 12 o'clock (see illustration).

74   Lubricate the bearing caps and install them in their original positions and facing the original directions. Don't install the thrust bearing caps (the ones closest to the sprocket and phaser) at this time. Install the bolts finger-tight.

75   Install the exhaust thrust bearing cap last. Install the bolts finger-tight. Don't install the intake camshaft thrust bearing cap at this time.

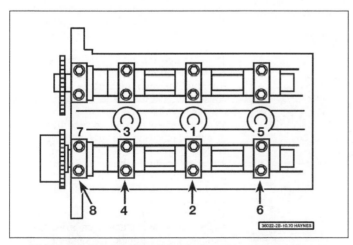

10.69 Intake camshaft bolt tightening sequence for the rear cylinder head - 2009 through 2012 models

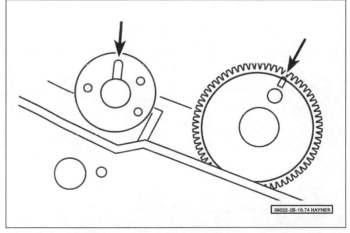

10.73 The mark on the exhaust camshaft sprocket and the slot on the intake camshaft must be aligned as shown before installing the phaser - 2009 through 2012 models

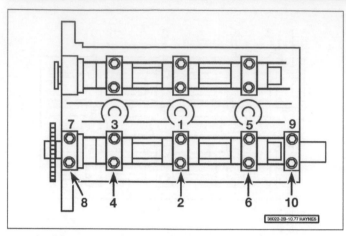

10.76 Exhaust camshaft bolt tightening sequence for the front cylinder head - 2009 through 2012 models

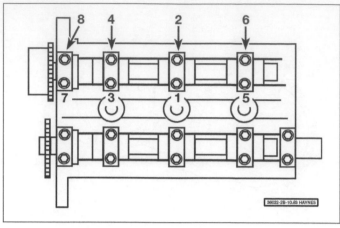

10.82 Intake camshaft bolt tightening sequence for the front cylinder head - 2009 through 2012 models

76   Tighten the exhaust camshaft bearing cap bolts to the torque listed in this Chapter's Specifications and in the correct sequence (see illustration).

77   Install the camshaft oil seal retainer with a new gasket.

78   Install a new camshaft oil seal using manufacturer special tools #303-463, 303-464 and 211-185. These tools can be purchased from dealers and from specialty automotive tool dealers.

79   Install the camshaft phaser and sprocket to the timing chain, sprocket and phaser, aligning the marks you made earlier with those on the timing chain.

80   Secure the camshaft with the ratchet in the D slot in the rear of the camshaft, then tighten the sprocket and phaser bolts to the torque listed in this Chapter's Specifications.

81   Lubricate the thrust bearing cap for the intake camshaft and install it now.

82   Tighten the intake camshaft bearing cap bolts in the proper sequence to the torque listed in this Chapter's Specifications (see illustration).

83   Again verify that the marks you made earlier on the timing chain and the sprocket and phaser are aligned and that the locking tool hasn't slipped out of place. If the align-

ment isn't correct, refer to Section 9, remove the front cover and set the alignments of all the timing components.

84   Remove the timing chain locking tool.

85   Install the manufacturer water pump pulley installation tool #303-458 onto the rear of the camshaft. Place the water pump pulley over the camshaft installation tool, then use manufacturer tools #211-185 and 303-459 to install the water pump pulley until it's flush with the end of the camshaft. Refer to Chapter 3 for additional information.

86   The remainder of installation is the reverse of removal.

11   Cylinder heads - removal and installation

Warning: *Wait until the engine is completely cool before beginning this procedure.*

Note: *The following instructions describe the steps necessary to remove both cylinder heads. If only one cylinder head requires removal, disregard the steps for the other cylinder head. If just the rear cylinder head requires removal, you only need to remove the timing chain, camshafts, and rocker arms from*

the rear cylinder head; the front can remain installed. However, if just the front cylinder head must be removed, you must first remove the rear timing chain for access to the front chain.

Removal

1   Relieve the fuel system pressure (see Chapter 5). Disconnect the cable from the negative battery terminal (see Chapter 5).

2   Drain the cooling system (see Chapter 1).

3   Remove the upper and lower intake manifolds (see Section 5).

4   Drain the engine oil (see Chapter 1). Remove the oil pan (see Section 12).

5   Remove the engine front cover (see Section 8).

6   Remove the timing chains, camshafts, rocker arms and lash adjusters (see Sections 9 and 10).

7   Remove the exhaust manifold(s) (see Section 6).

8   Remove the water pump (see Chapter 3) and the water pump housing (see illustration).

9   If you're working on the front cylinder head of a 2009 through 2012 models, remove the upper radiator support brackets. On all models, remove the oil dipstick and its tube.

10   Disconnect the EGR valve pipe that connects the EGR valve to the rear exhaust manifold.

11   Remove the hoses and any electrical connectors from the coolant bypass tube. Remove the two fasteners securing the coolant bypass tube (see illustration). Then remove the tube.

12   Loosen each cylinder head bolt, a little at a time, following the reverse order of the tightening sequence (see illustration 11.21a). When all cylinder head bolts are loose, remove and discard the bolts. New torque-to-yield cylinder head bolts must be used during installation.

13   Remove the cylinder head from the engine block and place it on a workbench. Remove and discard the cylinder head gasket. **Caution:** *If the cylinder head sticks to the block, pry only on a casting protrusion to prevent damaging the mating surfaces.*

11.8 Remove the water pump housing from the left bank (front) cylinder head

11.11 Remove the coolant bypass tube mounting bolts and remove the tube from the cylinder heads

11.21 Cylinder head bolt TIGHTENING sequence

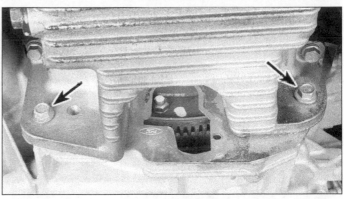

12.8 Remove the bolts that secure the oil pan to the transaxle

Installation

14   The mating surfaces of the cylinder head and the block must be perfectly clean before installing the cylinder head. Clean the surfaces with a scraper, but be careful not to gouge the aluminum.

Caution: *Be very careful when scraping on aluminum engine parts. Aluminum is soft and gouges easily. Severely gouged parts may require replacement.*

15   Check the mating surfaces of the block and the cylinder head for nicks, deep scratches, and other damage. If slight, they can be removed carefully with a file; if excessive, machining may be the only alternative to replacement.

16   If you suspect warpage of the cylinder head gasket surface, use a straightedge to check it for distortion. If the gasket mating surface of your cylinder head or block is out of specification or is severely nicked or scratched, consult an automotive machine shop for advice.

17   Clean the mating surfaces of the cylinder head and block with a clean shop towel and solvent as needed.

18   Ensure that the two locating dowels are in position in the cylinder block and that all cylinder head bolt holes are free of oil, corrosion, or other contamination.

19   Install new cylinder head gaskets on the block, over the locating dowels.

20   Carefully install the cylinder heads. Use caution when lowering the cylinder heads onto the cylinder block to prevent damage to the cylinder heads or block. Make sure the cylinder heads fit properly over the locating dowels in the block.

Note: *The method used for the cylinder head bolt tightening procedure is referred to as the "torque angle" or "torque-to-yield" method; follow the procedure exactly. Tighten the bolts using a torque wrench, then use a breaker bar and a special torque angle adapter (available at auto parts stores) to tighten the bolts the required angle.*

21   Install new cylinder head bolts and turn down by hand until snug. Using a torque wrench and an angle gauge, tighten the cylinder head bolts in the sequence shown to the torque listed in this Chapter's Specifications (see illustration).

Caution: *The cylinder head bolts are the torque-to-yield type and are stretched during tightening. Therefore, the original bolts must be discarded and new bolts installed during assembly.*

22   Install the rest of the parts in the reverse order of removal. Tighten fasteners to the torque values listed in this Chapter's Specifications.

23   Refer to Chapter 1 and replace the oil filter, then fill the engine with fresh engine oil. Also refill the cooling system.

24   Reconnect the battery (see Chapter 5).

25   Start the engine and check for leaks.

Note: *The Powertrain Control Module (PCM) must relearn its idle and fuel trim strategy for optimum driveability and performance.*

12   Oil pan - removal and installation

Removal

1   Disconnect the cable from the negative battery terminal (see Chapter 5).

2   Raise the vehicle and support it securely on jackstands.

3   Drain the engine oil (see Chapter 1). Reinstall the oil drain plug and tighten it to the torque listed in the Chapter 1 Specifications, using a new gasket if necessary.

4   Remove the oil filter (see Chapter 1).

5   Disconnect the oxygen sensor connector(s) (see Chapter 6).

6   Refer to Chapter 4 and remove the crossover and flexible exhaust pipe and converter assembly from the vehicle.

7   Remove the driveplate access cover.

8   Remove the bolts that secure the oil pan to the transaxle (see illustration).

9   Remove the oil pan fasteners and remove the oil pan. Note the location of any stud bolts.

10   Remove and discard the oil pan gasket. If necessary, remove the fasteners that secure the oil screen and pick-up tube and remove the screen and tube assembly.

11   Thoroughly clean the oil pan and cylinder block mating surfaces using brake system cleaner. The surfaces must be free of any residue that will keep the sealant from adhering properly. Clean the oil pan inside and out with solvent and dry with compressed air.

Installation

12   If removed, install a new O-ring seal onto the oil pick-up tube. Install the tube and screen assembly and tighten the retaining fasteners to the torque listed in this Chapter's Specifications. Use a new self-locking nut to secure the pick-up tube support bracket. Tighten the nut to the torque listed in this Chapter's Specifications.

13   Install a new gasket on the oil pan. Apply a 1/8-inch bead of RTV sealant to the oil pan gasket in the area of the engine front cover-to-cylinder block parting line.

14   Install the oil pan, being careful not to dislodge the pan gasket. Install the pan bolts and tighten by hand. Be sure to install any stud bolts in the locations noted during removal. Install the oil pan-to-transaxle bolts. Firmly push the oil pan against the transaxle and tighten the pan-to-transaxle bolts snugly. Then, tighten the oil pan bolts (see illustration) gradually and evenly, to the torque listed in this Chapter's Specifications.

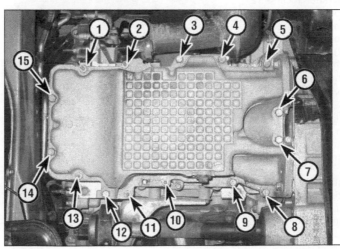

12.14 Oil pan bolt tightening sequence

13.3 Remove the oil pump retaining bolts and pull the oil pump off the crankshaft

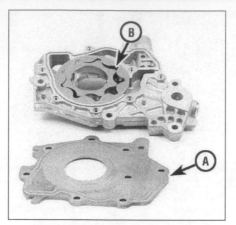

13.5 Remove the pump cover (A) and noting any identification marks on the rotors, remove the inner and outer rotors (B) from the pump body

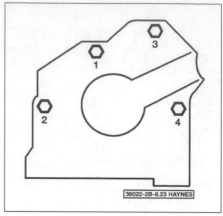

13.11 Oil pump bolt tightening sequence

15   The remainder of installation is the reverse of removal, noting the following items:
   a) *Tighten all fasteners to the torque values listed in this Chapter's Specifications.*
   b) *Always replace any self-locking nuts disturbed on removal.*
   c) *Refer to (Chapter 1) and fill the engine with fresh engine oil. Install a new oil filter.*
   d) *Refill the cooling system (see Chapter 1).*
   e) *Start the engine and check for leaks.*
16   Reconnect the battery (see Chapter 5).
Note: *The Powertrain Control Module (PCM) must relearn its idle and fuel trim strategy for optimum driveability and performance.*

13   Oil pump - removal, inspection and installation

Note: *Because of the difficulty of removing the oil pump, this job is best done as part of complete engine overhaul with the engine out of the vehicle.*

Removal

1   Remove the oil pan (see Section 12). Remove the oil pick-up tube and screen.
2   Remove the engine front cover (see Section 8), timing chains and crankshaft sprockets (see Section 9).
3   Loosen each of the four oil pump mounting bolts one turn. Then, gradually and evenly, loosen each bolt in several steps. When all bolts are loose, remove the bolts and oil pump (see illustration).

Inspection

4   Remove the oil pump cover from the oil pump body.
5   Note any identification marks on the rotors and withdraw the rotors from the pump body (see illustration).
6   Thoroughly clean and dry the components.

7   Inspect the rotors for obvious wear or damage. If either rotor, the pump body or the cover is scored or damaged, the complete oil pump assembly must be replaced.
8   If the oil pump components are in acceptable condition, dip the rotors in clean engine oil and install them into the pump body with any identification marks positioned as noted during disassembly.
9   Install the cover and tighten the screws securely.

Installation

10   Rotate the oil pump inner rotor so it aligns with the flats on the crankshaft. Install the oil pump over the crankshaft and fit it firmly against the cylinder block.
11   Install the oil pump bolts and tighten by hand until snug. Tighten the bolts gradually and evenly, in sequence, to the torque listed this Chapter's Specifications (see illustration).
12   Install the remainder of the components in the reverse order of removal.
13   Refer to Chapter 1 and fill the engine with fresh engine oil. Install a new oil filter. Refill the cooling system.

14   Start the engine and check for leaks.

14   Rear main oil seal - replacement

1   Refer to Chapter 2A for the crankshaft rear oil seal replacement procedure.

15   Driveplate - removal and installation

1   This procedure is essentially the same as the flywheel/driveplate removal procedure for the four-cylinder engine. Refer to Chapter 2A and follow the procedure outlined there. However, use the bolt torque listed in this Chapter's Specifications.

16   Engine mounts - inspection and replacement

1   This procedure is essentially the same as for the four cylinder engine. Refer to Chapter 2A and follow the procedure outlined there but refer to the illustrations in this Section (see illustrations).

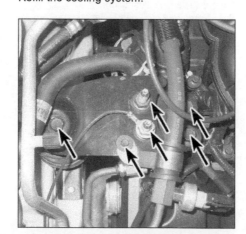

16.1a Remove the engine mount bracket bolts and separate the bracket from the engine and insulator

16.1b Remove the engine mount insulator bolts and separate the insulator from the engine compartment

Chapter 2 Part C
General engine overhaul procedures

Contents

Specifications

General

Displacement
  Four-cylinder models
    1.5L.. 91.54 cubic inches
    1.6L .... 97.64 cubic inches
    2.0L (2004 and earlier models)....................................... 121 cubic inches
    2.0L (2013 and later models)... 122 cubic inches
    2.3L.. 140 cubic inches
    2.5L.. 153 cubic inches
  V6 models
    3.0L.. 182 cubic inches
Bore and Stroke
  Four-cylinder models
    1.5L.. 3.10 x 3.0 inches (79.015 x 76.4 mm)
    1.6L .... 3.10 x 3.2 inches (79.01 x 81.4 mm)
    2.0L.. 3.34 x 3.46 inches (84.8 x 88.0 mm)
    2.3L.. 3.44 x 3.70 inches (87.5 x 94.0 mm)
    2.5L.. 3.50 x 3.94 inches (89.0 x 100.0 mm)
  V6 models .... 3.50 x 3.13 inches (89.0 x 79.5 mm)
Cylinder compression.. Lowest cylinder must be within 75 percent of highest cylinder
Oil pressure (at operating temperature)
  Four-cylinder models
    1.5L and 1.6L engines
      Hot @ 800 rpm.. 14.5 psi (100kPa)
      Hot @ 2000 rpm.. 29.0 psi (200kPa)
    2.0L, 2.3L and 2.5L engines
      2012 and earlier models.. 29 to 39 psi (200 to 268 kPa) @ 2000 rpm
      2013 and later models... 29 to 60 psi (200 to 414 kPa) @ 2000 rpm
  V6 models .... 29 to 45 psi (200 to 310 kPa) @ 2,000 rpm
Balance shaft assembly backlash (all 2006 and later four-cylinder
  engines except 1.5L and 1.6L models).. 0.00078 to 0.00470 inch (0.020 to 0.120 mm)

Torque specifications

	Ft-lbs (unless otherwise indicated)	Nm

Note: *One foot-pound (ft-lb) of torque is equivalent to 12 inch-pounds (in-lbs) of torque. Torque values below approximately 15 ft-lbs are expressed in inch-pounds, since most foot-pound torque wrenches are not accurate at these smaller values.*

	Ft-lbs (unless otherwise indicated)	Nm
Balance shaft assembly mounting bolts (all 2006 and later four-cylinder engines except 1.5L and 1.6L models)		
Step 1	18	24
Step 2	31	42
Connecting rod bearing cap bolts*		
Four-cylinder models		
2004 and earlier models		
Step 1	26	35
Step 2	Tighten an additional 90 degrees	
2005 and later 2.0L, 2.3L and 2.5L models		
Step 1	21	29
Step 2	Tighten an additional 90 degrees	
1.5L and 1.6L models		
Step 1	159 in-lbs	18
Step 2	Tighten an additional 45 degrees	
Step 3	Tighten an additional 45 degrees	
V6 models		
Step 1	17	23
Step 2	32	43
Main bearing cap or beam bolts (four-cylinder models) (in sequence - see illustration 10.19a)		
2004 and earlier models*		
Step 1	18	25
Step 2	Tighten an additional 60 degrees	
2005 and later 2.0L, 2.3L and 2.5L models*		
Step 1	44 in-lbs	5
Step 2	18	25
Step 3	Tighten an additional 90 degrees	
1.5L and 1.6L models*		
Step 1	22	30
Step 2	37	50
Step 3	Tighten an additional 45 degrees	
Step 4	Tighten an additional 45 degrees	
Main bearing beam bolts* (V6 models) (in sequence - see illustration 10.19b)		
Step 1 (bolts 1 through 8)..	18	25
Step 2 (bolts 9 through 16)..	30	40
Step 3 (bolts 1 through 16)..	Tighten an additional 90 degrees	
Step 4 (bolts 17 through 22)..	18	25

*Use NEW bolts

1   General information - engine overhaul

1   Included in this portion of Chapter 2 are general information and diagnostic testing procedures for determining the overall mechanical condition of your engine.

2   The information ranges from advice concerning preparation for an overhaul and the purchase of replacement parts and/or components to detailed, step-by-step procedures covering removal and installation.

3   The following Sections have been written to help you determine whether your engine needs to be overhauled and how to remove and install it once you've determined it needs to be rebuilt. For information concerning in-vehicle engine repair, see Chapter 2A or Chapter 2B.

4   The Specifications included in this Part are general in nature and include only those necessary for testing the oil pressure and checking the engine compression. Refer to Chapter 2A or Chapter 2B for additional engine Specifications.

5   It's not always easy to determine when, or if, an engine should be completely overhauled, because a number of factors must be considered.

6   High mileage is not necessarily an indication that an overhaul is needed, while low mileage doesn't preclude the need for an overhaul. Frequency of servicing is probably the most important consideration.

7   Excessive oil consumption is an indication that piston rings, valve seals and/or valve guides are in need of attention. Perform a cylinder compression check to determine the extent of the work required (see Section 3). Also check the vacuum readings under various conditions (see Section 4).

8   Check the oil pressure with a gauge installed in place of the oil pressure sending unit and compare it to this Chapter's Specifications (see Section 2). If it's extremely low, the bearings and/or oil pump are probably worn out.

9   Loss of power, rough running, knocking or metallic engine noises, excessive valve train noise and high fuel consumption rates may also point to the need for an overhaul, especially if they're all present at the same time.

10   An engine overhaul involves restoring the internal parts to the specifications of a new engine. During an overhaul, the piston rings are replaced and the cylinder walls are reconditioned (rebored and/or honed). If a rebore is done by an automotive machine shop, new oversize pistons will also be installed. The main bearings, connecting rod bearings and camshaft bearings are generally replaced with new ones and, if necessary, the crankshaft may be reground to restore the journals. Generally, the valves are serviced as well, since they're usually in less-than-perfect condition at this point.

Note: *Critical cooling system components such as the hoses, drivebelts, thermostat and water pump should be replaced with new parts when an engine is overhauled. The radiator should be checked carefully to ensure that it isn't clogged or leaking (see Chapter 3).*

11   Overhauling the internal components on today's engines is a difficult and time-consuming task which requires a significant amount of specialty tools and is best left to a professional engine rebuilder. A competent engine rebuilder will handle the inspection of your old parts and offer advice concerning the reconditioning or replacement of the original engine, never purchase parts or have machine work done on other components until the block has been thoroughly inspected by a professional machine shop.

2.2a On four-cylinder engines the oil pressure sending unit is located near the oil filter

2.2b On V6 engines the oil pressure sending unit is located on the front of the engine block behind the air conditioning compressor

3.6 Use a compression gauge with a threaded fitting for the spark plug hole, not the type that requires hand pressure to maintain the seal

2 Oil pressure check

1 Low engine oil pressure can be a sign of an engine in need of rebuilding. A "low oil pressure" indicator (often called an "idiot light") is not a test of the oiling system. Such indicators only come on when the oil pressure is dangerously low. Even a factory oil pressure gauge in the instrument panel is only a relative indication, although much better for driver information than a warning light. A better test is with a mechanical (not electrical) oil pressure gauge.

2 Locate the oil pressure sending unit on the engine block:

a) On 1.5L and 1.6L EcoBoost models, the oil pressure sending unit is located on the front side of the engine block, above the oil filter (see illustration).

b) On 2.0L Zetec models, the oil pressure sending unit is located on the rear side of the engine block, near the oil filter adaptor.

c) On 2.0L EcoBoost and 2.3L models, the oil pressure sending unit is located on the front side of the engine, on the oil filter adapter.

d) On 2.5L models, the oil pressure sending unit is located on the rear side of the engine, on the oil filter adapter.

e) On V6 engines, the sending unit is located near the oil filter housing behind the air conditioning compressor, on the front side of the block (see illustration).

3 On 1.5L and 1.6L engines, remove the charge air cooler pipe (see Chapter 4) and the oil filter (see Chapter 1) to access the oil pressure sending unit. Reinstall the filter and the charge air cooler pipe after the gauge has been installed.

Note: *The manufacturer recommends replacing the oil filter once it has been removed and reinstalled for the oil pressure test.*

4 Unscrew the oil pressure sending unit and screw in the hose for your oil pressure gauge. If necessary, install an adapter fitting. Use Teflon tape or thread sealant on the threads of the adapter and/or the fitting on the end of your gauge's hose.

5 Connect an accurate tachometer to the engine, according to the tachometer manufacturer's instructions.

6 Check the oil pressure with the engine running (normal operating temperature) at the specified engine speed, and compare it to this Chapter's Specifications. If it's extremely low, the bearings and/or oil pump are probably worn out.

3 Cylinder compression check

1 A compression check will tell you what mechanical condition the upper end of your engine (pistons, rings, valves, head gaskets) is in. Specifically, it can tell you if the compression is down due to leakage caused by worn piston rings, defective valves and seats or a blown head gasket.

Note: *The engine must be at normal operating temperature and the battery must be fully charged for this check.*

2 Begin by cleaning the area around the spark plugs before you remove them (compressed air should be used, if available). The idea is to prevent dirt from getting into the cylinders as the compression check is being done.

3 Remove all of the spark plugs from the engine (see Chapter 1).

4 On 2008 and earlier models, block the throttle wide open.

5 On models with a coil pack and spark plug wires, disconnect the primary (low voltage) wires from the coil pack (see Chapter 5). Remove the fuel pump relay (see Chapter 4). The relays are located in the power distribution center in the engine compartment.

6 Install a compression gauge in the spark plug hole (see illustration).

7 On 2009 and later models, have an assistant depress the accelerator and hold it to the floor during the next Step.

8 Crank the engine over at least seven compression strokes and watch the gauge. The compression should build up quickly in a

healthy engine. Low compression on the first stroke, followed by gradually increasing pressure on successive strokes, indicates worn piston rings. A low compression reading on the first stroke, which doesn't build up during successive strokes, indicates leaking valves or a blown head gasket (a cracked head could also be the cause). Deposits on the undersides of the valve heads can also cause low compression. Record the highest gauge reading obtained.

9 Repeat the procedure for the remaining cylinders and compare the results to this Chapter's Specifications.

10 Add some engine oil (about three squirts from a plunger-type oil can) to each cylinder, through the spark plug hole, and repeat the test.

11 If the compression increases after the oil is added, the piston rings are definitely worn. If the compression doesn't increase significantly, the leakage is occurring at the valves or head gasket. Leakage past the valves may be caused by burned valve seats and/or faces or warped, cracked or bent valves.

12 If two adjacent cylinders have equally low compression, there's a strong possibility that the head gasket between them is blown. The appearance of coolant in the combustion chambers or the crankcase would verify this condition.

13 If one cylinder is slightly lower than the others, and the engine has a slightly rough idle, a worn lobe on the camshaft could be the cause.

14 If the compression is unusually high, the combustion chambers are probably coated with carbon deposits. If that's the case, the cylinder head(s) should be removed and decarbonized.

15 If compression is way down or varies greatly between cylinders, it would be a good idea to have a leak-down test performed by an automotive repair shop. This test will pinpoint exactly where the leakage is occurring and how severe it is.

16 On 2008 and earlier models, don't forget to unblock the throttle.

4   Vacuum gauge diagnostic checks

1   A vacuum gauge provides inexpensive but valuable information about what is going on in the engine. You can check for worn rings or cylinder walls, leaking head or intake manifold gaskets, incorrect carburetor adjustments, restricted exhaust, stuck or burned valves, weak valve springs, improper ignition or valve timing and ignition problems.

2   Unfortunately, vacuum gauge readings are easy to misinterpret, so they should be used in conjunction with other tests to confirm the diagnosis.

3   Both the absolute readings and the rate of needle movement are important for accurate interpretation. Most gauges measure vacuum in inches of mercury (in-Hg). The following references to vacuum assume the diagnosis is being performed at sea level. As elevation increases (or atmospheric pressure decreases), the reading will decrease. For every 1,000 foot increase in elevation above approximately 2,000 feet, the gauge readings will decrease about one inch of mercury.

4   Connect the vacuum gauge directly to the intake manifold vacuum, not to ported (throttle body) vacuum. Be sure no hoses are left disconnected during the test or false readings will result.

5   Before you begin the test, allow the engine to warm up completely. Block the wheels and set the parking brake. With the transaxle in Park, start the engine and allow it to run at normal idle speed.

Warning: *Keep your hands and the vacuum gauge clear of the fans and drivebelt.*

6   Read the vacuum gauge; an average, healthy engine should normally produce about 17 to 22 in-Hg with a fairly steady needle (see illustration). Refer to the following vacuum gauge readings and what they indicate about the engine's condition:

7   A low steady reading usually indicates a leaking gasket between the intake manifold and cylinder head(s) or throttle body, a leaky vacuum hose, late ignition timing or incorrect camshaft timing. Check ignition timing with a timing light and eliminate all other possible causes, utilizing the tests provided in this Chapter before you remove the timing chain cover to check the timing marks.

8   If the reading is three to eight inches below normal and it fluctuates at that low reading, suspect an intake manifold gasket leak at an intake port or a faulty fuel injector.

9   If the needle has regular drops of about two-to-four inches at a steady rate, the valves are probably leaking. Perform a compression check or leak-down test to confirm this.

10   An irregular drop or down-flick of the needle can be caused by a sticking valve or an ignition misfire. Perform a compression check or leak-down test and read the spark plugs.

11   A rapid vibration of about four in-Hg vibration at idle combined with exhaust smoke indicates worn valve guides. Perform a leak-down test to confirm this. If the rapid vibration occurs with an increase in engine speed,

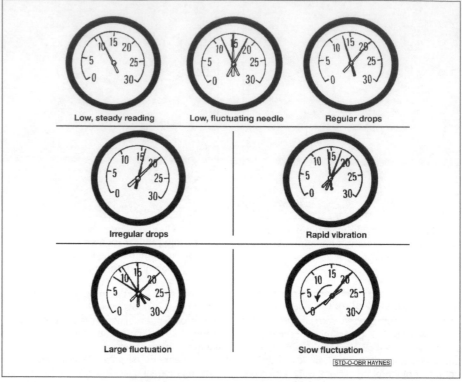

Low, steady reading   Low, fluctuating needle   Regular drops

Irregular drops   Rapid vibration

Large fluctuation   Slow fluctuation

STD-O-OBR HAYNES

4.6 Typical vacuum gauge readings

check for a leaking intake manifold gasket or head gasket, weak valve springs, burned valves or ignition misfire.

12   A slight fluctuation, say one inch up and down, may mean ignition problems. Check all the usual tune-up items and, if necessary, run the engine on an ignition analyzer.

13   If there is a large fluctuation, perform a compression or leak-down test to look for a weak or dead cylinder or a blown head gasket.

14   If the needle moves slowly through a wide range, check for a clogged PCV system, incorrect idle fuel mixture, throttle body or intake manifold gasket leaks.

15   Check for a slow return after revving the engine by quickly snapping the throttle open until the engine reaches about 2,500 rpm and let it shut. Normally the reading should drop to near zero, rise above normal idle reading (about 5 in-Hg over) and then return to the previous idle reading. If the vacuum returns slowly and doesn't peak when the throttle is snapped shut, the rings may be worn. If there is a long delay, look for a restricted exhaust system (often the muffler or catalytic converter). An easy way to check this is to temporarily disconnect the exhaust ahead of the suspected part and redo the test.

5   Engine rebuilding alternatives

1   The do-it-yourselfer is faced with a number of options when purchasing a rebuilt engine. The major considerations are cost, warranty, parts availability and the time required for the rebuilder to complete the proj-

ect. The decision to replace the engine block, piston/connecting rod assemblies and crankshaft depends on the final inspection results of your engine. Only then can you make a cost effective decision whether to have your engine overhauled or simply purchase an exchange engine for your vehicle.

2   Some of the rebuilding alternatives include:

Individual parts - If the inspection procedures reveal that the engine block and most engine components are in reusable condition, purchasing individual parts and having a rebuilder rebuild your engine may be the most economical alternative. The block, crankshaft and piston/connecting rod assemblies should all be inspected carefully by a machine shop first.

Short block - A short block consists of an engine block with a crankshaft and piston/connecting rod assemblies already installed. All new bearings are incorporated and all clearances will be correct. The existing camshafts, valve train components, cylinder head and external parts can be bolted to the short block with little or no machine shop work necessary.

Long block - A long block consists of a short block plus an oil pump, oil pan, cylinder head, valve cover, camshaft and valve train components, timing sprockets and chain or gears and timing cover. All components are installed with new bearings, seals and gaskets incorporated throughout. The installation of manifolds and external parts is all that's necessary.

Low mileage used engines - Some companies now offer low mileage used

6.7a Get an engine stand sturdy enough to firmly support the engine while you're working on it. Stay away from three-wheeled models; they have a tendency to tip over more easily, so get a four-wheeled unit

6.7b A clutch alignment tool is necessary if you plan to install a rebuilt engine mated to a manual transaxle

engines which is a very cost effective way to get your vehicle up and running again. These engines often come from vehicles which have been in totaled in accidents or come from other countries which have a higher vehicle turn over rate. A low mileage used engine also usually has a similar warranty like the newly remanufactured engines.

3   Give careful thought to which alternative is best for you and discuss the situation with local automotive machine shops, auto parts dealers and experienced rebuilders before ordering or purchasing replacement parts.

6   Engine removal - methods and precautions

1   If you've decided that an engine must be removed for overhaul or major repair work, several preliminary steps should be taken. Read all removal and installation procedures carefully prior to committing to this job.

2   Locating a suitable place to work is extremely important. Adequate work space, along with storage space for the vehicle, will be needed. If a shop or garage isn't available, at the very least a flat, level, clean work surface made of concrete or asphalt is required.

3   Cleaning the engine compartment and engine before beginning the removal procedure will help keep tools clean and organized.

4   An engine hoist will be necessary. Make sure the hoist is rated in excess of the combined weight of the engine and transaxle. Safety is of primary importance, considering the potential hazards involved in removing the engine from the vehicle.

5   A vehicle hoist will also be necessary for engine removal on all except 2004 and earlier four-cylinder models, since the subframe must be removed and the engine/transaxle assembly must be lowered from the engine compartment, then the vehicle is raised and the powertrain unit is removed from under

the vehicle. If the necessary equipment is not available, the engine will have to be removed by a qualified automotive repair facility.

6   If you're a novice at engine removal, get at least one helper. One person cannot easily do all the things you need to do to remove a big heavy engine and transaxle assembly from the engine compartment. Also helpful is to seek advice and assistance from someone who's experienced in engine removal.

7   Plan the operation ahead of time. Arrange for or obtain all of the tools and equipment you'll need prior to beginning the job (see illustrations). Some of the equipment necessary to perform engine removal and installation safely and with relative ease are (in addition to a vehicle hoist and an engine hoist) a heavy duty floor jack (preferably fitted with a transmission jack head adapter), complete sets of wrenches and sockets as described in the front of this manual, wooden blocks, plenty of rags and cleaning solvent for mopping up spilled oil, coolant and gasoline.

8   Plan for the vehicle to be out of use for quite a while. A machine shop can do the work that is beyond the scope of the home mechanic. Machine shops often have a busy schedule, so before removing the engine, consult the shop for an estimate of how long it will take to rebuild or repair the components that may need work.

7   Engine - removal and installation

Warning: *Gasoline is extremely flammable, so take extra precautions when you work on any part of the fuel system. Don't smoke or allow open flames or bare light bulbs near the work area, and don't work in a garage where a gas-type appliance (such as a water heater or clothes dryer) is present. Since gasoline is carcinogenic, wear fuel-resistant gloves when there's a possibility of being exposed to fuel, and, if you spill any fuel on your skin, rinse it off immediately with soap and water. Mop up any spills immediately and do not store fuel-soaked rags where they could ignite. The*

fuel system is under constant pressure, so, if any fuel lines are to be disconnected, the fuel pressure in the system must be relieved first (see Chapter 4 for more information). When you perform any kind of work on the fuel system, wear safety glasses and have a Class B type fire extinguisher on hand.
Warning: *The air conditioning system is under high pressure. DO NOT loosen any fittings or remove any components until after the system has been discharged. Air conditioning refrigerant must be properly discharged into an EPA-approved container at a dealer service department or an automotive air conditioning repair facility. Always wear eye protection when disconnecting air conditioning system fittings.*
Warning: *The engine must be completely cool before beginning this procedure.*

Removal

All engines
Note: *Except for 2001 through 2004 four-cylinder models, engine removal is a difficult and potentially dangerous job, especially for the do-it-yourself mechanic working at home. Because of the design of the vehicles, the manufacturer states that the engine and transaxle have to be lowered as a single assembly from the underside of the vehicle, not lifted out the top. With a floor jack and jackstands, the vehicle simply cannot be raised high enough and supported safely enough for the engine/transaxle assembly to slide out from underneath. Instead, the manufacturer states that removal of the engine/transaxle assembly must be performed with the vehicle on a vehicle hoist.*

1   Have the air conditioning system discharged and recovered by a licensed air conditioning technician.

2   Relieve the fuel system pressure (see Chapter 4).

3   Disconnect the cable from the negative battery terminal (see Chapter 5).

4   Remove the engine cover and the hood (see Chapter 11).

5   Remove the air intake duct and air filter housing (see Chapter 4).

6   Remove the snow shield (if equipped) from the accelerator cable bracket. Disconnect the accelerator cable and, if equipped, the cruise control cable. Detach the accelerator/cruise control cable bracket from the engine and position the bracket and the cable(s) aside (see Chapter 6).

7   Disconnect the fuel supply line fitting at the fuel rail, then plug the supply line and fuel rail sides of the fitting. Disconnect the vacuum line from the fuel pulsation damper. Disconnect the electrical connectors from the fuel injectors, detach any harness retainers and set the harness aside.

8   Remove the battery and the battery tray (see Chapter 5).

9   Remove the cover from the fuse box, disconnect the two battery cables and disconnect the electrical connector from the underside of the box. Disconnect the ground strap and disconnect the large electrical connector near the box.

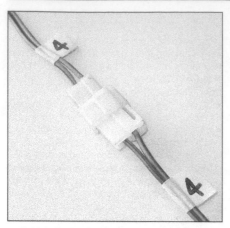

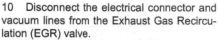

7.14 Label both ends of each wire and hose before disconnecting it

7.39 The engine hoist chains have been connected to the engine mount bracket at the right end (front) of the engine and to the lifting hook at the left end of the engine (V6 engine shown)

10   Disconnect the electrical connector and vacuum lines from the Exhaust Gas Recirculation (EGR) valve.

11   Disconnect the power brake booster vacuum line and the vacuum reservoir line.

12   Disconnect the electrical connectors and ground wire from the PCM (see Chapter 6).

13   Disconnect the large electrical connector that connects the main engine wiring harness to the vehicle harness.

14   Clearly label and disconnect any other vacuum lines, emissions hoses, wiring harness connectors, harness or connector retainer clips, ground straps and/or ground wires and fuel lines that are routed between the vehicle and the engine. Label both ends of each connector and/or harness with electrical tape or numbered labels (see illustration). Or, take photos or sketch the locations of components and brackets.

Note: *It's not necessary to disconnect every single connector, harness, harness retainer, vacuum line, emission control hose, etc., at this time. Harnesses and connectors that are routed between components on the engine itself or between the engine and the transaxle can be disconnected and removed after the engine/transaxle assembly is removed from the engine compartment.*

15   Loosen the front wheel lug nuts. Raise the vehicle and place it securely on jackstands. Remove the front wheels. Remove all engine under-covers.

16   Drain the engine oil and remove the oil filter (see Chapter 1).

17   Drain the cooling system (see Chapter 1), then disconnect and remove the upper and lower radiator hoses from the radiator, the two heater hoses from the firewall, the coolant expansion tank on 1.5L and 1.6L engines (see Chapter 3) and the coolant hoses from the throttle body, if applicable.

18   Drain the transaxle fluid (see Chapter 1). Use a new washer when installing the transaxle drain plug.

19   If equipped, remove the lateral support crossmember bolts and remove the lateral support crossmember. Remove the side frame support bolts and supports from each side of the frame (see Chapter 10).

20   Remove the front exhaust pipe (see Chapter 4) and the catalytic converter(s) (see Chapter 6).

Note: *On some models you might not be able to remove the downpipe or the intermediate pipe until after you have removed the lateral support crossmember.*

21   Disconnect the power steering pump hose support bracket bolts and the upper pump mounting bracket bolts (see Chapter 10).

22   If the vehicle is equipped with cruise control, remove the cruise control unit mounting bolts and set the cruise control unit aside.

23   Remove the accessory drivebelt (see Chapter 1).

24   On 2WD models, disconnect the shift linkage rods from the transaxle; on 4WD models, disconnect the shifter cables from the transaxle (see Chapter 7C).

2004 and earlier four-cylinder models

25   Locate the small support bracket on the lower back side of the engine, then remove the two nuts and remove the bracket.

26   Remove the right front lower splash shield bolts and remove the splash shield.

27   Loosen the coolant hose clamps and disconnect and remove the coolant hoses.

28   Disconnect the electrical connector from the air conditioning compressor clutch field coil. Disconnect both air conditioning lines from the compressor and plug both lines. Remove the air conditioning compressor (see Chapter 3).

29   Remove and discard both front axle wheel hub nuts (see Chapter 8).

30   Separate the lower control arms from the steering knuckles (see Chapter 10).

31   Disengage the outer end of each driveaxle assembly from the steering knuckle and the inner end of each driveaxle from the transaxle and remove the driveaxles, then remove the intermediate shaft bracket and the intermediate shaft as a single assembly (see Chapter 8).

32   Remove the bolt and nuts that secure the two shift linkage rods to the transaxle (see Chapter 7B).

33   Disconnect the block heater electrical connector (if applicable).

34   Remove the front transaxle mount through-bolt.

35   On 4WD models, remove the transfer case (see Chapter 7C).

36   Remove the engine-to-transaxle bolts that are accessible from below.

37   Disconnect the electrical connector from the power steering pressure sensor on the power steering pump, then remove the power steering hose bracket bolts. Remove the power steering pump mounting bolts and position the pump and power steering hoses out of the way (it's not necessary to disconnect the hoses from the pump or from the steering gear).

38   Remove the clip that secures the clutch release cylinder line to the line mounting bracket, then remove the clutch release cylinder mounting bolts and set the release cylinder and the clutch hydraulic line aside.

39   Support the engine with an engine hoist (see illustration). Roll the engine hoist into position and attach it to the engine with a couple pieces of heavy-duty chain. If the engine is equipped with lifting brackets, use them. If not, you'll have to fasten the chain to some substantial part of the engine - one that is strong enough to take the weight, but in a location that will provide good balance. If you're attaching the chain to a stud on the engine, or are using a bolt passing through the chain and into a threaded hole, place a washer between the nut or bolt head and the chain, and tighten the nut or bolt securely. Take up the slack in the chain, but don't lift the engine.

Warning: *DO NOT place any part of your body under the engine when it's supported only by a hoist or other lifting device.*

40   Raise the engine slightly and then inspect it thoroughly once more to make sure that nothing is still attached, then slowly raise the engine out of the engine compartment. Check carefully to make sure nothing is hanging up.

Warning: *When raising the engine, never place any part of your body underneath it or between the engine and the engine compartment.*

41   Remove the two bolts and two nuts that secure the rear transaxle mounting bracket and remove the mounting bracket.

42   Remove the five bolts that secure the front support insulator and mounting bracket and remove the insulator and mounting bracket.

43   Remove the five bolts that secure the left transaxle mounting bracket and remove the mounting bracket.

44   Remove the two nuts and bolt that secure the right upper engine mounting bracket and remove the mounting bracket. Note that the nut closer to the crank pulley also secures a ground wire. Don't forget to install this ground wire during reassembly.

45   Verify that nothing is still connecting the engine to the vehicle. If you find anything else, disconnect or remove it as necessary.

46   Carefully remove the engine and transaxle as a single assembly from the engine compartment.

47   Disconnect the electrical connectors from the alternator (see Chapter 5), the oxygen sensor, the Vehicle Speed Sensor (VSS), the Crankshaft Position (CKP) sensor and the EGR differential pressure feedback sensor (see Chapter 6), the ignition coil and the starter motor (see Chapter 5) and the back-up light switch. Disconnect the ground wires from the bellhousing and from the top of the transaxle. Detach the wiring harnesses for all of these components from any harness retainers, then remove and set aside the entire harness.

48   Remove the starter mounting bolts and remove the starter motor (see Chapter 5).

49   Remove the rest of the engine-to-transaxle bolts and remove the transaxle from the engine (see Chapter 7B).

50   Remove the flywheel (see Chapter 2A) and mount the engine on an engine stand (see illustration 6.3).

51   Inspect the engine and transaxle mounts (see Chapter 2A). If they're worn or damaged, replace them.

2005 and later four-cylinder models

Note: *Engine removal on these models is a difficult and potentially dangerous job, especially for a do-it-yourself mechanic working at home. Because of the vehicle's design, the manufacturer states that the engine and transaxle have to be lowered as a single assembly from the underside of the vehicle, not lifted out the top. With a floor jack and jackstands, the vehicle simply cannot be raised high enough and supported safely enough for the engine/transaxle assembly to slide out from underneath. Instead, the manufacturer states that removal of the engine/transaxle assembly must be performed with the vehicle on a vehicle hoist.*

52   On 2013 and later models, remove the cowl panels (see Chapter 11).

53   Remove the four lateral support member bolts and remove the lateral support member from the sub-frame.

54   Remove the driveaxles and the intermediate shaft (see Chapter 8).

55   On 4WD models, disconnect the driveshaft from the transfer case (see Chap-

ter 8), secure the driveshaft out of the way and remove the transfer case (see Chapter 7C).

56   On 2005 through 2008 models, press the two locking tabs to release the lower air intake duct from the upper air intake duct, then remove the alternator shield bolts and remove the alternator shield.

57   On 2009 through 2012 models, depress the locking tabs to release the alternator air duct and remove the duct.

58   Remove the air filter housing (see Chapter 4).

59   If applicable, unbolt and disconnect the ground wire that's routed from the engine to the subframe, under the air conditioning compressor.

60   Remove the cover from the fuse box and disconnect the two battery cables, the electrical connector from the underside of the box, the ground strap and the large electrical connector.

61   Detach the wiring harness retainers from the battery tray support bracket and set the harness aside.

62   Remove the coolant expansion tank (see Chapter 3).

63   On models with a manual transaxle, disconnect the clutch hydraulic line fitting, plug the line and the clutch release cylinder, disengage the hydraulic line from the spring clip and set the line aside. Remove the retaining clips and disconnect both transaxle control cables.

64   On models with an automatic transaxle, disconnect the transaxle electrical connector, disconnect the shift cable from the transaxle manual lever, detach the wiring harness retainer, remove the cable support bracket bolts and set the transaxle control cable and brackets aside. Disconnect the electrical connector from the Transmission Range (TR) sensor, detach the transaxle control harness retainers and set the harness aside.

65   On models with a manual transaxle, disconnect the Vehicle Speed Sensor (VSS) electrical connector, detach the retainer and set the VSS harness aside.

66   On models with an automatic transaxle, disconnect the fluid cooler tube and position it aside. Disconnect the Output Shaft Speed (OSS) sensor and the Turbine Shaft Speed (TSS) electrical connectors, remove the transaxle fluid cooler support bracket bolt and set the fluid cooler tube aside.

67   On models with a manual transaxle, disconnect the back-up light indicator switch electrical connector, detach the retainer and set the harness aside.

68   On models with an automatic transaxle, remove the OSS mounting bolt and remove the OSS.

69   On models with an automatic transaxle, detach the transaxle control harness from its retainer clip.

70   Disconnect the electrical connector from the block heater, if equipped. Detach the block heater harness retainers and set the harness aside.

71   Remove the radiator hoses retaining clips and disconnect the hoses then turn the the heater hose connectors while pushing

inwards on the connectors and disconnect the heater hoses.

72   On turbocharged models, disconnect and remove the charge air cooler tubes (see Chapter 4).

73   On 2013 and later models, remove the PCM (see Chapter 6).

74   Loosen the hose clamps and disconnect the upper radiator and coolant vent hoses. Remove the coolant vent hose support bracket retaining nuts, detach the brackets and set the coolant vent hose aside.

75   Disconnect the upper end of the heater hose support strap from the firewall, then disconnect the two heater hoses from the firewall.

76   Remove the nut from the power steering line support bracket, detach the bracket, and set the power steering line and bracket aside.

77   Disconnect the vacuum supply tube, the fuel vapor return tube and the vacuum reservoir tube and set them aside.

78   Remove the catalytic converter (see Chapter 6).

79   Remove the two upper power steering pump mounting bolts, then disconnect the lower radiator hose from the radiator and the air conditioning compressor electrical connector. Remove the four compressor mounting bolts, support the compressor with coat hanger wire and disconnect the Power Steering Pressure (PSP) sensor electrical connector. Remove the two lower power steering pump mounting bolts and remove the power steering pump.

80   Support the engine and transaxle from above with an engine hoist (see illustration 7.39). Roll the engine hoist into position and attach it to the engine with a couple pieces of heavy-duty chain. If the engine is equipped with lifting brackets, use them. If not, you'll have to fasten the chain to some substantial part of the engine - one that is strong enough to take the weight, but in a location that will provide good balance. If you're attaching the chain to a stud on the engine, or are using a bolt passing through the chain and into a threaded hole, place a washer between the nut or bolt head and the chain, and tighten the nut or bolt securely. Take up the slack in the chain, but don't lift the engine.

Warning: *DO NOT place any part of your body under the engine when it's supported only by a hoist or other lifting device.*

Note: *The chains must be long enough to allow the engine hoist to lower the engine and transaxle to the floor without the hoist arm contacting the vehicle.*

81   Remove the front engine mount through-bolt and the two bolts for the longitudinal engine support crossmember. Remove the rear nut and the engine support crossmember and remove the crossmember.

82   Remove the four lower transaxle-to-engine bolts.

Note: *If the bolts are different lengths, mark them to ensure that they are correctly reinstalled.*

83   Lower the engine and transaxle to the floor as a single assembly, then disconnect the hoist from the chains. Roll the hoist out of the way.

7.88 Remove the torque converter fasteners through the access hole under the transaxle driveplate - 2005 through 2012 models

84   Raise the vehicle until it clears the engine/transaxle assembly. Reconnect the hoist to the engine.

85   Remove the engine mounting bracket nuts and bolts from the engine mounting bracket and from the front and rear transaxle mounting brackets.

86   Disconnect the electrical connectors from the alternator (see Chapter 5), the oxygen sensor, the Vehicle Speed Sensor (VSS), the Crankshaft Position (CKP) sensor and the EGR differential pressure feedback sensor (see Chapter 6), the ignition coil and the starter motor (see Chapter 5) and the back-up light switch. Disconnect the ground wires from the bellhousing and from the top of the transaxle. Detach the wiring harnesses for all of these components from any harness retainers, then remove and set aside the entire harness.

87   On engines with a manual transaxle, remove the rest of the transaxle-to-engine bolts and separate and remove the transaxle from the engine. Remove the flywheel.

88   On engines with an automatic transaxle, remove the starter motor isolator, then remove the four torque converter nuts (see illustration). Remove the six remaining engine-to-transaxle bolts and separate and remove the transaxle from the engine. Remove the driveplate.

Note: *On 2013 and later models, the torque converter nuts are accessed through the starter opening.*

89   Using the engine hoist, raise the engine off the floor and install it on an engine stand for further disassembly.

V6 engines

Note: *Engine removal on these models is a difficult and potentially dangerous job, especially for a do-it-yourself mechanic working at home. Because of the vehicle's design, the manufacturer states that the engine and transaxle have to be lowered as a single assembly from the underside of the vehicle, not lifted out the top. With a floor jack and jackstands, the vehicle simply cannot be raised high enough and supported safely enough for the engine/transaxle assembly to slide out from underneath. Instead, the*

manufacturer states that removal of the engine/ transaxle assembly must be performed with the vehicle on a vehicle hoist.

90   Disconnect the electrical connector from the Manifold Absolute Pressure (MAP) sensor, if equipped.

91   Disconnect the electrical connector and EVAP hoses from the EVAP canister purge valve.

92   Unscrew and remove the EGR tube between the exhaust manifold and the EGR valve (see Chapter 6).

93   Disconnect the transaxle electrical connector, disconnect the shift cable from the transaxle manual lever, detach the wiring harness retainer, remove the cable support bracket bolts and set the transaxle control cable and brackets aside. Disconnect the electrical connector from the Transmission Range (TR) sensor, detach the transaxle control harness retainers and set the harness aside.

94   Loosen the coolant hose clamps and disconnect and remove the coolant hoses from the radiator and the thermostat housing.

95   Remove the bolts that secure the transmission fluid line support bracket, then disconnect and plug the two transmission lines.

96   Loosen the transaxle cooler hose clamps and disconnect the cooler hoses from the transaxle and from the cooler. Be sure to plug the cooler pipes on the transaxle and on the transaxle cooler.

97   Disconnect the electrical connector from the Power Steering Pressure (PSP) switch. Disconnect and plug the power steering hoses. Remove the power steering hose bracket bolt or nut and set the power steering hoses aside. On later models, a ground wire is secured by the bracket mounting nut; be sure to reattach it during reassembly.

98   Remove the dipstick, unbolt the dipstick tube and remove the tube.

99   Disconnect the block heater wiring harness, if applicable.

100 Unbolt the air conditioning compressor (see Chapter 3) and set the compressor and the air conditioning hoses aside. Be sure to safely secure the compressor with a coat hanger or a piece of sturdy wire.

101 On 4WD models, disconnect the driveshaft from the transfer case (see Chapter 7C).

102 Unbolt and remove the left and right wheel speed sensors from the steering knuckles. Unbolt the brackets for the speed sensor harnesses and set the speed sensors and harnesses aside.

103 Remove the left and right brake calipers from the steering knuckles (see Chapter 9).

104 Separate the left and right balljoints and the left and right outer tie-rod ends from the steering knuckles (see Chapter 10).

105 Disconnect the left and right stabilizer bar link rods from the front struts (see Chapter 10).

106 Remove the strut-to-steering knuckle nuts and bolts and separate the struts from the left and right steering knuckles (see Chapter 10).

107 Remove the pinch bolt from the lower steering shaft U-joint and separate the steering shaft from the steering gear (see Chapter 10).

108 Remove the torque converter inspection cover and remove the four torque converter nuts.

109 Remove the oil pan-to-transaxle bolts (three bolts on earlier models; two on later models).

110 Support the engine and transaxle from above with an engine hoist (see illustration 7.39). Roll the engine hoist into position and attach it to the engine with a couple pieces of heavy-duty chain. If the engine is equipped with lifting brackets, use them. If not, you'll have to fasten the chain to some substantial part of the engine - one that is strong enough to take the weight, but in a location that will provide good balance. If you're attaching the chain to a stud on the engine, or are using a bolt passing through the chain and into a threaded hole, place a washer between the nut or bolt head and the chain, and tighten the nut or bolt securely. Take up the slack in the chain, but don't lift the engine.

Warning: *DO NOT place any part of your body under the engine when it's supported only by a hoist or other lifting device.*

Note: *The chains must be long enough to allow the engine hoist to lower the engine and transaxle to the floor without the hoist arm contacting the vehicle.*

111 On 2001 through 2004 models:

a) *Remove the three bolts and two nuts from the engine mounting bracket and remove the mounting bracket.*

b) *Remove the five bolts from the transaxle mounting bracket and remove the mounting bracket.*

c) *Remove the two rear subframe bolts.*

d) *Remove the two subframe side nuts.*

e) *Remove the two bolts from the motor mount support bracket.*

112 On 2005 and later models:

a) *Remove the right transaxle mounting insulator bolt.*

Note: *The right transaxle mounting bracket insulator on 2009 and later models includes a damper that is also secured by this same bolt; be sure to install the damper when installing the insulator bolt.*

b) *Remove the bolt and nuts, then remove the right transaxle mounting bracket.*

c) *Remove the rear transaxle mounting bracket bolt.*

d) *Remove the front engine support bolt and nuts.*

113 Lower the engine and transaxle to the floor as a single assembly, then disconnect the hoist from the chains. Roll the hoist out of the way.

114 Raise the vehicle until it clears the engine/ transaxle assembly.

115 Reconnect the engine hoist. Detach the engine/transaxle from the engine/transaxle support, then lift the engine and transaxle up and remove the support.

116 Disconnect the electrical connectors for the oxygen sensors. To find the connectors, trace the sensor harnesses from the upstream and downstream sensors to their respective connectors. Some oxygen sensors use cable ties to secure their harnesses; if so, cut the cable ties.

Unscrew and remove the oxygen sensors.

117 Disconnect the electrical connectors for the Transmission Range (TR) sensor, Output Shaft Speed (OSS), Turbine Speed Sensor (SS) and the electrical connectors and harnesses for any other transaxle electronic controls. Remove the bolted clips and/or plastic retainers that secure the transaxle control harness support bracket(s), then detach the harness and set it aside.

118 Remove the two starter bolts, then remove the starter motor and starter harness as a single assembly.

119 On 4WD models:

a) *Disconnect the right (rear) oxygen sensor electrical connector.*

b) *Remove the bolt and pin-type retainer that secure the power take-off vent tube and set the vent tube aside.*

c) *Remove the three heat shield bolts and the heat shield*

d) *Remove the three nuts and the right (rear) catalytic converter.*

e) *Remove the Power Transfer Unit (PTU) support bracket bolts and remove the PTU support bracket.*

f) *Remove the four PTU mounting bolts and remove the PTU and the PTU heat shield.*

120 Disconnect the electrical connector from the knock sensor.

121 Disconnect the electrical connectors from the alternator and remove the alternator (see Chapter 5).

122 On 4WD models, remove the six exhaust manifold nuts from the right (rear) exhaust manifold, remove the manifold, then remove the exhaust manifold studs.

123 Remove the intermediate shaft support bracket bolts and remove the intermediate shaft support bracket.

124 Remove the upper transaxle-to-engine bolts (six on 2001 through 2004 models, seven on 2005 through 2007 models and five on 2008 and later models).

125 Separate the transaxle from the engine. Remove the driveplate and mount the engine on a stand.

Installation

126 Install the flywheel or driveplate (see Chapter 2A).

127 On models with a manual transaxle, install the clutch and pressure plate (see Chapter 8). We recommend installing a new clutch.

128 Reattach the transaxle to the engine (see Chapter 2A or Chapter 2B). Apply a dab of high-temperature grease to the input shaft and guide it into the crankshaft pilot bearing until the bellhousing is flush with the engine block. Be sure to tighten the engine-to-transaxle bolts to the torque listed in this Chapter's Specifications.

Caution: *DO NOT use the bolts to force the transaxle and engine together!*

129 To install the engine, reverse the removal procedure.

130 Refill the engine with coolant and oil and the transaxle with transaxle fluid (see Chapter 1). If you had to open the power steering system, be sure refill the power steering reservoir.

131 Run the engine and check for leaks and proper operation of all accessories.

132 Install the hood (see Chapter 11) and the engine cover.

133 Test drive the vehicle. The Powertrain Control Module (PCM) must relearn its idle and fuel trim strategy for optimum drivability and performance (see Chapter 5, Section 1).

134 Have the air conditioning system recharged and leak tested by the shop that discharged it.

8 Engine overhaul - disassembly sequence

1   It's much easier to remove the external components if it's mounted on a portable engine stand. A stand can often be rented quite cheaply from an equipment rental yard. Before the engine is mounted on a stand, the flywheel/driveplate should be removed from the engine.

2   If a stand isn't available, it's possible to remove the external engine components with it blocked up on the floor. Be extra careful not to tip or drop the engine when working without a stand.

3   If you're going to obtain a rebuilt engine, all external components must come off first, to be transferred to the replacement engine. These components include:

Clutch and flywheel (models with
    manual transaxle)
Driveplate (models with
    automatic transaxle)
Ignition system components
Emissions-related components
Engine mounts and mount brackets
Engine rear cover (spacer plate between
    flywheel/driveplate and engine block)
Intake/exhaust manifolds
Fuel injection components
Oil filter
Spark plug wires (or ignition coils) and
    spark plugs
Thermostat and housing assembly
Water pump

9.1 Before you try to remove the pistons, use a ridge reamer to remove the raised material (ridge) from the top of the cylinders

Note: *When removing the external components from the engine, pay close attention to details that may be helpful or important during installation. Note the installed position of gaskets, seals, spacers, pins, brackets, washers, bolts and other small items.*

4   If you're going to obtain a short block (assembled engine block, crankshaft, pistons and connecting rods), remove the timing belt, cylinder head, oil pan, oil pump pick-up tube, oil pump and water pump from your engine so that you can turn in your old short block to the rebuilder as a core. See *Engine rebuilding alternatives* for additional information regarding the different possibilities to be considered.

9 Pistons and connecting rods - removal and installation

Removal

Note: *Prior to removing the piston/connecting rod assemblies, remove the cylinder head and oil pan and, on 1.5L and 1.6L engines, the oil pump pickup tube and windage tray (see Chapter 2A or Chapter 2B).*

1   Use your fingernail to feel if a ridge has formed at the upper limit of ring travel (about 1/4-inch down from the top of each cylinder). If carbon deposits or cylinder wear have produced ridges, they must be completely removed with a special tool (see illustration). Follow the manufacturer's instructions provided with the tool. Failure to remove the ridges before attempting to remove the piston/connecting rod assemblies may result in piston breakage.

2   After the cylinder ridges have been removed, turn the engine so the crankshaft is facing up.

3   Before the main bearing cap assembly and connecting rods are removed, check the connecting rod endplay with feeler gauges. Slide them between the first connecting rod and the crankshaft throw until the play is removed (see illustration). Repeat this procedure for each connecting rod. The endplay is equal to the

9.3 Checking the connecting rod endplay (side clearance)

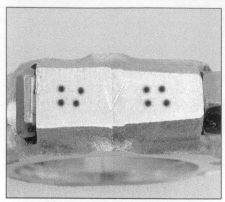

9.4 If the connecting rods and caps are not marked, use permanent ink or paint to mark the caps to the rods by cylinder number (for example, this would be the No. 4 connecting rod)

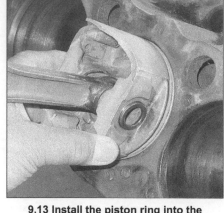

9.13 Install the piston ring into the cylinder then push it down into position using a piston so the ring will be square in the cylinder

9.14 With the ring square in the cylinder, measure the ring end gap with a feeler gauge

thickness of the feeler gauge(s). Check with an automotive machine shop for the endplay service limit (a typical end play limit should measure between 0.005 to 0.015 inch [0.127 to 0.369 mm]). If the play exceeds the service limit, new connecting rods will be required. If new rods (or a new crankshaft) are installed, the endplay may fall under the minimum allowable. If it does, the rods will have to be machined to restore it. If necessary, consult an automotive machine shop for advice.

4   Check the connecting rods and caps for identification marks. If they aren't plainly marked, use paint or marker to clearly identify each rod and cap (1, 2, 3, etc., depending on the cylinder they're associated with) (see illustration).

5   Loosen each of the connecting rod cap bolts 1/2-turn at a time until they can be removed by hand.

Note: *New connecting rod cap bolts must be used when reassembling the engine, but save the old bolts for use when checking the connecting rod bearing oil clearance.*

6   Remove the number one connecting rod cap and bearing insert. Don't drop the bearing insert out of the cap.

7   Remove the bearing insert and push the connecting rod/piston assembly out through the top of the engine. Use a wooden or plastic hammer handle to push on the upper bearing surface in the connecting rod. If resistance is felt, double-check to make sure that all of the ridge was removed from the cylinder.

8   Repeat the procedure for the remaining cylinders.

9   After removal, reassemble the connecting rod caps and bearing inserts in their respective connecting rods and install the cap bolts finger-tight. Leaving the old bearing inserts in place until reassembly will help prevent the connecting rod bearing surfaces from being accidentally nicked or gouged.

10   The pistons and connecting rods are now ready for inspection and overhaul at an automotive machine shop.

Piston ring installation

11   Before installing the new piston rings, the ring end gaps must be checked. It's assumed that the piston ring side clearance has been checked and verified correct.

12   Lay out the piston/connecting rod assem-

blies and the new ring sets so the ring sets will be matched with the same piston and cylinder during the end gap measurement and engine assembly.

13   Insert the top (number one) ring into the first cylinder and square it up with the cylinder walls by pushing it in with the top of the piston (see illustration). The ring should be near the bottom of the cylinder, at the lower limit of ring travel.

14   To measure the end gap, slip feeler gauges between the ends of the ring until a gauge equal to the gap width is found (see illustration). The feeler gauge should slide between the ring ends with a slight amount of drag. A typical ring gap should fall between 0.010 and 0.020 inch [0.25 to 0.50 mm] for compression rings and up to 0.030 inch [0.76 mm] for the oil ring steel rails. If the gap is larger or smaller than specified, double-check to make sure you have the correct rings before proceeding.

15   If the gap is too small, it must be enlarged or the ring ends may come in contact with each other during engine operation, which can cause serious damage to the engine. If necessary,

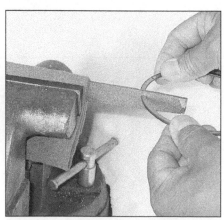

9.15 If the ring end gap is too small, clamp a file in a vise as shown and file the piston ring ends - be sure to remove all raised material

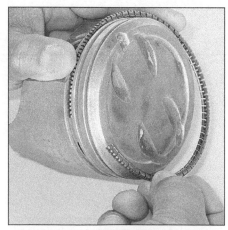

9.19a Installing the spacer/expander in the oil ring groove

9.19b DO NOT use a piston ring installation tool when installing the oil control side rails

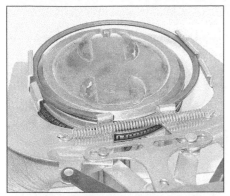

9.22 Use a piston ring installation tool to install the number 2 and the number 1 (top) rings - be sure the directional mark on the piston ring(s) is facing toward the top of the piston

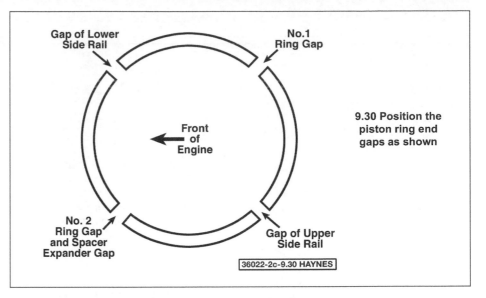

9.30 Position the piston ring end gaps as shown

Gap of Lower Side Rail

No.1 Ring Gap

Front of Engine

No. 2 Ring Gap and Spacer Expander Gap

Gap of Upper Side Rail

36022-2c-9.30 HAYNES

increase the end gaps by filing the ring ends very carefully with a fine file. Mount the file in a vise equipped with soft jaws, slip the ring over the file with the ends contacting the file face and slowly move the ring to remove material from the ends. When performing this operation, file only by pushing the ring from the outside end of the file towards the vise (see illustration).

16   Excess end gap isn't critical unless it's greater than 0.040 inch (1.01 mm). Again, double-check to make sure you have the correct ring type.

17   Repeat the procedure for each ring that will be installed in the first cylinder and for each ring in the remaining cylinders. Remember to keep rings, pistons and cylinders matched up.

18   Once the ring end gaps have been checked/corrected, the rings can be installed on the pistons.

19   The oil control ring (lowest one on the piston) is usually installed first. It's composed of three separate components. Slip the spacer/expander into the groove (see illustration). If an anti-rotation tang is used, make sure it's inserted into the drilled hole in the ring groove. Next, install the upper side rail in the same manner (see illustration). Don't use a piston ring installation tool on the oil ring side rails, as they may be damaged. Instead, place one end of the side rail into the groove between the spacer/expander and the ring land, hold it firmly in place and slide a finger around the piston while pushing the rail into the groove. Finally, install the lower side rail.

20   After the three oil ring components have been installed, check to make sure that both the upper and lower side rails can be rotated smoothly inside the ring grooves.

21   The number two (middle) ring is installed next. It's usually stamped with a mark which must face up, toward the top of the piston. Do not mix up the top and middle rings, as they have different cross-sections.

Note: *Always follow the instructions printed on the ring package or box - different manufacturers may require different approaches.*

22   Use a piston ring installation tool and make sure the identification mark is facing

the top of the piston, then slip the ring into the middle groove on the piston (see illustration). Don't expand the ring any more than necessary to slide it over the piston.

23   Install the number one (top) ring in the same manner. Make sure the mark is facing up. Be careful not to confuse the number one and number two rings.

24   Repeat the procedure for the remaining pistons and rings.

Installation

25   Before installing the piston/connecting rod assemblies, the cylinder walls must be perfectly clean, the top edge of each cylinder bore must be chamfered, and the crankshaft must be in place.

26   Remove the cap from the end of the number one connecting rod (refer to the marks made during removal). Remove the original bearing inserts and wipe the bearing surfaces of the connecting rod and cap with a clean, lint-free cloth. They must be kept spotlessly clean.

Connecting rod bearing oil clearance check

27   Clean the back side of the new upper bearing insert, then lay it in place in the connecting rod.

28   Make sure the tab on the bearing fits into the recess in the rod. Don't hammer the bearing insert into place and be very careful not to nick or gouge the bearing face. Don't lubricate the bearing at this time.

29   Clean the back side of the other bearing insert and install it in the rod cap. Again, make sure the tab on the bearing fits into the recess in the cap, and don't apply any lubricant. It's critically important that the mating surfaces of the bearing and connecting rod are perfectly clean and oil free when they're assembled.

30   Position the piston ring gaps at 90-degree intervals around the piston as shown (see illustration).

31   Lubricate the piston and rings with clean engine oil and attach a piston ring compres-

sor to the piston. Leave the skirt protruding about 1/4-inch to guide the piston into the cylinder. The rings must be compressed until they're flush with the piston.

32   Rotate the crankshaft until the number one connecting rod journal is at BDC (bottom dead center) and apply a liberal coat of engine oil to the cylinder walls.

33   With the mark on top of the piston facing the front (timing belt or chain end) of the engine, gently insert the piston/connecting rod assembly into the number one cylinder bore and rest the bottom edge of the ring compressor on the engine block.

34   Tap the top edge of the ring compressor to make sure it's contacting the block around its entire circumference.

35   Gently tap on the top of the piston with the end of a wooden or plastic hammer handle (see illustration) while guiding the end of the connecting rod into place on the crankshaft journal. The piston rings may try to pop out of the ring compressor just before entering the cylinder bore, so keep some downward pressure on the ring compressor. Work slowly, and if any resistance is felt as the

9.35 Use a plastic or wooden hammer handle to push the piston into the cylinder

ENGINE BEARING ANALYSIS

Debris

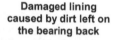

Babbitt bearing embedded with debris from machinings

Microscopic detail of debris

Microscopic detail of gouges

Overplated copper alloy bearing gouged by cast iron debris

Aluminum bearing embedded with glass beads

Microscopic detail of glass beads

Damaged lining caused by dirt left on the bearing back

Misassembly

Result of a lower half assembled as an upper - blocking the oil flow

Excessive oil clearance is indicated by a short contact arc

Polished and oil-stained backs are a result of a poor fit in the housing bore

Result of a wrong, reversed, or shifted cap

Overloading

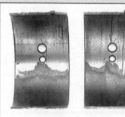

Damage from excessive idling which resulted in an oil film unable to support the load imposed

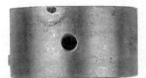

Damaged upper connecting rod bearings caused by engine lugging; the lower main bearings (not shown) were similarly affected

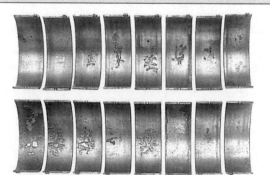

The damage shown in these upper and lower connecting rod bearings was caused by engine operation at a higher-than-rated speed under load

Misalignment

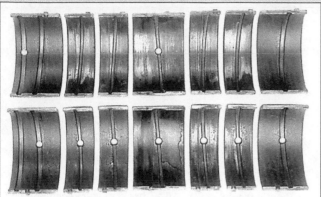

A warped crankshaft caused this pattern of severe wear in the center, diminishing toward the ends

A poorly finished crankshaft caused the equally spaced scoring shown

A bent connecting rod led to the damage in the "V" pattern

A tapered housing bore caused the damage along one edge of this pair

Lubrication

Result of dry start: The bearings on the left, farthest from the oil pump, show more damage

Result of a low oil supply or oil starvation

Severe wear as a result of inadequate oil clearance

Corrosion

Microscopic detail of corrosion

Corrosion is an acid attack on the bearing lining generally caused by inadequate maintenance, extremely hot or cold operation, or inferior oils or fuels

Microscopic detail of cavitation

Example of cavitation - a surface erosion caused by pressure changes in the oil film

Damage from excessive thrust or insufficient axial clearance

Bearing affected by oil dilution caused by excessive blow-by or a rich mixture

9.37 Place Plastigage on each connecting rod bearing journal parallel to the crankshaft centerline

9.41 Use the scale on the Plastigage package to determine the bearing oil clearance - be sure to measure the widest part of the Plastigage and use the correct scale; it comes with both standard and metric scales

10.1 Checking crankshaft endplay with a dial indicator

piston enters the cylinder, stop immediately. Find out what's hanging up and fix it before proceeding. Do not, for any reason, force the piston into the cylinder - you might break a ring and/or the piston.

36 Once the piston/connecting rod assembly is installed, the connecting rod bearing oil clearance must be checked before the rod cap is permanently installed.

37 Cut a piece of the appropriate size Plastigage slightly shorter than the width of the connecting rod bearing and lay it in place on the number one connecting rod journal, parallel with the journal axis (see illustration).

38 Clean the connecting rod cap bearing face and install the rod cap. Make sure the mating mark on the cap is on the same side as the mark on the connecting rod (see illustration 9.4).

39 Install the old rod bolts, at this time, and tighten them to the torque listed in this Chapter's Specifications.

Note: *Use a thin-wall socket to avoid erroneous torque readings that can result if the socket is wedged between the rod cap and the bolt head. If the socket tends to wedge itself between the fastener and the cap, lift up on it slightly until it no longer contacts the cap.*

Note: *DO NOT rotate the crankshaft at any time during this operation.*

40 Remove the fasteners and detach the rod cap, being very careful not to disturb the Plastigage.

41 Compare the width of the crushed Plastigage to the scale printed on the Plastigage envelope to obtain the oil clearance (see illustration). The connecting rod oil clearance is usually about 0.001 to 0.002 inch (0.025 to 0.05 mm). Consult an automotive machine shop for the clearance specified for the rod bearings on your engine.

42 If the clearance is not as specified, the bearing inserts may be the wrong size (which means different ones will be required). Before deciding that different inserts are needed, make sure that no dirt or oil was between the bearing inserts and the connecting rod or cap when the clearance was

measured. Also, recheck the journal diameter. If the Plastigage was wider at one end than the other, the journal may be tapered. If the clearance still exceeds the limit specified, the bearing will have to be replaced with an undersize bearing.

Caution: *When installing a new crankshaft always use a standard size bearing.*

Final installation

43 Carefully scrape all traces of the Plastigage material off the rod journal and/or bearing face. Be very careful not to scratch the bearing - use your fingernail or the edge of a plastic card.

44 Make sure the bearing faces are perfectly clean, then apply a uniform layer of clean moly-base grease or engine assembly lube to both of them. You'll have to push the piston into the cylinder to expose the face of the bearing insert in the connecting rod.

Caution: *Install new connecting rod cap bolts. Do NOT reuse old bolts - they have stretched and cannot be reused.*

45 Slide the connecting rod back into place on the journal, install the rod cap, install the new bolts and tighten them to the torque listed in this Chapter's Specifications. Again, work up to the torque in three steps.

46 Repeat the entire procedure for the remaining pistons/connecting rods.

47 The important points to remember are:

a) *Keep the back sides of the bearing inserts and the insides of the connecting rods and caps perfectly clean when assembling them.*

b) *Make sure you have the correct piston/rod assembly for each cylinder.*

c) *The mark on the piston must face the front (timing belt/chain end) of the engine.*

d) *Lubricate the cylinder walls liberally with clean oil.*

e) *Lubricate the bearing faces when installing the rod caps after the oil clearance has been checked.*

48 After all the piston/connecting rod assemblies have been correctly installed, rotate the crankshaft a number of times by hand to check for any obvious binding.

49 As a final step, check the connecting rod endplay, as described in Step 3. If it was correct before disassembly and the original crankshaft and rods were reinstalled, it should still be correct. If new rods or a new crankshaft were installed, the endplay may be inadequate. If so, the rods will have to be removed and taken to an automotive machine shop for resizing.

10 Crankshaft - removal and installation

Removal

Note: *The crankshaft can be removed only after the engine has been removed from the vehicle. It's assumed that the flywheel or driveplate, crankshaft pulley, timing belt, oil pan, oil pump body, oil filter and piston/connecting rod assemblies have already been removed. The rear main oil seal retainer must be unbolted and separated from the block before proceeding with crankshaft removal.*

1 Before the crankshaft is removed, measure the endplay. Mount a dial indicator with the indicator in line with the crankshaft and just touching the end of the crankshaft as shown (see illustration).

2 Pry the crankshaft all the way to the rear and zero the dial indicator. Next, pry the crankshaft to the front as far as possible and check the reading on the dial indicator. The distance traveled is the endplay. A typical crankshaft endplay will fall between 0.003 to 0.010 inch (0.076 to 0.254 mm). If it is greater than that, check the crankshaft thrust surfaces for wear after it's removed. If no wear is evident, new main bearings should correct the endplay.

3 If a dial indicator isn't available, feeler gauges can be used. Gently pry the crankshaft all the way to the front of the engine. Slip feeler gauges between the crankshaft and the front face of the thrust bearing or washer to determine the clearance (see illustration).

10.3 Checking the crankshaft endplay with feeler gauges at the thrust bearing journal

10.17 Place the Plastigage onto the crankshaft bearing journal as shown

4   Loosen the main bearing cap or main bearing beam bolts (four-cylinder engines) or lower cylinder block (bedplate) bolts 1/4-turn at a time each, until they can be removed by hand. Loosen the bolts in the reverse of the tightening sequence (see illustrations 10.19a and 10.19b)
5   If you're working on a four-cylinder engine, remove the main bearing caps or main bearing beam. If you're working on a V6 engine, gently tap the lower cylinder block with a soft-face hammer around its perimeter. Pull the lower cylinder block straight up and off the cylinder block. Try not to drop the bearing inserts if they come out with the assembly.
6   Carefully lift the crankshaft out of the engine. It may be a good idea to have an assistant available, since the crankshaft is quite heavy and awkward to handle. With the bearing inserts in place inside the engine block and main bearing caps or lower cylinder block, reinstall the main bearing caps or lower cylinder block onto the engine block and tighten the bolts finger-tight. If you're working on a four-cylinder engine, make sure you install the caps with the arrows pointing toward the front (timing belt end) of the engine.

Installation
7   Crankshaft installation is the first step in engine reassembly. It's assumed at this point that the engine block and crankshaft have been cleaned, inspected and repaired or reconditioned.
8   Position the engine block with the bottom facing up.
9   Remove the bolts and lift off the main bearing caps or lower cylinder block.
10   If they're still in place, remove the original bearing inserts from the block and from the main bearing cap assembly. Wipe the bearing surfaces of the block and main bearing cap assembly with a clean, lint-free cloth. They must be kept spotlessly clean. This is critical for determining the correct bearing oil clearance.

Main bearing oil clearance check
11   Without mixing them up, clean the back sides of the new upper main bearing inserts (with grooves and oil holes) and lay one in each main bearing saddle in the engine block. Each upper bearing (engine block) has an oil groove and oil hole in it. Clean the back sides

of the lower main bearing inserts and lay them in the corresponding location in the main bearing caps or the lower cylinder block. Make sure the tab on the bearing insert fits into the recess in the block or lower cylinder block.
Caution: *The oil holes in the block must line up with the oil holes in the upper bearing inserts.*
Caution: *Do not hammer the bearing insert into place and don't nick or gouge the bearing faces. DO NOT apply any lubrication at this time.*
Note: *The thrust bearing on the four-cylinder engine is located on the engine block number 3 journal. The thrust bearing on the V6 engine is located on the 4th journal on the lower cylinder block, and the thrust washer is located on the 4th journal (upper).*
12   Clean the faces of the bearing inserts in the block and the crankshaft main bearing journals with a clean, lint-free cloth.
13   Check or clean the oil holes in the crankshaft, as any dirt here can go only one way - straight through the new bearings.
14   Once you're certain the crankshaft is clean, carefully lay it in position in the cylinder block.
15   Before the crankshaft can be permanently installed, the main bearing oil clearance must be checked.
16   Cut several strips of the appropriate size of Plastigage. They must be slightly shorter than the width of the main bearing journal.
17   Place one piece on each crankshaft main bearing journal, parallel with the journal axis as shown (see illustration).
18   Clean the faces of the bearing inserts in the main bearing caps or the lower cylinder block. Hold the bearing inserts in place and install the caps or the lower cylinder block onto the crankshaft and cylinder block. DO NOT disturb the Plastigage.
19   Apply clean engine oil to all bolt threads prior to installation, then install all bolts finger-tight. Tighten the bolts in the sequence shown (see illustrations) progressing in steps, to the torque listed in this Chapter's Specifications. DO NOT rotate the crankshaft at any time during this operation.
Note: *Use the old bolts at this time.*

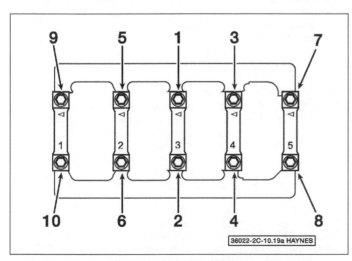

10.19a Main bearing cap or beam bolt tightening sequence - four-cylinder engines

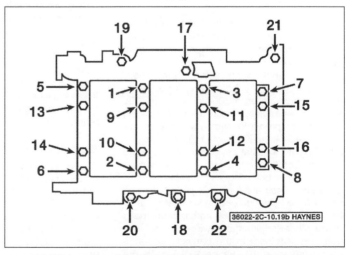

10.19b Lower cylinder block bolt tightening sequence - V6 engines

20   Remove the bolts in the reverse order of the tightening sequence and carefully lift the caps or the lower cylinder block straight up and off the block. Do not disturb the Plastigage or rotate the crankshaft.

21   Compare the width of the crushed Plastigage on each journal to the scale printed on the Plastigage envelope to determine the main bearing oil clearance (see illustration). Check with an automotive machine shop for the oil clearance for your engine.

22   If the clearance is not as specified, the bearing inserts may be the wrong size (which means different ones will be required). Before deciding if different inserts are needed, make sure that no dirt or oil was between the bearing inserts and the cap assembly or block when the clearance was measured. If the Plastigage was wider at one end than the other, the crankshaft journal may be tapered. If the clearance still exceeds the limit specified, the bearing insert(s) will have to be replaced with an undersize bearing insert(s).

Caution: *When installing a new crankshaft always install a standard bearing insert set.*

23   Carefully scrape all traces of the Plastigage material off the main bearing journals and/or the bearing insert faces. Be sure to remove all residue from the oil holes. Use your fingernail or the edge of a plastic card - don't nick or scratch the bearing faces.

Final installation

24   Carefully lift the crankshaft out of the cylinder block.

25   Clean the bearing insert faces in the cylinder block, then apply a thin, uniform layer of moly-base grease or engine assembly lube to each of the bearing surfaces. Be sure to coat the thrust faces as well as the journal face of the thrust bearing.

26   Make sure the crankshaft journals are clean, then lay the crankshaft back in place in the cylinder block.

27   Clean the bearing insert faces and apply the same lubricant to them. Clean the engine block and the bearing caps/lower cylinder block thoroughly. The surfaces must be free of oil residue.

28   On V6 engines, apply a bead of RTV to the engine block before installing the lower cylinder block. Be sure the bead is the correct thickness (see illustration).

29   On V6 engines, install the lower cylinder block onto the crankshaft and cylinder block. If you're working on a 2004 and earlier 2.0L four-cylinder engines, install the main bearing caps in their proper locations, with the arrows on the caps facing the front of the engine.

30   Prior to installation, apply clean engine oil to all bolt threads, wiping off any excess. Then install all bolts finger-tight.

Caution: *These are torque-to-yield bolts and can only be tightened once; use new bolts.*

31   Tighten the main bearing cap/beam bolts (four-cylinder engines) or lower cylinder block bolts (V6 engines) in the correct sequence (see illustration 10.19a or 10.19b), to the torque listed in this Chapter's Specifications.

10.21 Use the scale on the Plastigage package to determine the bearing oil clearance - be sure to measure the widest part of the Plastigage and use the correct scale; it comes with both standard and metric scales

32   Recheck the crankshaft endplay with a feeler gauge or a dial indicator. The endplay should be correct if the crankshaft thrust faces aren't worn or damaged and if new bearings have been installed.

33   Rotate the crankshaft a number of times by hand to check for any obvious binding. It should rotate with a running torque of 50 in-lbs or less. If the running torque is too high, correct the problem at this time.

34   Install a new rear main oil seal (see Chapter 2A or Chapter 2B).

---

11   Balance shaft assembly - inspection, removal and installation

Note: *This procedure applies to all 2006 and later four-cylinder engines except 1.5L and 1.6L models.*

1   The balance shaft assembly is bolted to the crankshaft main bearing support beam. A gear on the crankshaft meshes with a gear on the balance shaft assembly. When the engine is operating, the assembly smoothes out engine vibrations.

2   The balancer is a precision-machined assembly; there are no serviceable parts inside and it should not be disassembled. If the backlash is out of Specification (see Steps 5 through 8), the assembly must be replaced as a complete unit.

Inspection

3   When the balance shaft assembly is in place, the backlash between the drive gear (on the crankshaft) and the driven gear on the assembly can be checked. Remove the timing peg.

4   Attach a 5 mm Allen wrench to the top of the driveshaft with the long end of the wrench pointing straight up. Secure a dial indicator fixture to the engine so that the tip of the indi-

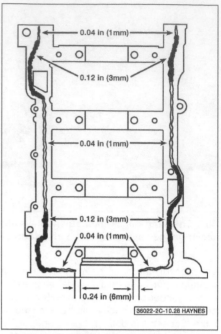

10.28 Apply RTV sealant to the cylinder block mating surface - make sure the bead of sealant is the correct diameter

cator is against the wrench, 3.15 inches (80 mm) from the centerline of the balance shafts (see illustration).

5   Use a pry tool against the crankshaft front counterweight and apply force to move the crankshaft back-and-forth. Record the measurements of the dial indicator. Measurements should be taken at the following degrees of engine rotation: 10, 30, 100, 190 and 210 degrees. Compare your results with the allowable range of backlash given in this Chapter's Specifications.

6   If the backlash is out of range, the assembly must be replaced.

Removal

7   Position the crankshaft at TDC (see Chapter 2A and illustration 11.9).

8   With the engine crankshaft-side-up on the engine stand, remove the four mounting bolts and lift the assembly straight up from the engine.

Installation

9   When installing the balance shaft assembly, the engine must be set to TDC for cylinder number 1. Before installing the assembly, rotate it to align the timing marks on both shafts (see illustration). Bolt the assembly to the engine and recheck that the timing marks are still aligned and that the crankshaft has not moved. Tighten the bolts in a criss-cross pattern to the torque listed in this Chapter's Specifications.

10   The crankshaft timing peg must remain installed to keep the engine at TDC until installation of the timing chain and sprockets is completed.

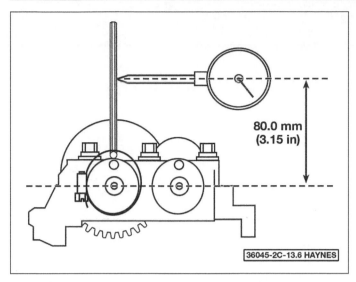

11.4 Method of checking the backlash in the balancer assembly

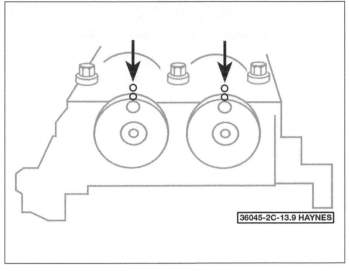

11.9 Remove/install the balance shaft assembly bolts only when the crankshaft is at TDC and these balancer dots align

12 Engine overhaul - reassembly sequence

1   Before beginning engine reassembly, make sure you have all the necessary new parts, gaskets and seals as well as the following items on hand:

Common hand tools
A 1/2-inch drive torque wrench
New engine oil
Gasket sealant
Thread locking compound

2   If you obtained a short block it will be necessary to install the cylinder head, the oil pump and pick-up tube, the oil pan, the water pump, the timing belt or timing chain and timing cover, and the valve cover (see Chapter 2A or Chapter 2B). In order to save time and avoid problems, the external components must be installed in the following general order:

Thermostat and housing cover
Water pump
Intake and exhaust manifolds
Fuel injection components
Emission control components
Spark plug wires and spark plugs
Ignition coils
Oil filter
Engine mounts and mount brackets
Clutch and flywheel (manual transaxle)
Driveplate (automatic transaxle)

13 Initial start-up and break-in after overhaul

Warning: *Have a fire extinguisher handy when starting the engine for the first time.*

1   Once the engine has been installed in the vehicle, double-check the engine oil and coolant levels.

2   On models with a coil pack and spark plug wires, disconnect the primary (low voltage) wires from the coil pack (see Chapter 5). Remove the fuel pump relay (see Chapter 4). The relays are located in the power distribution center in the engine compartment.

3   With the spark plugs out of the engine and the ignition system and fuel pump disabled, crank the engine until oil pressure registers on the gauge or the light goes out.

4   Install the spark plugs, hook up the plug wires and restore the ignition system and fuel pump functions.

5   Start the engine. It may take a few moments for the fuel system to build up pressure, but the engine should start without a great deal of effort.

6   After the engine starts, it should be allowed to warm up to normal operating temperature. While the engine is warming up, make a thorough check for fuel, oil and coolant leaks.

7   Shut the engine off and recheck the engine oil and coolant levels.

8   Drive the vehicle to an area with minimum traffic, accelerate from 30 to 50 mph, then allow the vehicle to slow to 30 mph with the throttle closed. Repeat the procedure 10 or 12 times. This will load the piston rings and cause them to seat properly against the cylinder walls. Check again for oil and coolant leaks.

9   Drive the vehicle gently for the first 500 miles (no sustained high speeds) and keep a constant check on the oil level. It is not unusual for an engine to use oil during the break-in period.

10   At approximately 500 to 600 miles, change the oil and filter.

11   For the next few hundred miles, drive the vehicle normally. Do not pamper it or abuse it.

12   After 2,000 miles, change the oil and filter again and consider the engine broken in.

Notes

Chapter 3
Cooling, heating and air conditioning systems

Contents

Specifications

General

Expansion tank cap pressure rating
    All models .... 17.4 to 21.7 psi (120 to 150 kPa)
Thermostat rating (opening temp to fully open)
    Four-cylinder non-turbocharged models, and 2004 and
        earlier V6 models .... 194 to 223 degrees F (90 to 106 degrees C)
    Four-cylinder turbocharged models
        1.5L and 1.6L models .... 180 to 207 degrees F (82 to 97 degrees C)
        2.0L models .... 176 to 207 degrees F (80 to 97 degrees C)
    V6 models .... 183 to 210 degrees F (84 to 99 degrees C)
Cooling system capacity .... See Chapter 1
Refrigerant type
    2016 and earlier models R-134a
    2017 models R-1234yf
Refrigerant capacity .... Refer to HVAC specification tag

Torque specifications

Note: *One foot-pound (ft-lb) of torque is equivalent to 12 inch-pounds (in-lbs) of torque. Torque values below approximately 15 ft-lbs are expressed in inch-pounds, since most foot-pound torque wrenches are not accurate at these smaller values.*

	Ft-lbs (unless otherwise indicated)	Nm
Condenser (A/C line) inlet and outlet nuts		
2007 and earlier models	71 in-lbs	8
2008 through 2016 models	133 in-lbs	15
2017 models	159 in-lbs	18
Engine oil cooler mounting bolt (four-cylinder engines)		
2004 and earlier 2.0L models	43	58
2005 to 2008 2.3L models		
Oil cooler bolt	25	34
Oil filter adapter bolts	18	25
1.5L and 1.6L engines	41	55
Cooling module (radiator/condenser assembly) mounting bolts		
(2013 and later models)	18	25
Radiator bracket support bolts (2012 and earlier models)	89 in-lbs	10
Thermostat fasteners		
All engines except 1.5L	89 in-lbs	10
1.5L engine		
Thermostat housing nut	159 in-lbs	18
Thermostat housing stud	44 in-lbs	5
Water pump bolts		
All engines (except 2004 and earlier 2.0L)	89 in-lbs	10
2004 and earlier 2.0L engine	144 in-lbs	16
Water pump pulley bolts		
1.5L and 1.6L models	89 in-lbs	10
1.6L, 2.5L, and 2013 and later 2.0L models	177 in-lbs	20
1.5L, 2.3L, and 2004 and earlier 2.0L models	18	25
Water pump drivebelt tensioner bolt (early V6 models)	89 in-lbs	10
A/C compressor		
Mounting bolts/nuts	18	25
Mounting stud (2013 and later models)	80 in-lbs	9
A/C line inlet/outlet fasteners		
2004 and earlier models	15	20
2005 to 2007 models		
Four-cylinder engines	71 in-lbs	8
V6 engine	15	20
2008 through 2016 models	133 in-lbs	15
2017 models	159 in-lbs	18

1 General information

Engine cooling system

1 The cooling system consists of a radiator, an expansion tank, a pressure cap (located on the expansion tank), a thermostat, a cooling fan and clutch, and a belt-driven water pump.

2 The expansion tank (referred to by the manufacturer as a "degas bottle") functions somewhat differently than a conventional recovery tank. Designed to separate any trapped air in the coolant, it is pressurized by the radiator and has a pressure cap on top. The radiator on these models does not have a pressure cap. When the thermostat is closed, no coolant flows in the expansion tank, but when the engine is fully warmed up, coolant flows from the top of the radiator through a small hose that enters the top of the expansion tank, where the air separates and the coolant falls into a coolant reservoir in the bottom of the tank, which is fed to the cooling system through a larger hose connected to the lower radiator hose.

Warning: *Unlike a conventional coolant recov-ery tank, the pressure cap on the expansion tank should never be opened after the engine has warmed up, because of the danger of severe burns caused by steam or scalding coolant.*

3 Coolant in the left side of the radiator cir-culates through the lower radiator hose to the water pump, where it is forced through cool-ant passages in the cylinder block. The cool-ant then travels up into the cylinder head, cir-culates around the combustion chambers and valve seats, travels out of the cylinder head past the open thermostat into the upper radia-tor hose and back into the radiator.

4 When the engine is cold, the thermostat restricts the circulation of coolant to the engine. When the minimum operating temperature is reached, the thermostat begins to open, allow-ing coolant to return to the radiator.

Transaxle cooling systems

5 2007 and earlier vehicles with automatic transaxles have a transaxle cooler that is built into the radiator. Some models also have an auxiliary cooler that is mounted in front of the air conditioning condenser. On 2008 through 2012 models, the automatic transaxle cooler is built into the air conditioning condenser. On 2013 and later models, the transmission uses a cooler-warmer that is mounted to the transmission, along with a system of associated coolant valves (see Chapter 7B, Section 7).

Engine oil cooling system

6 Besides the engine and transaxle cool-ing systems described above, engine heat is also dissipated through an external oil cooler that's integrated into the lubrication system. The oil cooler helps keep engine and oil tem-peratures within design limits under extreme load conditions.

7 The oil cooling system on these models consists of a cylindrical housing (oil cooler) mounted inline between the oil filter and the engine block, a heat exchanger inside the oil cooler and a pair of coolant hoses that deliver coolant from the radiator to the oil cooler housing.

Heating system

8 The heating system consists of the heater controls, the heater core, the heater blower assembly (which houses the blower

2.2 The cooling system pressure tester is connected in place of the pressure cap, then pumped up to pressurize the system

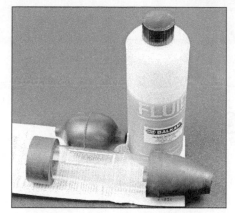

2.5a The combustion leak detector consists of a bulb, syringe and test fluid

2.5b Place the tester over the cooling system filler neck and use the bulb to draw a sample into the tester

motor and the blower motor resistor), and the hoses connecting the heater core to the engine cooling system. Hot engine coolant is circulated through the heater core. When the heater mode is activated, a flap door opens to expose the heater box to the passenger compartment. A fan switch on the heater controls activates the blower motor, which forces air through the core, heating the air.

Air conditioning system

9   The air conditioning system consists of the condenser, which is mounted in front of the radiator, the evaporator case assembly under the dash, a compressor mounted on the engine, and the plumbing connecting all of the above components.

10   A blower fan forces the warmer air of the passenger compartment through the evaporator core (sort of a radiator-in-reverse), transferring the heat from the air to the refrigerant. The liquid refrigerant boils off into low pressure vapor, taking the heat with it when it leaves the evaporator.

2   Troubleshooting

Coolant leaks

1   A coolant leak can develop anywhere in the cooling system, but the most common causes are:

a) A loose or weak hose clamp
b) A defective hose
c) A faulty pressure cap
d) A damaged radiator
e) A bad heater core
f) A faulty water pump
g) A leaking gasket at any joint that carries coolant

2   Coolant leaks aren't always easy to find. Sometimes they can only be detected when the cooling system is under pressure. Here's where a cooling system pressure tester comes in handy. After the engine has cooled completely, the tester is attached in place

of the pressure cap, then pumped up to the pressure value equal to that of the pressure cap rating (see illustration). Now, leaks that only exist when the engine is fully warmed up will become apparent. The tester can be left connected to locate a nagging slow leak.

Coolant level drops, but no external leaks

3   If you find it necessary to keep adding coolant, but there are no external leaks, the probable causes include:

a) A blown head gasket
b) A leaking intake manifold gasket (only on engines that have coolant passages in the manifold)
c) A cracked cylinder head or cylinder block

4   Any of the above problems will also usually result in contamination of the engine oil, which will cause it to take on a milkshake-like appearance. A bad head gasket or cracked head or block can also result in engine oil contaminating the cooling system.

5   Combustion leak detectors (also known as block testers) are available at most auto parts stores. These work by detecting exhaust gases in the cooling system, which indicates a compression leak from a cylinder into the coolant. The tester consists of a large bulb-type syringe and bottle of test fluid (see illustration). A measured amount of the fluid is added to the syringe. The syringe is placed over the cooling system filler neck and, with the engine running, the bulb is squeezed and a sample of the gases present in the cooling system are drawn up through the test fluid (see illustration). If any combustion gases are present in the sample taken, the test fluid will change color.

6   If the test indicates combustion gas is present in the cooling system, you can be sure that the engine has a blown head gasket or a crack in the cylinder head or block, and will require disassembly to repair.

Pressure cap

Warning: *Wait until the engine is completely cool before beginning this check.*

7   The cooling system is sealed by a spring-loaded cap, which raises the boiling point of the coolant. If the cap's seal or spring are worn out, the coolant can boil and escape past the cap. With the engine completely cool, remove the cap and check the seal; if it's cracked, hardened or deteriorated in any way, replace it with a new one.

8   Even if the seal is good, the spring might not be; this can be checked with a cooling system pressure tester (see illustration). If the cap can't hold a pressure within approximately 1-1/2 lbs of its rated pressure (which is marked on the cap), replace it with a new one.

9   The cap is also equipped with a vacuum relief spring. When the engine cools off, a vacuum is created in the cooling system. The vacuum relief spring allows air back into the system, which will equalize the pressure and prevent damage to the radiator (the radiator tanks could collapse if the vacuum is great enough). If, after turning the engine off and allowing it to cool down you notice any of the cooling system hoses collapsing, replace the pressure cap with a new one.

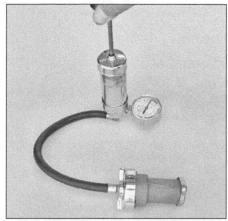

2.8 Checking the cooling system pressure cap with a cooling system pressure tester (radiator cap shown, expansion tank cap pressure check procedure is identical)

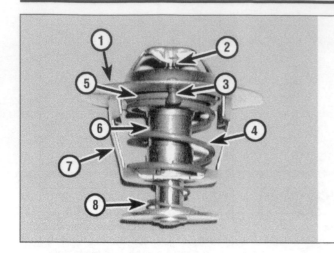

2.10 Typical thermostat:

1   Flange
2   Piston
3   Jiggle valve
4   Main coil spring
5   Valve seat
6   Valve
7   Frame
8   Secondary coil
     spring

2.28 The water pump weep hole is generally located on the underside of the pump

Thermostat

10   Before assuming the thermostat (see illustration) is responsible for a cooling system problem, check the coolant level (see Chapter 1), drivebelt tension (see Chapter 1) and temperature gauge (or light) operation.

11   If the engine takes a long time to warm up (as indicated by the temperature gauge or heater operation), the thermostat is probably stuck open. Replace the thermostat with a new one.

12   If the engine runs hot or overheats, a thorough test of the thermostat should be performed.

13   Definitive testing of the thermostat can only be made when it is removed from the vehicle. If the thermostat is stuck in the open position at room temperature, it is faulty and must be replaced.

Caution: *Do not drive the vehicle without a thermostat. The computer may stay in open loop and emissions and fuel economy will suffer.*

14   To test a thermostat, suspend the (closed) thermostat on a length of string or wire in a pot of cold water.

15   Heat the water on a stove while observing thermostat. The thermostat should fully open before the water boils.

16   If the thermostat doesn't open and close as specified, or sticks in any position, replace it.

Cooling fan

Electric cooling fan

17   If the engine is overheating and the cooling fan is not coming on when the engine temperature rises to an excessive level, unplug the fan motor electrical connector(s) and connect the motor directly to the battery with fused jumper wires. If the fan motor doesn't come on, replace the motor.

18   If the fan motor is okay, but it isn't coming on when the engine gets hot, the cooling fan control module or the Powertrain Control Module (PCM) might be defective.

19   These control circuits are fairly complex, and checking them should be left to a qualified automotive technician.

20   Check all wiring and connections to the fan motor. Refer to the wiring diagrams at the end of this manual.

21   If no obvious problems are found, the problem could be the Cylinder Head Temperature (CHT) sensor or the Powertrain Control Module (PCM). Have the cooling fan system and circuit diagnosed by a dealer service department or repair shop with the proper diagnostic equipment.

Belt-driven cooling fan

22   Disconnect the cable from the negative terminal of the battery and rock the fan back and forth by hand to check for excessive bearing play.

23   With the engine cold (and not running), turn the fan blades by hand. The fan should turn freely.

24   Visually inspect for substantial fluid leakage from the clutch assembly. If problems are noted, replace the clutch assembly.

25   With the engine completely warmed up, turn off the ignition switch and disconnect the negative battery cable from the battery. Turn the fan by hand. Some drag should be evident. If the fan turns easily, replace the fan clutch.

Water pump

26   A failure in the water pump can cause serious engine damage due to overheating.

Drivebelt-driven water pump

27   There are two ways to check the operation of the water pump while it's installed on the engine. If the pump is found to be defective, it should be replaced with a new or rebuilt unit.

28   Water pumps are equipped with weep (or vent) holes (see illustration). If a failure occurs in the pump seal, coolant will leak from the hole.

29   If the water pump shaft bearings fail, there may be a howling sound at the pump while it's running. Shaft wear can be felt with the drivebelt removed if the water pump pulley is rocked up and down (with the engine off). Don't mistake drivebelt slippage, which causes a squealing sound, for water pump bearing failure.

Timing chain or timing belt-driven water pump

30   Water pumps driven by the timing chain or timing belt are located underneath the timing chain or timing belt cover.

31   Checking the water pump is limited because of where it is located. However, some basic checks can be made before deciding to remove the water pump. If the pump is found to be defective, it should be replaced with a new or rebuilt unit.

32   One sign that the water pump may be failing is that the heater (climate control) may not work well. Warm the engine to normal operating temperature, confirm that the coolant level is correct, then run the heater and check for hot air coming from the ducts.

33   Check for noises coming from the water pump area. If the water pump impeller shaft or bearings are failing, there may be a howling sound at the pump while the engine is running.

Note: *Be careful not to mistake drivebelt noise (squealing) for water pump bearing or shaft failure.*

34   It you suspect water pump failure due to noise, wear can be confirmed by feeling for play at the pump shaft. This can be done by rocking the drive sprocket on the pump shaft up and down. To do this you will need to remove the tension on the timing chain or belt as well as access the water pump.

All water pumps

35   In rare cases or on high-mileage vehicles, another sign of water pump failure may be the presence of coolant in the engine oil. This condition will adversely affect the engine in varying degrees.

Note: *Finding coolant in the engine oil could indicate other serious issues besides a failed water pump, such as a blown head gasket or a cracked cylinder head or block.*

36   Even a pump that exhibits no outward signs of a problem, such as noise or leakage, can still be due for replacement. Removal for close examination is the only sure way to tell. Sometimes the fins on the back of the impeller can corrode to the point that cooling efficiency is diminished significantly.

Heater system

37   Little can go wrong with a heater. If the fan motor will run at all speeds, the electrical part of the system is okay. The three basic heater problems fall into the following general categories:

a) *Not enough heat*
b) *Heat all the time*
c) *No heat*

38   If there's not enough heat, the control valve or door is stuck in a partially open position, the coolant coming from the engine isn't hot enough, or the heater core is restricted. If the coolant isn't hot enough, the thermostat in the engine cooling system is stuck open, allowing coolant to pass through the engine so rapidly that it doesn't heat up quickly enough. If the vehicle is equipped with a temperature gauge instead of a warning light, watch to see if the engine temperature rises to the normal operating range after driving for a reasonable distance.

39   If there's heat all the time, the control valve or the door is stuck wide open.

40   If there's no heat, coolant is probably not reaching the heater core, or the heater core is plugged. The likely cause is a collapsed or plugged hose, core, or a frozen heater control valve. If the heater is the type that flows coolant all the time, the cause is a stuck door or a broken or kinked control cable.

Air conditioning system

41   If the cool air output is inadequate:
Inspect the condenser coils and fins to make sure they're clear.

a) *Check the compressor clutch for slippage.*
b) *Check the blower motor for proper operation.*
c) *Inspect the blower discharge passage for obstructions.*
d) *Check the system air intake filter for clogging.*

42   If the system provides intermittent cooling air:

a) *Check the circuit breaker, blower switch and blower motor for a malfunction.*
b) *Make sure the compressor clutch isn't slipping.*
c) *Inspect the plenum door to make sure it's operating properly.*
d) *Inspect the evaporator to make sure it isn't clogged.*
e) *If the unit is icing up, it may be caused by excessive moisture in the system, incorrect super heat switch adjustment or low thermostat adjustment.*

43   If the system provides no cooling air:

a) *Inspect the compressor drivebelt. Make sure it's not loose or broken.*
b) *Make sure the compressor clutch engages. If it doesn't, check for a blown fuse.*
c) *Inspect the wire harness for broken or disconnected wires.*
d) *If the compressor clutch doesn't engage, bridge the terminals of the A/C pressure switch(es) with a jumper wire; if the clutch*

3.1a Look for the evaporator drain hose on the firewall; make sure it isn't plugged

now engages, and the system is properly charged, the pressure switch is bad.
e) *Make sure the blower motor is not disconnected or burned out.*
f) *Make sure the compressor isn't partially or completely seized.*
g) *Inspect the refrigerant lines for leaks.*
h) *Check the components for leaks.*
i) *Inspect the receiver-drier/accumulator or expansion valve/tube for clogged screens.*

44   If the system is noisy:

a) *Look for loose panels in the passenger compartment.*
b) *Inspect the compressor drivebelt. It may be loose or worn.*
c) *Check the compressor mounting bolts. They should be tight.*
d) *Listen carefully to the compressor. It may be worn out.*
e) *Listen to the idler pulley and bearing and the clutch. Either may be defective.*
f) *The winding in the compressor clutch coil or solenoid may be defective.*
g) *The compressor oil level may be low.*
h) *The blower motor fan bushing or the motor itself may be worn out.*
i) *If there is an excessive charge in the system, you'll hear a rumbling noise in the high pressure line, a thumping noise in the compressor, or see bubbles or cloudiness in the sight glass.*
j) *If there's a low charge in the system, you might hear hissing in the evaporator case at the expansion valve, or see bubbles or cloudiness in the sight glass.*

---

3   Air conditioning and heating system - check and maintenance

---

Air conditioning system

Warning: *The air conditioning system is under high pressure. Do not loosen any hose fittings or remove any components until after the system has been discharged. Air conditioning refrigerant should be properly discharged into an EPA-approved recovery/recycling unit at a dealer service department or an automotive air conditioning*

3.1b On models where the drain hose isn't directly visible from the firewall, peel back the passenger's side carpeting to access the rubber drain hose (if equipped) - checking the hose or ports for blockage

repair facility. Always wear eye protection when disconnecting air conditioning system fittings.
Caution: *The models covered by this manual use environmentally friendly R-134a (2016 and earlier models) or R-1234yf (2017 models). This refrigerant (and its appropriate refrigerant oils) are not compatible with R-12 refrigerant system components, or each other, and must never be mixed or the components will be damaged.*
Caution: *When replacing entire components, additional refrigerant oil should be added equal to the amount that is removed with the component being replaced. Read the can before adding any oil to the system, to make sure it is compatible with the R-134a or R-1234yf system, as applicable.*

1   The following maintenance checks should be performed on a regular basis to ensure that the air conditioning continues to operate at peak efficiency.

a) *Inspect the condition of the compressor drivebelt. If it is worn or deteriorated, replace it (see Chapter 1).*
b) *Check the drivebelt tension (see Chapter 1).*
c) *Inspect the system hoses. Look for cracks, bubbles, hardening and deterioration. Inspect the hoses and all fittings for oil bubbles or seepage. If there is any evidence of wear, damage or leakage, replace the hose(s).*
d) *Inspect the condenser fins for leaves, bugs and any other foreign material that may have embedded itself in the fins. Use a fin comb or compressed air to remove debris from the condenser.*
e) *Make sure the system has the correct refrigerant charge.*
f) *Check the evaporator housing drain tube for blockage (see illustrations).*

2   It's a good idea to operate the system for about ten minutes at least once a month. This is particularly important during the winter months because long term non-use can cause hardening and subsequent failure of the seals. Note that using the Defrost function operates the compressor.

3.8 Insert a thermometer in the center vent, turn on the air conditioning system and wait for it to cool down; depending on the humidity, the output air should be 35 to 40 degrees cooler than the ambient air temperature

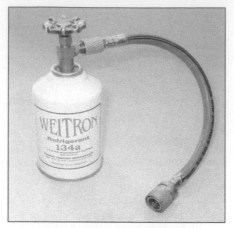

3.10 Automotive air conditioning charging kit

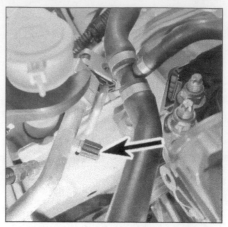

3.12 Location of the low-side charging port (2013 and later 1.6L model shown, other models similar)

3   If the air conditioning system is not working properly, proceed to Step 6 and perform the general checks outlined below.

4   Because of the complexity of the air conditioning system and the special equipment necessary to service it, in-depth troubleshooting and repairs beyond checking the refrigerant charge and the compressor clutch operation are not included in this manual. However, simple checks and component replacement procedures are provided in this Chapter. For more complete information on the air conditioning system, refer to the *Haynes Automotive Heating and Air Conditioning Manual.*

5   The most common cause of poor cooling is simply a low system refrigerant charge. If a noticeable drop in system cooling ability occurs, one of the following quick checks will help you determine if the refrigerant level is low.

Checking the refrigerant charge

6   Warm the engine up to normal operating temperature.

7   Place the air conditioning temperature selector at the coldest setting and put the blower at the highest setting.

8   Insert a thermometer in the center air distribution duct (see illustration) while operating the air conditioning system at its maximum setting - the temperature of the output air should be 35 to 40 degrees F below the ambient air temperature (down to approximately 40 degrees F). If the ambient (outside) air temperature is very high, say 110 degrees F, the duct air temperature may be as high as 60 degrees F, but generally the air conditioning is 35 to 40 degrees F cooler than the ambient air.

9   Further inspection or testing of the system requires special tools and techniques and is beyond the scope of the home mechanic.

Adding refrigerant

Caution: *Make sure any refrigerant, refrigerant oil or replacement component you purchase is designated as compatible*

with R-134a (2016 and earlier models) or R-1234yf (2017 models) systems.

10   Purchase an R-134a or R-1234yf automotive charging kit (as applicable) at an auto parts store (see illustration). A charging kit includes a can of refrigerant, a tap valve and a short section of hose that can be attached between the tap valve and the system low side service valve.

Caution: *Never add more than one can of refrigerant to the system. If more refrigerant than that is required, the system should be evacuated and leak tested.*

11   Back off the valve handle on the charging kit and screw the kit onto the refrigerant can, making sure first that the O-ring or rubber seal inside the threaded portion of the kit is in place.

Warning: *Wear protective eyewear when dealing with pressurized refrigerant cans.*

12   Remove the dust cap from the low-side charging port and attach the hose's quick-connect fitting to the port (see illustration). The low-side port is typically located on the larger diameter A/C line.

Warning: *DO NOT hook the charging kit hose to the system high side (smaller diameter A/C line)! The fittings on the charging kit are designed to fit only on the low side of the system.*

13   Warm up the engine and turn on the air conditioning. Keep the charging kit hose away from the fan and other moving parts.

Note: *The charging process requires the compressor to be running. If the clutch cycles off, you can put the air conditioning switch on High and leave the car doors open to keep the clutch on and compressor working. The compressor can be kept on during the charging by removing the connector from the pressure switch and bridging it with a paper clip or jumper wire during the procedure.*

14   Turn the valve handle on the kit until the stem pierces the can, then back the handle out to release the refrigerant. You should be able to hear the rush of gas. Keep the can upright at all times, but shake it occasionally. Allow stabilization time between each addition.

Note: *The charging process will go faster if*

you wrap the can with a hot-water-soaked rag to keep the can from freezing up.

15   If you have an accurate thermometer, you can place it in the center air conditioning duct inside the vehicle and keep track of the output air temperature. A charged system that is working properly should cool down to approximately 40 degrees F. If the ambient (outside) air temperature is very high, say 110 degrees F, the duct air temperature may be as high as 60 degrees F, but generally the air conditioning is 35 to 40 degrees F cooler than the ambient air.

16   When the can is empty, turn the valve handle to the closed position and release the connection from the low-side port. Reinstall the dust cap.

17   Remove the charging kit from the can and store the kit for future use with the piercing valve in the UP position, to prevent inadvertently piercing the can on the next use.

Heating systems

18   If the carpet under the heater core is damp, or if antifreeze vapor or steam is coming through the vents, the heater core is leaking. Remove it (see Section 14) and install a new unit (most radiator shops will not repair a leaking heater core).

19   If the air coming out of the heater vents isn't hot, the problem could stem from any of the following causes:

a)   *The thermostat is stuck open, preventing the engine coolant from warming up enough to carry heat to the heater core. Replace the thermostat (see Section 4).*

b)   *There is a blockage in the system, preventing the flow of coolant through the heater core. Feel both heater hoses at the firewall. They should be hot. If one of them is cold, there is an obstruction in one of the hoses or in the heater core, or the heater control valve is shut. Detach the hoses and back flush the heater core with a water hose. If the heater core is clear but circulation is impeded, remove the two hoses and flush them out with a water hose.*

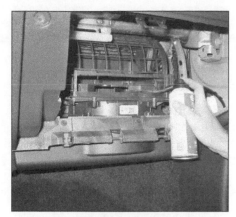

3.23 Insert the nozzle of the disinfectant can into the return-air intake behind the glove box

4.4a Disconnect the radiator hose from the thermostat housing cover - 2004 and earlier four-cylinder engine shown

4.4b Remove the spring clamp from the radiator hose at the thermostat housing cover - early V6 engine shown

c) If flushing fails to remove the blockage from the heater core, the core must be replaced (see Section 14).

Eliminating air conditioning odors

20   Unpleasant odors that often develop in air conditioning systems are caused by the growth of a fungus, usually on the surface of the evaporator core. The warm, humid environment there is a perfect breeding ground for mildew to develop.

21   The evaporator core on most vehicles is difficult to access, and factory dealerships have a lengthy, expensive process for eliminating the fungus by opening up the evaporator case and using a powerful disinfectant and rinse on the core until the fungus is gone. You can service your own system at home, but it takes something much stronger than basic household germ-killers or deodorizers.

22   Aerosol disinfectants for automotive air conditioning systems are available in most auto parts stores, but remember when shopping for them that the most effective treatments are also the most expensive. The basic procedure for using these sprays is to start by running the system in the RECIRC mode for ten minutes with the blower on its highest speed. Use the highest heat mode to dry out the system and keep the compressor from engaging by disconnecting the wiring connector at the compressor.

23   The disinfectant can usually comes with a long spray hose. Insert the nozzle into an intake port inside the cabin, and spray according to the manufacturer's recommendations (see illustration). Try to cover the whole surface of the evaporator core, by aiming the spray up, down and sideways. Follow the manufacturer's recommendations for the length of spray and waiting time between applications.

Automatic heating and air conditioning systems

24   Some vehicles are equipped with an optional automatic climate control sys-

tem. This system has its own computer that receives inputs from various sensors in the heating and air conditioning system. This computer, like the PCM, has self-diagnostic capabilities to help pinpoint problems or faults within the system. Vehicles equipped with automatic heating and air conditioning systems are very complex and considered beyond the scope of the home mechanic. Vehicles equipped with automatic heating and air conditioning systems should be taken to dealer service department or other qualified facility for repair.

4   Thermostat - replacement

Warning: Wait until the engine is completely cool before performing this procedure.
Note: The thermostat/thermostat housing on 1.5L, 1.6L, 2.3L, 2.5L, and 2013 and later 2.0L engine models are replaced as an assembly.

1   Disconnect the cable from the negative battery terminal (see Chapter 5).
2   Drain the cooling system (see Chap-

4.6a Remove the thermostat housing cover to access the thermostat - 2004 and earlier four-cylinder engine shown

ter 1). If the coolant is relatively new and still in good condition, save it and reuse it.

2004 and earlier four-cylinder and V6 models

3   Follow the lower radiator hose to the engine to locate the thermostat housing.
4   Loosen (or slide back) the hose clamp, then detach the hose from the fitting (see illustrations). If it's stuck, grasp it near the end with a pair of adjustable pliers and twist it to break the seal, then pull it off. If the hose is old or if it has deteriorated, cut it off and install a new one.
5   If the outer surface of the thermostat housing cover, which mates with the hose, is already corroded, pitted, or otherwise deteriorated, it might be damaged even more by hose removal. If it is, replace the thermostat housing cover.
6   Remove the thermostat cover mounting bolts, then pull off the thermostat cover (see illustrations). If the cover is stuck, tap it with a soft-faced hammer to jar it loose. Be prepared for some coolant to spill as the gasket seal is broken.

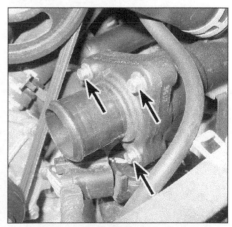

4.6b Remove the thermostat housing cover to access the thermostat - early V6 engine shown

4.7a Remove the thermostat from the housing - 2004 and earlier four-cylinder engine shown

4.7b Remove the thermostat from the housing - early V6 engine shown

7   Note how it's installed - which end is facing up or out - then remove the thermostat, noting that the jiggle valve on the thermostat is located in the 12 o'clock position (see illustrations).

2005 and later 2.0L, 2.3L, 2.5L and 3.0L models

Note: *The illustrations mentioned in Steps 3 through 7 are also relevant for this heading of the procedure*

Caution: *On 2013 and later models, shield the alternator from coolant spillage by covering it with waterproof plastic, to avoid damage*

8   On 2.3L engines, remove the drivebelt (see Chapter 1).
9   On 2008 and earlier 3.0L V6 models, remove the air intake tube (see Chapter 4).
10   Follow the lower radiator hose to the engine to locate the thermostat housing.
11   On 2.3L, 2.5L, and 2013 and later 2.0L engines, disconnect the heater hose from the

thermostat housing.
12   Slide back the clamp and disconnect the radiator hose from the thermostat housing.
13   On V6 models, remove the housing cover bolts and cover. Then, noting how it's installed, remove the thermostat. On all other models, remove the thermostat housing bolts and thermostat housing as an assembly.

1.5L and 1.6L models

14   On 1.5L models, remove the drivebelt (see Chapter 1) and A/C compressor (see Section 15). Position aside the A/C compressor and hang it securely out of the way with a length of wire, without disconnecting the refrigerant lines.
15   Remove the alternator (see Chapter 5).
16   On 1.6L models, follow the accompanying photos in order for the removal of the thermostat housing (see illustrations).
17   On 1.5L models, remove the single thermostat housing retaining nut, then unscrew

and remove the stud. Slide back the clamps, then disconnect the radiator and additional coolant hoses from the housing. Also detach the wiring harness clip from the underside of the housing. Using a slight twisting motion, remove the thermostat housing - replacing the O-ring upon installation.

All models

18   Remove all traces of old gasket material and sealant from the housing and cover with a gasket scraper, then with a rag saturated in brake cleaner.
19   Install a new rubber gasket(s) on the thermostat or thermostat housing (see illustrations), then install the thermostat in the housing, spring-end first (for models that apply). Make sure that you install the thermostat with the "jiggle valve" at 12 o'clock (see illustration 4.7a). If the thermostat housing cover uses a paper gasket, apply a slight film of gasket sealant on both sides, then position it on the thermostat housing with the holes in the gasket aligned with the bolt holes in the housing.
20   Install the thermostat housing or housing cover and bolts, then tighten the bolts to the torque listed in this Chapter's Specifications.
21   Attach the radiator hose(s) to the thermostat housing cover. Make sure that the hose clamps are still tight. If they aren't, replace them.
22   Installation of components is otherwise the reverse of removal.
23   Refill the cooling system (see Chapter 1).
24   Reconnect the battery. After you're done, the Powertrain Control Module (PCM) must relearn its idle and fuel trim strategy for optimum driveability and performance (see Chapter 5, Section 1).
25   Start the engine and allow it to reach normal operating temperature, then check for leaks and proper thermostat operation (as described in Steps 2 through 4).

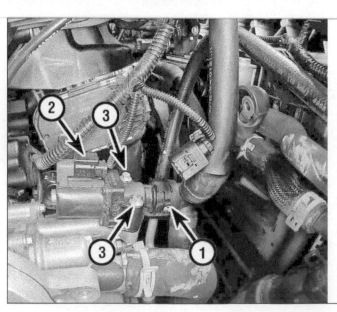

4.16a Remove the coolant shutoff solenoid valve (heater shutoff valve)

1   *Pry-out the quick release clip and disconnect the coolant hose fitting*
2   *Disconnect the electrical connector*
3   *Remove the mounting bolts*

4.16b Slide back the clamps, then disconnect the radiator and heater hoses from the thermostat housing

4.16c Remove the thermostat housing mounting bolts

4.19a Install a new rubber seal around the perimeter of the thermostat

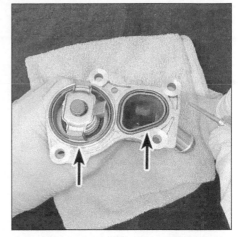

4.19b On 1.6L models, there are two new rubber seals on the housing

5   Engine cooling fans - replacement

Warning: *2005 through 2012 models have a Supplemental Restraint System, or airbag sensor, mounted in the center of the radiator support bracket. It will be necessary to disarm the system prior to performing any work around the radiator, fans or other components in this area (see Chapter 12).*

Warning: *To avoid possible injury or damage, DO NOT operate the engine with a damaged fan. Do not attempt to repair fan blades - replace a damaged fan with a new one.*

Note: *Always be sure to check for blown fuses before attempting to diagnose an electrical circuit problem.*

Note: *Four-cylinder models are equipped with a high, a medium and a low speed cooling fan relay. V6 models are equipped with a low speed relay, a medium speed relay and two high speed fan relays that are controlled by the PCM.*

Warning: *Wait until the engine is completely cool before performing this procedure.*

2008 and earlier models

1   Disconnect the cable from the negative battery terminal (see Chapter 5).
2   Disconnect the fan motor electrical connector(s).
3   Raise the vehicle and secure it on jackstands.
4   Drain the cooling system (see Chapter 1). If the coolant is relatively new and still in good condition, save it and reuse it.
5   On 2004 and earlier models, remove the hood latch (see Chapter 11).
6   Remove the upper radiator hose (see illustration 7.7). Loosen the hose clamp by squeezing the ends together. Hose clamp pliers work best, but regular pliers will work also. If the radiator hose is stuck, grasp it near the end with a pair of adjustable pliers and twist it to break the seal, then pull it off. If the hose is old or if it has deteriorated, cut it off and install a new one.
7   Remove the center support bolts (see illustrations) and lift the center support from the engine compartment.
8   Remove the left and right upper radiator

5.7a Remove the bolts from the center support (upper bolt shown) . . .

support brackets (see illustrations).
9   Tilt the radiator forward to make additional clearance for fan removal.

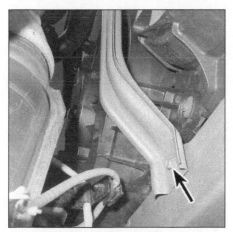

5.7b . . . and the lower section of the center support on 2008 and earlier models

5.8a Remove the right side radiator support bracket bolts . . .

5.8b . . . and the left side radiator support bracket bolts - 2008 and earlier models

10   Detach the wire harness from the fan shroud, remove the upper and lower fan shroud mounting bolts (see illustration) and then carefully lift the fan motor and shroud out of the engine compartment as a single assembly.

11   Remove the fan mounting screws.

12   To detach the fan blade from the motor, remove the nut or retaining clip from the motor shaft (see illustration). Remove the fan blade from the motor.

13   To detach the motor from the shroud, remove the retaining bolts. Remove the motor from the shroud.

14   Installation is the reverse of removal.

Note: *When reinstalling the fan assembly, make sure the rubber air shields around the shroud are still in place - without them, the cooling system may not work efficiently.*

15   Reconnect the battery and refill the cooling system (see Chapter 1). After you're done, the Powertrain Control Module (PCM) must relearn its idle and fuel trim strategy for optimum driveability and performance (see Chapter 5, Section 1).

16   Start the engine and allow it to reach normal operating temperature, then check for leaks and proper operation.

2009 through 2012 models

17   Raise the vehicle and support it securely on jackstands.

18   On four-cylinder models, refer to Chapter 1 and drain the engine coolant.

19   Refer to Chapter 11 and remove the front bumper cover.

20   Remove both airbag sensors (see Warning at the top). Disarm the airbag system before performing any work in the area of the cooling fans (see Chapter 12).

21   Remove the brackets from the top of the radiator.

22   Remove the nuts and bolts from the hood latch, then set the latch aside with the cable attached.

23   Detach the wiring harness clips from the radiator.

5.10 Remove the cooling fan shroud bolts - lower bolt not visible

24   Remove the bolt from the vertical center brace in front of the radiator.

25   On four-cylinder models, disconnect the wiring from the fan resistor, then disconnect the coolant recovery hose and push it to the side.

26   Disconnect the wiring from the cooing fans.

27   Remove the two cooling fan mounting bolts and carefully lift the fan assembly out of the engine compartment.

28   Installation is the reverse of removal.

2013 and later models

29   Remove the cooling module (see Section 9).

30   Remove the lower radiator support from the bottom of the cooling module assembly.

Note: *Some later models may not have any transmission cooler control valve hoses to disconnect*

31   Using pliers, disconnect the spring clamps and remove the hoses from the transmission fluid cooler control valve.

5.12 Use a flat-bladed screwdriver and drive the locking washer off the fan motor shaft

32   Lift the cooling fan shroud clips at each side of the radiator and lift the shroud and motor assembly from the radiator.

33   Installation is the reverse of removal.

6   Coolant expansion tank - removal and installation

Warning: *Wait until the engine is completely cool before beginning this procedure.*

2012 and earlier models

1   Drain the cooling system (see Chapter 1).

2   Slide back the hose clamps and disconnect the expansion tank hoses.

3   Remove the expansion tank mounting nuts (see illustration). Lift the tank out of the engine compartment.

4   Clean out the tank with soapy water and a brush to remove any deposits inside. Inspect the reservoir carefully for cracks. If you find a crack, replace the reservoir.

5   Installation is the reverse of removal.

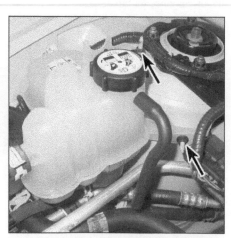

6.3 Location of the expansion tank mounting nuts - 2012 and earlier models

6.8a Lift expansion tank up and off the front mounting tab, using the screwdriver for leverage as necessary . . .

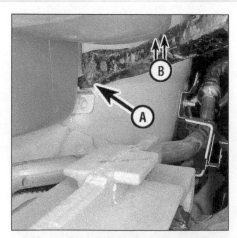

6.8b . . . then detach the tank from the rear tab, using a prybar with the end placed on the bracket (A) and prying up sharply to release the tank (B)

2013 and later models

Note: *The expansion tank design on 2017 models is slightly different - however, the replacement concept is identical*

6   Refer to Chapter 1 and drain the cooling system until no more coolant remains in the expansion tank. Clamp-off the lower tank hose to prevent coolant spillage, if necessary.

7   Slide back the hose clamps, then disconnect the hoses from the tank.

8   Lift expansion tank up and off of the mounting tabs (see illustrations).

9   Installation is the reverse of removal.

---

7   Radiator - removal and installation

Warning: *Wait until the engine is completely cool before beginning this procedure.*

Warning: *2005 through 2012 models have a Supplemental Restraint System, or airbag sensor, mounted in the center of the radiator support bracket. It will be necessary to disarm the system prior to performing any work around the radiator, fans or other components in this area (see Chapter 12)*

Removal

2007 and earlier models

1   Disconnect the cable from the negative battery terminal (see Chapter 5).

2   Raise the vehicle and place it securely on jackstands. Remove the lower splash shields.

3   Drain the cooling system (see Chapter 1). If the coolant is relatively new and in good condition, save it and reuse it.

4   Unbolt the cooling fan shrouds from the radiator. On 2005 through 2007 models, remove the cooling fan shrouds completely (see Section 5).

5   On 2004 and earlier models, remove the hood latch and the center support (see Section 5).

6   On 3.0L models, disconnect the lower radiator hose and expansion tank hose from the left side tank of the radiator (see illustration). On four-cylinder engines, disconnect the

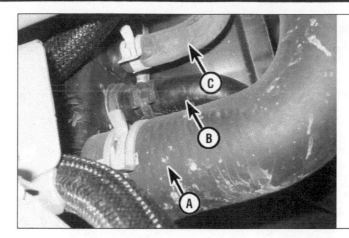

7.6 Detach the hoses from the lower left side of the radiator - 3.0L model shown

A   *Lower radiator hose*
B   *Transaxle fluid cooler hose*
C   *Expansion tank hose*

lower radiator and expansion tank hoses from the right side tank of the radiator. If the vehicle is equipped with an automatic transaxle, also disconnect the cooler hose from the fitting.

7   Lower the vehicle and disconnect the expansion tank hose and upper radiator hose from the right side of the radiator (see illustration). Also, if the vehicle is equipped with an automatic transaxle, disconnect the other fluid cooler hose from the upper left side of the radiator.

8   Remove the radiator support brackets from the radiator support beam and the radiator (see illustration).

9   Carefully lift out the radiator. Don't spill coolant on the vehicle or scratch the paint. Make sure the rubber radiator insulators that fit on the bottom of the radiator and into the sockets in the body remain in place in the body for proper installation of the radiator.

10   Remove bugs and dirt from the radiator with compressed air and a soft brush. Don't bend the cooling fins. Inspect the radiator for leaks and damage. If it appears damaged, replace it.

2008 through 2012 models

11   Refer to Chapter 1 and drain the engine coolant.

12   Remove the engine cooling fans (see Section 5).

13   Detach all hoses from the radiator.

14   Remove the two upper bolts that attach the a/c condenser to the radiator. Move the air conditioning condenser away from the radiator.

15   Carefully remove the radiator.

2013 and later models

16   Remove the cooling module (see Section 9).

17   Remove the lower radiator support from the bottom of the cooling module assembly.

18   Using pliers, disconnect the spring clamps and remove the hoses from the transmission fluid cooler control valve located on the right-hand side of the radiator (if equipped).

19   Pry outward on the cooling fan shroud clips at each side of the radiator, then slide the shroud and motor assembly upward and remove it from the radiator.

20   Pull the lower radiator hose spring clip (see illustration 9.13) until the end of the spring clip is snapped into the detent on the quick-connect coupling and disconnect the hoses from the radiator.

21   Remove the charge air cooler (CAC, intercooler) from the radiator (see Chapter 4).

22   Cut the foam seal between the radiator and the condenser using a utility knife.

23   Press the condenser retaining tabs in on both sides of the radiator and slide the condenser up to detach it from the radiator.

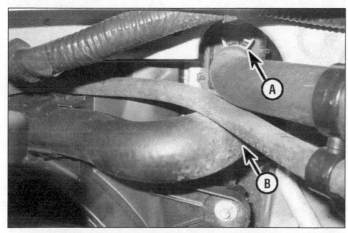

7.7 Detach the expansion tank hose (A) and the upper radiator hose (B) from the upper right side of the radiator - 3.0L model shown

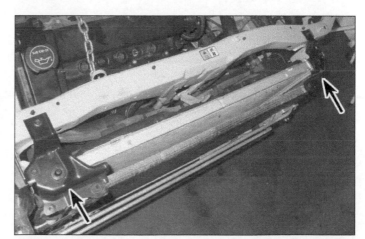

7.8 Remove the radiator support brackets and lift out the radiator (front bumper cover removed for clarity)

8.7 Remove the water pump bolts

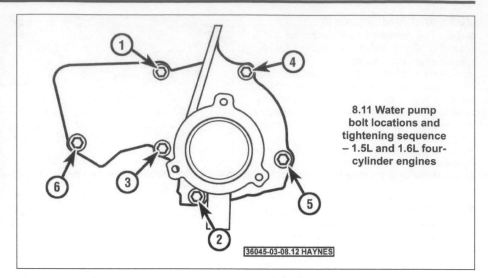

8.11 Water pump bolt locations and tightening sequence – 1.5L and 1.6L four-cylinder engines

Installation

24   Inspect the rubber insulators in the lower crossmember for cracks and deterioration. Make sure that they're free of dirt and gravel. When installing the radiator, make sure that it's correctly seated on the insulators before fastening the retaining brackets.

25   Installation is otherwise the reverse of the removal procedure. After installation, fill the cooling system with the correct mixture of antifreeze and water (see Chapter 1).

26   Reconnect the battery. After you're done, the Powertrain Control Module (PCM) must relearn its idle and fuel trim strategy for optimum driveability and performance (see Chapter 5, Section 1).

27   Start the engine and check for leaks. Allow the engine to reach normal operating temperature, indicated by the upper radiator hose becoming hot. Recheck the coolant level and add more if required.

28   If you're working on an automatic transaxle equipped vehicle, check and add fluid as needed.

8   Water pump - removal and installation

Warning: *Wait until the engine is completely cool before beginning this procedure.*

Removal

1   Disconnect the cable from the negative battery terminal (see Chapter 5).

2   Drain the cooling system (see Chapter 1).

Note: *Loosening the water pump pulley bolts before the drivebelt is removed will make removal of the pulley easier.*

3   Remove the drivebelt on four-cylinder models or the water pump drivebelt on V6 models (see Chapter 1).

Four-cylinder models

4   Loosen the right front wheel lug nuts. Raise the vehicle and secure it on jackstands.

5   Remove the right front wheel and the fender splash shield.

6   Remove the water pump pulley.

2.0L, 2.3L and 2.5L models

7   Remove the bolts attaching the water pump to the engine block and remove the pump from the engine (see illustration). If the water pump is stuck, gently tap it with a soft-faced hammer to break the seal. Have a drain pan positioned under the pump to catch coolant spilling out as the seal is broken.

8   Clean the bolt threads and the threaded holes in the engine and remove all corrosion and sealant. Remove all traces of old gasket material from the sealing surfaces or replace the O-ring.

1.5L and 1.6L models

9   Remove the cowl panel (Chapter 11), then remove the air filter outlet pipe(s) and turbocharger inlet pipe (see Chapter 4).

10   Remove the timing belt and tensioner (see Chapter 2A, Section 9).

11   Remove the water pump mounting bolts (see illustration) and remove the pump.

8.14 Remove the water pump drivebelt tensioner bolt (2004 and earlier models)

8.16 Remove the water pump bolts

V6 models

2008 and earlier models

12   Remove the air filter housing and intake pipe (see Chapter 4).

Note: *On models where it's necessary, removing the water pump pulley requires a special puller that is used to slide the press-fitted pulley off the water pump shaft. Use Ford special tool numbers: 303-456, 303-457, 303-009 or equivalent.*

13   If you're working on a 2006 (late build) to 2008 model, remove the water pump pulley.

14   On 2004 and earlier models, remove the water pump drivebelt tensioner (see illustration).

15   If equipped, disconnect the coolant hoses from the water pump.

16   Remove the water pump mounting bolts (see illustration).

17   Remove the gasket, then clean all the gasket and O-ring surfaces on the pump and the housing.

2009 through 2012 models

18   Remove the outlet air duct from the air filter housing (see Chapter 4).

19   Remove the water pump drivebelt (see Chapter 1).

20   Remove the lower radiator hose from the thermostat housing (see Section 4).

21   Remove the three mounting bolts, then separate the water pump from the housing. Discard the water pump gasket.

22   Clean the sealing surfaces of the housing and the pump. Remove all traces of old gasket.

All models

23   Compare the new pump to the old one to make sure that they're identical.

24   Apply a thin film of RTV sealant to hold the new gasket in place during installation.

Caution: *Make sure that the gasket is correctly positioned on the water pump and the mating surfaces are clean and free of old gasket material. Carefully mate the pump to the water pump housing.*

25   Install the water pump bolts and tighten them to the torque listed in this Chapter's Specifications. On 1.5L and 1.6L models, also follow the tightening sequence shown in Illustration 8.11.

26   The remainder of installation is the reverse of removal. Refill the cooling system (see Chapter 1).

27   Reconnect the battery. After you're done, the Powertrain Control Module (PCM) must relearn its idle and fuel trim strategy for optimum driveability and performance (see Chapter 5, Section 1).

28   Operate the engine and check for leaks.

---

9   Cooling module (2013 and later models) - removal and installation

---

Warning: *The air conditioning system is under high pressure. DO NOT loosen any fittings or remove any components until after the system has been discharged. Air conditioning refrigerant must be properly discharged into an EPA-approved container at a dealer service department oran automotive air conditioning repairfacility. Always wear eye protection when disconnecting air conditioning system fittings.*

1   Have the air conditioning system discharged by a dealer service department or by an automotive air conditioning shop before proceeding (see Warning above).

2   Drain the cooling system (see Chapter 1).

3   Raise the vehicle and secure it on jackstands, then remove the engine splash shield (see Chapter 1).

4   On 1.5L models, remove the left-hand headlight housing (see Chapter 12). On 2017 2.0L models, remove the air filter housing (see Chapter 4).

5   Remove the fasteners for the air conditioning lines (see illustration), then disconnect the lines and cap the openings to prevent

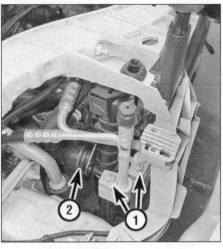

9.5 Remove the air conditioning line mounting nuts

1   *Air conditioning line mounting nuts*
2   *Coolant hose quick-connect coupling spring clip location*

foreign material from entering. Move the lines out the way.

6   Disconnect the coolant hose quick-connect coupling by prying-out the spring clip and pulling the coupling off the radiator stub (see illustration 9.4).

7   Disconnect the transmission fluid cooler control valve electrical connector (see Chapter 7B, Section 7).

8   Disconnect the coolant hose quick-connect coupling from the transmission fluid cooler control valve and release the hose from the retaining clip.

9   Disconnect the hose clamp and remove the lower hose from the thermostat housing (see illustration).

10   Disconnect the cooling fan motor electrical connector from the module (see illustration).

11   Squeeze the tabs and remove the coolant tube (see illustration).

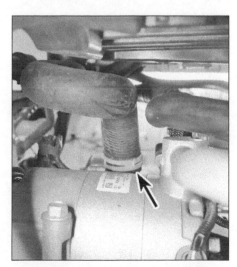

9.9 Use a pair of pliers to release the spring clamps and slide the clamps up the hose

9.10 Disconnect the electrical connector from the cooling fan module

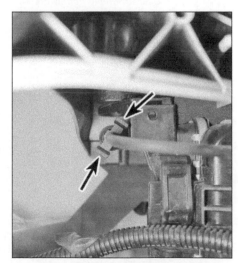

9.11 Depress the tabs inwards to release the coolant tube

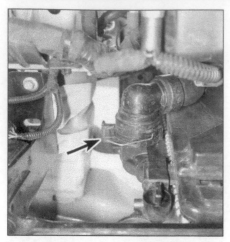

9.14 Pry the lower radiator hose spring clip out until the end of the clip is in the detent

9.15 Wiring harness connector location

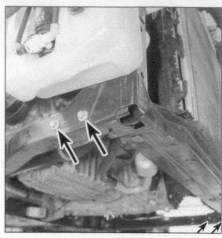

9.16a Remove the cooling module side mounting bolts . . .

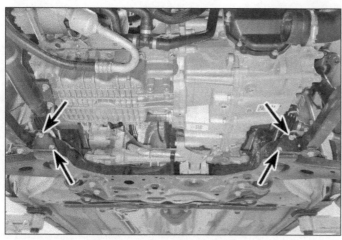

9.16b . . . support arm bolts . . .

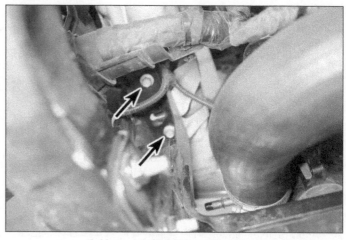

9.16c . . . left side lower bolts . . .

12   Remove the active grille shutter (see Section 10), if equipped.
Note: *Some vehicles are equipped with a active grille shutter assembly framework only and will not have a grille shutter actuator or a grille shutter linkage.*
13   Disconnect the quick-connect couplings from the charge air cooler (CAC, intercooler),

then disconnect the inlet and outlet hoses, (see Chapter 4) if equipped.
Note: *On 1.5L models, it will be necessary to completely remove the charge air cooler (intercooler).*
14   Pull the lower radiator hose spring clip (see illustration) until the end of the spring clip is snapped into the detent on the quick-con-

nect coupling and disconnect the lower radiator hose.
15   Disconnect the wiring harness electrical connector (see illustration), unclip the connector from the cooling module and move the harness out of the way.
16   Support the cooling module assembly with a floor jack and block of wood, then

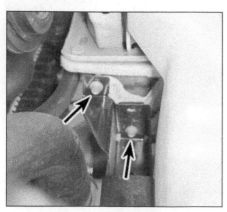

9.16d . . . and right side bolts

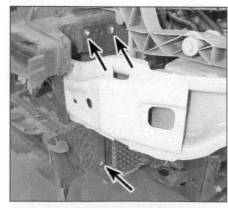

10.2 Lower air deflector fastener locations - left side shown, right side identical

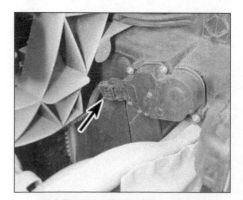

10.3 Depress the locking tab and disconnect the electrical connector to the grille shutter actuator

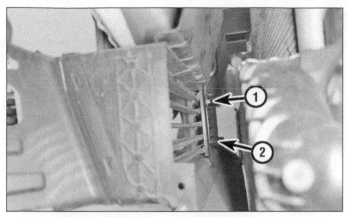

10.4a Location of the bridge link fasteners and retaining tabs
1 *Bridge link lower fastener, upper fastener identical*
2 *Bridge link lower retaining tab, upper tab identical*

10.4b Release the retaining tabs by squeezing it towards the center of the bridge link - upper retaining tab shown, lower tab release identical

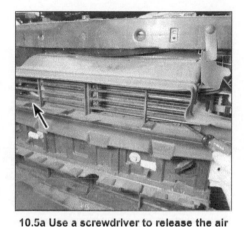

10.5a Use a screwdriver to release the air deflector tabs, pulling the upper part of the deflector rearward . . .

10.5b . . . then unhook the bottom of the deflector and slide it out from behind the grille shutter

10.6 Upper air deflector fastener locations - left side shown, right side identical

remove the cooling module mounting bolts (see illustrations).

17   With the help of an assistant, lower the floor jack and cooling module to the ground.

18   Installation is the reverse of removal, tightening the cooling module mounting bolts to the torque listed in this Chapter's Specifications.

19   Refill the cooling system (see Chapter 1).

20   Take the vehicle to the shop that discharged the air conditioning system and have the system properly recharged.

10 Active grille shutter (2013 and later models) - removal and installation

Note: *Although 2017 models have a slightly different design arrangement, the replacement procedure is similar to the one mentioned in this Section. Be sure to disconnect the shutter actuator electrical connector and wiring harness clips, located on the inside, before pulling off the active grille shutter assembly.*

Removal

1   Remove the front bumper cover (see Chapter 11).

2   Remove the radiator lower air deflector fasteners from both sides of the grille shutter (see illustration) and remove the deflectors.

3   Disconnect the electrical connector to the grille shutter actuator (see illustration).

4   Working on the back side of the grille shutter, remove the retainers, then release the retaining tabs by squeezing them together and separate the bridge link from the grille

shutter (see illustrations).

5   Release the locking tabs for the front bumper air deflector, then lift the deflector up and remove it from the vehicle (see illustrations).

6   Remove the radiator upper air deflector fasteners from both sides of the radiator (see illustration).

7   Remove the active grille shutter fasteners (see illustration).

10.7 Active grille shutter fastener locations

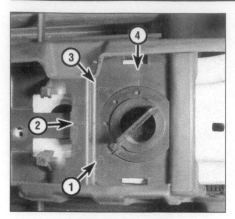

10.8 Stow knob details

1 *Lock position*   3 *Stow position*
2 *Assemble position*   4 *Ship position*

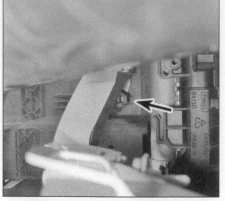

10.9 Disengage the retaining tab by placing into the center position and pulling it slightly forward - left side shown, right side identical

10.10 After rotating the stow knob to the "ship" position, remove the active grille shutter from the front of the vehicle

8   Rotate the stow knobs clockwise from the "lock" position to the "assemble" position (see illustration).
9   Disengage the retaining tabs (see illustration) and move the active grille shutter slightly forward.
10   Rotate the stow knobs clockwise from the "assemble" position to the "ship" position (see illustration 10.8) and remove the active grille shutter assembly from the vehicle (see illustration).

Installation

11   Install the active shutter grille to the front of the vehicle and rotate the stow knob counterclockwise from the "ship position" to the "assemble" position (see illustration 10.8).
12   Installation is the reverse of removal, with the following points: Once all the components are installed, rotate the stow knob counterclockwise from the "assemble" position to the "lock" position (see illustration 10.8), then install the bumper cover.

11   Coolant temperature sending unit - general information

1   The coolant temperature indicator system consists of a warning light or a temperature

gauge on the dash and a coolant temperature sending unit mounted on the engine. A Cylinder Head Temperature (CHT) sensor is used on all four-cylinder models and 2009 and later V6 models. An Engine Coolant Temperature (ECT) sensor is used on 2008 and earlier V6 models. These are information sensors for the Powertrain Control Module (PCM); they also function as the coolant temperature sending unit.
2   See Chapter 6 for the CHT or ECT sensor replacement procedure.

12   Blower motor resistor and blower motor - replacement

Warning: *The models covered by this manual are equipped with a Supplemental Restraint System (SRS), more commonly known as airbags. Always disable the airbag system before working in the vicinity of any airbag system component to avoid the possibility of accidental deployment of the airbag, which could cause personal injury (see Chapter 12).*

Blower motor resistor

1   On 2007 and earlier models, remove the trim panel from below the glove box

(see Chapter 11). On 2008 through 2012 models, remove the entire glove box. On 2013 and later models, remove the accelerator pedal position (APP) sensor (see Chapter 6).
2   Disconnect the electrical connector from the blower motor resistor, then remove the mounting screws (see illustration). Remove the resistor from the housing.
3   On 2013 and later models, the blower motor resistor is accessed from the drivers side footwell with the APP sensor removed. Disconnect the electrical connector(s) from the resistor (see illustration), then remove the mounting screw and guide the resistor out of the housing.
4   Installation is the reverse of removal.

Blower motor
2007 and earlier models

5   Remove the lower trim panel from below the glovebox (see Chapter 11).
6   Disconnect the blower motor electrical connector (see illustration).
7   Remove the blower motor mounting screws (see illustration) and remove the blower motor.
8   Installation is the reverse of removal.

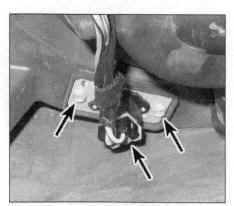

12.2 Unplug the blower motor resistor connector, then remove the mounting screw(s) - 2007 earlier models shown, later models (until 2013) similar

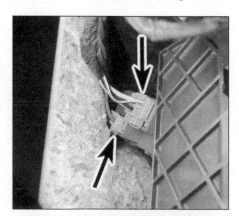

12.3 Blower motor resistor electrical connectors - 2013 and later models

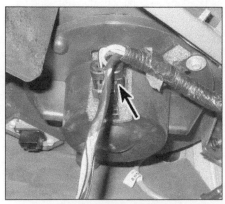

12.6 Location of the blower motor electrical connector - 2007 and earlier models

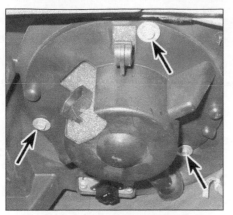

12.7 Location of the blower motor mounting screws - 2007 and earlier models

12.11 Pull off this small duct from the bottom of the blower motor - (2008 through 2012 models)

12.12 Rotate the blower motor counterclockwise to remove it - (2008 through 2012 models)

2008 through 2012 models

9 Refer to Chapter 11 and remove the glove box and the trim panel below it.

10 Disconnect the wiring from the blower motor.

11 Release the two vent duct clips, then pull the vent duct away from the heater core (see illustration).

12 Turn the blower motor counterclockwise until it is released from the main housing, then remove it (see illustration).

13 Installation is the reverse of removal.

2013 and later models

Caution: *The blower motor removal procedure, on 2013 and later models is complex. Do not attempt this procedure unless you are highly experienced with this type of repair. If you are at all unsure of your abilities, consult an expert.*

14 Disconnect the cable from the negative battery terminal (see Chapter 5, Section 1).

15 Remove the accelerator pedal position (APP) sensor (see Chapter 6).

16 Remove the glove box and end trim panels (see Chapter 11), then disconnect the electrical connectors (see illustration) to the Body Control Module (BCM), depress the retaining tabs at the bottom of the BCM and remove the BCM from the vehicle.

17 Once the BCM has been removed, remove

the BCM harness bracket nuts, then detach the harness retaining clips and remove the bracket.

18 If equipped with a front lighting module, disconnect the module electrical connector on the backside of the BCM bracket. Remove the adaptive front lighting module bracket fasteners and remove the assembly.

Note: *When removing the air inlet duct, there are three torx fasteners - two of which are hidden on the backside of the air inlet duct assembly. Use a torx socket and a long extension/ratchet to access these fasteners. A magnetic pickup tool and small mirror may also be useful tools.*

19 Disconnect the air inlet actuator electrical connector, then remove the air inlet duct assembly torx fasteners. Remove the air inlet duct assembly.

20 Working from the driver's side footwell, disconnect the electrical connector to the blower motor.

21 It will take the help of an assistant to remove the blower motor. Have your assistant work from the passenger's side of the vehicle while you work from the driver's side. Working from the driver's side, insert a flat blade screwdriver between the blower motor and the retaining clip to disengage the clip and keep the screwdriver in this position. Have the assistant reach through to the blower motor cage and turn the blower motor housing 45-degrees clockwise

(from his perspective). Once the blower motor is in this position it can be removed from the passenger's side of the vehicle.

22 Installation is the reverse of removal.

13 Heater/air conditioner control assembly - removal and installation

Warning: *The models covered by this manual are equipped with a Supplemental Restraint System (SRS), more commonly known as airbags. Always disable the airbag system before working in the vicinity of any airbag system component to avoid the possibility of accidental deployment of the airbag, which could cause personal injury (see Chapter 12).*

1 Disconnect the cable from the negative battery terminal (see Chapter 5).

2 Remove the dashboard center trim panels (see Chapter 11).

Note: *On 2013 and later models, it might make it easier to first pry-out and remove the vent, located directly above the control assembly, before removing the air conditioning control assembly trim panel.*

3 On 2013 and later models, pry the heater/air conditioner control assembly trim panel out (see illustration).

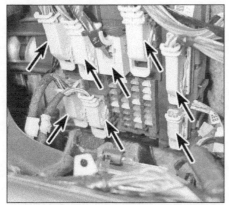

12.16 Disconnect the electrical connectors to the Body Control Module (BCM)

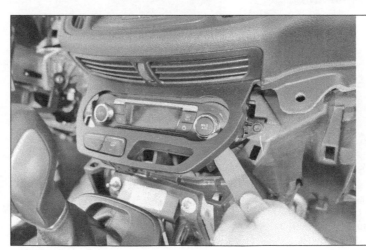

13.3 Pry the control assembly trim panel out - 2013 and later models

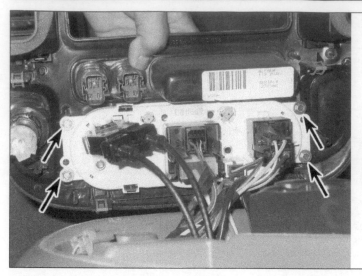

13.4 Location of the heater/ air conditioner control assembly mounting screws - 2007 and earlier model shown, 2008 through 2012 models similar

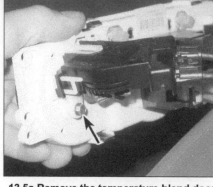

13.5a Remove the temperature blend door control cable mounting screw and separate the cable assembly from the control assembly - 2007 and earlier models

4   Remove the heater/air conditioner control assembly retaining screws (see illustration). On 2013 and later models, the screws retain the control assembly directly to the dash.

5   On 2007 and earlier models, disconnect the temperature blend door control cables, the blower motor switch and the temperature control switch connectors (see illustrations). On later models, disconnect the wiring harness (see illustration).

6   Installation is the reverse of removal.

14   Heater core - replacement

Warning: *The models covered by this manual are equipped with a Supplemental Restraint System (SRS), more commonly known as airbags. Always disarm the airbag system before working in the vicinity of any airbag system component to avoid the possibility of accidental deployment of the airbag, which could cause personal injury (see Chapter 12).*

Warning: *Wait until the engine is completely cool before beginning this procedure.*

2007 and earlier models

1   Disconnect the cable from the negative battery terminal (see Chapter 5).

2   Drain the cooling system (see Chapter 1), then disconnect the heater hoses from the heater core inlet and outlet pipes at the firewall (see illustration).

3   Remove the instrument panel (see Chapter 11).

4   Remove the lever from the heater blend door shaft (see illustration).

5   Remove the actuating rod from the lever (see illustration).

6   Remove the heater core cover screws (see illustration).

7   Pull the heater core out of the housing (see illustration).

8   Installation is the reverse of removal. Don't forget to reconnect the heater core inlet and outlet hoses at the firewall.

9   Refill the cooling system (see Chapter 1).

2008 through 2012 models

10   Have the air conditioning system discharged and recovered by an approved service facility.

11   Disconnect the cable from the negative battery terminal (see Chapter 1).

12   Refer to Chapter 1 and drain the engine coolant.

13   Remove the entire instrument panel (see Chapter 11).

14   Disconnect the air conditioning refrigerant lines from the engine side of the firewall (see illustration).

15   Disconnect the heater hoses from the heater core at the engine side of the firewall.

16   Remove the heater core housing mounting nuts from the instrument panel studs.

17   Pull the right side of the heater core housing to the rear, then slide it to the right to release it. Remove it from the vehicle.

18   Remove the seal from the heater core tubes.

19   Remove the bracket that secures the heater core, then slide the heater core out.

20   Installation is the reverse of removal.

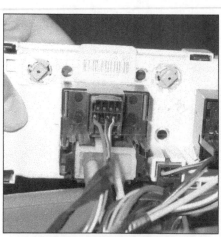

13.5b Release the tabs and disconnect the blower switch and temperature control switch connectors - 2007 and earlier models

13.5c Disconnect the wiring connectors from the back side of the control assembly - 2013 and later model shown, 2008 through 2012 models similar

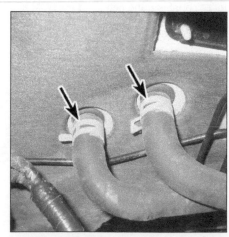

14.2 To disconnect the heater hoses from the heater core pipes, squeeze the tabs on the spring clamps together with a pair of pliers and slide them up the hoses

14.4 Carefully pry the lever off the blend door shaft - 2007 and earlier models

14.5 Remove the actuating rod from the lever - 2007 and earlier models

14.6 Remove the screws and detach the cover from the HVAC housing - 2007 and earlier models

Check for coolant leaks, then have the air conditioning system evacuated and recharged by the shop that discharged it.

2013 and later models

21   Have the air conditioning system discharged and recovered by an approved service facility.

22   Disconnect the cable from the negative battery terminal (see Chapter 1).

23   Refer to Chapter 1 and drain the engine coolant.

24   Remove the cowl panel (see Chapter 11) and the air filter outlet and turbo intake pipes (see Chapter 4).

25   Disconnect electrical connector to the pressure transducer, then remove the nut and the air conditioning refrigerant lines from the engine side of the firewall (see illustration).

26   Disconnect the heater hose quick-connect fittings from the heater core at the engine side of the firewall (see illustration). Be prepared for coolant loss.

27   Remove the entire instrument panel (see Chapter 11).

28   Remove the evaporator housing bracket

14.7 Slide the heater core from the housing - 2007 and earlier models

bolts and bracket from the heater/evaporator housing and instrument panel center brace.

29   Remove the instrument panel-to firewall seal, then remove the evaporator tube support bracket fastener.

30   Remove the heater core tube mounting bracket and the evaporator cover fasteners

14.14 The refrigerant lines attach to the air conditioning ovaporator at the firewall

(the cabin air filter door is mounted to this cover), then lift the cover off of the heater/evaporator housing.

31   On the opposite side of the heater/evaporator housing, remove the four air plenum cover fasteners and cover.

32   Remove the five lower heater core cover

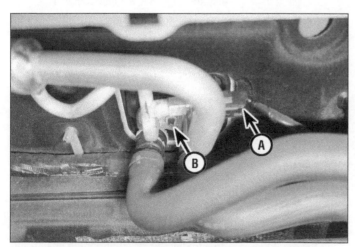

14.25 Disconnect electrical connector to the pressure transducer (A), then remove the nut (B) and the air conditioning refrigerant lines from the firewall

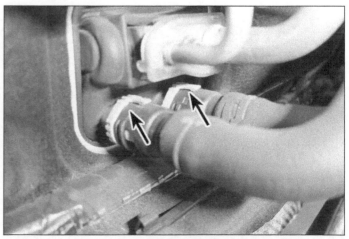

14.26 Disconnect the heater hose quick-connect fittings - rotate the collar counterclockwise, then pull off the hoses

15.6a Disconnect the connector from the compressor clutch field coil - 2007 and earlier models shown

15.6b Disconnect the two electrical connectors from the compressor - (1.5L, 1.6L and 2013 and later 2.0L models)

15.7 Remove the compressor mounting nut, then remove the stud - 1.5L and 1.6L models

fasteners and remove the heater core along with the cover from the heater/evaporator housing.

33 Installation is the reverse of removal. Check for coolant leaks, then have the air conditioning system evacuated and recharged by the shop that discharged it.

15 Air conditioning compressor - removal and installation

Warning: *The air conditioning system is under high pressure. DO NOT loosen any fittings or remove any components until after the system has been discharged. Air conditioning refrigerant must be properly discharged into an EPA-approved container at a dealer service department or an automotive air conditioning repair facility. Always wear eye protection when disconnecting air conditioning system fittings.*

Note: *If you are replacing the compressor, you must also replace the accumulator or receiver/drier (see Section 16), and the orifice tube (see Section 19) or the TXV valve (see Section 20).*

Removal

1 Have the air conditioning system discharged by a dealer service department or by an automotive air conditioning shop before proceeding (see Warning above).

2 Loosen the right front wheel lug nuts. Raise the vehicle and secure it on jackstands. Remove the right front wheel.

3 Remove the right front fender splash shield.

4 On 2013 and later models, remove the engine splash shield.

5 Remove the drivebelt (see Chapter 1).

6 Disconnect the electrical connector(s) from the compressor (see illustrations).

7 On 1.5L and 1.6L models, remove the compressor mounting nut (see illustration) then remove the stud using a reverse Torx socket.

Caution: *Don't allow the compressor to hang from the lines.*

8 Remove the compressor mounting bolts (see illustration), then lower the compressor for access to the refrigerant line fittings (if necessary).

9 Disconnect the refrigerant lines from the compressor (see illustration) and discard the O-rings.

10 Remove the compressor without allowing any oil to drain out of the compressor.

Installation

11 If you're installing a new compressor, make sure the compressor has the correct amount of oil as follows:

a) *Remove the compressor drain plug, then rotate the compressor shaft and allow the oil to drain out of the compressor into a container.*

b) *Remove the new compressor drain plug, then rotate the compressor shaft and allow the oil to drain into a clean container.*

c) *Measure the amount of oil from the old compressor and add the same amount of new oil into the new compressor.*

d) *Install the compressor drain plug and tighten the plug securely.*

12 The clutch may have to be transferred from the original to the new compressor.

15.8 Location of the air conditioning compressor bolts (V6 model shown)

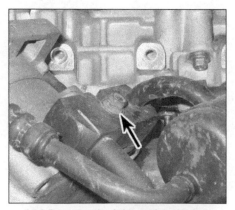

15.9 Remove the inlet and outlet line manifold bolt from the compressor and separate the lines from the compressor - early models shown, later models have separate lines and are mounted with nuts

16.3 Install the air conditioning line spring lock tool with the lip of the tool towards the spring inside the coupling - 2007 and earlier models

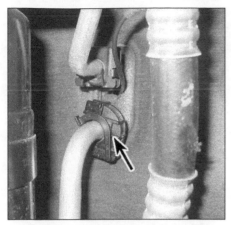

16.4 Remove the protective cover from the evaporator line coupler, then use the spring lock tool to disconnect the line - 2007 and earlier models

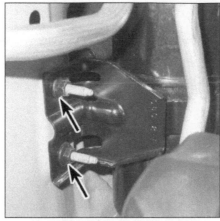

16.5 Location of the accumulator mounting nuts - 2007 and earlier models

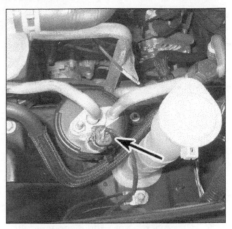

16.12 Wiring connections on the top of the receiver/drier on 2008 through 2012 models

13   Before reconnecting the inlet and outlet lines to the compressor, replace all manifold O-rings and lubricate them with refrigerant oil.
14   Installation is otherwise the reverse of removal. Tighten the compressor fasteners to the torque settings listed in this Chapter's Specifications.
15   Replace the accumulator or receiver/drier (see Section 16), and the orifice tube (see Section 19) or TXV valve (see Section 20).
16   Have the system evacuated, recharged and leak tested by the shop that discharged it.

16   Air conditioning accumulator or receiver/drier - removal and installation

Warning: *The air conditioning system is under high pressure. DO NOT loosen any fittings or remove any components until after the system has been discharged. Air conditioning refrigerant must be properly discharged into an EPA-approved container at a dealer service department or an automotive air conditioning repair facility. Always wear eye protection when disconnecting air conditioning system fittings.*

1   Have the air conditioning system discharged by a dealer service department or by an automotive air conditioning shop before proceeding (see Warning above).

Accumulator (2007 and earlier models)

2   Disconnect the air conditioning pressure cycling switch electrical connector (see Section 18).
3   Disconnect the condenser line from the accumulator (see illustration). Use special spring lock coupling tools to disconnect the air conditioning line. Cap the line to prevent contamination. Remove and discard the old O-ring.
4   Disconnect the line from the accumulator to the evaporator (see illustration). Cap the line to prevent contamination. Remove and

discard the old O-ring.
5   Remove the accumulator mounting bracket nuts (see illustration) and remove the accumulator.
6   Replace all old O-rings. Before installing the new O-rings, coat them with refrigerant oil.
7   Installation is otherwise the reverse of removal.
8   Take the vehicle to the shop that discharged it and have the system evacuated and recharged.

Receiver/drier

All 2008 and later models (except 2013 and later 2.5L and 1.6L engines)
9   Refer to Chapter 11 and remove the front bumper cover.
10   On 1.5L models, remove the cooling module (see Section 9) and condenser (see Section 17).
11   On 2008 through 2012 models, disconnect the refrigerant outlet line from the condenser. Discard the gasket and seal.
12   On 2008 through 2012 models, disconnect the wiring from the top of the receiver/drier (see illustration).
13   On 2008 through 2012 models, disconnect the outlet line from the receiver/drier. Detach the line from the receiver/drier bracket.

14   On 2.0L models, remove the a/c line retaining plate bolt from the bottom of the receiver/drier. Disconnect the a/c line.
15   On 1.5L models, remove the fasteners retaining the inlet/outlet line bracket to the receiver/drier, then disconnect the a/c line bracket.
16   Remove the mounting fasteners, then remove the receiver/drier.
17   Replace the O-rings as well as the gasket and seal, coating them with refrigerant oil before installation.
18   Installation is the reverse of removal.
19   Take the vehicle to the shop that discharged it and have the system evacuated and recharged.

2013 and later 1.6L and 2.5L engines
20   Raise the vehicle and support it securely on jackstands.
21   Remove the front bumper cover (see Chapter 11).
22   Remove the lower cooling module arm fasteners on both sides (see Section 9) and unbolt lower radiator crossmember from the cooling module arms.
23   Remove the threaded plug at the bottom left of the condenser-mounted receiver/drier (see illustration). With the plug removed, push the drier plug upwards and use a small screwdriver to remove the retaining ring.
24   Install an M5 bolt in the center of the receiver drier plug and pull the receiver drier plug out.
25   Using a 90-degree angle pick or needle-nose pliers, pull the desiccant bag (element) out of the receiver-drier.
26   Insert a new element into the receiver-drier.
27   Install the receiver/driver plug into the housing and remove the M5 bolt. Install the plug retaining ring into the inside housing groove.

16.23 Unscrew and remove the plug from the bottom left of the condenser-mounted receiver/drier

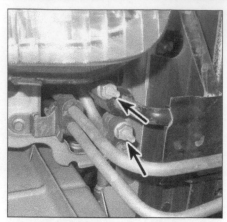

17.3 Remove the nuts and disconnect the inlet and outlet refrigerant lines from the condenser - 2007 and earlier models shown, 2008 and later models' location similar

17.4 On 2008 and later models, the automatic transaxle cooler lines attach to the top of the air conditioning condenser

17.6 Lift the condenser away from the radiator, being careful not to damage the fins on the radiator

28   Install the plug and tighten it securely.
29   Installation is the reverse of removal. Lubricate all air conditioning line fittings/O-rings with refrigerant oil when installing.
30   Have the system evacuated, recharged and leak tested by the shop that discharged it.

17   Air conditioning condenser - removal and installation

Warning: *The air conditioning system is under high pressure. DO NOT loosen any fittings or remove any components until after the system has been discharged. Air conditioning refrigerant must be properly discharged into an EPA-approved container at a dealer service department or an automotive air conditioning repair facility. Always wear eye protection when disconnecting air conditioning system fittings.*
Note: *The accumulator should be replaced if the condenser was damaged, causing the system to be open for some time (see Section 16).*
1   Have the air conditioning system discharged by a dealer service department or an automotive air conditioning shop before proceeding (see Warning above).
2   Remove the front bumper cover (see Chapter 11).

2012 and earlier models
3   Disconnect the refrigerant inlet and outlet lines from the condenser (see illustration). Cap the lines to prevent contamination.
4   On 2007 and earlier models, position aside the power steering cooler. On 2008 and later models with automatic transaxles, disconnect the transaxle cooler lines from the condenser (see illustration).
5   Remove the upper radiator support brackets (see illustration 17.4a).
6   On 2008 and later models, remove the condenser mounting bolts. Remove the condenser (see illustration). If you're going to install the same condenser, store it with the fittings facing up to limit oil loss.

7   If you're going to install a new condenser, pour one ounce of refrigerant oil of the correct type into it prior to installation.
8   Before reconnecting the refrigerant lines to the condenser, be sure to coat a pair of new O-rings with refrigerant oil, install them in the refrigerant line fittings and then tighten the condenser inlet and outlet nuts to the torque listed in this Chapter's Specifications.

2013 and later models
9   Remove the cooling module and radiator (see Section 7).
10   Cut the foam seal between the radiator and the condenser.
11   Press the condenser retaining tabs in on both sides of the radiator and lift the condenser up and off of the radiator.

All models
12   Installation is otherwise the reverse of removal. Lubricate all air conditioning line fittings/O-rings with refrigerant oil when installing.
13   Have the system evacuated, recharged and leak tested by the shop that discharged it.

18   Air conditioning pressure cycling switch - replacement

Note: *This switch is used only on 2007 and earlier models with accumulators. Later models with receiver/driers don't use a pressure cycling switch.*
Note: *The pressure cycling switch is located on top of the accumulator. The pressure cycling switch detects low refrigerant line pressure at 22 to 28 psi, switches the A/C system off, then back on again at 40 to 47 psi. If the pressure increases over 550 psi, the pressure cut-off switch, located in the high pressure line between the compressor and the accumulator, shuts the system off. A Schrader valve in the accumulator prevents refrigerant loss during pressure cycling switch replacement.*

1   Unplug the electrical connector from the pressure cycling switch (see illustration).
2   Unscrew the pressure cycling switch from the accumulator.
3   Lubricate the switch O-ring with clean refrigerant oil of the correct type.
4   Screw the new switch onto the accumulator threads until hand-tight, then tighten it securely.
5   Reconnect the electrical connector.

19   Air conditioning orifice tube (2007 and earlier models) - removal and installation

Warning: *The air conditioning system is under high pressure. DO NOT loosen any hose fittings or remove any components until the system has been discharged. Air conditioning refrigerant must be properly discharged into an EPA-approved recovery/recycling unit by a dealer service department or an automotive air conditioning repair facility. Always wear eye protection when disconnecting air conditioning system fittings.*
Note: *The orifice tube is located in the condenser to evaporator line. The orifice tube*

18.1 The air conditioning pressure cycling switch is mounted on top of the accumulator - 2007 and earlier models

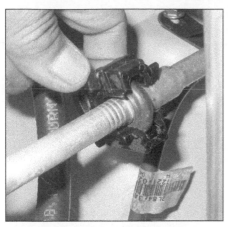

19.2a Remove the protective cover from the air conditioning line coupling

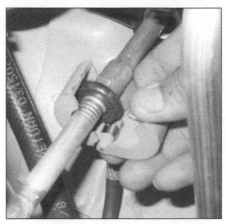

19.2b Install the special spring lock tool and release the coupler by pressing the tool into the spring

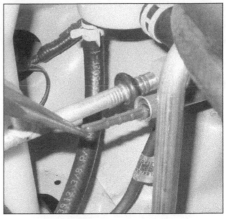

19.3 Remove the orifice tube using needle nose pliers

changes the high pressure liquid refrigerant into a low pressure liquid.

1   Have the air conditioning system discharged by a dealer service department or by an automotive air conditioning shop before proceeding (see Warning above).

2   Disconnect the refrigerant line from the condenser to the evaporator (see illustrations).

3   Using a pair of needle-nose pliers pull the orifice tube from the line (see illustration).

4   Installation is the reverse of removal. Be sure to remove and discard the old O-rings.

5   Take the vehicle back to the shop that discharged it. Have the system evacuated, recharged and leak tested.

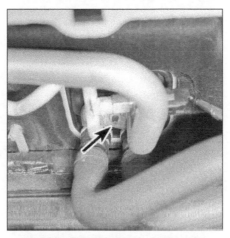

20.2 The TXV valve is located on the firewall, just above the heater hoses

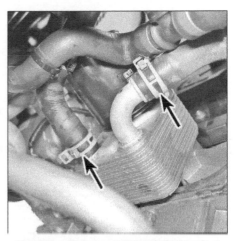

21.9 Use pliers to slide back the coolant hose spring clamps, then disconnect the coolant hoses

---

20   Air conditioning thermostatic expansion valve (TXV) (2008 and later models) - general information

Warning: *The air conditioning system is under high pressure. DO NOT loosen any hose fittings or remove any components until the system has been discharged. Air conditioning refrigerant must be properly discharged into an EPA-approved recovery/recycling unit by a dealer service department or an automotive air conditioning repair facility. Always wear eye protection when disconnecting air conditioning system fittings.*

1   There are several ways that air conditioning systems convert the high-pressure liquid refrigerant from the compressor to lower-pressure vapor. The conversion takes place at the air conditioning evaporator; the evaporator is chilled as the refrigerant passes through, cooling the airflow through the evaporator for delivery to the vents. The conversion is usually accomplished by a sudden change in the tubing size. Many vehicles have a removable controlled orifice in one of the AC pipes at the firewall.

2   2008 and later models use a thermostatic expansion valve (TXV) that accomplishes the same purpose as a controlled orifice. To remove the TXV, first have the air conditioning system discharged by a licensed air conditioning technician. Disconnect the electrical

connector to the pressure transducer, then unscrew the nut and disconnect the refrigerant lines from the TXV at the firewall (see illustration). Remove the two bolts securing the TXV valve, then remove the valve.

3   Installation is the reverse of removal. Replace the gasket seals when installing the TXV valve. Lubricate all a/c line fittings/O-rings with refrigerant oil.

---

21   Oil cooler - removal and installation

Warning: *Wait until the engine is completely cool before beginning this procedure.*

Note: *Oil coolers are used only on 2007 and earlier 2.0L and 2.3L models and all 1.5L and 1.6L models. They are also used on early V6 models.*

1   Raise the vehicle and place it securely on jackstands.

2   Remove the engine splash shield (see Chapter 1).

3   Drain the engine coolant and the engine oil (see Chapter 1).

4   Remove the oil filter (see Chapter 1).

Four-cylinder models

2007 and earlier 2.0L and 2.3L (and early V6) models

5   Disconnect the two coolant hoses from the oil cooler.

6   Remove the oil cooler mounting bolt and detach the oil cooler from the engine block.

7   Be sure to install a new O-ring.

8   Installation is the reverse of removal. Tighten the oil cooler mounting bolt to the torque listed in this Chapter's Specifications.

1.5L and 1.6L models

9   Slide back the hose clamps (see illustration) then disconnect the coolant hoses from the oil cooler.

10   Insert a Allen wrench into the center of the oil cooler adapter bolt and remove the adapter bolt.

11   Remove the oil cooler from the engine block and discard the O-ring.

12   Clean the block surface and install a new O-ring onto the oil cooler, lightly lubricate the O-ring with oil then install the cooler and tighten the adapter bolt to the torque listed in this Chapter's Specifications.

Notes

Chapter 4
Fuel and exhaust systems

Contents

Specifications

Fuel pressure
- 2001 — 35 to 65 psi (240 to 448 kPa)
- 2002
  - Four-cylinder engine — 35 to 65 psi (240 to 448 kPa)
  - V6 engine — 39 to 55 psi (270 to 380 kPa)
- 2003 and 2004 — 50 to 65 psi (345 to 448 kPa)
- 2005 through 2008 — 35 to 70 psi (240 to 485 kPa)
- 2009 through 2012 models — 38 to 65 psi (262 to 448 kPa)
- 2013 and later models
  - 1.5L turbocharged models (low pressure system) — 46 to 80 psi (320 to 550 kPa)
  - 1.6L and 2.0L turbocharged models (low pressure system) — 55 to 73 psi (380 to 500 kPa)
  - 2.5L models
    - 2013 through 2016 models — 50.8 to 60.2 psi (350 to 415 kPa)
    - 2017 and later models — 55.0 to 72.5 psi (379 to 500 kPa)

Fuel injector resistance
- 2001
  - Four-cylinder engine — 11.4 to 12.6 ohms
  - V6 engine — 10.3 to 17.3 ohms
- 2002 through 2004
  - Four-cylinder engine — 13.8 to 15.2 ohms
  - V6 engine — 13.1 to 14.5 ohms
- 2005 and later models — 11 to 18 ohms

Torque specifications

	Ft-lbs (unless otherwise indicated)	Nm

Note: *One foot-pound (ft-lb) of torque is equivalent to 12 inch-pounds (in-lbs) of torque. Torque values below approximately 15 ft-lbs are expressed in inch-pounds, since most foot-pound torque wrenches are not accurate at these smaller values.*

	Ft-lbs (unless otherwise indicated)	Nm
Throttle body-to-intake manifold bolts		
2004 and earlier four-cylinder models	62 in-lbs	7
2005 through 2012 four-cylinder models and		
2012 and earlier V6 models	89 in-lbs	10
2013 and later models		
1.5L and 1.6L models	106 in-lbs	12
2.0L and 2.5L models	89 in-lbs	10
Fuel rail mounting nuts or bolts		
2004 and earlier four-cylinder models and		
2012 and earlier V6 models	89 in-lbs	10
2005 through 2012 four-cylinder models	17	23
2013 and later 1.5L, 1.6L, and 2.5L models	17	23
2013 through 2016 2.0L models		
Step 1	71 in-lbs	8
Step 2	Tighten an additional 26 degrees	
2017 2.0L models		
Step 1	89 in-lbs	10
Step 2	Loosen to 0 in-lbs	0
Step 3	Wait 5 seconds	
Step 4	124 in-lbs	14
Step 5	Tighten an additional 30 degrees	
High-pressure fuel pump bolts (2013 and later models)		
1.5L models	80 in-lbs	9
1.6L models	115 in-lbs	13
2.0L models		
Step 1	44 in-lbs	5
Step 2	Tighten an additional 55 degrees	
High-pressure fuel line fittings (2013 and later models)		
1.5L and 1.6L models		
Step 1	15	21
Step 2	Wait 5 minutes	
Step 3	15	21
2.0L models		
Step 1	133 in-lbs	15
Step 2	Tighten an additional 30 degrees	
Fuel pulsation damper bolts (2012 and earlier four-cylinder models)	53 in-lbs	6
Turbocharger flange nuts		
1.6L models		
Step 1	18	25
Step 2	18	25
1.5L and 2.0L models		
Step 1	37	50
Step 2	37	50
Turbocharger coolant tube banjo bolts	21	28
Catalytic converter manifold flange nuts (2013 and later 2.5L models)	41	55

1   General information, precautions and troubleshooting

Fuel system warnings

Warning: *Gasoline is extremely flammable and repairing fuel system components can be dangerous. Consider your automotive repair knowledge and experience before attempting repairs, which may be better suited for a professional mechanic.*

a) *Don't smoke or allow open flames or bare light bulbs near the work area*
b) *Don't work in a garage with a gas-type appliance (water heater, clothes dryer)*
c) *Use fuel-resistant gloves. If any fuel spills on your skin, wash it off immediately with soap and water*
d) *Clean up spills immediately*
e) *Do not store fuel-soaked rags where they could ignite*
f) *Prior to disconnecting any fuel line, you must relieve the fuel pressure (see Section 2)*
g) *Wear safety glasses*
h) *Have a proper fire extinguisher on hand*

Fuel system

1   The fuel system consists of the fuel tank, electric fuel pump/fuel level sending unit (located in the fuel tank), fuel rail, fuel injectors and, on turbocharged models, a high-pressure fuel pump driven by one of the camshafts. The fuel injection system is a multi-port system; multi-port fuel injection uses timed impulses to inject the fuel directly into the intake port of each cylinder. The Powertrain Control Module (PCM) controls the injectors. The PCM monitors various engine parameters and delivers the exact amount of fuel required into the intake ports.

2   Fuel is circulated from the fuel pump to the fuel rail through fuel lines running along the underside of the vehicle. Various sections of the fuel line are either rigid metal or nylon, or flexible fuel hose. The various sections of the fuel hose are connected either by quick-connect fittings or threaded metal fittings.

Exhaust system

3   The exhaust system consists of the exhaust manifold(s), catalytic converter(s), muffler(s), tailpipe and all connecting pipes, flanges and clamps. The catalytic converter is an emission control device added to the exhaust system to reduce pollutants.

2.4a Remove the trim panel . . .

2.4b . . . and the scuff plate . . .

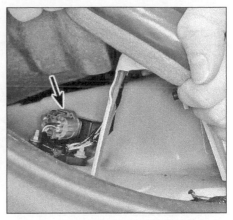

2.4c . . . then disconnect the electrical connector from the FPDM

Troubleshooting

Fuel pump

4   The fuel pump is located inside the fuel tank. Sit inside the vehicle with the windows closed, turn the ignition key to ON (not START) and listen for the sound of the fuel pump as it's briefly activated. You will only hear the sound for a second or two, but that sound tells you that the pump is working. Alternatively, have an assistant listen at the fuel filler cap.

5   If the pump does not come on, check the fuses and the fuel pump relay in the underhood fuse and relay box (see Chapter 12). If the fuse and relay are okay, check the Inertia switch (2012 and earlier models; see the next Step) and the wiring back to the fuel pump. If the fuse, relay and wiring are okay, the fuel pump is probably defective. If the pump runs continuously with the ignition key In the ON position, the Powertrain Control Module (PCM) is probably defective. Have the PCM checked by a professional mechanic.

6   A fuel-cut off system is used in case of an accident:

a) On 2012 and earlier models, a fuel pump inertia switch, located behind the right front kick panel in the passenger's footwell, is used. In the event of an impact of sufficient force, the switch will oen the circuit to the fuel pump and turn it off. If the fuel pump does not operate, depress the reset button on top of the switch and re-check.

b) On 2013 and later models, the restraint system sends a signal to the Fuel Pump Driver Module (FPDM) and shuts the module off. To reset the FPCM, turn the ignition key to the "OFF" position, then to the "ON" position and start the engine.

Fuel injection system

Note: The following procedure is based on the assumption that the fuel pump is working and the fuel pressure is adequate (see Section 3).

7   Check all electrical connectors that are related to the system. Check the ground wire connections for tightness.

8   Verify that the battery is fully charged (see Chapter 5).

9   Inspect the air filter element (see Chapter 1).

10   Check all fuses related to the fuel system (see Chapter 12).

11   Check the air induction system between the throttle body and the intake manifold for air leaks. Also inspect the condition of all vacuum hoses connected to the intake manifold and to the throttle body.

12   Remove the air intake duct from the throttle body and look for dirt, carbon, varnish, or other residue in the throttle body, particularly around the throttle plate. If it's dirty, clean it with carb cleaner, a toothbrush and a clean shop towel.

13   With the engine running, place an automotive stethoscope against each injector, one at a time, and listen for a clicking sound that indicates operation.

Warning: Stay clear of the drivebelt and any rotating or hot components.

14   If you can hear the injectors operating, but the engine is misfiring, the electrical circuits are functioning correctly, but the injectors might be dirty or clogged. Try a commercial injector cleaning product (available at auto parts stores). If cleaning the injectors doesn't help, replace the injector(s).

15   If an injector is not operating (it makes no sound), disconnect the injector electrical connector and measure the resistance across the injector terminals with an ohmmeter. Compare this measurement to the other injectors. If the resistance of the non-operational injector is quite different from the other injectors, replace it.

16   If the injector is not operating, but the resistance reading is within the range of resistance of the other injectors, the PCM or the circuit between the PCM and the injector might be faulty.

2   Fuel pressure relief procedure

Warning: Gasoline is extremely flammable, so take extra precautions when you work on any part of the fuel system. Don't smoke or allow open flames or bare light bulbs near the work area, and don't work in a garage where a gas-type appliance (such as a water heater or a clothes dryer) is present. Since gasoline is carcinogenic, wear latex gloves when there's a possibility of being exposed to fuel, and, if you spill any fuel on your skin, rinse it off immediately with soap and water. Mop up any spills immediately and do not store fuel-soaked rags where they could ignite. The fuel system is under constant pressure, so, if any fuel lines are to be disconnected, the fuel pressure in the system must be relieved first. When you perform any kind of work on the fuel system, wear safety glasses and have a Class B type fire extinguisher on hand.

Caution: After the fuel pressure has been relieved, it's a good idea to lay a shop towel over any fuel connection to be disassembled, to absorb the residual fuel that may leak out when servicing the fuel system.

Note: The fuel system referred to in this Chapter is defined as the fuel tank and tank-mounted fuel pump/fuel gauge sender unit, the fuel filter, the fuel injectors and the metal pipes and flexible hoses of the fuel lines between these components. All these contain fuel, which will be under pressure while the engine is running and/or while the ignition is switched on.

1   The fuel system will remain under pressure for some time after the ignition has been switched off, and it must be relieved before any fuel lines are disconnected.

2   Remove the fuel filler cap or depress the flap in the filler opening, as applicable, to relieve any built-up vapor pressure in the fuel tank.

2012 and earlier models

3   Remove the rear seat and the fuel pump access cover (see Section 5). Disconnect the fuel pump electrical connector. Start and run the engine. The engine will run briefly, then stall. Turn the engine over once or twice to verify that all fuel system pressure has been released, then switch off the ignition.

2013 and later models

Low-pressure side (all models)

4   Remove the right rear door scuff plate trim panel and scuff plate, then peel back the carpet to expose the Fuel Pump Driver Module (FPDM) (see illustrations). Disconnect

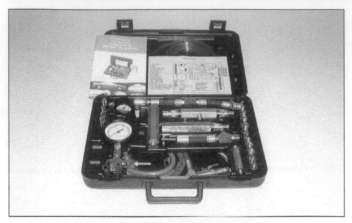

3.1 This fuel pressure testing kit contains all the necessary fittings and adapters, along with the fuel pressure gauge, to test most automotive systems

3.2a On V6 models, connect the fuel pressure gauge to the Schrader valve on the fuel rail

3.2b On turbocharged models, remove the engine cover, disconnect the fuel feed line from the high-pressure fuel pump line and connect a fuel pressure gauge between the line and the pump (1.6L turbocharged engine shown)

High-pressure side (turbocharged models)

6   Relieve the low-side fuel system pressure, then wait two hours before proceeding.

7   Place a flare nut wrench on the high-pressure fuel tube, then wrap the wrench and nut with a shop towel to absorb any remaining fuel pressure while loosening the nut (see Section 6).

8   Loosen the high pressure fuel tube-to-fuel injection pump flare nut and release the remaining pressure.

---

3   Fuel pressure - check

Warning: *Gasoline is extremely flammable, so take extra precautions when you work on any part of the fuel system. See the Warning in Section 2.*

Fuel pressure check

1   To check the fuel pressure on a V6 engine, you'll need a fuel pressure gauge with an adapter that fits the Schrader valve on the left end of the fuel rail. Four-cylinder engines do not have a Schrader valve, so you will need to obtain adapters with which you can tee into the fuel system at a suitable fuel line fitting (see illustration).

2   Relieve the fuel pressure (see Section 2). If you're using a pressure gauge with a bleeder valve, make sure that the valve is closed. On V6 models, connect a fuel pressure gauge to the Schrader valve test port (see illustration). On four-cylinder models, you'll have to tee into the fuel system at the connection between the fuel supply line and the fuel rail (non-turbocharged models) or between the fuel supply line and the high-pressure fuel pump (see illustration).

3   Start the engine and allow it to idle. Note the gauge reading as soon as the pressure stabilizes, and compare it with the pressure listed in this Chapter's Specifications.

a) If the pressure is lower than specified, check the fuel lines and hoses for

kinks, blockages and leaks. Remove the fuel pump module (see Section 5) and check the fuel strainer for restrictions. If all these components are okay, the fuel filter could be clogged or the fuel pump module could be defective.

b) If the fuel pressure is higher than specified, replace the fuel pump module (the pressure regulator, part of the fuel pump module, is not available separately).

4   Turn off the engine. Verify that the fuel pressure stays at the specified level for five minutes after the engine is turned off.

5   Relieve the fuel pressure (see Section 2), then disconnect the fuel pressure gauge. Be sure to cover the fitting with a rag before loosening it. Mop up any spilled gasoline.

6   Start the engine and verify that there are no fuel leaks.

---

4   Fuel lines and fittings - general information and disconnection

Warning: *Gasoline is extremely flammable, so take extra precautions when you work on any part of the fuel system. See Fuel system warnings in Section 1.*

1   Relieve the fuel pressure before servicing fuel lines or fittings (see Section 2). Disconnect the cable from the negative battery terminal (see Chapter 5) before proceeding.

2   The fuel supply line connects the fuel pump in the fuel tank to the fuel rail on the engine. The Evaporative Emissions Control (EVAP) system lines connect the fuel tank to the EVAP canister and connect the canister to the intake manifold.

3   Whenever you're working under the vehicle, be sure to inspect all fuel and evaporative emission lines for leaks, kinks, dents and other damage. Always replace a damaged fuel or EVAP line immediately.

4   If you find signs of dirt in the lines during disassembly, disconnect all lines and blow them out with compressed air. Inspect the fuel strainer on the fuel pump pick-up unit for damage and deterioration.

the FPDM electrical connector. Start and run the engine. The engine will run briefly, then stall. Turn the engine over once or twice to verify that all fuel system pressure has been released, then switch off the ignition.

Warning: *This procedure will merely relieve the increased pressure necessary for the engine to run. Remember that fuel will still be present in the system components, and take precautions accordingly before disconnecting any of them. Disconnect the cable from the negative terminal of the battery before performing any work on the fuel system.*

5   Reconnect the fuel pump or FPDM electrical connector when your work is completed. Note that once the fuel system has been depressurized and drained (even partially), it will take significantly longer to restart the engine - perhaps several seconds of cranking - before the system is refilled and pressure restored.

Disconnecting Fuel Line Fittings

Two-tab type fitting; depress both tabs with your fingers, then pull the fuel line and the fitting apart

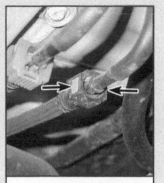

On this type of fitting, depress the two buttons on opposite sides of the fitting, then pull it off the fuel line

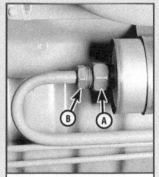

Threaded fuel line fitting; hold the stationary portion of the line or component (A) while loosening the tube nut (B) with a flare-nut wrench

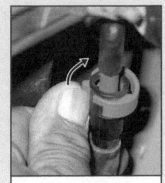

Plastic collar-type fitting; rotate the outer part of the fitting

Metal collar quick-connect fitting; pull the end of the retainer off the fuel line and disengage the other end from the female side of the fitting . . .

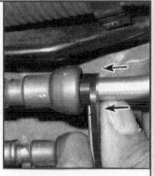

. . . insert a fuel line separator tool into the female side of the fitting, push it into the fitting and pull the fuel line off the pipe

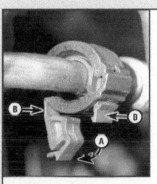

Some fittings are secured by lock tabs. Release the lock tab (A) and rotate it to the fully-opened position, squeeze the two smaller lock tabs (B) . . .

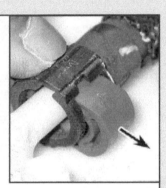

. . . then push the retainer out and pull the fuel line off the pipe

Spring-lock coupling; remove the safety cover, install a coupling release tool and close the tool around the coupling . . .

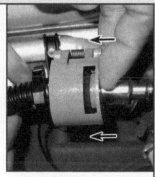

. . . push the tool into the fitting, then pull the two lines apart

Hairpin clip type fitting: push the legs of the retainer clip together, then push the clip down all the way until it stops and pull the fuel line off the pipe

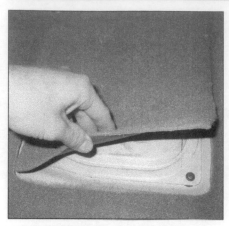

5.4a Pull back the carpeting covering the fuel pump/fuel level sending unit access cover . . .

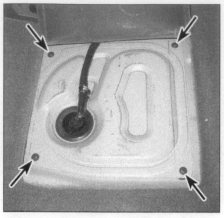

5.4b . . . remove the access cover screws and remove the cover

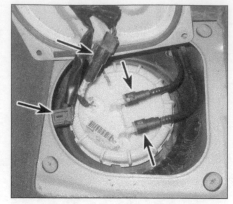

5.5 Disconnect the electrical connectors for the fuel pump/fuel level sending unit module and for the fuel pressure sensor, then disconnect the fuel and vapor line connection (see Section 4)

Steel tubing

5   It is critical that the fuel lines be replaced with lines of equivalent type and specification.
6   Some steel fuel lines have threaded fittings. When loosening these fittings, hold the stationary fitting with a wrench while turning the tube nut.

Plastic tubing

7   When replacing fuel system plastic tubing, use only original equipment replacement plastic tubing.
Caution: *When removing or installing plastic fuel line tubing, be careful not to bend or twist it too much, which can damage it. Also, plastic fuel tubing is NOT heat resistant, so keep it away from excessive heat.*

Flexible hoses

8   When replacing fuel system flexible hoses, use only original equipment replacements.
9   Don't route fuel hoses (or metal lines) within four inches of the exhaust system or

within ten inches of the catalytic converter. Make sure that no rubber hoses are installed directly against the vehicle, particularly in places where there is any vibration. If allowed to touch some vibrating part of the vehicle, a hose can easily become chafed and it might start leaking. A good rule of thumb is to maintain a minimum of 1/4-inch clearance around a hose (or metal line) to prevent contact with the vehicle underbody.

---

5   Fuel pump module/fuel level sending unit - removal and installation

Warning: *Gasoline is extremely flammable, so take extra precautions when you work on any part of the fuel system. See the Warning in Section 1.*
Warning: *Fuel will pour out when the fuel pump module is removed on these models if the fuel level is above 3/4-full. Drive the vehicle until the fuel is below this level or siphon*

the excess fuel through the filler neck. Siphoning kits are available at most automotive supply stores. Never siphon by mouth!
Note: *On 2012 and earlier models, the fuel pump/fuel level sending unit module can be replaced without removing the fuel tank. On 2013 and later models, the fuel tank must be removed.*
Note: *The fuel level sender can be separated from the fuel pump module and replaced on 2009 and later models.*
1   Relieve the residual pressure in the fuel system (see Section 2), and equalize tank pressure by removing the fuel filler cap.
2   Disconnect the cable from the negative battery terminal (see Chapter 5, Section 1).

2012 and earlier models

3   Fold the rear seat cushion forward (see Chapter 11, Section 25 if necessary).
4   Remove the carpeting covering the fuel pump/fuel level sending unit module access cover, then remove the access cover (see illustrations). To reduce the risk of introduc-

5.8a To unscrew the fuel pump/fuel level sending unit module retaining ring, turn it counterclockwise; if it's tight, loosen it with a hammer and a brass punch (2004 and earlier models)

5.8b 2005 and later models have a metal lock ring. A special tool is available to turn it, but a pair of large pliers will work

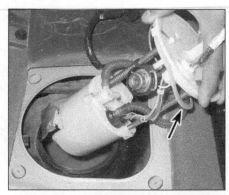

5.9 Carefully remove the fuel pump/fuel level sending unit module from the fuel tank; angle the pump as shown so that the fuel level sending unit float arm clears the edge of the access hole. Be sure to inspect the neoprene gasket for wear and damage while the pump is out (2012 and earlier models shown)

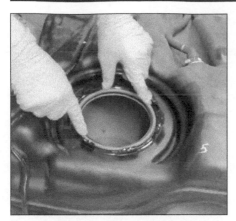

5.10 Install a new pump O-ring

6.8 Remove the high-pressure fuel pump insulator (1.6L shown, others similar)

6.9 Remove the coil-on-plug bracket (1.6L models)

ing water, dirt and dust into the tank while it is open, use a vacuum cleaner to clean the carpeting surrounding the access cover hole. Also blow (or wipe) off any dirt from the top of the fuel pump/fuel level sending unit and from the surface of the tank surrounding the module.

5   Disconnect the fuel pump/fuel level sending unit module electrical connector (see illustration). Also disconnect the electrical connector for the fuel pressure sensor. Set the connectors and the electrical harnesses aside.

2013 and later models

6   Remove the fuel tank (see Section 8).

All models

7   Disconnect the fuel supply and EVAP lines from the fuel pump/fuel level sending unit module. Use a rag to soak up any spilled fuel.
8   Unscrew the fuel pump/fuel level sending unit module retaining ring by turning it counterclockwise using a special lock ring tool. If the ring is tight, use a hammer and a brass punch to loosen it (see illustrations).
9   Remove the fuel pump/fuel level sending unit module, taking care not to bend the sending unit float arm (see illustration). If only the fuel level sender needs to be replaced

on 2009 and later models, disconnect its wiring harness from the fuel pump module, then depress the locking tabs and remove the sender.
10   Install a new O-ring gasket in the pump opening (see illustration).
11   Align the fuel pump/fuel level sending unit module with its hole in the tank and carefully insert it into the tank. Make sure that you don't damage the float arm during installation. If the float arm is bent, the fuel level gauge reading will be incorrect.
12   The remainder of installation is the reverse of removal. on models with a metal lock ring, make sure the raised portions of the ring are properly engaged with raised portions of the tank.
13   Reconnect the battery and perform any necessary relearn/initialization procedures (see Chapter 5).

---

6   High-pressure fuel pump - removal and installation

Note: *2013 and later 1.5L, 1.6L and 2.0L models are equipped with Direct Injection and are equipped with a high-pressure fuel pump.*

Removal

1   If equipped, remove the engine cover.
2   Relieve the fuel system pressure (see Section 2).
3   Disconnect the negative battery cable (see Chapter 5).
4   On 1.6L models, remove the ignition coils (see Chapter 5, Section 5).
5   On 1.6L and 2.0L models, remove the turbocharger piping between the air filter and turbocharger inlet (see Section 14, illustrations 17.25a and 17.25b).
6   On 2.0L models, remove the cowl cover (see Chapter 11, Section 12).
7   Disconnect the high-pressure fuel pump electrical connector.
8   Remove the high-pressure fuel pump insulator (cover) on the back side of the valve cover (see illustration).
9   On 1.6L models, disconnect the electrical connectors and hoses and remove the coil-on-plug bracket bolts and bracket from the top of the valve cover (see illustration).
10   Remove the noise insulator from the valve cover (see illustration).
11   Disconnect the fuel supply line at the high-pressure fuel pump pipe (see Section 3). If necessary, remove the fuel supply line feed pipe at the high-pressure fuel pump (see illustration).

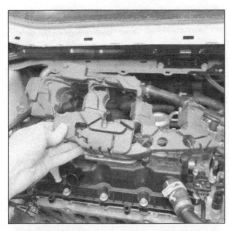

6.10 Remove the noise insulator from the fuel rail (1.6L shown, others similar)

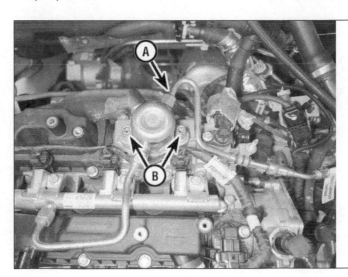

6.11 Identifying the fuel supply line at the pump (A) and high-pressure fuel pump bolts (B) (1.6L shown, others similar)

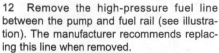

6.12 Remove the high-pressure fuel line between the pump and fuel rail (1.6L shown, others similar)

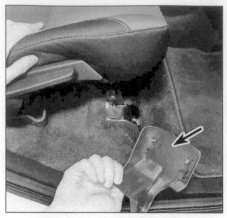

7.1 Remove the trim cover from the seat hinge

12   Remove the high-pressure fuel line between the pump and fuel rail (see illustration). The manufacturer recommends replacing this line when removed.

Note: *Wrap a rag around the fittings when loosening to soak up any spilled fuel.*

13   On 2.0L models, remove the high-pressure fuel pump bracket bolts and bracket.

14   Loosen the fuel pump bolts (see illustration 6.11), alternating one turn at a time until the bolts and pump can be removed. The bolts must be replaced after they have been removed.

Caution: *The fuel pump is under heavy spring pressure - if the bolts are not loosened evenly and one complete turn at a time the pump or housing will be damaged.*

Installation

15   Replace the O-ring for the pump and inspect the fuel pump tappet located in the cylinder head.

16   Before installation, the cam lobe for the high-pressure fuel pump drive tappet must be at the bottom of its travel before installing the high-pressure fuel pump. Rotate the engine using the crankshaft pulley bolt until the tappet is at the bottom of its travel.

17   Lubricate the tappet with clean engine oil before installing.

18   Lubricate the new O-ring with clean engine oil when installing the high-pressure fuel pump. Tighten the bolts a little at a time to the torque listed in this Chapter's Specifications.

19   Install the new high-pressure fuel line between the pump and fuel rail and tighten to the torque listed in this Chapter's Specifications.

20   The remainder of installation is reverse of removal.

21   Start the vehicle and check for fuel leaks.

---

7   Fuel Pump Driver Module (FPDM) - replacement

Note: *2013 and later models are equipped with a FPDM. The FPDM is located under the carpet below the right (passenger's) rear seat.*

1   Remove the right rear seat hinge cover (see illustration).

2   Remove the passenger's rear sill plate (see illustration).

3   Remove the right rear seat hinge bolt and fold the hinge up and out of the way to allow access to the FPDM.

4   Peel back the carpet to expose the FPDM (see illustration).

5   Disconnect the FPDM electrical connector.

6   Remove the nuts and the FPDM from the vehicle.

7   Installation is reverse of removal.

---

8   Fuel tank - removal and installation

Warning: *Gasoline is extremely flammable, so take extra precautions when you work on any part of the fuel system. See the Warning in Section 2.*

Warning: *Before disconnecting or opening any part of the fuel system, relieve the residual pressure (see Section 2), and equalize the pressure inside the fuel tank by removing the fuel filler cap.*

1   There is no fuel tank drain plug, so it's easier to remove the fuel tank when it's nearly empty. If that's not possible, try to siphon out the fuel in the tank before removing the tank.

7.2 Remove the passenger's rear door sill plate

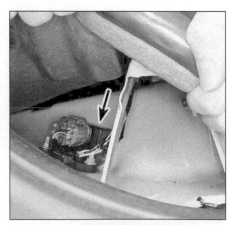

7.4 Pull back the carpet to access the FPDM

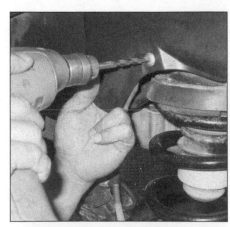

8.7 Drill out the rivets that attach the fuel filler hose shield and then remove the shield - 2008 and earlier models

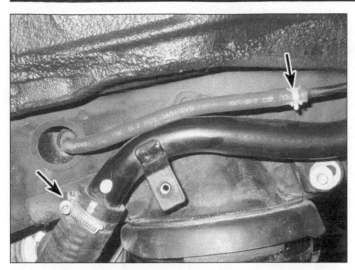

8.8 Loosen the hose clamps and disconnect the fuel filler hose and the fuel filler vent hose - 2008 and earlier models

8.11 Support the fuel tank with a transmission jack (shown) or a floor jack; if you're using a floor jack, be sure to put a sturdy piece of plywood between the jack head and the fuel tank to protect the tank; to detach the fuel tank straps, remove the bolts at rear end of each strap

2   Relieve the fuel system pressure (see Section 2).

3   Disconnect the cable from the negative battery terminal (see Chapter 5, Section 1).

4   On 2012 and earlier models, working inside the vehicle, disconnect the electrical connectors from the fuel pump/fuel level sending unit module and from the fuel pressure sensor (see Section 5). Also disconnect the fuel line(s) (some models have two, others only have one). If you don't need to siphon any fuel from the tank, it's not necessary to remove the module to remove the tank (as long as everything is disconnected from the module!).

5   If there's still a lot of fuel in the tank, siphon or hand-pump the remaining fuel from the tank now. We recommend that you siphon the fuel out of the tank through the hole in the top of the tank for the fuel pump/fuel level sending unit module, but you can also siphon it out through the fuel filler neck or through the fuel filler neck pipe (at the tank), after you

have disconnected the fuel filler neck hose from the fuel tank.

Warning: *Don't start the siphoning action by mouth! Use a siphoning kit (available at most auto parts stores).*

2008 and earlier models

6   Loosen the left rear wheel lug nuts, raise the vehicle and place it securely on jackstands. Remove the left rear wheel.

7   Drill out the rivets that attach the fuel filler hose shield (see illustration) and then remove the shield.

8   Disconnect the fuel tank filler hose and the fuel tank filler vent hose from the fuel tank filler tube (see illustration).

9   Remove the muffler (see Section 16).

10   On 4WD models, remove the driveshaft, remove the differential mounting bracket-to-subframe mounting bolt, remove the differential-to-bracket bolts, loosen the differential mounting bracket-to-subframe mounting bolts, rotate the brackets and position the axle

assembly to the rear (see Chapter 8).

11   Support the fuel tank with a transmission jack (see illustration) or a floor jack. If you're going to use a floor jack, place a sturdy piece of plywood between the jack head and the tank to protect the tank.

12   Disconnect the EVAP hose (see illustration).

13   Disconnect the pin-type retainer that attaches the parking brake to the left fuel tank strap (see illustration) and reposition the parking brake cable so that it's out of the way.

14   On 4WD models, disconnect the wiring harness clip from the rear of the left fuel tank strap.

2009 through 2012 models

15   Raise the vehicle and support it securely on jackstands.

16   Remove the muffler and resonator. On 4WD models, remove the rear driveshaft.

17   Detach the filler hose from the fuel tank.

8.12 Disconnect this EVAP hose - 2008 and earlier models

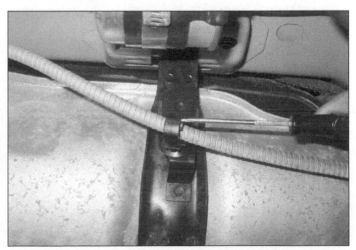

8.13 Detach this pin-type retainer from the front of the left fuel tank strap and reposition the parking brake cable so that it's out of the way - 2008 and earlier models

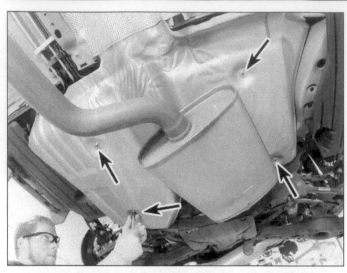

8.25a Remove the heat shield fasteners and the heat shield

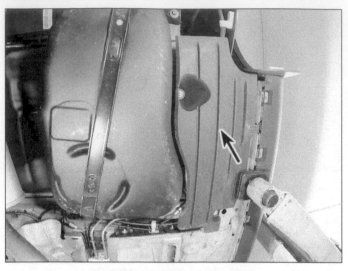

8.25b Remove the splash shield

8.25c Remove the clips from the bracket for the heat shield before removing the bracket

18   Disconnect the parking brake retainer and the wiring harness retainer from the left fuel tank support strap.
19   Place a floor jack under the fuel tank with a piece of plywood to protect the tank. Raise the jack enough to take weight off of the support straps.

2013 and later models

20   Raise the vehicle and support it securely on jackstands.
21   Remove the underbody panels to allow access to the fuel tank.
22   Unbolt the rear portion of the exhaust system from the front exhaust pipe and remove it.
23   If equipped, remove the bolts and bracket from the rear of the vehicle chassis tunnel.
24   On AWD models, remove the driveshaft (see Chapter 8, Section 9).
25   Remove the fuel tank heat shield and small splash shield (see illustrations).
26   Disconnect the fuel line and fuel filler pipe vent quick-disconnect fittings.
27   Disconnect the EVAP lines at the canister (see Chapter 6, Section 28) and disconnect the electrical connector.
28   Loosen the clamp and disconnect the fuel tank filler hose from the tank or the filler pipe (see illustration).
29   If equipped, remove the push-pin retainers from the tank straps.
30   Place a floor jack under the fuel tank with a piece of plywood to protect the tank (see illustration). Raise the jack enough to take weight off of the support straps.

All models

31   Remove the fuel tank strap bolts (see illustration 8.11 and 8.30) and straps. On 2012 and earlier models, swing the fuel tank straps down and out of the way (the straps are hinged at the front; there are no front strap bolts).
32   Lower the tank enough to have a look at the top of the tank and unclip or detach any remaining cables, connectors, hoses or lines (see illustration).
33   Lower the tank the rest of the way.
34   Installation is the reverse of the removal procedure.
Note: *Replace the O-ring seal whenever the module is removed.*

8.28 Disconnect the fuel tank filler hose

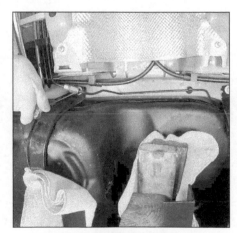

8.30 Support the tank and loosen the strap bolts

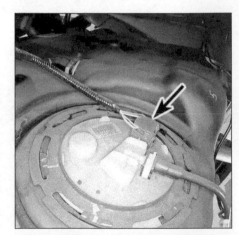

8.32 On 2013 and later models, disconnect the fuel pump electrical connector before lowering the tank

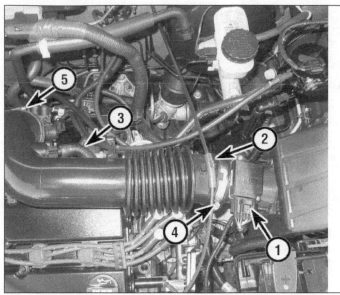

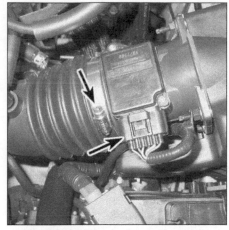

9.2a Air intake duct installation details (four-cylinder models)

1 Mass Air Flow (MAF) sensor electrical connector
2 Accelerator cable clip
3 Fresh air intake hose
4 Hose clamp at MAF sensor
5 Hose clamp at throttle body

9.2b On 2008 and earlier V6 models, disconnect the electrical connector from the Mass Air Flow (MAF) sensor and loosen the air intake duct hose clamp screw

35 Turn the key to Run and check for leaks.

36 On 2012 and earlier models, after you've reconnected the battery, the Powertrain Control Module (PCM) must relearn its idle and fuel trim strategy for optimum driveability and performance (see Chapter 5, Section 1 for this procedure).

9 Air filter housing - removal and installation

Air filter housing

1 Disconnect the cable from the negative battery terminal (see Chapter 5).

2012 and earlier models

2 Disconnect the electrical connector from the Mass Air Flow (MAF) sensor (see illustrations).

3 On 2008 and earlier four-cylinder models, detach the accelerator cable from the air intake duct (see illustration 8.2a). On V6 mod-

9.3 On V6 models, disconnect the crankcase vent hose from the air intake duct - 2008 and earlier models shown

els, disconnect the crankcase vent hose from the air intake duct (see illustration).

4 Loosen the air intake duct hose clamp

9.4a Release the clips to access the air filter - 2009 and later V6 model shown

screw at the MAF sensor end of the air intake duct and disconnect the air intake duct from the air filter housing (see illustrations).

9.4b Remove these pins on 2009 and later V6 models to remove the air filter housing (the housing is also secured by ballstuds and grommets at the bottom)

9.4c Disconnect the MAF sensor (A) and the throttle body hose clamp (B) - 2009 and later V6 models

9.5a To detach the air filter housing from the vehicle, remove this bolt - 2008 and earlier models

9.5b To disengage the air filter housing from these two grommets, pull it straight up (air filter housing cover and air filter element removed for clarity)

9.6 When installing the air filter housing, make sure that the air intake mouth of the housing is correctly seated

9.10 Remove the screws (1) and disconnect the duct (2) (2014 and later 1.6L shown)

installing the air filter housing. If they're cracked, torn or otherwise damaged, replace them. Installation is otherwise the reverse of removal. Make sure that the air filter housing locator pins are correctly seated in their grommets, and that the air intake mouth of the filter housing is correctly seated inside the fresh air inlet duct (see illustration).

7   After you've reconnected the battery, the Powertrain Control Module (PCM) must relearn its idle and fuel trim strategy for optimum driveability and performance (see Chapter 5, Section 1 for this procedure).

2013 and later models

8   On 2013 1.6L and 2.0L models and 2013 and later 2.5L models, loosen the intake air hose at the air filter housing and disconnect the hose.

9   Disconnect the MAF sensor connector.

10   On 2014 and later 1.5L, 1.6L, and 2.0L models, remove the screws attaching the intake air duct and hose to the air filter housing (this includes the MAF) and disconnect the duct from the housing (see illustration).

5   On 2008 and earlier models, remove the air filter housing mounting bolt (see illustration). On all models, to disengage the air filter housing from its grommets (see illus-

tration), pull it straight up. You'll also have to pull the housing to the rear far enough to disengage it from the fresh air inlet duct.

6   Inspect the rubber grommets before

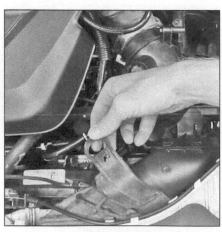

9.11 Pull up on the tab to detach the intake air duct

9.12 Pull up on the housing to disconnect from the rubber bushings and remove (2014 and later 1.6L shown)

9.17 To detach the fresh air inlet duct from the vehicle body, remove this bolt - 2008 and earlier models

9.21 To detach the left inner fender splash shield, remove these fasteners (the shield uses a combination of screws and pin-type retainers; if any of them are damaged or missing, be sure to replace them when installing the shield)

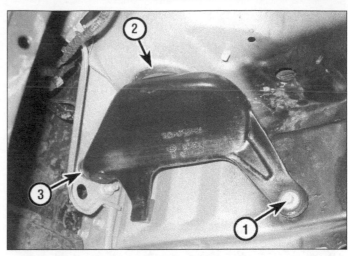

9.22 To detach the fresh air inlet duct resonator from the vehicle body, remove the bolt (1), then disengage the resonator from the fresh air inlet insulator (2) and from the locator pin grommet (3); inspect both the insulator and the grommet and make sure that they're in good condition

11   Unhook the rubber retaining flap from the intake duct tube (see illustration).
12   Pull up on the rear of the air filter housing to pull the ballstud from the rubber bushing, then pull the housing rearward to disconnect the front retainer from the rubber bushing (see illustration).
13   Pull the intake ducts from the front duct work and remove the air filter housing from the vehicle.
14   Installation is reverse of removal.

Fresh air inlet duct (2008 and earlier models)

15   Remove the battery and the battery tray (see Chapter 5, Section 3).
16   Remove the air filter housing (see Steps 1 through 6).
17   Remove the fresh air inlet duct retaining bolt (see illustration) and remove the inlet duct.
18   Installation is the reverse of removal.

19   After you've reconnected the battery, the Powertrain Control Module (PCM) must relearn its idle and fuel trim strategy for optimum driveability and performance (see Chapter 5, Section 1 for this procedure).

Fresh air inlet duct resonator (2008 and earlier models)

20   Loosen the left front wheel lug nuts. Raise the front of the vehicle and place it securely on jackstands. Remove the left front wheel.
21   Remove the pin-type retainers and screws, then remove the left inner fender splash shield (see illustration).
22   Remove the resonator retaining bolt (see illustration) and remove the resonator.
23   Inspect the rubber insulator between the resonator and the fresh air inlet duct and the rubber grommet for the locator pin. If either is damaged or worn, replace it.

24   Installation is otherwise the reverse of removal.

---

10 Accelerator cable - removal and installation

Note: *This procedure applies only to 2008 and earlier models. Later models use an electronic throttle body that is controlled by the Powertrain Control Module (PCM) with input from the Accelerator Pedal Position (APP) sensor (see Chapter 6).*

1   Disconnect the cable from the negative battery terminal (see Chapter 5).
2   On V6 models, remove the engine cover (see illustration).
3   On four-cylinder models, detach the accelerator cable from the air intake duct (see illustration 8.2a). On V6 models, detach the accelerator cable bracket from the intake manifold (see illustration).

10.2 To detach the engine cover from a V6 model, remove these three nuts

10.3 On V6 models, remove this bolt to detach the accelerator cable bracket from the intake manifold

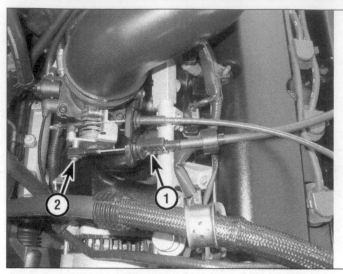

10.4 To disengage the accelerator cable housing (1) from the cable bracket, rotate the cable housing 90 degrees to the left or right and pull the housing out of its slot in the bracket; to disconnect the cable from the throttle body, remove the retainer clip (2) and pull the cable end off the pin on the linkage (four-cylinder models)

10.7 Pull out the cable from the engine compartment side, then pry out the grommet and remove the cable and grommet from the vehicle

4   Disengage the accelerator cable from the cable bracket (see illustration).

5   Disconnect the accelerator cable from the throttle body.

6   Working inside the passenger's compartment, disengage the cable from the accelerator pedal.

7   Returning to the engine compartment, pull the cable through the firewall until the cable's lower end plug reaches the grommet, then remove the grommet and pull the cable through the hole (see illustration).

8   Installation is the reverse of the removal procedure.

9   Have an assistant depress and release the accelerator pedal while you verify that the throttle valve moves smoothly and easily from the fully closed to the fully open position and back again.

10   After you've reconnected the battery, the Powertrain Control Module (PCM) must relearn its idle and fuel trim strategy for optimum driveability and performance (see Chapter 5, Section 1 for this procedure).

11   Throttle body - inspection, removal and installation

Removal

Warning: *Wait until the engine is completely cool before beginning this procedure.*

1   Disconnect the cable from the negative battery terminal (see Chapter 5).

2008 and earlier models

2   On four-cylinder models, detach the accelerator cable from the air intake duct. On V6 models, disconnect the crankcase vent hose from the air intake duct (see illustration 8.3).

3   Disconnect the air intake duct from the throttle body (see illustration).

4   Disconnect the accelerator cable from the throttle body (see Section 10).

5   If the vehicle is equipped with cruise control, disconnect the cruise control cable from the throttle body.

6   Disconnect the electrical connector from the Throttle Position (TP) sensor (see illustration).

7   Disconnect the electrical connector from the Idle Air Control (IAC) valve (see illustration). On V6 models, detach the IAC harness from the throttle body mounting stud located near the TP sensor.

8   On V6 models, remove the transaxle vent tube bracket from the throttle body.

9   Remove the throttle body mounting bolts (see illustrations) and then remove the throttle body from the manifold.

10   Remove the throttle body gasket and discard it.

11   If necessary, clean the throttle body as outlined in Step 2.

2009 through 2012 models

12   Remove the outlet duct from the air filter housing and the throttle body.

13   Disconnect the electrical connector from the throttle body (see illustration).

14   Remove the throttle body mounting bolts, lift off the throttle body and discard the gasket.

11.3 To detach the air intake duct from the throttle body, loosen this hose clamp screw (V6 model shown, four-cylinder models similar)

11.6 Disconnect the electrical connector from the Throttle Position (TP) sensor (V6 model shown, four-cylinder models similar)

11.7 Disconnect the electrical connector from the Idle Air Control (IAC) valve (V6 model shown, four-cylinder models similar)

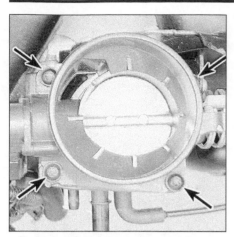

11.9a To detach the throttle body from the air intake manifold on a four-cylinder model, remove these four bolts

11.9b To detach the throttle body from the air intake manifold on a V6 model, remove these two bolts - 2008 and earlier models

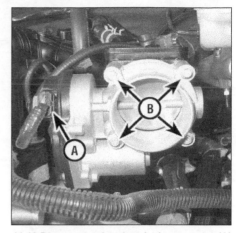

11.13 Disconnect the electrical connector (A) from the throttle body, then remove the mounting bolts (B)

15   Installation is the reverse of removal. Tighten the bolts to the torque listed in this Chapter's Specifications.

2013 and later 1.5L, 1.6L and 2.0L models

16   Raise and support the front of the vehicle on jack stands.
17   Remove the engine undercover.
18   On 1.5L models, remove the intercooler intake pipe (see Section 15).
19   On 1.6L and 2.0L models, working between the engine and the radiator, loosen the clamp and disconnect the throttle body inlet hose from between the throttle body and intercooler (see illustration).
Note: *Disconnect any quick-disconnect fittings and remove any brackets that may be securing the intake hose to allow it to be positioned out of the way for throttle body replacement.*
20   On all models, disconnect the throttle body electrical connector (**see illustration**).
21   Remove the four throttle body bolts (see

illustration) and remove the throttle body from the vehicle.
22   Inspect the throttle body O-ring and replace if damaged.

2013 and later 2.5L models

23   Remove the engine cover.
24   Loosen the clamp and disconnect the intake air hose from the throttle body.
25   Disconnect the throttle body electrical connector.
26   Remove the four throttle body bolts and remove the throttle body from the vehicle.
27   Inspect the throttle body O-ring and replace if damaged.

Installation

28   Using a new gasket, install the throttle body and tighten the throttle body mounting bolts to the torque listed in this Chapter's Specifications.
29   The remainder of installation is the reverse of removal.
30   On 2012 and earlier models, after you've

reconnected the battery, the Powertrain Control Module (PCM) must relearn its idle and fuel trim strategy for optimum driveability and performance (see Chapter 5, Section 1 for this procedure).
31   Start the engine and verify that the throttle body operates correctly and that there are no air leaks.

---

12 Fuel pulsation damper - removal and installation

Warning: *Gasoline is extremely flammable, so take extra precautions when you work on any part of the fuel system. See the Warning in Section 2.*
Note: *This procedure applies only to 2008 and earlier vehicles. Later models don't use a fuel pulsation damper.*
1   Relieve the fuel system pressure (see Section 2).
2   Disconnect the cable from the negative battery terminal (see Chapter 5).

11.19 Disconnect the throttle body inlet hose (1.6L model shown)

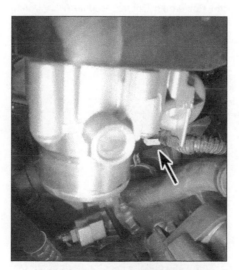

11.20 Disconnect the throttle body connector (1.6L model shown)

11.21 Remove the bolts and the throttle body (1.6L model shown)

12.4 Disconnect the vacuum line from the pulsation damper (V6 model shown) - 2008 and earlier models

12.6a On V6 models, remove the pulsation damper snap-ring . . .

12.6b . . . and remove the pulsation damper - 2008 and earlier models

3   On V6 models, remove the throttle body (see Section 11).

4   Disconnect the vacuum line from the pulsation damper (see illustration).

5   On four-cylinder models, remove the two bolts that attach the pulsation damper to the fuel rail and then remove the damper.

6   On V6 models, remove the fuel pulsation damper snap-ring and then remove the damper (see illustrations).

7   Before installing the pulsation damper, be sure to remove the old O-ring and discard it. Install a new O-ring on the pulsation damper (even if you're planning to reuse the old damper).

8   Installation is the reverse of removal. On four-cylinder models, be sure to tighten the pulsation damper bolts to the torque listed in this Chapter's Specifications.

9   After you've reconnected the battery, the Powertrain Control Module (PCM) must relearn its idle and fuel trim strategy for optimum driveability and performance (see Chapter 5, Section 1 for this procedure).

10   Start the engine and verify that there are no fuel leaks.

6   Disconnect the vacuum line from the fuel pulsation damper.

7   Disconnect the electrical connectors from the fuel injectors.

8   Remove the fuel rail mounting bolts.

9   Remove the fuel rail and injectors as a single assembly.

10   Remove the clamp that secures each fuel injector to the fuel rail and pull out the injector (see illustration).

11   Remove the old O-rings from each injector (see illustration) and discard them. Always install new O-rings on the injectors before reassembling the injectors and the fuel rail.

12   Installation is otherwise the reverse of removal. To ensure that the new injector O-rings are not damaged when the injectors are installed into the fuel rail and into the intake manifold, lubricate them with clean engine oil. And be sure to tighten the fuel rail mounting bolts to the torque listed in this Chapter's Specifications.

13   After you've reconnected the battery, the Powertrain Control Module (PCM) must

relearn its idle and fuel trim strategy for optimum driveability and performance (see Chapter 5, Section 1 for this procedure).

14   Start the engine and verify that there are no fuel leaks.

V6 models

15   Remove the upper intake manifold (see Chapter 2B, Section 5).

16   Disconnect the vacuum line from the fuel pulsation damper (see illustration 13.4).

17   Disconnect the fuel supply line from the fuel rail.

18   Disconnect the electrical connectors from the injectors (see illustration) and detach the fuel injector harness from the brackets at the left and right ends of the vee. Then trace the harness to each electrical component to which it's connected and disconnect the electrical connectors from those components and set the harness aside.

19   Remove the fuel rail mounting bolts (see illustration).

20   Carefully disengage the injectors from the lower intake manifold and lift the fuel rail

---

13   Fuel rail and injectors - removal and installation

1   Relieve the fuel system pressure (see Section 2) and equalize tank pressure by removing the fuel filler cap (or depressing the filler flap, as applicable).

2   Disconnect the cable from the negative battery terminal (see Chapter 5).

2008 and earlier models
Four-cylinder models

3   Remove the air intake duct.

4   Disconnect the accelerator cable from the throttle lever cam. If equipped, also disconnect the cruise control cable.

5   Disconnect the fuel supply line from the fuel rail.

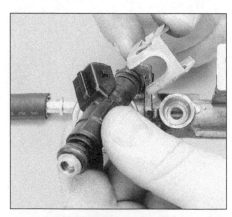

13.10 To disengage an injector from the fuel rail on a four-cylinder model, remove this clamp and pull the injector straight out of the fuel rail (wiggle and pull at the same time if the O-ring is stuck to the injector bore)

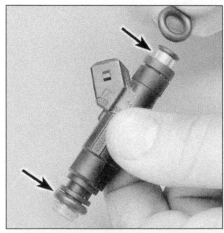

13.11 Always remove and discard the old injector O-rings and install new O-rings on each injector (four-cylinder models)

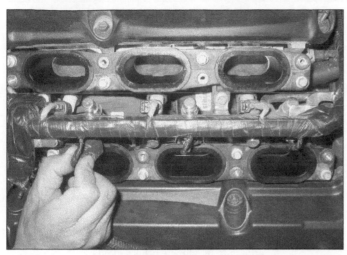

13.18 Disconnect the electrical connectors from the injectors

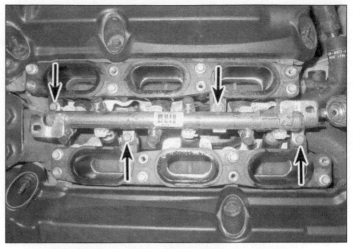

13.19 To detach the fuel rail from the engine,
remove these four bolts (V6 models)

and all six injectors from the engine as a single assembly.

21   Remove the injectors from the fuel rail (see illustration).

22   Remove and discard the old injector O-rings (see illustration). Always install new O-rings on the injectors before reassembling the injectors and the fuel rail.

23   Installation is the reverse of the removal procedure. To ensure that the new injector O-rings are not damaged when the injectors are installed into the fuel rail and into the intake manifold, lubricate them with clean engine oil. Be sure to tighten the fuel rail mounting bolts to the torque listed in this Chapter's Specifications.

24   After you've reconnected the battery, the Powertrain Control Module (PCM) must relearn its idle and fuel trim strategy for optimum driveability and performance (see Chapter 5, Section 1 for this procedure).

25   Start the engine and verify that there are no fuel leaks.

2009 through 2012 models

Four-cylinder models

26   Disconnect the fuel supply line from the fuel rail at the quick-disconnect fitting.

27   Disconnect the wiring from each fuel injector, then disconnect the other wiring harnesses clips from the fuel rail.

28   Detach the capacitor and move it to the side.

29   Remove the fuel rail mounting bolts and lift the rail off along with the fuel injectors.

30   Remove the clamp that secures each injector to the fuel rail and pull out the injector.

31   Remove the O-rings from the injectors and discard them. Install new O-rings, lubricating them with clean engine oil.

32   Installation is the reverse of removal. Tighten the fuel rail mounting fasteners to the torque listed in this Chapter's Specifications.

V6 models

33   Refer to Chapter 2B, Section 5 and remove the upper intake manifold.

34   Disconnect the fuel supply line from the fuel rail at the quick-disconnect fitting.

35   Detach all of the wiring harness retainers from the fuel rail.

36   Disconnect the wiring from each fuel injector.

37   Remove the fuel rail mounting bolts and lift the rail off along with the fuel injectors.

38   Remove the clamp that secures each injector to the fuel rail and pull out the injector.

39   Remove the O-rings from the injectors and discard them. Install new O-rings, lubricating them with clean engine oil.

40   Installation is the reverse of removal. Tighten the fuel rail mounting fasteners to the torque listed in this Chapter's Specifications.

2013 and later models

41   Disconnect the negative battery cable (see Chapter 5).

1.5L models

42   Remove the engine cover.

43   Disconnect the quick-disconnect fittings for the tubes running across the top of the engine and position the tubes out of the way.

44   Remove the ignition coils (see Chapter 5, Section 5).

45   Disconnect the Fuel Rail Pressure (FRP) sensor connector and detach harness from the valve cover.

46   Disconnect the fuel injector harness main connector, then disconnect the fuel injector connectors and remove the harness.

Note: Use compressed air to blow the area around the fuel injectors off to prevent debris from entering the engine.

47   Remove the high-pressure fuel line between the pump and fuel rail.

48   Remove the fuel rail bolts and pull the fuel rail with the injectors from the engine.

Note: Use a slide hammer to remove any injectors that are stuck in the head.

49   Remove the fuel injectors from the fuel rail and replace the O-rings and Teflon seals.

50   Installation is reverse of removal.

Caution: Do not lubricate the teflon rings when installing the fuel rail to the engine.

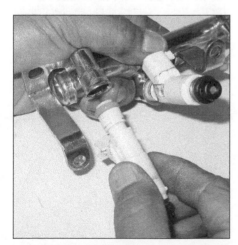

13.21 To detach an injector from the fuel rail, simply wiggle it side to side a little to loosen the O-ring and then pull it straight out as shown (V6 models)

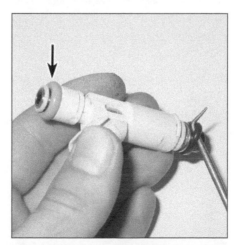

13.22 Remove the old O-rings from each injector and discard them; always install new O-rings when installing the injectors (V6 models)

13.53 Disconnect the fuel injectors (A) and the fuel rail pressure sensor (B) connectors (1.6L shown, others similar)

13.55 Remove the fuel rail bolts and remove the fuel rail from the engine (1.6L shown, others similar)

51   Tighten all fasteners to the torque listed in this Chapter's Specifications.

1.6L and 2.0L models

52   Remove the high-pressure fuel line between the pump and fuel rail (see Section 6).

53   Disconnect the fuel injector connectors and Fuel Rail Pressure (FRP) sensor if equipped (see illustration).

54   On 2.0L engines, remove the noise insulator from the fuel rail (see illustration 6.10).

Note: *Use compressed air to blow the area around the fuel injectors off to prevent debris from entering the engine.*

55   On all models, remove the fuel rail bolts and pull the fuel rail with the injectors from the engine (see illustration).

Note: *Use a slide hammer to remove any injectors that are stuck in the head.*

56   Remove the fuel injectors from the fuel rail and replace the O-rings and Teflon seals (see Step 66).

57   Installation is reverse of removal.

Caution: *DO NOT lubricate the teflon rings when installing the fuel rail to the engine.*

58   Tighten all fasteners to the torque listed in this Chapter's Specifications.

2.5L models

59   Disconnect the fuel rail supply lines (see Section 4).

60   Detach the wiring harness retainers from the fuel rail cover.

61   Disconnect the fuel injectors connectors.

62   Remove the nut and electrical connector bracket from the fuel rail and position out of the way.

63   Remove the fuel rail mounting bolts and lift the rail off along with the fuel injectors.

64   Remove the injectors and replace the injector O-rings.

65   Installation is the reverse of removal. Tighten the fuel rail mounting fasteners to the torque listed in this Chapter's Specifications.

DI injector removal and seal replacement

Removal

Note: *The fuel injectors may remain in the fuel rails when the rail is removed, but normally they remain in the cylinder heads and require the use of a removal tool.*

66   On all models, remove each fuel injector retaining clip with a pair of needle-nose pliers, then remove the injector from its bore in the fuel rail. Remove and discard the upper injector O-rings. Repeat this procedure for each injector.

Note: *Even if you only removed the fuel rail assembly to replace a single injector or a leaking O-ring, replace all of the fuel injector retaining clips and O-rings.*

67   If any injectors stick in the cylinder head, use special tool #310-206 attached to a slide hammer to remove the injector(s).

68   Remove the old combustion chamber Teflon sealing ring, upper O-ring and support ring from each injector (see illustration).

Caution: *Be extremely careful not to damage the groove for the seal or the rib in the floor of the groove. If you damage the groove or the rib, you must replace the injector.*

69   Before installing the new Teflon seal on each injector, thoroughly clean the groove for the seal and the injector shaft. Remove all combustion residue and varnish with a clean shop rag.

Teflon seal installation using the special tools

70   The manufacturer recommends that you use the tools included in the special injector

13.68 To remove the Teflon sealing ring, cut it off with a hobby knife (be careful not to scratch the injector groove)

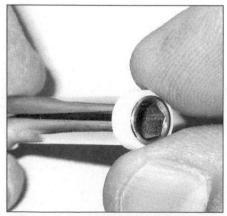

13.73 Slide the new Teflon seal onto the end of a socket that's the same diameter as the end of the fuel injector . . .

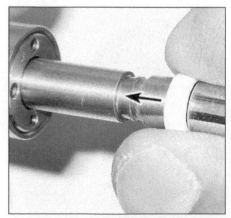

13.74 . . . align the socket with the end of the injector and slide the seal onto the injector and into its mounting groove

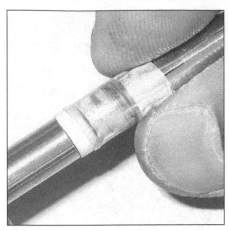

13.75a Use the socket to push a short section of plastic tubing onto the end of the injector and over the new seal . . .

13.75b . . . then leave the plastic tubing in place for several hours to compress the new seal

13.76 Note that the upper O-ring (1) is installed above the support ring (2)

tool set to install the Teflon lower seals on the injectors: Install the special seal assembly cone on the injector, install the special sleeve on the injector and use the sleeve to push on the assembly cone, which pushes the Teflon seal into place on its groove. Do not use any lubricants to do so.

71   Pushing the Teflon seal into place in its groove expands it slightly. There are three sizing sleeves in the special tool set with progressively smaller inside diameters. Using a clockwise rotating motion of about 180 degrees, install the slightly larger sleeve onto the injector and over the Teflon seal until the sleeve hits its stop, then carefully turn the sleeve counterclockwise as you pull it off the injector. Use the slightly smaller sizing sleeve the same way, followed by the smallest sizing ring. The seal is now sized. Repeat this step for each injector.

Teflon seal installation without special tools

72   If you don't have the special injector tool set, the Teflon seal can be installed using this method: First, find a socket that is equal or very close in diameter to the diameter of the end of the fuel injector.

73   Work the new Teflon seal onto the end of the socket (see illustration).

74   Place the socket against the end of the injector (see illustration) and slide the seal from the socket onto the injector. Do not use any lubricants to do so. Continue pushing the seal onto the injector until it seats into its mounting groove.

75   Because the inside diameter of the seal has to be stretched open to fit over the bore of the socket and the injector, its outside diameter is now slightly too large - it is no longer flush with the surface of the injector. It must be shrunk back to its original size. To do so, push a piece of plastic tubing with an interference fit onto the end of the socket; a plastic straw that fits tightly on the injector will work. After pushing the plastic tubing onto the socket about an inch, snip off the rest of the tubing, then use the socket to push the tubing onto the end of the injector (see illustration), sliding it onto the injector until

it completely covers the new seal (see illustration). Leave the tubing on for a few hours, then remove it. The seal should now be shrunk back its original outside diameter, or close to it.

Installation

76   Install a new support ring on the top of the injector, then lubricate the new upper O-ring with clean engine oil and install it on the injector. Do not oil the new Teflon seal. Note that the O-ring is installed above the support ring (see illustration).

77   Thoroughly clean the injector bores with a small nylon brush.

78   Lubricate the injector O-rings with clean motor oil before installing into the fuel rail.

79   Insert each injector into its bore in the fuel rail. Install the new retaining clips on the injectors.

80   Install the fuel rail.

81   Reconnect the cable to the negative battery terminal (see Chapter 5, Section 1), then turn the ignition switch to Run (but don't operate the starter). This activates the fuel pump for about two seconds, which builds up fuel pressure in the fuel lines and the fuel rail. Repeat this step two or three times, then check the fuel lines, fuel rails and injectors for fuel leaks.

---

14   Turbocharger - removal and installation

Warning: *Wait until the engine is completely cool before beginning this procedure.*
Note: *2013 and later 1.5L, 1.6L and 2.0L models are equipped with a turbocharger.*
1   Disconnect the negative battery cable (see Chapter 5).
2   Raise and support the vehicle on jackstands.
3   Remove the engine cover.
4   Drain the cooling system (see Chapter 1).

1.5L models

5   Remove the turbocharger piping between the air filter and turbocharger inlet.

6   Remove the intercooler piping between the turbocharger and intercooler. Use care to disconnect all connectors and harnesses.
7   Remove the exhaust bracket bolts at the front subframe.
8   Remove the catalytic converter support bracket.
9   Remove the battery and battery tray (see Chapter 5, Section 3).
10   Remove the catalytic converter flange nuts at the turbocharger and secure the converter out of the way. Discard the flange gasket.
11   Disconnect the turbocharger coolant lines from the turbocharger and secure out of the way. Discard the O-rings and sealing washers.
12   Remove the two nuts and one bolt and remove the turbocharger heat shield.
13   Remove the turbocharger oil supply line bolts at the turbocharger and block and discard the O-ring and sealing rings. Replace the turbocharger oil supply filter in the block.
14   Disconnect the wastegate vacuum hose.
15   Remove the bolts and the turbocharger oil return line from the vehicle. Discard the O-rings.
16   Support the engine using a floor jack, use a block of wood between the floor jack and oil pan to prevent damage.
17   Remove the front engine mount (see Chapter 2A, Section 18) to allow the engine to be lowered to remove the turbocharger.
Caution: *Detach the vacuum lines and harness from the valve cover so they are not damaged when the engine is lowered.*
18   Remove the four turbocharger exhaust flange nuts and remove the turbocharger from the vehicle through the engine compartment.
19   Replace the turbocharger mounting flange gasket with a new one.
20   If the turbocharger is to be replaced, transfer the inlet duct to the new turbocharger.
21   Installation is reverse of removal. Use new sealing washers and O-rings on the coolant and oil supply lines.
22   Tighten all fasteners to the torque listed in this Chapter's Specifications.
23   Refill the cooling system (see Chapter 1).

14.25a Loosen the clamps and remove the turbocharger inlet piping . . .

14.25b . . . and loosen the clamp to disconnect the inlet pipe from the turbocharger (1.6L model shown)

14.26 Remove the turbocharger outlet pipe (1.6L model shown)

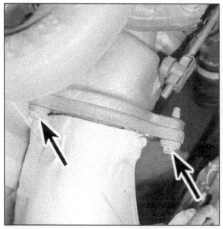

14.31 Remove the bolts and disconnect the converter from the turbocharger (1.6L model shown)

1.6L models

24   Remove the cowl panel (see Chapter 11, Section 12).

25   Remove the turbocharger piping between the air filter and turbocharger inlet (see illustrations).

26   Loosen the clamps and remove the turbocharger outlet pipe (see illustration).

27   On FWD models, remove the driveaxle bearing bracket nuts and bracket.

28   Remove the bolts and position the driveaxle bearing bracket out of the way.

29   On AWD models, remove the transfer case (see Chapter 7C, Section 4).

30   Disconnect the O2 sensor connector.

31   Remove the three bolts and disconnect the catalytic converter from the turbocharger exhaust outlet flange and exhaust pipe flange (see illustration).

32   Remove the bolts and front crossmember brace (see illustration).

33   Remove the exhaust bracket and catalytic converter assembly.

34   Remove the coolant tube bracket bolts (see illustration).

35   Remove the coolant tube banjo bolts at the turbocharger (see illustration) and discard the sealing rings.

36   Remove the clamps and disconnect the coolant tubes from the coolant hoses.

37   Remove the bolts and the turbocharger heat shield.

38   Remove the bolt and disconnect the turbocharger oil supply line from the engine block (see illustration). Inspect the O-ring and replace as necessary.

39   Remove the bolts at the block and the turbocharger and remove the turbocharger oil return line from the vehicle (see illustration). Replace the gasket, inspect the O-ring and replace as necessary.

40   Disconnect the turbocharger electrical connectors at the wastegate and vacuum valve and disconnect the vacuum supply line at the vacuum valve (see illustrations).

41   Remove the three turbocharger flange mounting nuts (see illustrations) and remove the turbocharger from the vehicle.

42   Replace the turbocharger mounting flange gasket with a new one.

43   If the turbocharger is to be replaced, transfer the oil supply line and inlet duct (see illustration) to the new turbocharger.

44   Installation is reverse of removal. Use new sealing washers on the coolant and oil supply line banjo bolts.

45   Tighten all fasteners to the torque listed in this Chapter's Specifications.

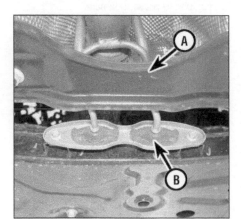

14.32 Remove the front crossmember brace (A) and the exhaust bracket (B) (1.6L model shown)

14.34 Remove the coolant tube bracket bolts (1.6L model shown)

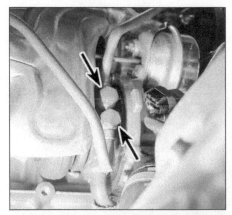

14.35 Remove the coolant tube banjo bolts at the turbocharger (1.6L model shown)

14.38 Disconnect the turbocharger oil
supply line from the block
(1.6L model shown)

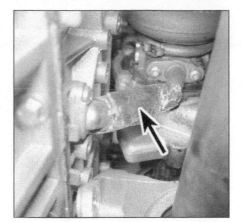

14.39 Remove the turbocharger oil
return line (1.6L model shown)

14.40a Disconnect the wastegate
electrical connector . . .

46   Refill the cooling system (see Chapter 1).

2.0L models

47   Disconnect both turbocharger down pipe oxygen sensor connectors before and after the catalytic converter.
48   Disconnect the catalytic converter from the exhaust pipe flange.
49   Remove the bolts and front crossmember brace.
50   Remove the exhaust bracket and hanger assembly.
51   Remove the catalytic converter support bracket.
52   Remove the lower transaxle mount bolt.
53   Remove the catalytic converter band clamp at the turbocharger flange and remove the catalytic converter.
Caution: *Use care not to damage the sealing surface of the converter or the flange or exhaust leaks may occur.*
54   Remove the front passenger's wheel.
55   Loosen the clamps and remove the turbocharger outlet pipe.
56   Remove the turbocharger piping between the air filter and turbocharger inlet.
57   Remove the clamps and disconnect the coolant tubes from the coolant hoses.

58   Remove the coolant tube banjo bolts at the turbocharger and discard the sealing rings.
59   Remove the bolts and the turbocharger heat shield.
60   Disconnect the turbocharger electrical connectors and hoses.
61   Remove the turbocharger oil supply line banjo bolts at the turbocharger and block and discard the sealing rings. Replace the turbocharger oil supply filter in the block.
62   Remove the bolts and the turbocharger oil return line from the vehicle. Replace the gasket, inspect the O-ring and replace as necessary.
63   On AWD models, support the engine using a floor jack, use a block of wood between the floor jack and oil pan to prevent damage.
64   Remove the front engine mount (see Chapter 2A, Section 18) to allow the engine to be lowered to remove the turbocharger.
Caution: *Detach the vacuum lines and harness from the valve cover so they are not damaged when the engine is lowered.*
65   On all models, remove the four turbocharger exhaust flange nuts and remove the turbocharger from the vehicle.
Note: *On FWD models, the turbocharger is removed from under the vehicle. On AWD mod-*

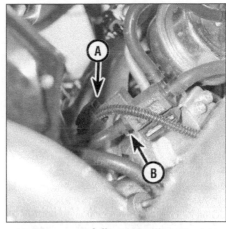

14.40b . . . and disconnect the vacuum
valve electrical connector (A) and vacuum
supply line (B) (1.6L model shown)

els, the turbocharger is removed through the engine compartment.
66   Replace the turbocharger mounting flange gasket with a new one.
67   If the turbocharger is to be replaced, transfer the inlet duct to the new turbocharger.

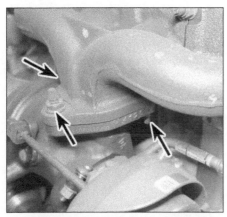

14.41a Remove the turbocharger flange
mounting nuts, one is accessed
from above . . .

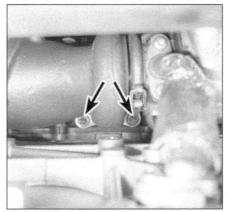

14.41b . . . and two are accessed from
below (1.6L model shown)

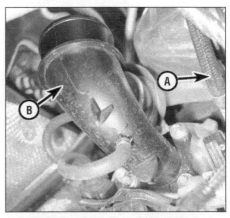

14.43 Transfer the oil supply line (A) and
inlet duct (B) to the new turbocharger if
necessary (1.6L model shown)

15.27 Support the bracket that is mounted under the intercooler and attached to the frame and firewall

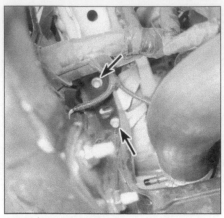

15.28a Remove the driver's side bolts . . .

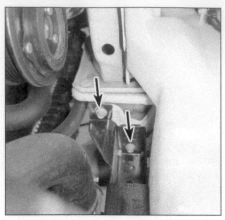

15.28b . . . and the passenger's side bolts to lower the bracket (not all bolts are shown)

68   Installation is reverse of removal. Use new sealing washers and O-rings on the coolant and oil supply lines.
69   Tighten all fasteners to the torque listed in this Chapter's Specifications.
70   Refill the cooling system (see Chapter 5).

15   Intercooler and piping - replacement

Warning: *Wait until the engine is completely cool before beginning this procedure.*
Note: *The air conditioning system is under high pressure. DO NOT loosen any fittings or remove any components until after the system has been discharged. Air conditioning refrigerant must be properly discharged into an EPA-approved container at a dealer service department or an automotive air conditioning repair facility. Always wear eye protection when disconnecting air conditioning system fittings.*
Note: *2013 and later 1.5L, 1.6L and 2.0L models are equipped with an intercooler. The intercooler is also referred to as the charge air cooler.*

Intercooler
1   Disconnect the negative battery cable (see Chapter 5).

1.5L models
Note: On 1.5L models, the intercooler is an air-to-water style located in the front of the engine near the cooling fan. The following procedure details positioning the engine in such a way that the intercooler can be removed from the engine.
2   Have the air conditioning system discharged by a dealer service department or by an automotive air conditioning shop before proceeding (see Warning above).
3   Raise and support the front of the vehicle on jackstands. Remove the engine undercover.
4   Drain the engine coolant (see Chapter 1).
5   Disconnect the AC lines at the AC compressor (see Chapter 3, Section 15).
6   Disconnect the AC pressure transducer switch connector.

7   Remove the nuts and bracket and disconnect the exhaust from the down pipe.
8   Remove the roll-restrictor mount bolt (see Chapter 7B, Section 10).
9   On AWD models, disconnect the driveshaft from the transfer case (see Chapter 8, Section 9) and secure out of the way.
10   Detach the AC line from the body at the front of the engine compartment.
11   Remove the nut and bracket from the top right corner of the intercooler.
12   Place reference marks for installation and remove the hood latch assembly from the core support.
13   Disconnect the coolant inlet and outlet quick-disconnect fittings at the intercooler.
14   Ensure any additional harnesses, connectors and hoses are disconnected and positioned out of the way for intercooler removal.
15   Remove the two stud bolts and ten bolts from the intercooler assembly.
16   Remove the two front motor mount-to-chassis bolts.
17   Place a floor jack under the oil pan with a wooden block between the jack pad and the pan to prevent damage.
18   Slowly jack the engine up about one inch.
19   Push at the top of the engine to rotate the engine back enough to pull the intercooler out of the housing to remove from the vehicle.
20   Inspect the intercooler housing O-ring and replace if damaged.
21   Installation is reverse of removal.
22   Refill the cooling system (see Chapter 1).
23   Have the system evacuated, recharged and leak-tested by the shop that discharged it.

1.6L and 2.0L models
24   Raise and support the vehicle on jackstands.
25   Remove the engine undercover.
26   On models, equipped with active grille shutters, remove the bumper cover and the grille shutter assembly (see Chapter 3).
27   Support the bracket that is mounted under the intercooler and attached to the frame and firewall (see illustration).
28   Remove the eight bolts attaching the

bracket to the vehicle (see illustrations) and lower the bracket about one inch (the lower radiator rubber mounts are located in this bracket).
29   Disconnect the electrical connector and quick-disconnect fittings from the intercooler (see illustration).
30   Loosen the clamps and disconnect the intercooler hoses from the inlet and outlet tubes.
31   Lift the radiator and intercooler up to remove the two bolts attaching the intercooler and brackets to the radiator.
32   Rotate the bottom of the intercooler towards the front of the vehicle and down off of the tabs at the top and remove the intercooler from the vehicle.
33   Installation is reverse of removal.

Intercooler intake pipe (1.5L models)
Note: *On 1.6L and 2.0L models, the intercooler intake pipe runs under the engine and is self-explanatory for replacement.*
34   Remove the cowl (see Chapter 11, Section 12).
35   Remove the engine cover.
36   Disconnect the turbocharger bypass valve electrical connector.
37   Disconnect the quick-disconnect fitting for the crankcase pressure sensor tube from the rear of the engine.
38   Disconnect the crankcase pressure sensor electrical connector.
39   Remove the clamp, disconnect the crankcase pressure sensor tube from the front of the engine and remove the tube from the vehicle.
40   Disconnect the turbocharger bypass hose from the bypass valve.
41   Remove the brake booster hose and retainer from the intercooler intake pipe.
42   Disconnect the Intake Air Temperature (IAT) sensor connector from the intercooler intake pipe.
43   Disconnect and detach the turbocharger boost pressure sensor hoses and electrical connector from the intercooler intake pipe.

15.29 Disconnect the connector (A) and quick-disconnect fitting (B) from the intercooler (1.6L model shown)

16.4 A typical rubber exhaust hanger; to remove an exhaust hanger, simply disengage it from the bracket attached to the floorpan and from the bracket attached to the exhaust pipe, catalytic converter, muffler, etc. If the hanger is difficult to remove, use a large prybar as a lever to pry it off

44   Disconnect the quick-disconnect fitting for the fuel vapor lines at the rear of the engine.
45   Loosen the clamp for the intercooler intake pipe at the turbocharger.
46   Remove the intercooler intake pipe bracket bolt below the turbocharger bypass valve.
47   Loosen the clamp for the intercooler intake pipe at the front of the engine.
48   Remove the intercooler intake pipe from the vehicle.
49   Installation is reverse of removal.

Intercooler coolant pump (1.5L models)

50   Raise and support the vehicle on jackstands.
51   Remove the engine undercover.
52   Ensure the cooling system pressure is released. After releasing any pressure, tighten the coolant expansion tank cap.
53   Locate the intercooler pump near the oil filter (attached to the oil pan). Clamp off the coolant hoses to the pump to prevent coolant loss. If the proper clamps are not available, drain the cooling system before disconnecting the hoses from the pump.
54   Disconnect the pump electrical connector.
55   Remove the two bolts attaching the pump to the oil pan. To remove the pump, move the pump downwards to pull the tab out of the retainer at the top.
56   Installation is reverse of removal. Top off the coolant level as necessary.

Intercooler radiator (1.5L models)

57   Raise and support the vehicle on jackstands.
58   Remove the engine undercover.
59   On models, equipped with active grille shutters, remove the bumper cover and the grille shutter assembly (see Chapter 3).
60   Disconnect the electrical connector from the intercooler radiator.
61   Remove the clamps and disconnect the three intercooler radiator coolant hoses.
62   Remove the two bolts attaching the intercooler radiator to the radiator.

63   Rotate the bottom of the intercooler radiator towards the front of the vehicle and down off of the tabs at the top and remove the intercooler radiator from the vehicle.
64   Installation is reverse of removal.

---

16   Exhaust system - inspection and component replacement

Inspection

Warning: *Inspect and repair exhaust system components only after allowing the exhaust components to cool completely. This applies particularly to the catalytic converter, which operates at very high temperatures. Also, when working under the vehicle, make sure it is securely supported on jackstands.*

1   The exhaust system consists of the exhaust manifold(s), the catalytic converter(s), the exhaust pipes, the resonator (V6 models), the muffler and all brackets, hangers and clamps. Inspect the exhaust system regularly to ensure that it remains safe and quiet. Look for any damaged or bent parts, open seams, holes, loose connections, excessive corrosion or other defects which could allow exhaust fumes to enter the vehicle. Also check the catalytic converter(s) when you inspect the exhaust system. Inspect the catalytic converter heat shield(s) for cracks, dents and loose or missing fasteners. If a heat shield is damaged, the converter might also be damaged. Have the converter inspected by a dealer service department. Damaged or deteriorated exhaust system components should not be repaired; they should be replaced with new parts.
2   Before trying to disassemble any exhaust components, spray the fasteners with a penetrating oil to help ease removal. If the exhaust system components are extremely corroded or rusted together, welding equipment will probably be required to remove them. The convenient way to accomplish this is to have a muffler repair shop remove the corroded sections with a cutting torch. If, however, you want to save money by doing it yourself (and you don't have a welding outfit with a cutting torch), simply cut

off the old components with a hacksaw. If you have compressed air, special pneumatic cutting chisels can also be used. If you decide to tackle the job at home, be sure to wear safety goggles to protect your eyes from metal chips and work gloves to protect your hands.
3   Here are some simple guidelines to follow when repairing the exhaust system:

a) *Work from the back to the front when removing exhaust system components.*
b) *Apply penetrating oil to the exhaust system component fasteners to make them easier to remove.*
c) *Use new gaskets, hangers and clamps when installing exhaust systems components.*
d) *Apply anti-seize compound to the threads of all exhaust system fasteners at reassembly.*
e) *Be sure to allow sufficient clearance between newly installed parts and all points on the underbody to avoid overheating the floor pan and possibly damaging the interior carpet and insulation. Pay particularly close attention to the catalytic converter and heat shield.*

Component replacement
Rubber exhaust hangers

4   The exhaust system is attached to the body with mounting brackets and rubber hangers. Because of the intense heat generated by the exhaust system, these hangers should be inspected regularly and often. Anytime you must raise the vehicle to perform any under-vehicle service, make sure that you inspect the rubber exhaust hangers. Look for cracks, tears and deterioration. If a hanger is worn or damaged, replace it (see illustration). If a hanger breaks or becomes disconnected from its hanger bracket

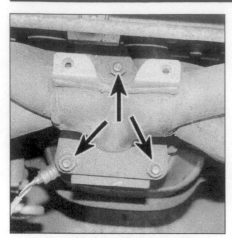

16.9 To detach the upper end of the catalytic converter from the exhaust manifold on a four-cylinder model, remove these three nuts

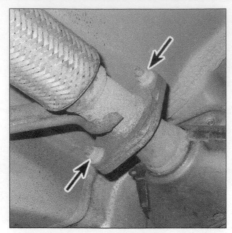

16.18 To disconnect the front end of the catalytic converter from the crossover and flexible exhaust pipe, remove these two flange nuts

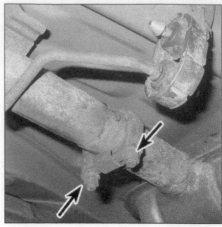

16.19 To disconnect the rear end of the catalytic converter from the exhaust pipe behind it, loosen the two nuts on this clamp and slide the clamp to the rear

or from the bracket on the part that it supports, the exhaust system will transmit excessive noise and vibration to the vehicle body.

Catalytic converters

Warning: *The catalytic converters get extremely hot during operation, and can remain very hot for hours after the engine has been turned off. Make sure that a converter has cooled down before you touch it.*

2012 and earlier four-cylinder models

5   Four-cylinder models are equipped with a three-way catalytic converter (TWC) that's bolted to the exhaust manifold
Note: *See Chapter 6 for more information on the function and operation of the catalytic converter.*
6   Remove the exhaust manifold heat shield (see Chapters 2A and 2B).
7   Disconnect the electrical connectors for the upstream and downstream oxygen sensors and then remove the upstream sensor

(see Chapter 6, Section 12).
8   Disconnect the Exhaust Gas Recirculation (EGR) pipe (see Chapter 6, Section 29).
9   Remove the three nuts from the exhaust manifold flange (see illustration).
10   Raise the vehicle and place it securely on jackstands.
11   Remove the downstream oxygen sensor (see Chapter 6, Section 12).
12   Remove the two bolts from the clamp that attaches the catalyst to the engine block. (It's not necessary to unbolt the clamp bracket from the engine block.)
13   Remove (and discard) the nuts from the flange at the lower end of the short section of exhaust pipe below the converter. Discard the old flange gasket.
14   Remove the catalytic converter assembly.
15   Installation is the reverse of removal. Apply anti-seize compound to the threads of all bolts and studs. Be sure to use new fasteners at both ends of the converter and a new flange gasket. Tighten all fasteners securely.

2012 and earlier V6 models

16   V6 models are equipped with three converters: an upstream "fast-light-off" catalyst directly below each exhaust manifold (and an integral part of the manifold), and a conventional three-way catalytic converter (TWC) located behind the junction between the two exhaust pipes for the front and rear cylinder heads. The following procedure applies to the larger catalyst underneath the vehicle. To replace an upstream converter, refer to Chapters 2A and 2B.
17   Raise the vehicle and place it securely on jackstands.
18   Remove the nuts from the flange located in front of the catalytic converter (see illustration).
19   Loosen the clamp located behind the converter (see illustration).
20   Disengage the two catalytic converter hanger brackets from their rubber exhaust hangers and then remove the catalytic converter.

16.32 To disconnect the front flange of the crossover and flexible exhaust pipe from the exhaust manifold/catalytic converter for the front cylinder head, remove these two nuts

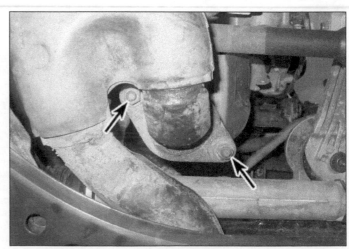

16.33 To disconnect the rear flange of the crossover and flexible exhaust pipe from the exhaust manifold/catalytic converter for the rear cylinder head, remove these two nuts

21   Be sure to inspect both rubber hangers (see Step 1). Installation is the reverse of removal. Apply anti-seize compound to the threads of all bolts and studs. Be sure to use new fasteners at both ends of the converter and a new flange gasket. Tighten all fasteners securely.

2013 and later 2.5L models

Note: *For catalytic converter removal procedures on 1.5L, 1.6L and 2.0L models, refer to Section 14.*

22   Raise and support the front of the vehicle of jackstands.
23   Remove the front subframe (see Chapter 10, Section 21).
24   Disconnect the O2 sensor harnesses and detach from the converter heat shield. If necessary, remove the O2 sensors.
25   Remove the four bolts and converter heat shield.
26   Remove the two converter bracket bolts.
27   Remove the nuts and disconnect the converter pipe from the exhaust pipe. Discard the flange gasket.
28   Remove the seven nuts attaching the converter manifold flange to the cylinder head and remove the converter from the vehicle. Discard the flange gasket.

Note: *The converter and exhaust manifold are one unit and not serviceable separately.*

29   Installation is reverse of removal. Use new flange gaskets.
30   Tighten all fasteners to the torque listed in this Chapter's Specifications.

Crossover and flexible exhaust pipe (2012 and earlier V6 models)

31   Raise the vehicle and place it securely on jackstands.
32   Remove the nuts that attach the forward crossover flange to the front exhaust manifold/catalytic converter (see illustration). Discard the nuts and the old flange gasket.
33   Remove the nuts that attach the rear crossover flange to the rear exhaust manifold/catalytic converter (see illustration). Discard the nuts and the old flange gasket.

16.37 The resonator is a small muffler that helps the muffler to reduce exhaust noise (V6 models only)

34   Remove the nuts that attach the flange behind the flexible exhaust pipe to the front end of the downstream catalytic converter (see illustration 16.18). Discard the nuts and the old flange gasket.
35   Remove the crossover and flexible exhaust pipe assembly.
36   Installation is the reverse of removal. Be sure to use new flange nuts and flange gaskets. Apply anti-seize compound to the threads of all bolts and studs. Tighten all fasteners securely.

Resonator (2012 and earlier V6 models)

37   The resonator (see illustration) is a small auxiliary muffler that helps the main muffler reduce exhaust noise. (Usually, the resonator reduces noise in one frequency, the muffler reduces noise in another frequency.)
38   Raise the vehicle and place it securely on jackstands.
39   Loosen the clamp ahead of the resonator (see illustration 16.19) and slide the clamp out of the way. Discard the nuts.
40   Remove the two nuts from the flange

behind the resonator (see illustration). Discard the nuts.
41   Remove the resonator.
42   Installation is the reverse of removal. Be sure to use new flange nuts and a new flange gasket at the front, and new nuts for the clamp at the rear of the resonator. Apply anti-seize compound to the threads of all bolts and studs. Tighten all fasteners securely.

Muffler

2012 and earlier models

43   Raise the vehicle and place it securely on jackstands.
44   Remove the two nuts from the flange behind the resonator (see illustration 15.31) and discard the nuts. (On 2.0L models, there's no resonator, just a straight exhaust pipe, but the flange is still there.)
45   Disengage the muffler hanger brackets from the front and rear muffler exhaust hangers (see illustrations) and remove the muffler.
46   Inspect the condition of the rubber exhaust hangers. Replace them if they're worn or damaged.

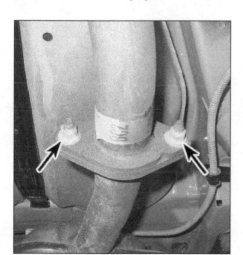

16.40 To disconnect the resonator flange from the muffler flange, remove these two nuts

16.45a Front muffler hanger (all models)

16.45b Rear muffler hanger (all models)

16.49 Remove the nuts and disconnect the pipe at the catalytic converter

16.50 Cut the pipe along here

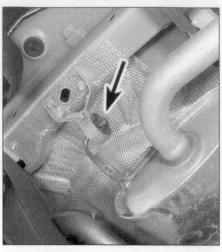

16.51 Disconnect the exhaust hangers

47   Installation is the reverse of removal. Be sure to use new fasteners and tighten all fasteners securely.

2013 and later models

Note: *Removal of the muffler and tailpipe assembly requires cutting of the one-piece assembly for removal and installing clamps at the cuts during installation.*

48   Raise and support the vehicle on jackstands.

49   Remove the nuts and disconnect the exhaust pipe from the catalytic converter flange (see illustration). Discard the flange gasket.

50   Carefully cut the tailpipe before the first bend after the flange, and before the last bend before the rear muffler (see illustration).

51   Disconnect the muffler and tailpipe from the exhaust hangers and remove from the vehicle (see illustration).

52   To install, attach the muffler and tailpipe to the hangers and install clamps where the cuts were made, to join the exhaust back together.

53   Use a new flange gasket.

Chapter 5
Engine electrical systems

Contents

Specifications

General

Firing order
Four-cylinder engines
    All except 1.5L 1-3-4-2
    1.5L 1-2-4-3
V6 engine 1-4-2-5-3-6
Cylinder numbering (from drivebelt end to transaxle end)
Four-cylinder engine 1-2-3-4
V6 engine
    Rear bank 1-2-3
    Front bank 4-5-6
Ignition timing .... Not adjustable

Torque specifications

Note: *One foot-pound (ft-lb) of torque is equivalent to 12 inch-pounds (in-lbs) of torque. Torque values below approximately 15 ft-lbs are expressed in inch-pounds, since most foot-pound torque wrenches are not accurate at these smaller values.*

	Ft-lbs (unless otherwise indicated)	Nm
Alternator bolts/nuts		
1.5L and 1.6L engines	35	48
Alternator idler pulley bolt	18	25
2.0L engine		
2004 and earlier models		
Upper bolt	18	25
Lower bolts	35	48
2013 and later models	35	48
2.3L engine	35	48
2.5L engine		
2009 through 2012 models	35	48
2013 and later models	18	25
3.0L V6 engine	35	48
Starter bolts		
1.5L engine	35	48
1.6L engine	26	35
2.0L engine		
2004 and earlier models	20	27
2013 and later models	26	35
2.3L engine	26	35
2.5L engine		
2009 through 2012 models	26	35
2013 and later models	35	48
3.0L V6 engine	20	27

1 General information, precautions and battery disconnection/reconnection

1   The engine electrical systems include all ignition, charging and starting components. Because of their engine-related functions, these components are discussed separately from chassis electrical devices such as the lights, the instruments, etc. (which are included in Chapter 12).

Precautions

2   Always observe the following precautions when working on the electrical system:

a) *Be extremely careful when servicing engine electrical components. They are easily damaged if checked, connected or handled improperly.*

b) *Never leave the ignition switched on for long periods of time when the engine is not running.*

c) *Never disconnect the battery cables while the engine is running.*

d) *Maintain correct polarity when connecting battery cables from another vehicle during jump starting - see the* Booster battery (jump) starting *Section at the front of this manual.*

e) *Always disconnect the negative cable from the battery before working on the electrical system, but read the following battery disconnection procedure first.*

3   It's also a good idea to review the safety-related information regarding the engine electrical systems located in the *Safety first!* Section at the front of this manual, before beginning any operation included in this Chapter.

Battery disconnection/ reconnection

Caution: *If the vehicle is equipped with the CD6 audio unit, when reconnecting the battery, make sure the battery is not disconnected or power is not interrupted from the audio unit for 30 seconds. If power is interrupted during the first 30 seconds, permanent damage to the CD6 audio unit will result.*

4   The battery is located in the engine compartment on all vehicles covered by this manual. To disconnect the battery for service procedures that require battery disconnection, simply disconnect the cable from the negative battery terminal. On 2013 and later models, the air filter housing must be removed to access the battery terminals. The alternative is to disconnect the negative battery terminal attached to the vehicle body (see illustration). Make sure that you isolate the cable to prevent it from coming into contact with the battery negative terminal (or the body on 2013 an later models, if it was disconnected from there).

5   Some vehicle systems (radio, alarm system, power door locks, windows, etc.) require battery power all the time, either to enable their operation or to maintain control unit memory (Powertrain Control Module, auto-

1.4 Disconnect the negative battery cable attached to the body (2013 and later models)

matic transaxle control module, etc.), which would be lost if the battery were to be disconnected. So before you disconnect the battery, note the following points:

a) *Before connecting or disconnecting the cable from the negative battery terminal, make sure that you turn the ignition key and the lighting switch to their Off positions. Failure to do so could damage semiconductor components.*

b) *On a vehicle with power door locks, it is a wise precaution to remove the key from the ignition and to keep it with you, so that it does not get locked inside if the power door locks should engage accidentally when the battery is reconnected!*

c) *After the battery has been disconnected, then reconnected (or a new battery has been installed) on vehicles with an automatic transaxle, the Transaxle Control Module (TCM) will need some time to relearn its adaptive strategy. As a result, shifting might feel firmer than usual. This is a normal condition and will not adversely affect the operation or service life of the transaxle. Eventually, the TCM will complete its adaptive learning process and the shift feel of the transaxle will return to normal.*

d) *The engine management system's PCM will lose some of the information stored in its "map" (program) when the battery is disconnected. Whenever the battery has been disconnected, the Powertrain Control Module (PCM) must relearn its idle and fuel trim strategy before the vehicle returns to its optimal driveability and performance. To initiate this process, refer to* Relearning the idle and fuel trim strategy after battery disconnection *below.*

e) *On 2008 through 2012 models, the cluster display needs to be calibrated. After the battery is reconnected, rotate the instrument panel dimmer switch from the lowest to the dome On position.*

f) *On 2008 through 2012 models, the cluster display needs to be calibrated. After the battery is reconnected, rotate the instrument panel dimmer switch from the lowest to the dome On position.*

g) *On models with One-Touch Open and Close power windows, the power window system must be initialized (see below).*

Memory savers

6   Devices known as "memory savers" (typically, small 9-volt batteries) can be used to avoid some of the above problems. A memory saver is usually plugged into the cigarette lighter, and then you can disconnect the vehicle battery from the electrical system. The memory saver will deliver sufficient current to maintain security alarm codes and - maybe, but don't count on it! - PCM memory. It will also run "unswitched" (always on) circuits such as the clock and radio memory, while isolating the car battery in the event that a short circuit occurs while the vehicle is being serviced.

Warning: *If you're going to work around any airbag system components, disconnect the battery and do not use a memory saver. If you do, the airbag could accidentally deploy and cause personal injury.*

Caution: *Because memory savers deliver current to operate unswitched circuits when the battery is disconnected, make sure that the circuit that you're going to service is actually open before working on it!*

Relearning the idle and fuel trim strategy after battery disconnection

Caution: *Failure to perform the following procedure after battery reconnection could adversely affect the idle quality of the engine until it eventually relearns its idle trim.*

7   With the vehicle stationary, apply the parking brake.

8   Put the shift lever in Park, turn off all accessories and then start the engine.

9   Allow the engine to warm up to its normal operating temperature.

10 Allow the engine to idle for at least one minute.

11 Turn on the air conditioning system and allow the engine to idle for at least another minute.

12 Drive the vehicle at least 10 miles.

13 The vehicle should now have relearned its idle and fuel trim strategy.

Battery Monitoring System (BSM) relearn

14 On 2013 and later models, when the current battery is replaced with a new one, the BSM relearn procedure must be performed using a scan tool.

One-Touch Open and Close power windows - initialization

15 Start the engine, raise the window until it is fully closed, then release the switch.

16   Operate the window switch in the Up position for at least one second.

17   Lower the window completely, release the switch, then raise the window until it is closed.

18   Repeat on the other front door window.

2   Troubleshooting

Ignition system

1   If a malfunction occurs in the ignition system, do not immediately assume that any particular part is causing the problem. First, check the following items:

a) *Make sure that the cable clamps at the battery terminals are clean and tight.*

b) *Test the condition of the battery (see Steps 15 through 19). If it doesn't pass all the tests, replace it.*

c) *Check the ignition coil or coil pack connections.*

d) *Check any relevant fuses in the engine compartment fuse and relay box (see Chapter 12). If they're burned, determine the cause and repair the circuit.*

Check

Warning: *Because of the high voltage generated by the ignition system, use extreme care when performing a procedure involving ignition components.*

Note: *The ignition system components on these vehicles are difficult to diagnose. In the event of ignition system failure that you can't diagnose, have the vehicle tested at a dealer service department or other qualified auto repair facility.*

Note: *You'll need a spark tester for the following test. Spark testers are available at most auto supply stores.*

2   If the engine turns over but won't start, verify that there is sufficient ignition voltage to fire the spark plugs as follows.

3   Remove a coil and install the tester between the boot at the lower end of the coil and the spark plug (see illustration).

4   Crank the engine and note whether or not the tester flashes.

Caution: *Do NOT crank the engine or allow it to run for more than five seconds; running the engine for more than five seconds may set a Diagnostic Trouble Code (DTC) for a cylinder misfire.*

5   If the tester flashes during cranking, the coil is delivering sufficient voltage to the spark plug to fire it. Repeat this test for each cylinder to verify that the other coils are OK.

6   If the tester doesn't flash, remove a coil from another cylinder and swap it for the one being tested. If the tester now flashes, you know that the original coil is bad. If the tester still doesn't flash, the PCM or wiring harness is probably defective. Have the PCM checked out by a dealer service department or other qualified repair shop (testing the PCM is beyond the scope of the do-it-yourselfer because it requires expensive special tools).

2.3 Spark tester

7   If the tester flashes during cranking but a misfire code (related to the cylinder being tested) has been stored, the spark plug could be fouled or defective.

Charging system

8   If a malfunction occurs in the charging system, do not automatically assume the alternator is causing the problem. First check the following items:

a) *Check the drivebelt tension and condition, as described in Chapter 1. Replace it if it's worn or deteriorated.*

b) *Make sure the alternator mounting bolts are tight.*

c) *Inspect the alternator wiring harness and the connectors at the alternator and voltage regulator. They must be in good condition, tight and have no corrosion.*

d) *Check the fusible link (if equipped) or main fuse in the underhood fuse/relay box. If it is burned, determine the cause, repair the circuit and replace the link or fuse (the vehicle will not start and/or the accessories will not work if the fusible link or main fuse is blown).*

e) *Start the engine and check the alternator for abnormal noises (a shrieking or squealing sound indicates a bad bearing).*

f) *Check the battery. Make sure it's fully charged and in good condition (one bad cell in a battery can cause overcharging by the alternator).*

g) *Disconnect the battery cables (negative first, then positive). Inspect the battery posts and the cable clamps for corrosion. Clean them thoroughly if necessary (see Chapter 1). Reconnect the cables (positive first, negative last).*

Alternator - check

9   Use a voltmeter to check the battery voltage with the engine off. It should be at least 12.6 volts (see illustration 2.16).

10   Start the engine and check the battery voltage again. It should now be approximately 13.5 to 15 volts.

2.16 To test the open circuit voltage of the battery, touch the black probe of the voltmeter to the negative terminal and the red probe to the positive terminal of the battery; a fully charged battery should be at least 12.6 volts

11   If the voltage reading is more or less than the specified charging voltage, the voltage regulator is probably defective, which will require replacement of the alternator (the voltage regulator is not replaceable separately). Remove the alternator and have it bench tested (most auto parts stores will do this for you).

12   The charging system (battery) light on the instrument cluster lights up when the ignition key is turned to ON, but it should go out when the engine starts.

13   If the charging system light stays on after the engine has been started, there is a problem with the charging system. Before replacing the alternator, check the battery condition, alternator belt tension and electrical cable connections.

14   If replacing the alternator doesn't restore voltage to the specified range, have the charging system tested by a dealer service department or other qualified repair shop.

Battery - check

15   Check the battery state of charge. Visually inspect the indicator eye on the top of the battery (if equipped with one); if the indicator eye is black in color, charge the battery as described in Chapter 1. Next perform an open circuit voltage test using a digital voltmeter.

Note: *The battery's surface charge must be removed before accurate voltage measurements can be made. Turn on the high beams for ten seconds, then turn them off and let the vehicle stand for two minutes.*

16   With the engine and all accessories Off, touch the negative probe of the voltmeter to the negative terminal of the battery and the positive probe to the positive terminal of the battery (see illustration). The battery voltage should be 12.6 volts or slightly above. If the battery is less than the specified voltage, charge the battery before proceeding to the next test. Do not proceed with the battery load test unless the battery charge is correct.

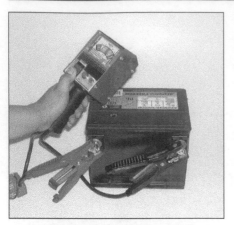

2.18 Connect a battery load tester to the battery and check the battery condition under load following the tool manufacturer's instructions

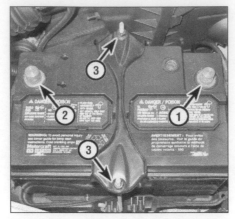

3.3a To remove the battery from the battery tray, disconnect the negative battery cable (1), then the positive cable (2); then remove the two hold-down nuts (3) and remove the hold-down bracket (2012 and earlier models shown)

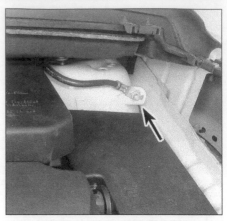

3.3b On 2013 and later models, disconnect the battery negative cable from the stud on the strut tower and remove it with the battery

17   Disconnect the negative battery cable, then the positive cable from the battery.

18   Perform a battery load test. An accurate check of the battery condition can only be performed with a load tester (see illustration). This test evaluates the ability of the battery to operate the starter and other accessories during periods of high current draw. Connect the load tester to the battery terminals. Load test the battery according to the tool manufacturer's instructions. This tool increases the load demand (current draw) on the battery.

19   Maintain the load on the battery for 15 seconds and observe that the battery voltage does not drop below 9.6 volts. If the battery condition is weak or defective, the tool will indicate this condition immediately.

Note: *Cold temperatures will cause the minimum voltage reading to drop slightly. Follow the chart given in the manufacturer's instructions to compensate for cold climates. Minimum load voltage for freezing temperatures (32 degrees F) should be approximately 9.1 volts.*

Starting system

The starter rotates, but the engine doesn't

20   Remove the starter (see Section 7).

Check the overrunning clutch and bench test the starter to make sure the drive mechanism extends fully for proper engagement with the flywheel ring gear. If it doesn't, replace the starter.

21   Check the flywheel ring gear for missing teeth and other damage. With the ignition turned off, rotate the flywheel so you can check the entire ring gear.

The starter is noisy

22   If the solenoid is making a chattering noise, first check the battery (see Steps 15 through 19). If the battery is okay, check the cables and connections.

23   If you hear a grinding, crashing metallic sound when you turn the key to Start, check for loose starter mounting bolts. If they're tight, remove the starter and inspect the teeth on the starter pinion gear and flywheel ring gear. Look for missing or damaged teeth.

24   If the starter sounds fine when you first turn the key to Start, but then stops rotating the engine and emits a zinging sound, the problem is probably a defective starter drive that's not staying engaged with the ring gear. Replace the starter.

The starter rotates slowly

25   Check the battery (see Steps 15 through 19).

26   If the battery is okay, verify all connections (at the battery, the starter solenoid and motor) are clean, corrosion-free and tight. Make sure the cables aren't frayed or damaged.

27   Check that the starter mounting bolts are tight so it grounds properly. Also check the pinion gear and flywheel ring gear for evidence of a mechanical bind (galling, deformed gear teeth or other damage).

The starter does not rotate at all

28   Check the battery (see Steps 15 through 19).

29   If the battery is okay, verify all connections (at the battery, the starter solenoid and motor) are clean, corrosion-free and tight. Make sure the cables aren't frayed or damaged.

30   Check all of the fuses in the underhood fuse/relay box.

31   Check that the starter mounting bolts are tight so it grounds properly.

32   Check for voltage at the starter solenoid "S" terminal when the ignition key is turned to the start position. If voltage is present, replace

3.4a On 2013 and later models, disconnect the BSM connector . . .

3.4b . . . then detach the harness and position it out of the way

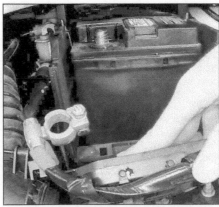

3.5 Detach the BJB from the battery box (2013 and later model shown)

3.6 Remove the battery hold-down bracket nuts (2013 and later models)

3.7 Loosen the clamp nut and detach the negative cable and Battery Monitoring Sensor from the battery (2013 and later models)

3.8 Battery tray mounting bolts - 2012 and earlier models

the starter/solenoid assembly. If no voltage is present, the problem could be the starter relay, the Transmission Range (TR) sensor (see Chapter 6), or with an electrical connector somewhere in the circuit (see the wiring diagrams at the end of this manual). Also, on many modern vehicles, the Powertrain Control Module (PCM) and the Body Control Module (BCM) control the voltage signal to the starter solenoid; on such vehicles a special scan tool is required for diagnosis.

3 Battery and battery tray - removal and installation

Warning: *Hydrogen gas is produced by the battery, so keep open flames and lighted cigarettes away from it at all times. Always wear eye protection when working around a battery. Rinse off spilled electrolyte immediately with large amounts of water.*

Battery

1   On 2013 and later models, remove the air filter housing components as needed to access the battery (see Chapter 4, Section 9).

2   On all models, remove the battery cover and/or the protective cover from the positive battery terminal.
3   Disconnect the negative cable first, then the positive cable second (see illustrations).
Warning: *Always disconnect the negative cable first and connect it last, or you might accidentally short the battery with the tool you're using to loosen the cable clamps.*
4   On 2013 and later models, disconnect the Battery Monitoring Sensor (BSM) connector and detach the wiring harness from the Battery Junction Box (BJB) and secure the harness to the side (see illustrations).
5   Release the clips and slide the BJB upward to remove from the battery box (see illustration).
6   On all models, remove the hold-down nuts and the hold-down bracket (see illustration 3.3a and the accompanying illustration).
7   Remove the battery. On 2013 and later models, the negative battery cable can now be disconnected (see illustration).

Battery tray

2012 and earlier models

8   If you need to access components located underneath the battery tray, remove

the bolts that secure the battery tray (see illustration), disconnect any components attached to the battery tray, and remove the tray.

2013 and later models

9   Unbolt the brake master cylinder remote fluid reservoir from the cowl panel and position it aside (see Chapter 9).
10   Detach the transaxle fluid cooler control valve from the side of the battery try; it is attached with a push-pin in the rear and a screw in the front (see illustration).
11   Remove the fasteners, pull up on the tray and guide it out from under the cowl (see illustration).

All models

12   Installation is the reverse of removal.
Warning: *When connecting the battery cables, always connect the positive cable first and the negative cable last to avoid a short circuit caused by the tool used to tighten the cable clamps.*
13   On 2012 and earlier models, after you're done, the Powertrain Control Module (PCM) must relearn its idle and fuel trim strategy for optimum driveability and performance (see Section 1 for this procedure).
14   On 2008 through 2012 models, the cluster display needs to be calibrated. After the battery is reconnected, rotate the instrument panel dimmer switch from the lowest to the dome On position.
15   On 2013 and later models, if a new battery has been installed, perform the BSM relearn procedure (see Section 1).
16   On models with One-Touch Open and Close power windows, the power window system must be initialized (see Section 1).

4 Battery cables - replacement

1   When removing the cables, always disconnect the cable from the negative battery terminal first and hook it up last, or you might accidentally short out the battery with the tool you're using to loosen the cable clamps. Even

3.10 Transaxle fluid cooler control valve fasteners

3.11 Battery tray mounting fastener locations

5.2 Disconnect the electrical connector from the coil pack

5.3 Disconnect the spark plug wires from the coil pack

if you're only replacing the cable for the positive terminal, be sure to disconnect the negative cable from the battery first.

2   Disconnect the old cables from the battery, then trace each of them to their opposite ends and disconnect them. Be sure to note the routing of each cable before disconnecting it to ensure correct installation.

3   If you are replacing any of the old cables, take them with you when buying new cables. It is vitally important that you replace the cables with identical parts.

4   Clean the threads of the solenoid or ground connection with a wire brush to remove rust and corrosion. Apply a light coat of battery terminal corrosion inhibitor or petroleum jelly to the threads to prevent future corrosion.

5   Attach the cable to the solenoid or ground connection and tighten the mounting nut/bolt securely.

6   Before connecting a new cable to the battery, make sure that it reaches the battery post without having to be stretched.

7   Connect the cable to the positive battery terminal first, then connect the ground cable to the negative battery terminal.

8   See Section 1 and perform any necessary relearn/initialization procedures.

---

5   Ignition coils - replacement

2004 and earlier four-cylinder engines

1   Disconnect the cable from the negative battery terminal (see Section 1).

2   Disconnect the ignition coil pack electrical connector (see illustration).

3   Disconnect all the spark plug wires from the ignition coil pack (see illustration). Pull on the spark plug wire boots, not on the wires.

4   Remove the four bolts that secure the coil pack to its mounting bracket and remove the coil pack (see illustration).

5   Installation is the reverse of the removal procedure with the following additions:

a) Before installing the spark plug wire connector into the ignition coil, coat the entire interior of the rubber boot with silicone dielectric compound.

b) Insert each spark plug wire into the proper terminal of the ignition coil. Push the wire into the terminal and make sure the boots are fully seated and both locking tabs are engaged properly.

c) After you're done, the Powertrain Control Module (PCM) must relearn its idle and fuel trim strategy for optimum driveability and performance (see Section 1 for this procedure as well as other procedures that need to be performed).

2005 and later four-cylinder models and all V6 engines

6   Disconnect the cable from the negative battery terminal (see Section 1).

7   Remove the engine cover (see illustration).

8   On V6 engines, to remove an ignition coil

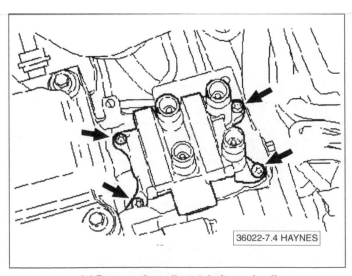

5.4 Remove the coil pack bolts and coil

5.7 To detach the engine cover, remove these three nuts (V6 models)

5.10a Disconnect the electrical connector from the ignition coil (2012 and earlier models shown, others similar)

5.10b On later models, slide out the lock, then unplug the electrical connector

5.11a To detach an ignition coil from the valve cover, remove this bolt (2012 and earlier models shown, others similar)

from the rear valve cover, remove the upper intake manifold (see Chapter 2B, Section 5).

9   On 2013 and later 2.0L turbocharged models, remove the intake air duct above the valve cover.

10   On all models, disconnect the electrical connector from the ignition coil (see illustrations).

11   Remove the coil mounting bolt(s) (see illustrations).

12   Remove the ignition coil (see illustration).

13   Coat the inside of the spark plug boot with silicone dielectric compound (see illustration).

14   Installation is otherwise the reverse of removal.

15   After you're done, the Powertrain Control Module (PCM) must relearn its idle and fuel trim strategy for optimum driveability and performance (see Section 1 for this procedure as well as other procedures that need to be performed).

6   Alternator - removal and installation

2007 and earlier models

Four-cylinder models

1   Disconnect the cable from the negative battery terminal (see Section 1).

2   Remove the alternator drivebelt (see Chapter 1, Section 24).

3   Disconnect the electrical connectors from the alternator.

4   Raise the front of the vehicle and place it securely on jackstands.

5   Remove the lower alternator bolts.

6   Unscrew the upper alternator bolt. (There's not enough clearance to remove the upper alternator bolt; it comes off with the alternator.)

7   Move the alternator to the rear, then lift it up and remove it from the engine compartment.

8   Before installing the alternator, insert the upper alternator bolt into its hole in the alternator mounting boss (you can't install the bolt once the alternator is in place).

9   Installation is otherwise the reverse of removal.

10   After you've reconnected the battery, the Powertrain Control Module (PCM) must relearn its idle and fuel trim strategy for optimum driveability and performance (see Section 1 for this procedure).

V6 models

11   Disconnect the cable from the negative battery terminal (see Section 1).

12   Loosen the driveaxle hub nut (see Chapter 8) and the wheel lug nuts. Raise the front of the vehicle and place it securely on jackstands. Remove the wheel.

5.11b On 1.5L and 1.6L engines, each coil is retained by two bolts

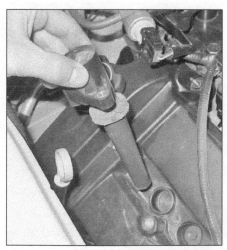

5.12 To remove an ignition coil from the valve cover, pull it straight up (2012 and earlier models shown, others similar)

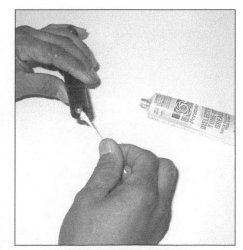

5.13 Before pushing the spark plug boot back onto the spark plug, coat the inside of the boot with silicone dielectric compound

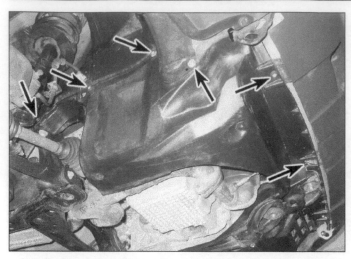

6.13 To detach the right engine splash shield from the vehicle, remove these fasteners (V6 models)

6.16a To detach the alternator splash shield, remove this push fastener . . .

13   Remove the right lower engine splash shield (see illustration).

14   Remove the right driveaxle (see Chapter 8, Section 8). (You will also need to remove the intermediate shaft before you can remove the alternator, but don't actually remove it at this time, or gear lubricant will leak out of the transaxle!)

15   Remove the drivebelt (see Chapter 1, Section 24).

16   Remove the alternator splash shield (see illustrations).

17   Detach the wiring harness clips (see illustration) and position the harness out of the way.

18   Remove the upper alternator mounting bolt (see illustration).

19   Remove the lower alternator mounting bolts and position the alternator out of the way so that you can remove the alternator mounting bracket.

20   Remove the alternator mounting bracket (see illustration).

21   Disconnect the electrical connectors from the alternator (see illustration).

22   Remove the driveaxle intermediate shaft (see Chapter 8, Section 8). When you detach the inner splined end of the intermediate shaft from the transaxle, some gear lubricant will leak out, so put a drain pan underneath to catch the spilled fluid.

23   Remove the alternator (see illustration) and then install the intermediate shaft immediately to prevent the loss of any more transaxle gear lube.

24   Be sure to check the fluid level inside the transaxle and refill as necessary (see Chapter 1).

25   Installation is otherwise the reverse of removal.

2008 models

26   Disconnect the cable from the negative battery terminal (see Section 1).

27   Raise the vehicle and support it securely on jackstands.

28   Remove the right lower splash shield.

29   Refer to Chapter 1, Section 24 and remove the drivebelt.

Four-cylinder models

30   Remove the lower air duct from the alternator by pushing the lock tab.

31   Remove the wiring harness guides from the alternator shield.

32   Move the alternator cover to the side and disconnect the wiring from the alternator.

33   Remove the two mounting nuts from the alternator shield, then remove the pin retainer from the bottom of the shield. Remove the shield.

34   Remove the alternator bolt, stud nuts and studs.

35   Remove the alternator upper air duct by removing its three screws.

36   Remove the alternator from the top of the engine compartment.

37   Installation is the reverse of removal.

6.16b . . . and this one, then remove the splash shield (2007 and earlier V6 models)

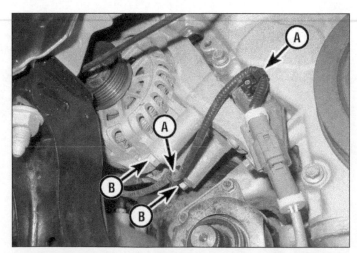

6.17 Detach the two wiring harness clips (A) and position the harness out of the way; to detach the lower part of the alternator from its mounting bracket, remove these two bolts (B) (2007 and earlier V6 models)

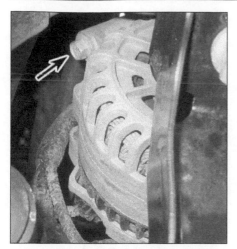

6.18 Remove the upper alternator bolt (2007 and earlier V6 models)

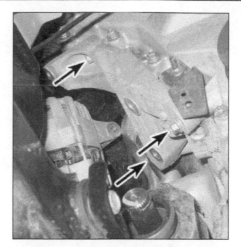

6.20 To detach the alternator mounting bracket from the engine block, remove these bolts (2007 and earlier V6 models)

6.21 Disconnect the electrical connector from the alternator, remove the nut from the battery positive (B+) terminal and disconnect the cable from the terminal (V6 models)

V6 models

38   Disconnect the electrical connectors from the alternator.
39   Disconnect the wire from the battery terminal of the alternator.
40   Unbolt the air conditioning compressor and use wire to secure it out of the way.
41   Loosen the two alternator nuts, then remove the two lower alternator studs.
42   Remove the upper alternator mounting bolt, then lift the alternator out.
43   Installation is the reverse of removal.

2009 through 2012 models

44   Raise the vehicle and support it securely on jackstands.
45   Disconnect the cable from the negative battery terminal (see Section 1).
46   Remove the right lower splash shield.
47   Refer to Chapter 1, Section 24 and remove the drivebelt from the alternator pulley.

Four-cylinder models

48   Remove the wiring harness guide from the bottom alternator mounting stud, then remove the mounting bolt and both stud nuts.
49   Working from above, remove the lower air duct by pushing in its lock tab.
50   Move the cover, then disconnect the wiring from the battery terminal of the alternator.
51   Disconnect the wiring harness from the alternator, then remove the alternator.
52   Installation is the reverse of removal.

V6 models

53   Disconnect the wiring harness from the alternator.
54   Move the cover, then disconnect the wire from the battery terminal of the alternator.
55   Unbolt the air conditioning compressor and use wire to secure it out of the way.
56   Loosen the two alternator nuts, then remove the two lower alternator studs (see illustration).

57   Remove the upper alternator mounting bolt, then lift the alternator out.
58   Installation is the reverse of removal.

2013 and later models

59   Raise and support the front of the vehicle on jackstands.
60   Disconnect the negative battery cable (see Section 1).

1.5L and 1.6L engines

Note: *The alternator pulley on the 1.5L and 1.6L engines is a clutch style.*

61   **1.5L models:** Remove the right (passenger's-side) driveaxle (see Chapter 8, Section 8).
62   Remove the roll-restrictor mount bolt at the subframe bracket.
63   Place a floor jack under the oil pan to support the engine. Place a wood block between the jack and oil pan to protect the oil pan.

6.23 To remove the alternator, lift it out through the space normally occupied by the right driveaxle and intermediate shaft (2007 and earlier V6 models)

6.56 Alternator lower mounting studs and nuts on a 2009 through 2012 V6 models

**6.68 Disconnect the B+ cable
(2013 and later 1.6L model shown)**

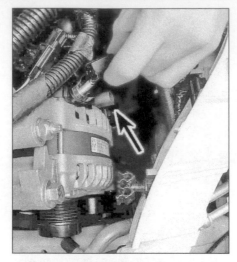

**6.69 Disconnect the alternator connector
(2013 and later 1.6L model shown)**

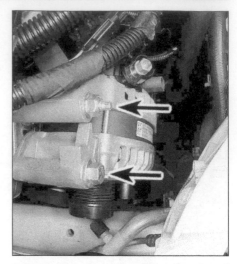

**6.70a Remove the upper bolt, nut and stud
(2013 and later 1.6L model shown) . . .**

64 Detach the coolant expansion tank from its mounting tabs and position it aside (see Chapter 3).

65 Remove the front engine mount (see Chapter 2A).

66 Jack up the engine approximately two inches (50 mm).

67 Remove the accessory drivebelt (see Chapter 1, Section 24).

68 Remove the nut and disconnect the battery B+ cable from the alternator (see illustration).

69 Disconnect the electrical connector from the alternator (see illustration).

70 Remove the two bolts, one nut and one stud attaching the alternator to the engine (see illustrations).

71 Rotate the alternator so that the pulley is facing downwards and the upper mounting holes are positioned pointing toward the right headlight, then carefully lift the alternator out of the engine compartment (see illustration).

72 If replacing the alternator, remove the idler pulley and transfer to the new alternator (see illustration).

73 Installation is reverse of removal.

2.0L and 2.5L engines

74 Remove the accessory drivebelt (see Chapter 1, Section 24).

75 Unbolt the air conditioning compressor and position it aside (see Chapter 3, Section 15).
Warning: *The air conditioning system is under pressure - DO NOT disconnect the hoses.*

76 Remove the nut and disconnect the battery B+ cable from the alternator. Disconnect the electrical connector from the alternator.

77 Remove the three bolts attaching the alternator to the engine (remove the AC bracket with the alternator).

78 Remove the alternator from the vehicle.

79 Remove the AC bracket from the alternator (if replacing the alternator).

80 Installation is reverse of removal.

All models

81 After you've reconnected the battery, the Powertrain Control Module (PCM) must relearn its idle and fuel trim strategy for optimum driveability and performance (see Section 1 for this procedure as well as other procedures that need to be performed).

---

7   Starter motor - removal and installation

2008 and earlier models
Four-cylinder models

1 Disconnect the cable from the negative battery terminal (see Section 1).

2 Remove the three starter motor mounting bolts.

3 Raise the front of the vehicle and place it securely on jackstands.

**6.70b . . . then remove the lower bolt
(2013 and later 1.6L model shown)**

**6.71 Remove the alternator from the vehicle with the pulley facing down
(2013 and later 1.6L model shown)**

**6.72 Remove the idler pulley
(2013 and later 1.6L model shown)**

7.12 To gain access to the starter motor on V6 models, disconnect these coolant hoses and position them out of the way (have some rags handy to catch any residual coolant that might drip onto the engine) - 2008 and earlier models

7.13 Disconnect the battery starter cable from the big terminal and the starter control cable from the smaller terminal (V6 models)

4   Remove the bolts from the support bracket for the intermediate shaft center bearing (see Chapter 8, Section 8).
5   It's not necessary to remove either the right driveaxle or the intermediate shaft, but detaching the center bearing support bracket will give you enough room to pull out the starter.
6   Disconnect the battery cable (the large cable) from the starter motor.
7   Disconnect the starter control electrical connector from the starter motor solenoid terminals.
8   Remove the starter.

V6 models

Warning: *Wait until the engine is completely cool before beginning this procedure.*
9   Disconnect the cable from the negative battery terminal (see Section 1).
10   Remove the air intake duct (see Chapter 4, Section 9), then remove the air filter housing and Mass Air Flow (MAF) sensor as a single assembly (see Chapter 6, Section 9). (It's not necessary to detach the MAF sensor from the air filter housing.)

11   Drain the cooling system (see Chapter 1, Section 22).
12   Remove the coolant hoses located above the starter motor (see illustration).
13   Disconnect the battery cable (the large cable) and the starter control electrical connector from the starter motor solenoid terminals (see illustration).
14   Remove the starter motor mounting bolts (see illustration) and remove the starter from the engine.

2009 through 2012 models

15   Disconnect the cable from the negative battery terminal (see Section 1).

Four-cylinder models

16   Raise the vehicle and support it securely on jackstands. Remove the right front fender splash shield.
17   Disconnect both wiring harnesses from the starter, then move them to the side.
18   Disconnect the ground wire and move it to the side.
19   Remove the two bolts, then lower the

starter motor from the front of the engine compartment.

V6 models

20   Working on top of the transaxle, disconnect both wiring harnesses from the starter, then move them to the side.
21   Disconnect the ground wire and move it to the side.
22   Remove the two bolts (vertical position) and lift the starter motor from the transaxle.

2013 and later models

23   Raise and support the front of the vehicle on jackstands.
24   Remove the engine undercover.
25   Disconnect the negative battery cable.

1.5L and 1.6L engines

26   Loosen the clamps and remove the bracket nut to remove the charge air cooler pipe from the front of the engine to allow access to the starter for replacement.
27   Remove the nuts and disconnect the wires at the starter and position the harness out of the way (see illustration).

7.14 To detach the starter motor from the engine, remove these two bolts (V6 models)

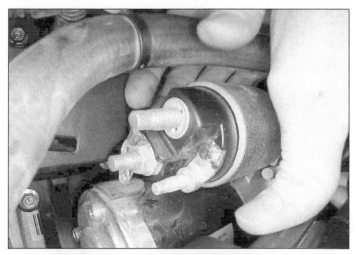

7.27 Disconnect the wires at the starter solenoid (2013 and later 1.6L model shown)

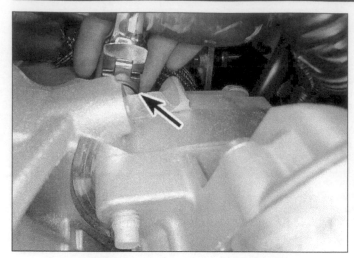

7.28 Remove this upper starter bolt (2013 and later 1.6L model shown)

7.29 Remove these two lower starter bolts from below (2013 and later 1.6L model shown)

28   From the engine compartment, remove the upper starter bolt nearest the engine block (see illustration).

29   From under the vehicle, remove the remaining two starter bolts (the lower has a ground cable attached) (see illustration).

30   Remove the starter from the vehicle.

31   Installation is reverse of removal. Ensure the ground cable is attached to the lower starter bolt at installation.

2.0L and 2.5L engines

32   On 2.0L models, loosen the clamps and remove the bracket nuts to remove the charge air cooler pipe from under the engine to allow access to the starter for replacement.

33   On all models, detach the coolant hose and wiring harness from the starter motor bracket.

34   Remove the nuts and disconnect the wires at the starter.

35   Remove the nuts attaching the starter motor bracket to the starter motor mounting bolts/studs and remove the bracket.

36   Working under the vehicle, disconnect the oil pressure sensor connector near the oil filter housing.

37   Remove the starter motor mounting bolts/studs and remove the starter from the vehicle.

38   Installation is reverse of removal.

All models

39   Installation is the reverse of removal.

40   After you've reconnected the battery, the Powertrain Control Module (PCM) must relearn its idle and fuel trim strategy for optimum driveability and performance (see Section 1 for this procedure as well as other procedures that need to be performed).

Chapter 6
Emissions and engine control systems

Contents

Specifications

Torque specifications

Note: *One foot-pound (ft-lb) of torque is equivalent to 12 inch-pounds (in-lbs) of torque. Torque values below approximately 15 ft-lbs are expressed in inch-pounds, since most foot-pound torque wrenches are not accurate at these smaller values.*

	Ft-lbs (unless otherwise noted)	Nm
Camshaft Position (CMP) sensor mounting bolt		
Four-cylinder engines		
2012 and earlier models	15	20
2013 and later models		
1.5L models	80 in-lbs	9
1.6L models	71 in-lbs	8
2.0L models	62 in-lbs	7
2.5L models	53 in-lbs	6
V6 engine	89 in-lbs	10
Crankshaft Position (CKP) sensor mounting bolt		
Four-cylinder engines		
2012 and earlier models	62 in-lbs	7
2013 and later models		
1.5L models	89 in-lbs	10
1.6L models	71 in-lbs	8
2.0L models	62 in-lbs	7
2.5L models	53 in-lbs	6
V6 engine	89 in-lbs	10

Torque specifications (continued)	Ft-lbs (unless otherwise noted)	Nm

Note: *One foot-pound (ft-lb) of torque is equivalent to 12 inch-pounds (in-lbs) of torque. Torque values below approximately 15 ft-lbs are expressed in inch-pounds, since most foot-pound torque wrenches are not accurate at these smaller values.*

Cylinder Head Temperature (CHT) sensor		
2012 and earlier models	89 in-lbs	10
2013 and later models		
1.5L models	89 in-lbs	10
2.0L models	97 in-lbs	11
2.5L models	106 in-lbs	12
Exhaust Gas Recirculation (EGR) valve mounting bolts		
Four-cylinder engine		
2012 and earlier models	80 in-lbs	9
2013 and later 2.5L models	177 in-lbs	20
V6 engine	18	25
Idle Air Control (IAC) valve mounting bolts		
2.0L four-cylinder and 3.0L V6 engines..................................	89 in-lbs	10
2.3L four-cylinder engine..	35 in-lbs	4
Knock sensor retaining nut/bolt		
2012 and earlier models		
Four-cylinder engine..	15	20
V6 engine..	18	25
2013 and later models		
1.5L and 1.6L models	159 in-lbs	18
2.0L and 2.5L models	177 in-lbs	20
Output Shaft Speed (OSS) sensor mounting bolt..................................	120 in-lbs	13
Oxygen sensors (all oxygen sensors, all models)..................................	30	40
Transmission Range (TR) sensor mounting bolts	108 in-lbs	12
Turbine Shaft Speed (TSS) sensor mounting bolt..................................	120 in-lbs	13
Vehicle Speed Sensor (VSS) (2WD and 4WD)..................................	62 in-lbs	7
Fuel rail pressure sensor		
1.5L models	24	32.5
1.6L and 2.0L models		
Step 1	53 in-lbs	6
Step 2	Tighten an additional 5 degrees	
Step 3	Loosen 90 degrees	
Step 4	53 in-lbs	6
Step 5	Tighten an additional 21 degrees	

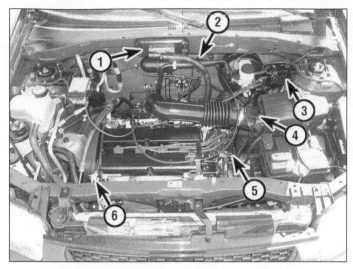

**1.1a Emissions and engine control component locations
(2.0L Zetec four-cylinder model)**

1   *Powertrain Control Module (PCM)*
2   *EGR vacuum regulator valve*
3   *Evaporative Emissions Control (EVAP) system canister purge valve*
4   *Mass Air Flow (MAF) sensor*
5   *Differential pressure feedback EGR system sensor*
6   *Oxygen sensor electrical connectors*

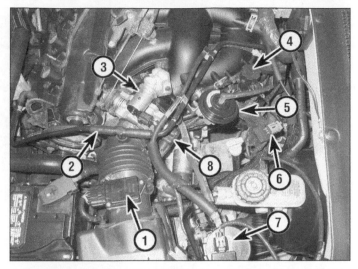

**1.1b Emissions and engine control component locations, left side
(2002 V6 model shown, other V6 models similar)**

1   *Mass Air Flow (MAF) sensor*
2   *Crankcase vent tube*
3   *Idle Air Control (IAC) valve*
4   *EGR vacuum regulator valve*
5   *Exhaust Gas Recirculation (EGR) valve*
6   *Differential pressure feedback EGR system sensor*
7   *Evaporative Emissions Control (EVAP) system canister purge valve*
8   *Throttle Position (TP) sensor*

1   General information

1   The emission control systems and components (see illustrations) are an integral part of the engine management system, which is called the Multiport Fuel Injection (MFI) system (see Chapter 4 for more information on the MFI system). The MFI system also includes all the government-mandated diagnostic features of the second generation of on-board diagnostics, which is known as On-Board Diagnostics II (OBD-II).

2   At the center of the MFI and OBD-II operations is the Powertrain Control Module (PCM), or on-board computer. The PCM controls not only engine operation but also the automatic transaxle. The PCM receives engine operating information as input signals from a number of sensors and provides output commands to actuators such as the ignition coil(s), the fuel injectors and various solenoids and relays for the engine and automatic transaxle. Other emission controls included in the overall engine control system are:

a) *Catalytic converter(s)*
b) *Evaporative Emission Control (EVAP) system*
c) *Exhaust Gas Recirculation (EGR) system*
d) *Positive Crankcase Ventilation (PCV) system*

3   The Sections in this Chapter include general descriptions and component replacement procedures for most of the information sensors and output actuators, as well as other components that are part of the systems listed

above. Refer to Chapter 4 for more information on the air intake, fuel and exhaust systems, and to Chapter 5 for information on the ignition system. Also refer to Chapter 1 for scheduled maintenance procedures for emission-related systems and components.

4   The procedures in this Chapter are intended to be practical, affordable and within the capabilities of the home mechanic. The diagnosis of most engine and emission control functions and driveability problems requires specialized tools, equipment and training. When checking or servicing become too difficult or require special test equipment, consult a dealer service department or other qualified repair shop.

5   Although engine and emission control systems are very sophisticated on late-model vehicles, you can perform many checks and

do most of the regular maintenance at home with common tune-up and hand tools and relatively inexpensive meters. Because of the Federally mandated warranty that covers the emission control system, check with a dealer about warranty coverage before working on any emission-related systems. After the warranty has expired, you may wish to perform some of the component replacement procedures in this Chapter to save money. Remember, the most frequent cause of emission and driveability problems is a loose electrical connector, or a broken wire or vacuum hose, so always check the electrical connections, the electrical wiring and the vacuum hoses first.

6   Pay close attention to any special precautions given in this Chapter. Remember that illustrations of various systems might

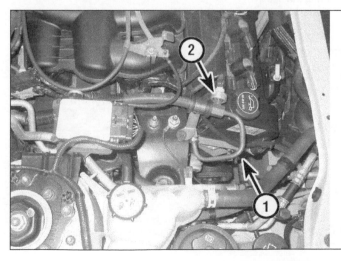

1.1c Emission control component locations, right side (2012 and earlier V6 models)

1   *Camshaft Position (CMP) sensor*
2   *Power Steering Pressure (PSP) switch*

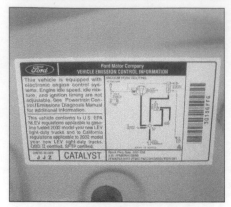

1.7 The VECI label in the engine compartment identifies the emission control devices on a particular engine and includes a vacuum hose routing diagram

2.4a Simple code readers are an economical way to extract trouble codes when the CHECK ENGINE light comes on

2.4b Hand-held scan tools like these can extract computer codes and also perform diagnostics

not exactly match the system installed on the vehicle on which you're working because of changes made by the manufacturer during production or from year to year.

7   A Vehicle Emission Control Information (VECI) label (see illustration) is located in the engine compartment (on the underside of the hood on our project vehicle, although it might be located somewhere else, like the upper radiator crossmember, on your vehicle). This label contains emission-control and engine tune-up specifications and adjustment information. It also includes a vacuum hose routing diagram for emission-control components. When servicing the engine or emission systems, always check the VECI label in your vehicle. If any information in this manual contradicts what you read on the VECI label on your vehicle, always defer to the information on the VECI label.

2   On-Board Diagnostic (OBD) system and Diagnostic Trouble Codes (DTCs)

OBD system general description

1   All models are equipped with the second generation OBD-II system. This system consists of an on-board computer known as the Powertrain Control Module (PCM), and information sensors, which monitor various functions of the engine and send data to the PCM. This system incorporates a series of diagnostic monitors that detect and identify fuel injection and emissions control systems faults and store the information in the computer memory. This updated system also tests sensors and output actuators, diagnoses drive cycles, freezes data and clears codes. This powerful diagnostic computer must be accessed using the new OBD-II scan tool and 16 pin Data Link Connector (DLC) located under the driver's dash area.

2   The PCM is the "brain" of the electronically controlled fuel and emissions system. It receives data from a number of sensors and

other electronic components (switches, relays, etc.). Based on the information it receives, the PCM generates output signals to control various relays, solenoids (i.e., fuel injectors) and other actuators. The PCM is specifically calibrated to optimize the emissions, fuel economy and driveability of the vehicle.

3   It isn't a good idea to attempt diagnosis or replacement of the PCM or emission control components at home while the vehicle is under warranty. Because of a Federally mandated warranty which covers the emissions system components and because any owner-induced damage to the PCM, the sensors and/or the control devices may void this warranty, take the vehicle to a dealer service department if the PCM or a system component malfunctions.

Scan tool information

4   Because extracting the Diagnostic Trouble Codes (DTCs) from an engine management system is now the first step in troubleshooting many computer-controlled systems and components, a code reader, at the very least, will be required (see illustration). More powerful scan tools can also perform many of the diagnostics once associated with expen-

sive factory scan tools (see illustration). If you're planning to obtain a generic scan tool for your vehicle, make sure that it's compatible with OBD-II systems. If you don't plan to purchase a code reader or scan tool and don't have access to one, you can have the codes extracted by a dealer service department or an independent repair shop.
Note: *Some auto parts stores even provide this service.*

Accessing the DTCs

5   The Diagnostic Trouble Codes (DTCs) can only be accessed with a code reader or scan tool. Professional scan tools are expensive, but relatively inexpensive generic code readers or scan tools (see illustrations 2.4a and 2.4b) are available at most auto parts stores. Simply plug the connector of the scan tool into the diagnostic connector (see illustrations). Then follow the instructions included with the scan tool to extract the DTCs.

6   Once you have outputted all of the stored DTCs, look them up on the accompanying DTC chart.

7   After troubleshooting the source of each DTC, make any necessary repairs or replace the defective component(s).

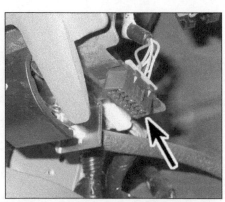

2.5a On 2012 and earlier models, the Data Link Connector (DLC) is located under the dash, near the hood release lever

2.5b On 2013 and later models, the Data Link Connector is located behind this panel on the left side of the instrument panel, marked "OBD"

Clearing the DTCs

8   Clear the DTCs with the code reader or scan tool in accordance with the instructions provided by the tool's manufacturer.

Diagnostic Trouble Codes

9   The accompanying tables are a list of the Diagnostic Trouble Codes (DTCs) that can be accessed by a do-it-yourselfer working at home (there are many more DTCs available to dealerships with proprietary scan tools and software, but those codes cannot be accessed by a generic scan tool). If, after you have checked and repaired the connectors, wire harness and vacuum hoses (if applicable) for an emission-related system, component or circuit, the problem persists, have the vehicle checked by a dealer service department or a qualified repair shop.

Diagnostic Trouble Codes

Code	Probable cause
P0001	Fuel volume regulator - control circuit open
P0002	Fuel volume regulator - control circuit low
P0003	Fuel volume regulator - control circuit high
P000A	Intake A camshaft position slow response (bank 1)
P000B	Exhaust B camshaft position slow response (bank 1)
P0010	Intake camshaft position actuator circuit open (bank 1)
P0011	"A" Camshaft position - timing over-advanced (bank 1)
P0012	"A" Camshaft position - timing over-retarded (bank 1)
P0013	"B" Camshaft position - actuator circuit malfunction (bank 1)
P0014	"B" Camshaft position - timing over-advanced (bank 1)
P0015	"B" Camshaft position - timing over-retarded (bank 1)
P0016	Crankshaft position/camshaft position, bank 1, sensor A - correlation
P0017	Crankshaft position/camshaft position, bank 1, sensor B - correlation
P0018	Crankshaft position/camshaft position, bank 2, sensor A - correlation
P0019	Crankshaft position/camshaft position, bank 2, sensor B - correlation
P0020	Intake camshaft position actuator circuit open (bank 2)
P0021	Intake camshaft position timing - over-advanced (bank 2)
P0022	Intake camshaft position timing - over-retarded (bank 2)
P0023	"B" Camshaft position - actuator circuit (bank 2)
P0024	"B" Camshaft position - timing over-advanced or system performance (bank 2)
P0025	"B" Camshaft position - timing over-retarded (bank 2)
P0030	O2 heater control circuit (bank 1, sensor 1)
P0031	O2 heater control circuit low (bank 1, sensor 1)
P0032	O2 heater control circuit high (bank 1, sensor 1)

Code	Probable cause
P0036	O2 heater control circuit (bank 1 sensor 2)
P0037	O2 heater control circuit low (bank 1, sensor 2)
P0038	O2 heater control circuit high (bank 1, sensor 2)
P0050	O2 heater control circuit (bank 2, sensor 1)
P0053	O2 heater resistance (bank 1, sensor 1)
P0054	O2 heater resistance (bank 1, sensor 2)
P0059	O2 heater resistance (bank 2, sensor 1)
P0060	O2 heater resistance (bank 2, sensor 2)
P0068	MAP/MAF - throttle position correlation
P0071	Ambient air temperature sensor range/performance
P0072	Ambient air temperature sensor circuit low input
P0073	Ambient air temperature sensor circuit high input
P0074	Ambient air temperature sensor circuit intermittent
P007B	Charge air cooler - temperature sensor range/performance
P007C	Charge air cooler - temperature sensor low (bank 1)
P007D	Charge air cooler - temperature sensor high (bank 1)
P0087	Fuel rail/system pressure - too low
P0088	Fuel rail/system pressure - too high
P008A	Low pressure fuel system pressure too low
P008B	Low pressure fuel system pressure too high
P0096	Intake air temperature (IAT) sensor 2 - circuit range/performance
P0097	Intake air temperature (IAT) sensor 2 - circuit low input
P0098	Intake air temperature (IAT) sensor 2 - circuit high input
P00BA	Low fuel pressure - limited power
P00DF	Charge air cooler temperature sensor - range/performance
P00E0	Charge air cooler temperature sensor - Low
P00E1	Charge air cooler temperature sensor - Hi
P00E2	Charge air cooler temperature sensor - Intermittent/Erratic
P0100	Mass air flow or volume air flow circuit
P0101	Mass air flow or volume air flow circuit, range or performance problem
P0102	Mass air flow or volume air flow circuit, low input

Diagnostic Trouble Codes

Code	Probable cause
P0103	Mass air flow or volume air flow circuit, high input
P0104	Mass air flow or volume air flow circuit, Intermittent
P0112	Intake air temperature circuit, low input
P0113	Intake air temperature circuit, high input
P0114	Intake air temperature circuit, Intermittent
P0117	Engine coolant temperature circuit, low input
P0118	Engine coolant temperature circuit, high input
P0119	Engine coolant temperature circuit, Intermittent
P0120	Throttle / pedal position sensor (switch A)
P0121	Throttle / pedal position sensor - range/performance (switch A)
P0122	Throttle / pedal position sensor - Low (switch A)
P0123	Throttle / pedal position sensor - High (switch A)
P0124	Throttle / pedal position sensor - Intermittent (switch A)
P0125	Insufficient coolant temp for closed loop fuel control
P0127	Intake air temperature too high
P0128	Coolant thermostat - coolant temperature below thermostat regulating temperature
P012B	Turbocharger inlet pressure - range/performance (sensor A)
P012C	Turbocharger inlet pressure - Low (sensor A)
P012D	Turbocharger inlet pressure - High (sensor A)
P012E	Turbocharger inlet pressure - Intermittent/erratic (sensor A)
P013A	O2 sensor slow response, bank 1, sensor 2
P013B	O2 sensor slow response, bank 1, sensor 2
P013C	O2 sensor slow response, bank 2, sensor 2
P013E	O2 sensor delayed response, bank 1 sensor 2
P0140	O2 sensor circuit - no activity detected (bank 1, sensor 2)
P0148	Fuel deli\very error
P014A	Sensor delayed response, bank 2, sensor 2
P0150	O2 sensor circuit malfunction (bank 2, sensor 1)
P0157	O2 sensor circuit, low voltage (bank 2, sensor 2)
P0159	O2 sensor circuit, slow response (bank 2, sensor 2)

Code	Probable cause
P016A	O2 sensor not ready, bank 1, sensor 1
P017C	Cylinder head temperature sensor - Low
P017D	Cylinder head temperature sensor - High
P017E	Cylinder head temperature sensor - Intermittent/erratic
P0181	Fuel temperature sensor A circuit - range/performance
P0182	Fuel temperature sensor A circuit, low input
P018C	Fuel pressure sensor B, circuit low
P018D	Fuel pressure sensor B, circuit high
P020x	Injector circuit open, cylinder x (x = cylinder number)
P0217	Engine overheating condition
P0218	Transmission overheating condition
P0221	Throttle / pedal position sensor - range/performance (switch B)
P0222	Throttle / pedal position sensor - Low (switch B)
P0223	Throttle / pedal position sensor - High (switch B)
P0231	Fuel pump secondary circuit - Low
P0232	Fuel pump secondary circuit - High
P0234	Turbocharger - Overboost condition (sensor A)
P0236	Turbocharger boost sensor - range/performance (sensor A)
P0237	Turbocharger boost sensor - Low (sensor A)
P0238	Turbocharger boost sensor - High (sensor A)
P023A	Charge air cooler coolant pump - Open
P023B	Charge air cooler coolant pump - Low
P023C	Charge air cooler coolant pump - High
P0243	Turbocharger wastegate (actuator A)
P0244	Turbocharger wastegate actuator A - range/performance (actuator A)
P0245	Turbocharger wastegate actuator A - Low (actuator A)
P0246	Turbocharger wastegate actuator A - High (actuator A)
P0247	Turbocharger wastegate (actuator B)
P0248	Turbocharger wastegate actuator A - range/performance (actuator B)
P0249	Turbocharger wastegate actuator A - Low (actuator B)
P0250	Turbocharger wastegate actuator A - Low (actuator B)

Diagnostic Trouble Codes

Code	Probable cause
P025E	Turbocharger boost sensor - intermittent/erratic (sensor A)
P0261	Cylinder no. 1 injector circuit, low
P0262	Cylinder no. 1 injector circuit, high
P0264	Cylinder no. 2 injector circuit, low
P0265	Cylinder no. 2 injector circuit, high
P0267	Cylinder no. 3 injector circuit, low
P0268	Cylinder no. 3 injector circuit, high
P026A	Charge air cooler efficiency below threshold
P0271	Cylinder no. 4 injector circuit, high
P027A	Fuel pump module B control circuit open
P027B	Fuel pump module control circuit range/performance
P028D	Charge air cooler cooling fan control circuit - Low
P028E	Charge air cooler cooling fan control circuit - High
P0297	Vehicle overspeed condition
P0298	Engine oil over temperature condition
P0299	Turbocharger underboost condition (sensor A)
P030x	Misfire detected, cylinder x (x = cylinder number)
P0313	Misfire detected with low fuel
P0315	Crankshaft position system variation not learned
P0316	Misfire detected on first startup (first 1000 revolutions)
P0322	Crankshaft Position (CKP) sensor/engine speed (RPM) sensor - no signal
P0325	Knock sensor no. 1 circuit malfunction (bank 1 or single sensor)
P0327	Knock sensor no. 1 circuit, low input (bank 1 or single sensor)
P0328	Knock sensor no. 1 circuit, high input (bank 1 or single sensor
P0332	Knock sensor no. 2 circuit, low input (bank 2)
P0333	Knock sensor no. 2 circuit, high input (bank 2)
P0335	Crankshaft position sensor A circuit malfunction
P0336	Crankshaft position sensor A circuit - range or performance problem
P0341	Camshaft position sensor "A", circuit - range or performance problem
P0344	Camshaft position sensor "A", circuit - intermittent

Code	Probable cause
P0346	Camshaft position sensor "A", range/performance problem (bank 2)
P0349	Camshaft position sensor "A", circuit - intermittent (bank 2)
P0365	Camshaft position sensor "B" circuit (bank 1)
P0366	Camshaft position sensor "B" circuit range/performance (bank 1)
P0390	Camshaft position sensor "B" circuit
P0391	Camshaft position sensor "B" circuit range/performance (bank 2)
P0394	Camshaft position sensor "B" circuit intermittent (bank 2)
P0444	Evaporative emission control system, open purge control valve circuit
P0450	Evaporative emission control system, pressure sensor malfunction
P0454	Evaporative emission control system, pressure sensor intermittent
P0458	Evaporative emission control system, purge control valve - circuit low
P0459	Evaporative emission control system, purge control valve - circuit high
P0461	Fuel level sensor circuit, range or performance problem
P0496	Evaporative emission system - high purge flow
P0497	Evaporative emission system - low purge flow
P0512	Starter request circuit
P051B	Crankcase pressure sensor - range/performance
P051C	Crankcase pressure sensor - Low
P051D	Crankcase pressure sensor - High
P053A	PCV heater control circuit open
P053F	Cold start fuel pressure performance
P0562	System voltage low
P0563	System voltage high
P0571	Cruise control/brake switch A, circuit malfunction
P0572	Cruise control/brake switch A, circuit low
P0573	Cruise control/brake switch A, circuit high
P0579	Cruise control system, multi-function input "A" - circuit range/performance
P0581	Cruise control system, multi-function input "A" - circuit high
P0600	Serial communication link malfunction
P0604	Internal control module, random access memory (RAM) error
P0607	Control module performance

Diagnostic Trouble Codes

Code	Probable cause
P060A	Internal control module monitoring processor performance
P060B	Internal control module A/D processing performance
P060C	Internal control module main processor performance
P060D	Internal control module accelerator position performance
P0610	Control module - vehicle options error
P0616	Starter relay - circuit low
P0617	Starter relay - circuit high
P061A	Internal control module torque performance
P061B	Internal control module torque calculation performance
P061C	Internal control module engine speed performance
P061D	Internal control module engine air mass performance
P061F	Internal control module throttle actuator controller performance
P062C	Internal control module vehicle speed performance
P062F	PCM Internal control module EEPROM error
P062F	TCM Internal control module EEPROM error
P0620	Generator control circuit malfunction
P0622	Generator lamp F, control circuit malfunction
P0625	Generator field terminal - circuit low
P0626	Generator field terminal - circuit high
P0627	Fuel pump control - circuit open
P0630	VIN not programmed or mismatch - ECM/PCM
P0642	Engine control module (ECM), knock control - defective
P0643	Sensor reference voltage A - circuit high
P064D	Internal control module O2 Sensor processor performance, bank 1
P064E	Internal control module O2 Sensor processor performance, bank 2
P0652	Sensor reference voltage B - circuit low
P0653	Sensor reference voltage B - circuit high
P0657	Actuator supply voltage - circuit open
P065B	Generator control circuit range/performance
P065C	Generator mechanical performance

Code	Probable cause
P0660	Intake manifold tuning valve control circuit (bank 1)
P0663	Intake manifold tuning valve control circuit (bank 2)
P0685	ECM power relay, control - circuit open
P0686	ECM power relay control - circuit low
P0687	Engine, control relay - short to ground
P0689	ECM power relay sense - circuit low
P0690	ECM power relay sense - circuit high
P0691	Engine coolant blower motor 1 - short to ground
P0692	Engine coolant blower motor 1 - short to positive
P06B6	Internal control module knock sensor processor 1 performance
P06B8	Internal control module non-volatile random access memory (NVRAM) error
P06D1	Internal control module coil control module performance
P0705	Transmission range sensor, circuit malfunction (PRNDL input)
P0706	Transmission range sensor circuit, range or performance problem
P0707	Transmission range sensor circuit, low input
P0708	Transmission range sensor circuit, high input
P0815	Upshift switch circuit
P0830	Clutch pedal switch "A" circuit
P0833	Clutch pedal switch B circuit
P0840	Transmission fluid pressure sensor/switch "A" circuit
P0A3B	Generator over-temperature
P0A5A	Generator current sensor circuit range/performance
P0A5B	Generator current sensor circuit low
P0A5C	Generator current sensor circuit high
P1001	KOER not able to complete, KOER aborted
P100F	Wastegate control pressure/BARO correlation
P1011	Wastegate control sensor - range/performance
P1012	Wastegate control sensor - Low
P1013	Wastegate control sensor - High
P1014	Wastegate control sensor - Intermittent/erratic
P1015	Wastegate control pressure - Lower than expected

Diagnostic Trouble Codes

Code	Probable cause
P1016	Wastegate control pressure - Higher than expected
P101F	Cylinder head temperature sensor - Out of self-test range (sensor 1)
P1021	Cylinder head temperature sensor - range/performance (sensor 2)
P1022	Cylinder head temperature sensor - Low (sensor 2)
P1023	Cylinder head temperature sensor - High (sensor 2)
P1024	Cylinder head temperature sensor - Intermittent/erratic (sensor 2)
P1025	Cylinder head temperature sensor - Out of self-test range (sensor 2)
P1026	Engine coolant temperature sensor 1 / Cylinder head temperature sensor 2 - correlation
P1060	Excessive camshaft chain wear
P1061	Excessive camshaft chain wear - Forced limited engine speed

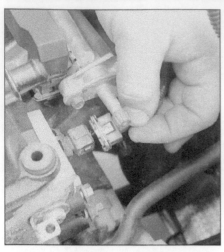

3.3 Disconnect the electrical connector from the CMP sensor (2004 and earlier four-cylinder models shown)

3.5 Remove the CMP sensor retaining bolt, then remove the CMP sensor from the cylinder head (four-cylinder models)

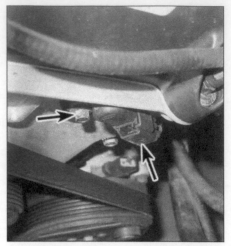

3.9 On V6 models, the CMP sensor is located at the right end of the front cylinder head; to remove the CMP sensor, disconnect the electrical connector and remove the retaining bolt

3　Camshaft Position (CMP) sensor - replacement

2012 and earlier models
Four-cylinder models

1　Disconnect the cable from the negative battery terminal (see Chapter 5, Section 1).
2　Locate the CMP sensor at the left rear corner of the cylinder head.
3　Disconnect the electrical connector from the CMP sensor (see illustration).
4　Remove the CMP sensor retaining bolt.
5　Remove the CMP sensor (see illustration).
6　Installation is the reverse of removal. Be sure to tighten the CMP sensor bolt to the torque listed in this Chapter's Specifications.
7　After you've reconnected the battery, the Powertrain Control Module (PCM) must relearn its idle and fuel trim strategy for optimum driveability and performance (see Chapter 5, Section 1 for this procedure).

V6 models
8　Disconnect the cable from the negative battery terminal (see Chapter 5, Section 1).
9　Locate the CMP sensor at the right end of the front cylinder head (see illustration).
10　Disconnect the electrical connector from the CMP sensor.
11　Remove the CMP sensor retaining bolt.
12　Remove the CMP sensor.
13　Installation is the reverse of removal. Be sure to tighten the CMP sensor bolt to the torque listed in this Chapter's Specifications.
14　After you've reconnected the battery, the Powertrain Control Module (PCM) must relearn its idle and fuel trim strategy for optimum driveability and performance (see Chapter 5, Section 1 for this procedure).

2013 and later models
1.5L, 1.6L and 2.0L models
Note: *The CMP sensors are located on top of the valve cover on the driver's end (rear of engine).*

15　Remove the engine cover.
16　To remove the left (LH) CMP sensor on 1.6L models, locate it on the left front corner of the valve cover and disconnect the electrical connector (see illustration).
17　To remove the left (LH) CMP sensor on 2.0L models, locate it near the oil filler cap and disconnect the electrical connector.
18　To remove the right (RH) CMP sensor on 1.6L and 2.0L models, remove the turbocharger piping between the air filter and turbocharger inlet (see illustration).
19　Disconnect the electrical connector.
20　Remove the bolt and pull either CMP out of the valve cover.
21　If the sensor is to be reused, inspect the O-ring and replace if damaged.
22　Installation is reverse of removal. Lubricate the O-ring with clean engine oil before installing.

2.5L models
23　Remove the engine cover.
24　Locate the CMP sensor on top of the valve cover on the driver's front corner.
25　Disconnect the CMP electrical connector.
26　Remove the bolt and pull the CMP out of the valve cover.
27　If the sensor is to be reused, inspect the O-ring and replace if damaged.
28　Installation is reverse of removal. Lubricate the O-ring with clean engine oil before installing.

4　Clutch Pedal Position (CPP) switch - replacement

Note: *Only 2012 and earlier models are equipped with a manual transaxle and clutch pedal requiring a CPP switch.*

1　Disconnect the cable from the negative battery terminal (see Chapter 5, Section 1).

3.16 Locating the left CMP sensor (2013 and later 1.6L model shown)

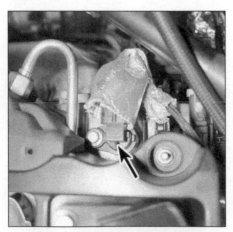

3.18 Locating the right CMP sensor (2013 and later 1.6L model shown)

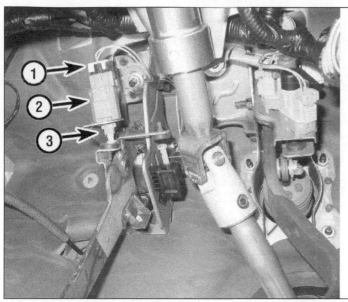

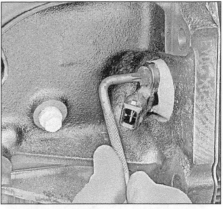

4.2 On models with a manual transaxle, the Clutch Pedal Position (CPP) switch is located at the top of the clutch pedal

1   Electrical connector
2   Clutch Pedal Position (CPP) switch
3   Locknut

5.5 On 2004 and earlier 2.0L four-cylinder models, the CKP sensor is located on the front side of the block, near the flywheel; to remove the CKP sensor, disconnect the electrical connector and remove the mounting bolt

2   Locate the CPP switch at the top of the clutch pedal (see illustration).
3   Disconnect the electrical connector from the CPP switch.
4   Unscrew the CPP switch locknut.
5   Remove the CPP switch.
6   Installation is the reverse of removal.
7   After you've reconnected the battery, the Powertrain Control Module (PCM) must relearn its idle and fuel trim strategy for optimum driveability and performance (see Chapter 5, Section 1 for this procedure).

5   Crankshaft Position (CKP) sensor - replacement

2012 and earlier models

2.0L four-cylinder engine

1   Disconnect the cable from the negative battery terminal (see Chapter 5, Section 1).
2   Remove the exhaust manifold (see Chapter 2A, Section 7).
3   Locate the CKP sensor on the front of the engine near the flywheel end of the block.
4   Disconnect the electrical connector from the CKP sensor.
5   Unscrew the CKP sensor mounting bolt (see illustration).
6   Remove the CKP sensor.
7   Installation is the reverse of removal. Be sure to tighten the CKP sensor mounting bolt to the torque listed in this Chapter's Specifications.
8   After you've reconnected the battery, the Powertrain Control Module (PCM) must relearn its idle and fuel trim strategy for optimum driveability and performance (see Chapter 5, Section 1 for this procedure).

V6 models

9   Disconnect the cable from the negative battery terminal (see Chapter 5, Section 1).
10   Loosen the right front wheel lug nuts. Raise

the front of the vehicle and place it securely on jackstands. Remove the right front wheel.
11   Remove the lower right splash shield.
12   Locate the CKP sensor (see illustrations) next to the crankshaft pulley.
13   Disconnect the electrical connector from the CKP sensor.
14   Remove the CKP sensor mounting bolt.
15   Remove the CKP sensor.
16   Installation is the reverse of removal. Be sure to tighten the CKP sensor mounting bolt to the torque listed in this Chapter's Specifications.
17   After you've reconnected the battery, the Powertrain Control Module (PCM) must relearn its idle and fuel trim strategy for optimum driveability and performance (see Chapter 5, Section 1 for this procedure).

2.3L four-cylinder engine

Note: *The CKP sensor for this engine must be replaced with a new one whenever it is*

removed because there is a special alignment tool that must be used to install the sensor that can only be obtained with the purchase of a new sensor - the tool is not sold separately.
18   Disconnect the cable from the negative battery terminal (see Chapter 5, Section 1).
19   The CKP sensor is mounted right next to the crankshaft pulley.
20   Disconnect the electrical connector to the CKP sensor.
21   Set the engine at TDC (see Chapter 2A, Section 3)
22   With the special timing pin and crankshaft pulley alignment tool in place, remove the mounting bolts for the CKP sensor.
23   To install the CKP sensor, install the mounting bolts - but do not tighten them.
24   Adjust the CKP sensor by using the CKP sensor alignment tool, then tighten the mounting bolts securely.

5.12a On V6 models, the CKP sensor is located near the crankshaft pulley; to remove the CKP sensor, disconnect the electrical connector and remove the sensor mounting bolt - 2004 and earlier model shown

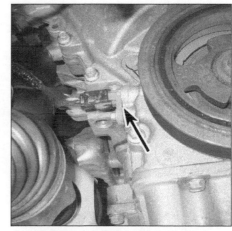

5.12b The sensor on later model V6 engines is located to the rear of the crankshaft pulley

5.28 Disconnect the CKP electrical connector (1) and remove the bolt (2) (2013 and later 1.6L model shown)

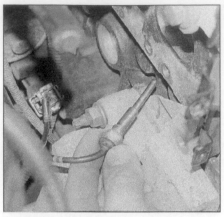

6.4 To remove the CHT sensor from a 2004 and earlier 2.0L four-cylinder engine, disconnect the electrical connector, then unscrew the sensor. On 2.3L/2.5L four-cylinder engines, the sensor is located between the number 2 and number 3 spark plugs

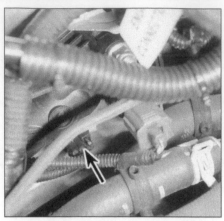

6.9 On 1.6L models, the CHT sensor is located at the left end of the cylinder head, under the water passage

2013 and later models

Removal

25 Raise and support the front of the vehicle on jackstands.
26 Remove the engine undercover.
27 Remove the passenger's inner splash shield (remove the wheel as necessary).
28 Locate the CKP sensor mounted to the rear of the crankshaft pulley, and disconnect the electrical connector (see illustration).
29 Remove the bolt and remove the CKP sensor from the engine.

Installation

1.5L and 1.6L models

30 Installation is reverse of removal.
31 Using a scan tool, perform the Misfire Monitor Neutral Profile Correction procedure and follow the on-screen instructions.

2.0L and 2.5L models

32 Remove the cylinder block hex plug on the side of the block behind where the CKP sensor mounts.
33 Insert the timing peg special tool 303-507 into the hole and rotate the engine (by hand) until the crankshaft makes contact with the special tool. When the crankshaft makes contact, the engine is at Top Dead Center (TDC).
Note: *DO NOT try and force the crankshaft to turn any further after it hits the special tool.*
34 With the engine at TDC and the special tool still installed, install the CKP sensor to the engine and tighten the bolts finger-tight. Install the CKP alignment special tool 303-1521 over the CKP sensor and tighten the bolts.
35 Connect the CKP sensor electrical connector.
36 Remove the timing peg special tool and install the hex plug.
37 The remainder of installation is reverse of removal.
38 Using a scan tool, perform the Misfire Monitor Neutral Profile Correction procedure and follow the on-screen instructions.

6 Cylinder Head Temperature (CHT) sensor - replacement

2012 and earlier models

Note: *This procedure does not apply to 2004 and earlier V6 models.*
Note: *On 2004 and earlier 2.0L four-cylinder engines, the sensor is mounted near the alternator. On all other four-cylinder engines, it's in the center top of the cylinder head. On later V6 engines, it's mounted in the left side of the rear cylinder head.*
1 Disconnect the cable from the negative battery terminal (see Chapter 5, Section 1).
2 If you're working on a 2.0L four-cylinder engine, remove the alternator (see Chapter 5, Section 6).
3 Disconnect the CHT sensor electrical connector.
4 Unscrew and remove the CHT sensor (see illustration).
5 Installation is the reverse of removal. Be sure to tighten the CHT sensor to the torque listed in this Chapter's Specifications.
6 After you've reconnected the battery, the Powertrain Control Module (PCM) must relearn its idle and fuel trim strategy for optimum driveability and performance (see Chapter 5, Section 1 for this procedure).

2013 and later models

Note: *On 1.5L and 2.0L models, the CHT sensor is located on the rear of the cylinder head, on the exhaust side. On 2.5L models, the CHT is located on the top of the cylinder head, between cylinder no. 2 and No.3 ignition coils. On 1.6L models, the CHT sensor is located on the driver's end of the cylinder head.*
7 Remove the engine cover.
8 On turbocharged models, remove the turbocharger piping between the air filter and turbocharger inlet.

9 On all models, disconnect the CHT sensor electrical connector (see illustration). First remove the protective cover to access the connector (if equipped).
10 Unscrew the CHT sensor from the cylinder head.
Note: *The manufacturer recommends to discard and use a NEW sensor when removed.*
11 Install the new sensor and tighten it to the torque listed in this Chapter's Specifications.

7 Engine Coolant Temperature (ECT) sensor - replacement

Warning: *Wait until the engine is completely cool before beginning this procedure.*

2012 and earlier models

Note: *This procedure applies to 2008 and earlier V6 models only. 2009 through 2012 V6 and four-cylinder models don't use an ECT sensor; they use a CHT sensor.*
1 Disconnect the cable from the negative battery terminal (see Chapter 5, Section 1).
2 Drain the cooling system (see Chapter 1, Section 22).
3 Remove the air intake duct (see Chapter 4, Section 10).
4 Locate the ECT sensor on the thermostat housing (see illustration).
5 Disconnect the electrical connector from the ECT sensor.
6 Remove the ECT sensor retaining clip (see illustration) by pulling it straight up.
7 Remove the ECT sensor.
8 Remove the ECT sensor O-ring and discard it. Always use a new O-ring when installing the ECT sensor, even if you're planning to reuse the old ECT sensor.
9 Installation is the reverse of removal.
10 Refill the cooling system (see Chapter 1, Section 22).
11 After you've reconnected the battery, the Powertrain Control Module (PCM) must relearn its idle and fuel trim strategy for optimum driveability and performance (see Chapter 5, Section 1 for this procedure).

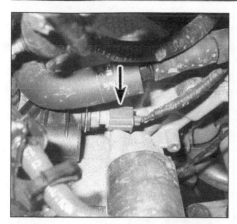

7.4 On 2008 and earlier V6 models, the ECT sensor is located on the thermostat housing; to remove the ECT sensor, disconnect the electrical connector and remove the retaining clip, then pull out the sensor

7.6 To remove the ECT sensor retaining clip, pull it straight up

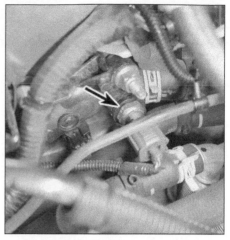

7.17 Locating the ECT sensor (2013 and later 1.6L model shown)

2013 and later models

Note: *The following procedure applies to 1.5L, 2.0L and 1.6L models. 2.5L models use a CHT sensor ONLY. 1.5L, 1.6L and 2.0L models use BOTH an ECT and CHT sensor.*

12   Remove the engine cover.
13   Drain the cooling system (see Chapter 1).
14   On 2.0L models, remove the battery and battery tray (see Chapter 5, Section 3).
15   On all models, remove the turbocharger piping between the air filter and turbocharger inlet.
16   2.0L models, remove the two nuts attaching the EVAP purge valve and tube bracket to the engine.
17   On all models, locate the ECT sensor on the coolant piping on the rear of the cylinder head, above the transaxle (see illustration). Disconnect the electrical connector.
18   Remove the ECT sensor retaining clip (see illustration 7.6) by pulling it straight out.
Note: *Have the NEW ECT sensor ready to install to reduce loss of coolant. If the same sensor is to be reinstalled, check the O-ring for damage and replace as necessary.*
19   Remove the existing ECT sensor and immediately install the new one. Insert the sensor retaining clip.
20   Remaining installation is reverse of removal. Top off the coolant as necessary.

8   Knock sensor - replacement

2004 and earlier 2.0L four-cylinder models

1   Disconnect the cable from the negative battery terminal (see Chapter 5, Section 1).
2   Raise the front of the vehicle and place it securely on jackstands.
3   Locate the knock sensor on the back side of the block, near the alternator.
4   The knock sensor electrical connector is

8.16 On 2006 and earlier V6 models, the knock sensor is located on the rear side of the block, near the transaxle; to detach the knock sensor from the block, trace the electrical lead up to the connector (see next illustration) and disconnect it, then remove the knock sensor retaining bolt and remove the sensor

attached to a bracket by two push fasteners. Detach the connector from the bracket and then disconnect it.
5   Remove the knock sensor retaining nut.
6   Remove the knock sensor.
7   Installation is the reverse of removal. Be sure to tighten the knock sensor retaining nut to the torque listed in this Chapter's Specifications.
8   After you've reconnected the battery, the Powertrain Control Module (PCM) must relearn its idle and fuel trim strategy for optimum driveability and performance (see Chapter 5, Section 1 for this procedure).

2.3L and 2009 through 2012 2.5L four-cylinder models

9   Disconnect the cable from the negative battery terminal (see Chapter 5, Section 1).
10   Remove the intake manifold (see Chapter 2A, Section 6).

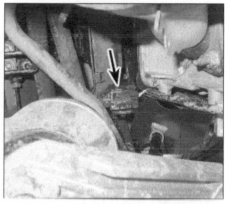

8.17 Trace the electrical lead from the knock sensor up to the sensor electrical connector and disconnect it

11   Disconnect the wiring harness from the sensor.
12   Remove the knock sensor mounting bolt, then remove the sensor.
13   Installation is the reverse of removal. Tighten the knock sensor bolt to the torque listed in this Chapter's Specifications.

2008 and earlier V6 models

14   Disconnect the cable from the negative battery terminal (see Chapter 5, Section 1).
15   Raise the front of the vehicle and place it securely on jackstands.
16   Locate the knock sensor on the rear side of the block (see illustration).
17   Disconnect the knock sensor electrical connector (see illustration).
18   Remove the knock sensor retaining bolt.
19   Remove the knock sensor.
20   Installation is the reverse of removal. Be sure to tighten the knock sensor retaining bolt to the torque listed in this Chapter's Specifications.
21   After you've reconnected the battery, the Powertrain Control Module (PCM) must relearn its idle and fuel trim strategy for optimum driveability and performance (see Chapter 5, Section 1 for this procedure).

8.29a Locating the front knock sensor
(2013 and later 1.6L model shown)

8.29b Identifying
the crankcase oil
separator hose
fitting (A) and
the rear knock
sensor (B) (2013
and later 1.6L
model shown)

2009 through 2012 V6 models

Note: *These vehicles have two knock sensors. One is attached to the left end of the rear cylinder head. The other is in the top of the engine block under the lower intake manifold*

22   Disconnect the cable from the negative battery terminal (see Chapter 5, Section 1).

23   If you're working on the engine block-mounted sensor, refer to Chapter 2B and remove the lower intake manifold.

24   Disconnect the sensor wiring harness.

25   Remove the mounting bolt, then remove the sensor.

26   Installation is the reverse of removal. Tighten the mounting bolt to the torque listed in this Chapter's Specifications.

2013 and later models

1.6L and 2.0L engines

Note: *The two knock sensors are located on the radiator side of the engine block.*

27   Locate and disconnect the knock sensor connectors. Detach the connectors from the tabs on the intake manifold.

28   On 1.6L models, disconnect the crankcase vent hose at the oil separator (see Section 30).

29   Remove the bolt attaching the knock

sensor to the block and remove the knock sensor(s) (see illustrations).

30   Installation is reverse of removal. Tighten the knock sensor bolt(s) to the torque listed in this Chapter's Specifications.

1.5L and 2.5L engines

Note: *The 1.5L engine has two knock sensors, the 2.5L engine has a single sensor. The knock sensors are located on the radiator side of the engine block.*

31   Remove the intake manifold (see Chapter 2A, Section 6).

32   Remove the bolt attaching the knock sensor to the block and remove the knock sensor(s).

33   Installation is reverse of removal. Tighten the knock sensor bolt(s) to the torque listed in this Chapter's Specifications.

---

9   Mass Air Flow (MAF) sensor - replacement

---

1   For the Mass Airflow/Intake Air Temperature (MAF/IAT) sensor on 1.5L, 1.6L and 2013 and later 2.0L four-cylinder engines, see Section 10.

2004 and earlier 2.0L engine and 2008 and earlier V6 engine

2   Disconnect the cable from the negative battery terminal (see Chapter 5, Section 1).

3   Remove the air intake duct (see illustration).

4   Disconnect the electrical connector from the MAF sensor (see illustration 9.3).

5   If necessary, remove the half of the air filter housing that includes the MAF sensor (see Chapter 4, Section 9).

6   Remove the four MAF sensor retaining nuts (see illustration) and remove the MAF sensor.

7   Installation is the reverse of removal. Be sure to tighten the MAF sensor retaining nuts securely.

8   After you've reconnected the battery, the Powertrain Control Module (PCM) must relearn its idle and fuel trim strategy for optimum driveability and performance (see Chapter 5, Section 1 for this procedure).

2.3L and 2.5L four-cylinder engines and 2009 and later V6 engines

Note: *2013 and later 2.5L models are equipped with BOTH a MAF and a MAP sensor.*

9.3 Loosen the air intake duct hose clamp screw and pull off the duct, then disconnect the electrical connector from the MAF sensor (V6 model shown, four-cylinder models similar)

9.6 To detach the MAF sensor from the air filter housing cover, remove these nuts (V6 model shown, four-cylinder models similar)

10.5 Pry up the anti-rotation tang, then twist the MAF/IAT 45 degrees counterclockwise to remove it

11.4 The OSS sensor is located on the left side of the automatic transaxle

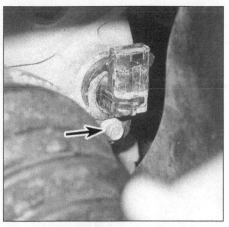

11.6 To detach the OSS sensor from the transaxle, remove this bolt

9   Disconnect the sensor wiring harness.
10   Remove the two mounting screws, then pull the sensor out of the air duct.
11   Installation is the reverse of removal.

10   Mass Air Flow/Intake Air Temperature (MAF/IAT) sensor - replacement

Note: *2013 and later 1.5L, 1.6L and 2.0L models are equipped with an MAF/IAT sensor, located in the intake air duct.*

1   Remove the engine cover.
2   Locate the MAF/IAT sensor in the intake air duct.
3   Disconnect the sensor electrical connector.
4   For MAF/IAT sensors attached with screws, remove the screws and remove the IAT sensor from the intake duct, near the air filter housing.
5   For MAF/IAT sensors that rotate and lock into the intake air duct, pry up the tang, rotate the sensor 45-degrees counterclockwise and pull it out of the duct (see illustration).
6   Installation is reverse of removal.

11   Output Shaft Speed (OSS) sensor - replacement

Note: *This sensor is used only on models with an automatic transaxle. The following procedure applies to 2007 and earlier models. On 2008 and later models, the OSS is located in the transaxle valve body and should be serviced by a qualified technician or repair facility.*

1   Disconnect the cable from the negative battery terminal (see Chapter 5, Section 1).
2   Raise the front of the vehicle and place it securely on jackstands.
3   Remove the lower left splash shield.
4   Locate the OSS sensor on the left side of the transaxle (see illustration).
5   Disconnect the electrical connector from the OSS sensor (see illustration 11.4).

6   Remove the OSS sensor retaining bolt (see illustration) and remove the OSS sensor.
7   Remove the old O-ring from the OSS sensor and discard it. Be sure to use a new O-ring when installing the OSS sensor (even if you're planning to reuse the old OSS sensor).
8   Installation is the reverse of removal. Be sure to tighten the OSS sensor retaining bolt to the torque listed in this Chapter's Specifications.
9   After you've reconnected the battery, the Powertrain Control Module (PCM) must relearn its idle and fuel trim strategy for optimum driveability and performance (see Chapter 5, Section 1 for this procedure).

12   Oxygen sensors - general information and replacement

General information

1   An oxygen sensor is a galvanic battery that produces a very small voltage output in response to the amount of oxygen in the exhaust gases. This voltage signal is the "input" side of the feedback loop between the oxygen sensor and the Powertrain Control Module (PCM). Without it, the PCM would be unable to correct the injector on-time (which determines the air/fuel ratio) to maintain the "perfect" (known as stoichiometric) air/fuel ratio of 14.7: 1 that the catalyst needs for optimal operation.
2   All vehicles covered by this manual have On-Board Diagnostics II (OBD-II) engine management systems, which means they have the ability to verify the accuracy of the basic feedback loop between the oxygen sensor and the PCM. They accomplish this by using an oxygen sensor ahead of the catalytic converter and another oxygen sensor behind the catalytic converter. By comparing the amount of oxygen in the post-catalyst exhaust gas to the oxygen content of the exhaust gas before it enters the catalyst, the PCM can determine the efficiency of the converter.

Four-cylinder models

3   Four-cylinder vehicles covered by this manual have one heated oxygen sensor upstream (ahead of the catalytic converter) and a Catalyst Monitor Sensor (after the catalyst).

V6 models

4   There are four heated oxygen sensors, one upstream and one downstream sensor for each cylinder bank. On these models, there are three catalysts: a smaller "fast-light-off" cat right below each exhaust manifold (the catalysts are actually an integral part of the exhaust manifolds) and another larger downstream catalyst. The oxygen sensors are located upstream and downstream in relation to the two smaller catalysts.

All models

5   The upstream and downstream oxygen sensors on all models are heated to speed up the warm-up time during which the sensors are unable to produce an accurate voltage signal. The circuit for each oxygen sensor heater is controlled by the PCM, which opens the ground side of the circuit to shut off the heater as soon as the sensor reaches its normal operating temperature.
6   Special care must be taken whenever a sensor is serviced.

 a) *Oxygen sensors have a permanently attached pigtail and an electrical connector that cannot be removed. Damaging or removing the pigtail or electrical connector will render the sensor useless.*
 b) *Keep grease, dirt and other contaminants away from the electrical connector and the louvered end of the sensor.*
 c) *Do not use cleaning solvents of any kind on an oxygen sensor.*
 d) *Oxygen sensors are extremely delicate. Do not drop a sensor or throw it around or handle it roughly.*
 e) *Make sure that the silicone boot on the sensor is installed in the correct position. Otherwise, the boot might melt and it might prevent the sensor from operating correctly.*

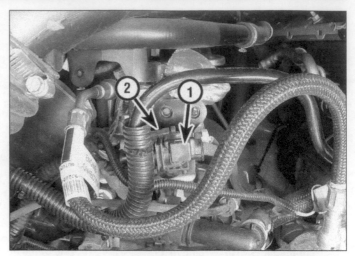

12.9a On early four-cylinder models, the electrical connectors for the oxygen sensors are located at the right front corner of the valve cover (2.0L Zetec model shown)

1   *Electrical connector for upstream oxygen sensor*
2   *Electrical connector for downstream oxygen sensor*

12.9b On 1.5L and 1.6L models, the upstream oxygen sensor electrical connectors are located on a bracket at the left end of the cylinder head

1   *Upstream O2 sensor electrical connector*
2   *Downstream O2 sensor electrical connector*

Replacement

Note: *Because it is installed in the exhaust manifold or pipe, both of which contract when cool, an oxygen sensor might be very difficult to loosen when the engine is cold. Rather than risk damage to the sensor or its mounting threads, start and run the engine for a minute or two, then shut it off. Be careful not to burn yourself during the following procedure.*

Four-cylinder models

Upstream oxygen sensor

Note: *This oxygen sensor is also known as O2 sensor, bank 1, sensor 1.*

7   Disconnect the cable from the negative battery terminal (see Chapter 5, Section 1).

8   On some models, certain components must be removed for access to the sensor:

a) *On 2004 and earlier 2.0L models, remove the exhaust manifold heat shield.*

b) *On 1.5L and 1.6L models, remove the battery and battery tray (see Chapter 5).*

12.10 Upstream oxygen sensor location - 1.5L and 1.6L models

9   Disconnect the upstream oxygen sensor electrical connector (see illustrations). Trace the electrical harness from the upstream oxygen sensor back to its connector to ensure that you're unplugging the correct connector.

10   Using an oxygen sensor socket (available at most automotive retailers), unscrew the upstream oxygen sensor (see illustration). If the sensor is difficult to loosen, spray some penetrant onto the sensor threads and allow it to soak in for awhile.

11   If you're going to install the old sensor, apply anti-seize compound to the threads of the sensor to facilitate future removal. If you're going to install a new oxygen sensor, it's not necessary to apply anti-seize compound to the threads. The threads on new sensors already have anti-seize compound on them.

12   Installation is otherwise the reverse of removal. Be sure to tighten the sensor to the torque listed in this Chapter's Specifications.

13   After you've reconnected the battery, the Powertrain Control Module (PCM) must relearn its idle and fuel trim strategy for optimum driveability and performance (see Chapter 5, Section 1 for this procedure).

Downstream oxygen sensor

Note: *The downstream O2 may also be referred to as the Catalyst Monitor Sensor (CMS). It is located at the bottom of the catalytic converter.*

Note: *This oxygen sensor is also known as O2 sensor, bank 1, sensor 2.*

14   Disconnect the cable from the negative battery terminal (see Chapter 5, Section 1). On 1.5L and 1.6L models, remove the battery and battery tray (see Chapter 5) for access to the oxygen sensor electrical connector.

15   Disconnect the oxygen sensor electrical connector (see illustration 12.9a and 12.9b). Trace the electrical harness from the downstream oxygen sensor back to its connector.

16   Raise the front of the vehicle and place it securely on jackstands.

17   Unscrew the downstream oxygen sensor (see illustrations), which is located right below the catalytic converter. If the sensor is difficult to loosen, spray some penetrant onto the sensor threads and allow it to soak in for awhile.

18   If you're going to install the old sensor, apply anti-seize compound to the threads of the sensor to facilitate future removal. If you're going to install a new oxygen sensor, it's not necessary to apply anti-seize compound to the threads. The threads on new sensors already have anti-seize compound on them.

19   Installation is otherwise the reverse of removal. Be sure to tighten the sensor to the torque listed in this Chapter's Specifications.

20   After you've reconnected the battery, the Powertrain Control Module (PCM) must relearn its idle and fuel trim strategy for optimum driveability and performance (see Chapter 5, Section 1 for this procedure).

V6 models

Upstream oxygen sensor (front cylinder bank)

Note: *This oxygen sensor is also known as O2 sensor, bank 2, sensor 1.*

21   Disconnect the cable from the negative battery terminal (see Chapter 5, Section 1).

22   Disconnect the upstream oxygen sensor electrical connector (see illustration).

23   Locate the upstream oxygen sensor for the front cylinder bank (see illustration). Using an oxygen sensor socket (available at most automotive retailers), unscrew the upstream oxygen sensor. If the sensor is difficult to loosen, spray some penetrant around the sensor and allow it to soak in for awhile.

24   If you're going to install the old sensor, apply anti-seize compound to the threads of the sensor to facilitate future removal. If

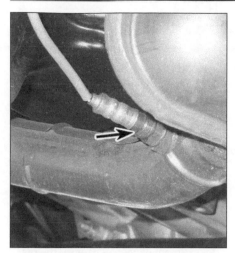

12.17a The downstream oxygen sensor on four-cylinder models is located right below the catalytic converter (2.0L Zetec engine shown)

12.17b Downstream oxygen sensor location - 1.5L and 1.6L models

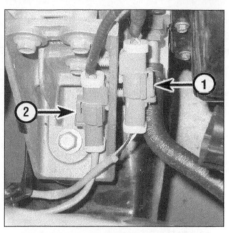

12.22 The electrical connectors for the upstream oxygen sensor (1) and the downstream sensor (2) for the front cylinder bank are located at the front engine mount (water pump removed for clarity, but you can disconnect either connector without removing the pump) (V6 models)

you're going to install a new oxygen sensor, it's not necessary to apply anti-seize compound to the threads. The threads on new sensors already have anti-seize compound on them.

25   Installation is otherwise the reverse of removal. Be sure to tighten the sensor to the torque listed in this Chapter's Specifications.

26   After you've reconnected the battery, the Powertrain Control Module (PCM) must relearn its idle and fuel trim strategy for optimum driveability and performance (see Chapter 5, Section 1 for this procedure).

Downstream oxygen sensor (front cylinder bank)

Note: *This oxygen sensor is also known as O2 sensor, bank 2, sensor 2.*

27   Disconnect the cable from the negative battery terminal (see Chapter 5, Section 1).

28   Disconnect the downstream oxygen sensor electrical connector (see illustration 12.22).

29   Raise the front of the vehicle and place it securely on jackstands.

30   Locate the downstream oxygen sensor for the front cylinder bank (see illustration), then unscrew the downstream oxygen sensor. If the sensor is difficult to loosen, spray some penetrant onto the sensor threads and allow it to soak in for awhile.

31   If you're going to install the old sensor, apply anti-seize compound to the threads of the sensor to facilitate future removal. If you're going to install a new oxygen sensor, it's not necessary to apply anti-seize compound to the threads. The threads on new sensors already have anti-seize compound on them.

32   Installation is otherwise the reverse of removal. Be sure to tighten the sensor to the torque listed in this Chapter's Specifications.

33   After you've reconnected the battery, the Powertrain Control Module (PCM) must relearn its idle and fuel trim strategy for optimum driveability and performance (see Chapter 5, Section 1 for this procedure).

Upstream oxygen sensor (rear cylinder bank)

Note: *This oxygen sensor is also known as O2 sensor, bank 1, sensor 1.*

34   Disconnect the cable from the negative battery terminal (see Chapter 5, Section 1).

35   Disconnect the upstream oxygen sensor electrical connector (see illustration), which is located behind the rear cylinder head, near the heater hoses.

36   Raise the front of the vehicle and place it securely on jackstands.

37   Locate the upstream oxygen sensor on the rear cylinder head exhaust manifold, right above the EGR pipe.

12.23 The upstream oxygen sensor for the front cylinder bank is located on the left side of the exhaust manifold, right above the catalytic converter; use an oxygen sensor socket to remove it (V6 models)

12.30 The downstream oxygen sensor for the front cylinder bank is located between the catalytic converter and the exhaust pipe flange

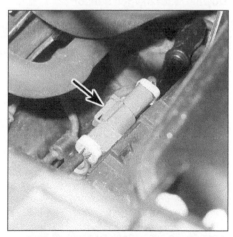

12.35 The electrical connector for the upstream oxygen sensor for the rear cylinder bank is located behind the rear cylinder head, near the heater hoses; if you find it difficult to unplug the connector, detach it from the plastic wiring harness protector, then unplug it (V6 models)

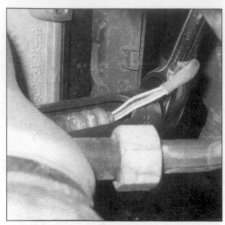

12.38 Use an oxygen sensor socket to unscrew the upstream oxygen sensor for the rear cylinder bank (V6 models)

12.45 The electrical connector for the downstream oxygen sensor for the rear cylinder bank is located near the crankshaft pulley

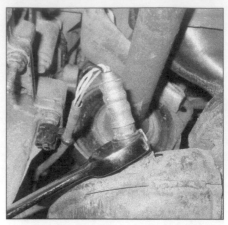

12.46 You should be able to remove the downstream oxygen sensor for the rear cylinder bank with a regular wrench (alternator removed for clarity)

38   Using an oxygen sensor socket, unscrew the oxygen sensor (see illustration). If the sensor is difficult to loosen, spray some penetrant onto the sensor threads and allow it to soak in awhile.

39   If you're going to install the old sensor, apply anti-seize compound to the threads of the sensor to facilitate future removal. If you're going to install a new oxygen sensor, it's not necessary to apply anti-seize compound to the threads. The threads on new sensors already have anti-seize compound on them.

40   Installation is otherwise the reverse of removal. Be sure to tighten the sensor to the torque listed in this Chapter's Specifications.

41   After you've reconnected the battery, the Powertrain Control Module (PCM) must relearn its idle and fuel trim strategy for optimum driveability and performance (see Chapter 5, Section 1 for this procedure).

**Downstream oxygen sensor
(rear cylinder bank)**

Note: *This oxygen sensor is also known as O2 sensor, bank 1, sensor 2.*

42   Disconnect the cable from the negative battery terminal (see Chapter 5, Section 1).

43   Loosen the right front wheel lug nuts. Raise the vehicle and place it securely on jackstands. Remove the right front wheel.

44   Remove the right lower engine splash shield.

45   Disconnect the downstream oxygen sensor electrical connector (see illustration).

46   Using a wrench, unscrew and remove the downstream oxygen sensor (see illustration). If the sensor is difficult to loosen, spray some penetrant onto the sensor threads and allow it to soak in for awhile.

Note: *In the accompanying photo, the alternator has been removed for clarity. You shouldn't have to actually remove the alternator to remove the downstream sensor, but it is a very tight fit, so if the sensor proves difficult to remove, you might have to remove the alternator (see Chapter 5, Section 6) to give yourself more room to work.*

47   If you're going to install the old sensor, apply anti-seize compound to the threads of the sensor to facilitate future removal. If you're going to install a new oxygen sensor, it's not necessary to apply anti-seize compound to the threads. The threads on new sensors already have anti-seize compound on them.

48   Installation is otherwise the reverse of removal. Be sure to tighten the downstream oxygen sensor to the torque listed in this Chapter's Specifications.

49   After you've reconnected the battery, the Powertrain Control Module (PCM) must relearn its idle and fuel trim strategy for optimum driveability and performance (see Chapter 5, Section 1 for this procedure).

13   Power Steering Pressure (PSP) switch - replacement

Note: *This procedure applies to 2004 and earlier V6 models only.*

1   Disconnect the cable from the negative battery terminal (see Chapter 5, Section 1).

2   Locate the PSP switch on the power steering pressure line, behind the oil filler cap (see illustration).

3   Disconnect the electrical connector from the PSP switch.

4   Using a wrench, unscrew the PSP switch and remove it.

5   Installation is the reverse of removal.

6   After you've reconnected the battery, the Powertrain Control Module (PCM) must relearn its idle and fuel trim strategy for optimum driveability and performance (see Chapter 5, Section 1 for this procedure).

14   Throttle Position (TP) sensor - replacement

Note: *This procedure applies to 2008 and earlier V6 models and to 2003 through 2008 four-cylinder models (it is not serviceable on 2001*

and 2002 2.0L Zetec models). 2009 and later models are throttle-by-wire and the TP sensor is part of the throttle body/motor assembly and is non-serviceable.

1   Disconnect the cable from the negative battery terminal (see Chapter 5, Section 1).

2   On V6 models, remove the engine cover (see illustration).

3   Disconnect the TP sensor electrical connector (see illustration).

4   Remove the TP sensor mounting screws and remove the TP sensor.

5   When installing the TP sensor screws on a four-cylinder model, be sure to hand-start the screws (they're self-threading, and failure to hand-start them could cause damage to the throttle body). Installation is otherwise the reverse of removal.

6   After you've reconnected the battery, the Powertrain Control Module (PCM) must relearn its idle and fuel trim strategy for optimum driveability and performance (see Chapter 5, Section 1 for this procedure).

13.2 The PSP switch is located on the power steering pressure line, near the oil filler cap

14.2 To detach the engine cover from a V6 engine, remove these fasteners

14.3 To remove the TP sensor, disconnect the electrical connector and remove the TP sensor mounting screws (2008 and earlier V6 model shown, 2003 through 2008 four-cylinder models similar)

15 Transmission Range (TR) sensor - replacement

Warning: *Wait until the engine is completely cool before beginning this procedure.*

Note: *This procedure applies to 2008 and earlier V6 models only. On 2009 and later models the TR sensor is located within the transaxle valve body and should be serviced by a qualified technician or repair facility.*

Note: *A special tool (manufacturer tool no.307-351, or equivalent) is required for this procedure - without it, chances are you won't be able to adjust the sensor properly, and a trouble code will then be set (along with driveability problems). If you don't have access to the special tool, have this procedure performed at a dealer service department or other qualified repair shop.*

Removal

1   Set the parking brake, then place the shift lever in Neutral.
2   Remove the battery and the battery tray (see Chapter 5, Section 3).
3   Remove the air filter housing and air intake duct (see Chapter 4, Section 9).

4   Drain the engine coolant (see Chapter 1, Section 22).
5   Remove the thermostat housing (see Chapter 3, Section 4).
6   Disconnect the electrical connector from the TR sensor (see illustration).
7   Unscrew the bolts and remove the sensor.

Installation

8   Fit the special tool onto the sensor, engaging the two prongs on the tool with the two notches in the rotating ring of the sensor. The large tang on the other end of the tool fits into the notch in the sensor housing.
9   Place the sensor onto the transaxle, making sure the projections on the rotating ring of the sensor align with the slots in the shaft.
10 Install the sensor mounting bolts and tighten them to the torque listed in this Chapter's Specifications. Remove the special tool, then reconnect the electrical connector.
11 The remainder of installation is the reverse of the removal procedure.
12 Refill the cooling system (see Chapter 1, Section 22).

13 After you've reconnected the battery, the Powertrain Control Module (PCM) must relearn its idle and fuel trim strategy for optimum driveability and performance (see Chapter 5, Section 1 for this procedure).

16 Turbine Shaft Speed (TSS) sensor - replacement

Note: *This procedure applies to models with an automatic transaxle.*

1   Disconnect the cable from the negative battery terminal (see Chapter 5, Section 1).
2   Loosen the left front wheel lug nuts. Raise the vehicle and place it securely on jackstands. Remove the left front wheel.
3   Remove the lower left splash shield.
4   Locate the TSS sensor on the left side of the transaxle (see illustrations).
5   Disconnect the TSS sensor electrical connector.
6   Put a drain pan under the TSS sensor to catch any spilled automatic transmission fluid. Then remove the TSS sensor mounting bolt and remove the sensor.

15.6 Disconnect the electrical connector from the TR sensor, then remove the mounting bolts

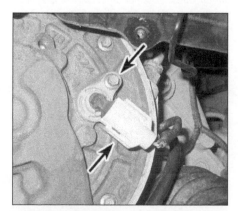

16.4a The TSS sensor is located on the left side of the automatic transaxle; to remove the TSS sensor, disconnect the electrical connector and remove the sensor mounting bolt (2012 and earlier models shown)

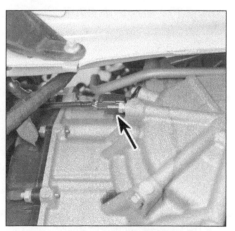

16.4b Locating the TSS sensor on 2013 and later models

7   Remove the old TSS sensor O-ring and discard it. Be sure to use a new O-ring when installing the TSS sensor (whether you're reusing the old sensor or installing a new one).

8   Installation is the reverse of removal. Be sure to tighten the TSS sensor mounting bolt to the torque listed in this Chapter's Specifications. Tighten the wheel lug nuts to the torque listed in the Chapter 1 Specifications.

9   On 2012 and earlier models, after you've reconnected the battery, the Powertrain Control Module (PCM) must relearn its idle and fuel trim strategy for optimum driveability and performance (see Chapter 5, Section 1 for this procedure).

17   Vehicle Speed Sensor (VSS) - replacement

Note: *This procedure applies to models with a manual transaxle.*

1   Disconnect the cable from the negative battery terminal (see Chapter 5, Section 1).

2   On 2002 4WD models, raise the vehicle and place it securely on jackstands. (The VSS is accessed from underneath the vehicle on these models.)

3   Locate the VSS on top of the rear part of the transaxle, near the inner CV joints.

4   Disconnect the VSS electrical connector.

5   Remove the VSS hold-down bolt.

6   Remove the VSS.

7   Installation is the reverse of removal. Be sure to tighten the VSS hold-down bolt to the torque listed in this Chapter's Specifications.

8   After you've reconnected the battery, the Powertrain Control Module (PCM) must relearn its idle and fuel trim strategy for optimum driveability and performance (see Chapter 5, Section 1 for this procedure).

18   Manifold Absolute Pressure (MAP) sensor - replacement

2005 through 2008 2.3L engine and 2009 through 2009 V6 engine

1   Disconnect the wiring harness from the sensor (see illustration).

2   Remove the mounting bolt(s) and remove the sensor.

Note: *Screws on four-cylinder models require a T25 Torx bit.*

3   Installation is the reverse of removal.

2013 and later models

Note: *On 1.5L and 1.6L models, the MAP sensor is attached to the top of the intake manifold. On 2.0L and 2.5L models, the MAP sensor is attached to the side of the intake manifold, near the radiator. 2.5L models are equipped with BOTH a MAP and MAF sensor.*

4   Remove the engine cover.

5   Locate the MAF sensor and disconnect the electrical connector (see illustration).

18.1 Location of the Manifold Absolute Pressure sensor - 2009 through 2012 V6 engines

6   Remove the mounting bolt and remove the sensor.

7   Installation is the reverse of removal.

19   Turbocharger boost pressure/ intercooler temperature sensor - replacement

Note: *2013 and later 1.5L, 1.6L and 2.0L models are equipped with a turbocharger. On 1.5L models, the boost pressure sensor and temperature sensor are separate sensors. On 1.5L models, the boost pressure sensor is mounted in the intercooler piping running along the top of the engine and the intercooler temperature sensor is mounted on the driver's (left) side of the intercooler. On 1.6L and 2.0L models, the boost pressure sensor and temperature sensor are one sensor. On 1.6L models, the turbocharger boost pressure/intercooler temperature sensor is located on the top side of the outlet of the intercooler on the driver's (left) side of the vehicle. On 2.0L models, the turbocharger boost pressure/intercooler temperature sensor is located on the top side of the outlet of the intercooler on the passenger's (right) side of the vehicle.*

1.5L engine

Turbocharger boost sensor

1   Remove the engine cover.

2   Locate the boost pressure sensor in the intercooler piping near the intake air duct boot (the sensor mounted on a square flange) and disconnect the electrical connector.

3   Remove the single screw and remove the boost pressure sensor from the intercooler piping.

Intercooler temperature sensor

Warning: *The intercooler on 1.5L models is air-to-water. When the sensor is changed, ensure the engine is cool and the cooling system pressure is released by opening and closing the expansion tank cap.*

18.5 Locating the MAP sensor (2013 and later 1.6L model shown)

4   Raise and support the front of the vehicle on jackstands.

5   Remove the engine under cover.

6   Locate the intercooler temperature sensor at the left (driver's) end of the intercooler and disconnect the electrical connector.

7   Position a drain pan under the left end of the intercooler. Remove the sensor retaining clip by pulling it straight out.

Note: *Have the NEW sensor sensor ready to install to reduce loss of coolant. If the same sensor is to be reinstalled, check the O-ring for damage and replace as necessary.*

8   Remove the existing sensor and immediately install the new one. Insert the sensor retaining clip.

9   Remaining installation is reverse of removal. Top off the coolant as necessary (see Chapter 1).

1.6L and 2.0L engines

10   Raise and support the front of the vehicle on jackstands.

11   Remove the engine undercover.

12   Disconnect the boost pressure/temp sensor electrical connector (see illustration).

13   Remove the screw and remove the boost pressure/temp sensor from the intercooler.

14   Installation is reverse of removal.

20   Turbocharger wastegate control solenoid - replacement

Note: *The following procedure applies to 2017 1.5L and 2.0L models.*

1   Remove the engine cover.

2   On 1.5L models, remove the intercooler piping between the turbocharger and intercooler. Use care to disconnect all connectors and harnesses.

3   Remove the turbocharger piping between the air filter and turbocharger inlet.

4   On 1.5L models, locate the wastegate control solenoid mounted to the vacuum pump on the rear of the cylinder head, above the transaxle and disconnect the electrical connector.

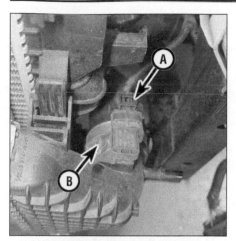

19.12 Disconnect the boost sensor connector (A) and remove the screw (B) (2013 and later 1.6L model shown, bumper cover removed for clarity)

22.4 Locating the fuel rail pressure sensor (1.6L model shown)

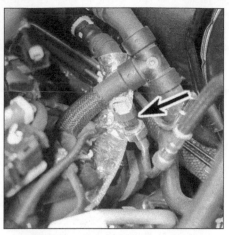

23.3 The fuel pressure sensor is located on the fuel feed line near the firewall (2013 and later models)

5   On 2.0L models, locate the wastegate control solenoid mounted on about the middle of the engine front cover and disconnect the electrical connector.

6   On all models, place identifying marks on the vacuum lines (for proper installation) and disconnect them from the solenoid.

7   Remove the two nuts (1.5L) or two bolts (2.0L) and remove the solenoid from the pipe.

8   Installation is reverse of removal.

21 Turbocharger wastegate vacuum sensor - replacement

Note: *The following procedure applies to 2017 2.0L models ONLY.*

1   Remove the engine cover.

2   Locate the wastegate vacuum sensor below the wastegate control solenoid mounted on about the middle of the engine front cover and disconnect the electrical connector.

3   Disconnect the vacuum line from the sensor.

4   Unscrew the vacuum sensor retaining bolt and remove the sensor.

5   Installation is reverse of removal.

22 Fuel Rail Pressure (FRP) sensor - replacement

Warning: *Gasoline is extremely flammable, so take extra precautions when you work on any part of the fuel system. Don't smoke or allow open flames or bare light bulbs near the work area, and don't work in a garage where a gas-type appliance (such as a water heater or a clothes dryer) is present. Since gasoline is carcinogenic, wear latex gloves when there's a possibility of being exposed to fuel, and, if you spill any fuel on your skin, rinse it off immediately with soap and water. Mop up any spills immediately and do*

not store fuel-soaked rags where they could ignite. The fuel system is under constant pressure, so, if any fuel lines are to be disconnected, the fuel pressure in the system must be relieved first. When you perform any kind of work on the fuel system, wear safety glasses and have a Class B type fire extinguisher on hand.

Note: *1.5L, 1.6L and 2013 and later 2.0L models are equipped with an FRP sensor. On 1.5L and 1.6L models, the FRP is accessible from the top of the engine. On 2.0L models, the FRP is accessible from the front of the engine, next to the intake manifold.*

1   Remove the engine cover.

2   Relieve the fuel system pressure (see Chapter 4, Section 2).

3   Disconnect the negative battery cable (see Chapter 5).

4   Locate and disconnect the FRP sensor connector (see illustration).

5   Unscrew the FRP sensor from the fuel rail.

6   Installation is reverse of removal.

Note: *Lubricate the O-ring with gasoline before installation.*

7   Tighten the FRP sensor to the torque listed in this Chapter's Specifications.

8   Start the vehicle and check for fuel leaks.

23 Fuel pressure sensor - replacement

Warning: *Gasoline is extremely flammable, so take extra precautions when you work on any part of the fuel system. Don't smoke or allow open flames or bare light bulbs near the work area, and don't work in a garage where a gas-type appliance (such as a water heater or a clothes dryer) is present. Since gasoline is carcinogenic, wear latex gloves when there's a possibility of being exposed to fuel, and, if you spill any fuel on your skin, rinse it off immediately with soap and water. Mop up any spills immediately and do*

not store fuel-soaked rags where they could ignite. The fuel system is under constant pressure, so, if any fuel lines are to be disconnected, the fuel pressure in the system must be relieved first. When you perform any kind of work on the fuel system, wear safety glasses and have a Class B type fire extinguisher on hand.

Note: *The fuel pressure sensor is installed in the fuel line and accessible between the brake master cylinder reservoir and cowl panel. It monitors the pressure from the fuel pump.*

1   Relieve the fuel system pressure (see Chapter 4, Section 2).

2   Remove the battery and the battery tray (see Chapter 5).

3   Locate the fuel pressure sensor and disconnect the electrical connector (see illustration).

4   Unscrew the sensor from the fuel line.

Caution: *Hold the fuel line while removing the sensor to prevent damage to the line.*

5   Installation is reverse of removal.

24 Powertrain Control Module (PCM) - replacement

Caution: *To avoid electrostatic discharge damage to the PCM, handle the PCM only by its case. Do not touch the electrical terminals during removal and installation. If available, ground yourself to the vehicle with an anti-static ground strap, available at computer supply stores.*

Note: *The PCM is a highly reliable component and rarely requires replacement. Because the PCM is the most expensive part of the engine management system, you should be absolutely positive that it has failed before replacing it. If in doubt, have the system tested by an experienced driveability technician at a dealer service department or other qualified repair shop. If the PCM is to be replaced, ensure all of the customer's keys and remotes are with the vehicle for programming and/or PATS reset.*

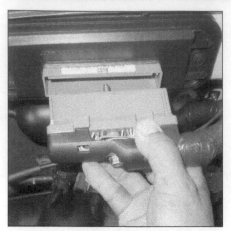

24.3a To disconnect the electrical connector, loosen the bolt, then pull the connector straight out (2008 and earlier models)

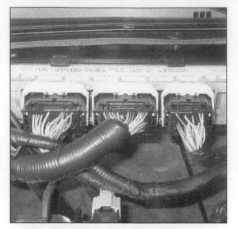

24.3b The Powertrain Control Module (PCM) is on the upper firewall on all models (later models have multiple connectors)

24.4 To remove the PCM cover, remove these two nuts

1   Disconnect the cable from the negative terminal of the battery (see Chapter 5, Section 1).

2012 and earlier models

2   The PCM is installed in the center of the upper firewall, and can be accessed from the engine compartment side.
3   Loosen the electrical connector retaining bolt (2008 and earlier models), then disconnect the electrical connector(s) from the PCM (see illustrations).
4   Remove the PCM cover nuts and remove the cover (see illustration).
5   Carefully remove the PCM from its receptacle.
6   Installation is the reverse of removal.
7   After you've reconnected the battery, the Powertrain Control Module (PCM) must relearn its idle and fuel trim strategy for optimum driveability and performance (see Chapter 5, Section 1 for this procedure).

2013 and later models

Caution: *Removal and installation of the same PCM does not require new keys or programming of the existing keys, however a parameter reset of the Passive Anti-Theft System (PATS) using a scan tool is required if replacing the PCM. If the PCM is to be replaced, the existing PCM data must be downloaded by a qualified technician or repair facility using a scan tool and then uploaded to the NEW PCM for the vehicle to operate correctly.*

8   Disconnect the negative battery terminal (see Chapter 5, Section 1).
9   Loosen the left front wheel lug nuts, then raise and support the front of the vehicle on jackstands.
10   Remove the left front wheel.
11   Remove the left inner fender splash shield (see Chapter 11, Section 10).
12   Remove the plastic retainer screws and the PCM cover (see illustration).
13   Remove the PCM retainers and pull the

PCM out from the vehicle with the connectors still connected (see illustration).
14   Pull the latches on the PCM connectors and disconnect the connectors.
15   Remove the PCM from the vehicle.
16   Installation is reverse of removal.

25   Idle Air Control (IAC) valve - replacement

Note: *This procedure applies only to 2008 and earlier models. 2009 and later models have an electronically operated throttle body that does not require an idle air control valve.*

Four-cylinder models

1   Disconnect the cable from the negative terminal of the battery (see Chapter 5, Section 1).
2   Raise the front of the vehicle and place it securely on jackstands.
3   Locate the IAC valve on the underside of the intake manifold (see illustration).

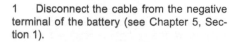

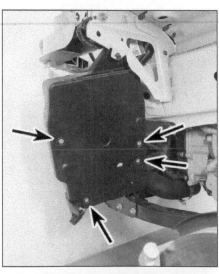

24.12 Remove the screws and the PCM cover (2013 and later models shown)

24.13 Remove the retainers and pull the PCM out (2013 and later models shown)

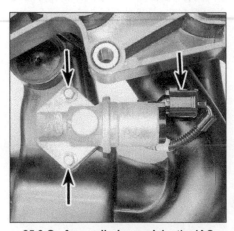

25.3 On four-cylinder models, the IAC valve is located underneath the intake manifold (manifold removed for clarity); to detach the IAC valve from the manifold, disconnect the electrical connector and remove the two mounting bolts

25.13 Disconnect the IAC valve electrical connector (2008 and earlier V6 models)

25.14 To detach the IAC valve from the intake manifold, remove these two bolts

26.2 The solenoid valves that control the variable camshaft timing are located on top of each valve cover (V6 shown)

4   Disconnect the electrical connector from the IAC valve.
5   Remove the IAC mounting bolts.
6   Remove the IAC valve and discard the old gasket.
7   Using a new gasket, install the IAC valve and tighten the bolts to the torque listed in this Chapter's Specifications.
8   Reconnect the IAC valve electrical connector.
9   Lower the vehicle and reconnect the cable to the negative battery terminal.
10   After you've reconnected the battery, the Powertrain Control Module (PCM) must relearn its idle and fuel trim strategy for optimum driveability and performance (see Chapter 5, Section 1 for this procedure).

V6 models
11   Disconnect the cable from the negative terminal of the battery (see Chapter 5, Section 1).
12   The IAC valve is located on top of the intake manifold, near the throttle body.
13   Disconnect the electrical connector (see illustration) from the IAC valve.

26.8 Locate the VCT oil control solenoids (2013 and later 1.6L model shown)

14   Remove the IAC valve mounting bolts (see illustration).
15   Remove the IAC valve.
16   Remove and discard the old IAC valve gasket.
17   Install a new IAC valve gasket. Make sure that the gasket is fully seated in its groove before installing the IAC valve.
18   Install the IAC valve. Be sure to tighten the IAC valve mounting bolts to the torque listed in this Chapter's Specifications.
19   The remainder of installation is the reverse of removal. Reconnect the cable to the negative battery terminal.
20   After you've reconnected the battery, the Powertrain Control Module (PCM) must relearn its idle and fuel trim strategy for optimum driveability and performance (see Chapter 5, Section 1 for this procedure).

26   Variable Camshaft Timing (VCT) oil control solenoid(s) (2009 and later models) - replacement

V6 engine
1   Remove the valve cover of the faulty sensor (see Chapter 2A, Section 4 or Chapter 2B, Section 4).
2   Disconnect the wiring harness from the solenoid (see illustration).
3   Remove the mounting bolt(s), then remove the sensor.
4   Installation is the reverse of removal.

Four-cylinder engines
1.5L and 1.6L models
5   Remove the engine cover.
6   On 1.6L models, remove the bolt and stud attaching the air intake to the valve cover and move out of the way.
7   On 1.6L models, disconnect the crankcase vent tube from the valve cover and intake manifold and remove.
8   On all models, locate the VCT oil control solenoids at the front of the valve cover (see illustration).

9   Disconnect the solenoid electrical connectors.
10   Remove the bolt and remove the solenoid from the cylinder head.
11   Installation is reverse of removal.

2.0L and 2.5L models
12   Remove the valve cover (see Chapter 2A, Section 4).
13   Locate the VCT oil control solenoids at the front of the cylinder head.
Note: *2.0L models have two solenoids, 2.5L models only have one solenoid.*
14   Remove the bolt and remove the solenoid from the cylinder head.
15   Installation is reverse of removal.

27   Catalytic converters

Note: *Because of a Federally mandated extended warranty which covers emission-related components such as the catalytic converter, check with a dealer service department before replacing the converter at your own expense.*

Check
1   The test equipment for a catalytic converter (a "loaded-mode" dynamometer and a 5-gas analyzer) is expensive. If you suspect that the converter on your vehicle is malfunctioning, take it to a dealer or authorized emission inspection facility for diagnosis and repair.
2   Whenever you raise the vehicle to service underbody components, inspect the converter for leaks, corrosion, dents and other damage. Carefully inspect the welds and/or flange bolts and nuts that attach the front and rear ends of the converter to the exhaust system. If you note any damage, replace the converter.
3   Although catalytic converters don't break too often, they can become clogged or even plugged up. The easiest way to check for a restricted converter is to use a

vacuum gauge to diagnose the effect of a blocked exhaust on intake vacuum.

a) *Connect a vacuum gauge to an intake manifold vacuum source (see Chapter 2A or 2B).*

b) *Warm the engine to operating temperature, place the transaxle in Park (automatic models) or Neutral (manual models) and apply the parking brake.*

c) *Note the vacuum reading at idle and jot it down.*

d) *Quickly open the throttle to near its wide-open position and then quickly get off the throttle and allow it to close. Note the vacuum reading and jot it down.*

e) *Do this test three more times, recording your measurement after each test.*

f) *If your fourth reading is more than one in-Hg lower than the reading that you noted at idle, the exhaust system might be restricted (the catalytic converter could be plugged, OR an exhaust pipe or muffler could be restricted).*

Replacement

4   See Chapter 4, Section 16 for the catalytic converter replacement procedures.

---

28   Evaporative Emissions Control (EVAP) system - component replacement

---

EVAP canister purge valve

2012 and earlier models

Note: *The illustrations accompanying this procedure depict a 2004 and earlier system. Later models are similar.*

1   Disconnect the cable from the negative terminal of the battery (see Chapter 5, Section 1).

2   The EVAP canister purge valve is located in the engine compartment, on a bracket near

28.3 Disconnect the electrical connector and the vacuum line from the EVAP canister purge valve; be sure to cap the vacuum line to prevent dirt and moisture from entering the line while it's disconnected (V6 model shown, four-cylinder models similar)

the master cylinder.

3   Disconnect the purge valve electrical connector (see illustration).

4   Disconnect the vacuum line from the purge valve. Cap the vacuum line to prevent dirt or moisture from entering it while it's disconnected.

5   Disconnect the EVAP canister return line and outlet line (see illustration). Cap the tubes to prevent dirt, dust and moisture from entering the EVAP system while the lines are open.

6   Remove the two purge valve mounting nuts (see illustration) and remove the purge valve from its mounting bracket (it's not necessary to remove the bracket).

7   Installation is the reverse of removal.

8   After you've reconnected the battery,

28.5 To disconnect the EVAP canister lines, squeeze the two tabs on the fitting together and pull off the line (V6 model shown, four-cylinder models similar)

the Powertrain Control Module (PCM) must relearn its idle and fuel trim strategy for optimum driveability and performance (see Chapter 5, Section 1 for this procedure).

2013 and later 1.5L, 1.6L and 2.0L models

Note: *The canister purge valve on 2013 and later 1.5L, 1.6L and 2.0L models is part of the EVAP hose assembly, and not serviceable.*

9   Remove the engine cover.

10   Locate the EVAP canister purge valve and disconnect the electrical connector (see illustration).

11   Disconnect the EVAP hose assembly fittings and remove the hose assembly from the vehicle (see illustrations).

12   On other models, additional fittings may require disconnection and there may be support brackets requiring removal of

28.6 To detach the EVAP canister purge valve from its mounting bracket, remove these two nuts (V6 model shown, four-cylinder models similar)

28.10 Disconnect the EVAP purge valve connector (A) and disconnect the EVAP fitting (B) (2013 and later 1.6L model shown, others similar)

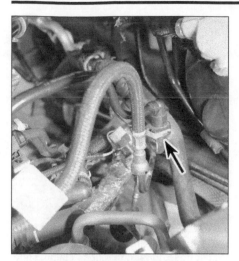

28.11a Disconnect the EVAP fitting near the brake booster . . .

28.11b . . . disconnect the EVAP fitting above the throttle body . . .

28.11c . . . disconnect the EVAP fitting at the intercooler (2013 and later 1.6L model shown, others similar)

fasteners to remove the EVAP hose assembly from the vehicle.

13   Installation is reverse of removal.

2013 and later 2.5L models

Note: *The canister purge valve on 2013 and later 2.5L models is bolted to the intake manifold, behind the throttle body.*

14   Remove the engine cover.

15   Locate the EVAP canister purge solenoid on the intake manifold and disconnect the electrical connector.

16   Disconnect the EVAP fitting from the valve.

17   Remove the bolts and pull the valve out of the intake manifold.

18   Installation is reverse of removal.

Fuel tank pressure sensor

Note: *The illustrations accompanying this procedure depict a 2004 and earlier system. Later models are similar.*

19   Disconnect the cable from the negative terminal of the battery (see Chapter 5, Section 1).

20   Remove the rear seat cushion (see Chapter 11).

21   Pull back the carpeting and remove the fuel pump/fuel level sending unit module access cover (see illustrations 5.4a and 5.4b in Chapter 4).

22   Disconnect the electrical connector from the fuel tank pressure sensor (see illustration).

23   Pull the fuel tank pressure sensor out from under the floorpan and out the two crimped hose clamps (see illustration).

24   Note which way the fuel tank pressure sensor is oriented before disconnecting the inlet and outlet hoses.

25   Installation is the reverse of removal. Be sure to secure the hoses to the sensor with new hose clamps (it's not necessary to

use crimp-style clamps).

26   After you've reconnected the battery, the Powertrain Control Module (PCM) must relearn its idle and fuel trim strategy for optimum driveability and performance (see Chapter 5, Section 1 for this procedure).

EVAP canister, check valve, dust separator and canister vent solenoid

2012 and earlier models

Note: *The illustrations accompanying this procedure depict a 2004 and earlier system. Later models are similar.*

27   Disconnect the cable from the negative terminal of the battery (see Chapter 5, Section 1).

28   Raise the vehicle and place it securely on jackstands.

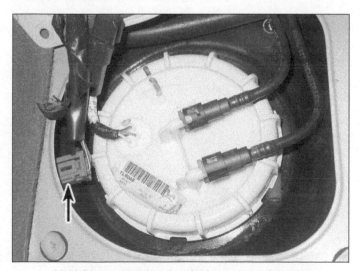

28.22 Disconnect the electrical connector from the fuel tank pressure sensor

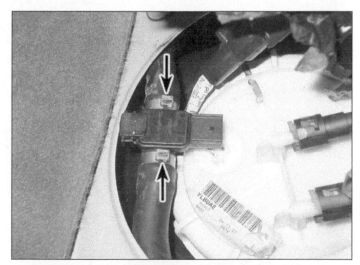

28.23 Pull the fuel tank pressure sensor out from under the floorpan and cut the two crimped hose clamps with a pair of diagonal cutters

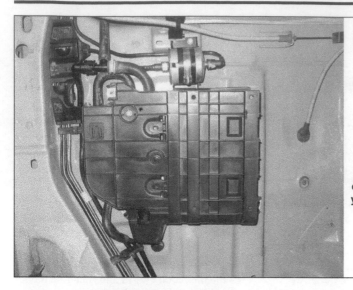

28.29a The EVAP canister, check valve, dust separator and canister vent solenoid, and the hoses connecting them, are located underneath the vehicle, near the left side; to replace any of these components, you must first remove the entire assembly (2012 and earlier models shown)

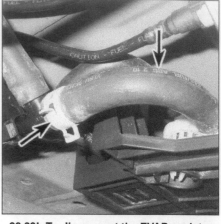

28.29b To disconnect the EVAP canister outlet hose, loosen this hose clamp, slide it back and pull off the hose

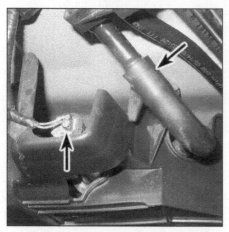

28.30 Disconnect the canister vent solenoid electrical connector and the canister vent hose

28.31 To detach the EVAP canister mounting bracket from the underside of the vehicle, remove the four nuts (left side shown)

29   On the back side of the EVAP canister assembly, disconnect the EVAP canister outlet hose (see illustrations).
30   On the front side of the EVAP canister assembly, disconnect the canister vent solenoid electrical connector and the canister vent hose (see illustration).
31   Remove the EVAP canister mounting bracket nuts and bolts (see illustration) and lower the EVAP canister, its mounting bracket and the rest of the EVAP components as a single assembly.
32   Remove the remote canister vent hoses, the check valve, the dust separator and the dust separator-to-canister vent solenoid hose (see illustration). To disengage the dust separator from the EVAP canister mounting bracket, simply pull it straight up to slide it off the retaining tab on the mounting bracket.
33   To detach the EVAP canister from its mounting bracket, pry up the right side of the canister and then slide the two mounting tabs on the left side of the canister out from their slots in the mounting bracket.
34   If you're replacing the EVAP canister or the canister vent solenoid, separate the two components. After you've separated the canister and the vent solenoid, inspect the solenoid O-ring and replace it if it's damaged or worn.
35   Installation is the reverse of removal.
36   After you've reconnected the battery, the Powertrain Control Module (PCM) must relearn its idle and fuel trim strategy for optimum driveability and performance (see Chapter 5, Section 1 for this procedure).

2013 and later models

37   Raise and support the vehicle on jackstands.
38   Disconnect the negative battery cable (see Chapter 5, Section 1).
39   To remove the canister vent solenoid without removing the EVAP canister, detach the vent hose, disconnect the solenoid electrical connector and rotate the solenoid clockwise and pull out of the canister (see illustration).
Note: *On AWD model, remove the shield under the rear differential before removing the solenoid.*
40   On AWD models, remove the fuel tank (see Chapter 4, Section 8).
41   On all models, disconnect the vent hose (see illustration).
42   On all models, disconnect the EVAP canister quick disconnect fittings.

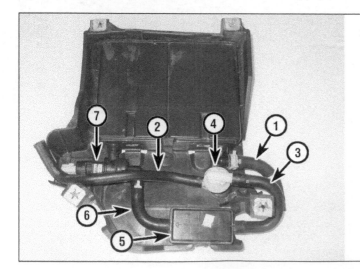

28.32 EVAP canister and related EVAP components

1   *EVAP canister outlet hose*
2   *Remote canister vent hose*
3   *Remote canister vent hose*
4   *EVAP check valve*
5   *Dust separator*
6   *Dust separator-to-canister vent solenoid hose*
7   *Canister vent solenoid*

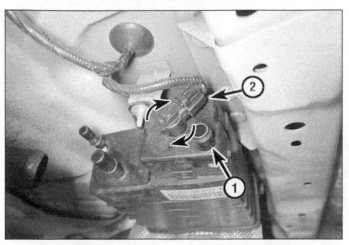

28.39 Disconnect the vent hose (1) then the vent solenoid connector (2) and rotate the solenoid clockwise to remove from the canister (2013 and later models shown)

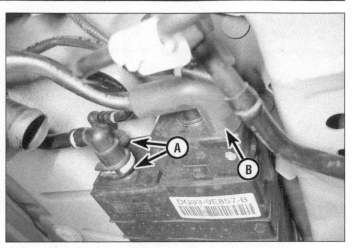

28.41 Disconnect the vent hose (A) and quick disconnect fittings (B) (2013 and later model shown)

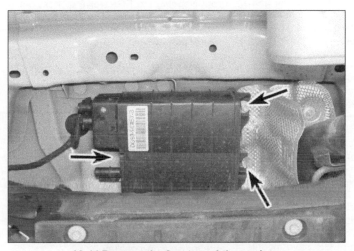

28.44 Remove the 3 nuts and the canister (2013 and later model shown)

28.46 Release the tabs to access and replace the canister filter (2013 and later model shown)

43   Disconnect the vent solenoid electrical connector.
44   Remove the three nuts and remove the EVAP canister from the vehicle (see illustration).
45   If the canister is to be replaced, remove the canister vent solenoid as described earlier and transfer it to the new canister.
46   To replace the canister filter, remove the vent solenoid and disengage the tabs to access and replace the filter (see illustration).
47   Installation is reverse of removal.

29   Exhaust Gas Recirculation (EGR) system - component replacement

EGR valve and pipe

2.0L four-cylinder models

1   Raise the vehicle and place it securely on jackstands.
2   Remove the right lower splash shield

(see illustration 6.13 in Chapter 5).
3   Unscrew the tube nut fitting and disconnect the EGR pipe from the exhaust manifold. This fitting might be difficult to loosen because of the intense heat to which it has been repeatedly subjected. If so, spray some penetrant on the threads of the fitting and then try again later. (The penetrant manufacturer will specify how long you should wait for the penetrant to loosen things up.)
4   In the engine compartment, remove the air intake duct (see Chapter 4, Section 9).
5   Unscrew the big nut at the upper end of the EGR pipe and disconnect the pipe from the EGR valve.
Caution: *The manufacturer says that anytime you disconnect the EGR pipe from the EGR valve, you MUST replace the old EGR pipe. Do NOT reuse the old EGR pipe! The manufacturer also says that you must inspect the (aluminum) threads inside the EGR valve for damage. If the threads are damaged, you MUST replace the EGR valve.*
6   Remove the two EGR pipe bracket bolts.

7   Carefully remove the differential pressure feedback EGR sensor from the EGR pipe.
8   Remove the EGR pipe and discard it.
9   Disconnect the vacuum line from the EGR valve.
10   Remove the two EGR valve mounting bolts.
11   Remove the EGR valve and remove and discard the old gasket.
12   Scrape all exhaust deposits and old gasket material from the EGR valve mounting surface on the manifold and, if you plan to use the same valve, from the mounting surface of the valve. Carefully remove carbon deposits from the valve port and pintle with a small screwdriver. If the EGR passage contains excessive deposits, clean it out with a small scraper and a vacuum. Make sure that all loose particles are removed to keep them from clogging the EGR valve or from being drawn into the engine.
Caution: *Do not try to wash the valve in solvent or sandblast it.*

29.25 Disconnect the vacuum line from the EGR valve (early V6 models)

29.26 To disconnect the EGR pipe from the EGR valve, unscrew this big nut and then carefully pull out the pipe just enough to clear the valve (early V6 models)

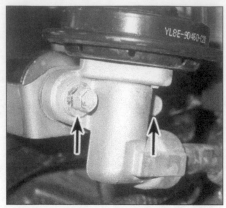

29.27 To detach the EGR valve from its mounting bracket, remove these two bolts (early V6 models)

13   Installation is the reverse of removal. Be sure to tighten the EGR valve mounting bolts to the torque listed in this Chapter's Specifications. And don't forget to use a new EGR pipe and tighten the EGR pipe fittings securely.

2.3L four-cylinder models

Warning: *Wait until the engine is completely cool before beginning this procedure.*

14   The EGR valve is located on the left side of the cylinder head. Engine coolant passes through the valve.

15   Drain the engine coolant.

16   Remove the upper radiator hose from the thermostat outlet, then remove the coolant hose from the EGR valve.

17   Remove the two EGR mounting bolts and remove the valve. Discard the old gasket.

18   Installation is the reverse of removal.

2.5L four-cylinder models

19   Refer to Chapter 1 and drain the cooling system.

20   Remove the outlet duct from the air filter housing.

21   Disconnect the wiring harness from the EGR valve.

22   Detach the coolant hose from the EGR valve.

23   Remove the two mounting bolts, then remove the EGR valve. Discard the gasket.

24   Installation is the reverse of removal. Be sure to use a new gasket.

2008 and earlier V6 models

25   Disconnect the vacuum line from the EGR valve (see illustration).

26   Disconnect the EGR pipe from the EGR valve (see illustration).

27   Remove the bolts securing the EGR valve to its mounting bracket on the intake manifold (see illustration) and remove the EGR valve. Remove and discard the old gasket.

28   Remove the differential pressure feedback EGR sensor from the EGR pipe.

29   Disconnect the EGR pipe from the exhaust manifold (see illustration).

30   Scrape all exhaust deposits and old gasket material from the EGR valve mounting surface on the mounting bracket at the intake manifold and, if you plan to use the same valve, from the mounting surface of the valve. Carefully remove carbon deposits from the valve port and pintle with a small screwdriver. If the EGR passage contains excessive deposits, clean it out with a small

scraper and a vacuum. Make sure that all loose particles are removed to keep them from clogging the EGR valve or from being drawn into the engine.

Caution: *Do not try to wash the valve in solvent or sandblast it.*

31   Installation is the reverse of removal. Attach the EGR pipe nuts loosely at the valve and at the exhaust manifold until the EGR valve is tightened on the intake manifold. Tighten the pipe fittings securely. Be sure to tighten the EGR valve mounting bolts to the torque listed in this Chapter's Specifications.

2009 through 2012 V6 models

32   Disconnect the wiring harness from the EGR valve.

33   Detach the EGR tube from the EGR valve.

34   Remove the two mounting bolts, then remove the EGR valve. Discard the gasket.

35   Installation is the reverse of removal. Be sure to use a new gasket.

2013 and later 2.5L models

Note: *The EGR valve is located on the rear of the cylinder head, above the transaxle and near the throttle body.*

36   Drain the cooling system (see Chapter 1, Section 22).

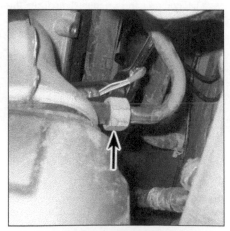

29.29 To disconnect the EGR pipe from the exhaust manifold, unscrew this big nut

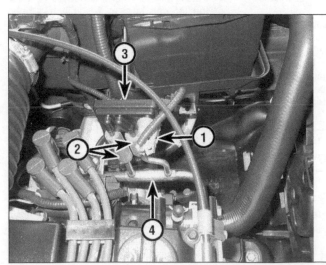

29.44a On early four-cylinder models, the differential pressure feedback EGR system sensor is located at the left end of the cylinder head, near the ignition coil

1   Electrical connector
2   Rubber hoses
3   Differential pressure feedback EGR system sensor
4   EGR pipe

29.44b Disconnect the electrical connector from the differential pressure feedback EGR system sensor (early V6 models)

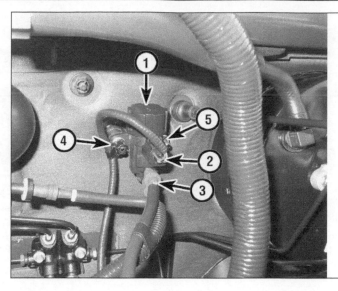

29.50 EGR vacuum regulator valve (2004 and earlier four-cylinder models)

1 EGR vacuum regulator valve
2 Electrical connector
3 Vacuum hose connector
4 Right mounting nut (remove harness clip from stud first, then remove nut)
5 Left mounting nut

37 Remove the intake air tube between the throttle body and the air filter housing.
38 Disconnect the EGR connector.
39 Disconnect the coolant hose to gain access to the EGR valve bolts.
40 Disconnect the small EGR coolant hose from the EGR valve.
41 Remove the two EGR bolts and remove the EGR valve from the engine.
42 Installation is reverse of removal. Refill the coolant.

Differential pressure feedback EGR system sensor

Note: This procedure applies only to early vacuum-operated models.
43 Disconnect the cable from the negative terminal of the battery (see Chapter 5, Section 1).
44 Disconnect the electrical connector from the differential pressure feedback EGR system sensor (see illustrations).
45 Carefully disconnect the differential pres-sure feedback EGR system sensor from the two hoses that connect it to the two small metal pipes welded to the EGR pipe.
46 While the differential pressure feedback EGR system sensor is removed, inspect the condition of the two small hoses. Look for cracks, tears and deterioration. If the hoses are worn or damaged, replace them.
47 Installation is the reverse of removal.
48 After you've reconnected the battery, the Powertrain Control Module (PCM) must relearn its idle and fuel trim strategy for optimum driveability and performance (see Chapter 5, Section 1 for this procedure).

EGR vacuum regulator valve

Note: This procedure applies only to early vacuum-operated models.
49 Disconnect the cable from the negative terminal of the battery (see Chapter 5, Section 1).
50 On four-cylinder models, locate the EGR vacuum regulator valve on the firewall (see illus-tration), to the right of the brake master cylinder. On V6 models, it's on the intake manifold.
51 Disconnect the electrical connector from the EGR vacuum regulator valve (V6 models, see accompanying illustration).
52 Disconnect the vacuum hoses from the EGR vacuum regulator valve.
53 On V6 models, detach the PCM wiring harness bracket and set it aside.
54 Remove the EGR vacuum regulator valve mounting nuts or bolts (V6 models, see accompanying illustration) and remove the valve from the firewall (four-cylinder models) or from the intake manifold (V6 models).
55 Installation is the reverse of removal. Be sure to tighten the vacuum regulator valve mounting nuts or bolts securely.
56 After you've reconnected the battery, the Powertrain Control Module (PCM) must relearn its idle and fuel trim strategy for optimum driveability and performance (see Chapter 5, Section 1 for this procedure).

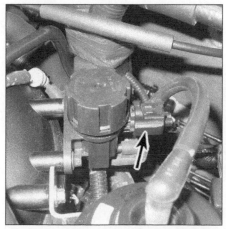

29.51 Disconnect the electrical connector from the EGR vacuum regulator valve (2004 and earlier V6 model shown, four-cylinder models similar)

29.54 To detach the EGR vacuum regulator valve, remove these two bolts; be careful not to damage the threads on the stud (on the right bolt) that secures the PCM wiring harness bracket (2004 and earlier V6 model shown, four-cylinder models similar, except that there's no stud on either bolt)

30.1 The fresh-air inlet hose on four-cylinder models connects the air intake duct (1) to the crankcase via a pipe on the valve cover (2) (valve cover pipe not visible in this photo; it's below the accordion pleats in the air intake duct)

30.7 The PCV hose, or crankcase ventilation hose, begins at the PCV valve (which is located on the left end of the crankcase vent oil separator) and ends here at the intake manifold, downstream from the throttle plate in the throttle body, so that the pressure differential caused by intake manifold vacuum will pull the crankcase vapors into the manifold (four-cylinder models)

30 Positive Crankcase Ventilation (PCV) system - component replacement

Four-cylinder models

Fresh air inlet hose

1   Disconnect the fresh air inlet hose from the air intake duct (see illustration).

2   Remove the air intake duct (see Chapter 4, Section 9).

3   Disconnect the fresh air inlet hose from the pipe at the left rear corner of the valve cover.

4   Thoroughly clean the fresh air inlet hose and inspect the condition of the hose. Look for cracks, tears and other deterioration. If it's damaged or worn, replace it. Also inspect the grommet at the air intake duct. Make sure that it's not cracked or torn. If it's worn or damaged, replace it. A leaking fresh air inlet hose or grommet will result in an unmetered air leak ("false air") that will lean out the air/fuel mixture ratio, which will cause a lean misfire and other driveability problems.

5   Installation is the reverse of removal.

PCV hose (crankcase ventilation hose)

6   Disconnect the PCV hose from the PCV valve, which is located at the upper left corner of the crankcase vent oil separator, which in turn is located on the front side of the engine block, behind the exhaust manifold and the crankcase vent oil separator.

7   Disconnect the PCV hose from the intake manifold (see illustration).

8   Thoroughly clean the PCV hose and inspect the condition of the hose. Look for cracks, tears and other deterioration. If it's damaged or worn, replace it. Also inspect the grommet at the air intake duct. Make sure that it's not cracked or torn. If it's worn or damaged, replace it. A leaking PCV hose will result in an unmetered air leak ("false air") that will lean out the air/fuel mixture ratio, which will cause a lean misfire and other driveability problems.

9   Installation is the reverse of removal.

PCV valve

10   See Chapter 1, Section 29.

Crankcase vent oil separator

2012 and earlier models

Note: *The PCV valve and oil separator on the 2.3L and 2.5L four-cylinder engine is located under the intake manifold, requiring its removal (see Chapter 2A, Section 6).*

11   The crankcase vent oil separator is located on the front side of the engine block, behind the exhaust manifold and the catalytic converter. To access the separator, remove the catalytic converter (see *Exhaust system - inspection and component replacement* in Chapter 4, Section 16) and remove the exhaust manifold (see Chapter 2A or 2B).

12   Disconnect the PCV hose from the PCV valve or remove the PCV valve from the crankcase vent oil separator.

13   Remove the crankcase vent oil separator mounting bolts and remove the separator (see illustrations). Remove and discard the old oil separator gasket.

14   Remove the PCV valve from the oil separator if you haven't already done so. Then wash out the oil separator with a suitable solvent. Blow it out with compressed air (be sure to wear safety goggles during this procedure!) and make sure that the passages are clear.

15   Install a new crankcase vent oil separator gasket (see illustration).

16   Installation is otherwise the reverse of removal. Be sure to tighten the oil separator mounting bolts to the torque listed in this Chapter's Specifications.

2013 and later models

Note: *Similar to some 2012 and earlier 4-cylinder models, 2013 and later models utilize an oil sep-*

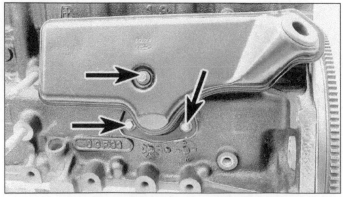

30.13a To detach the crankcase vent oil separator from the engine block on a four-cylinder model, remove these three bolts . . .

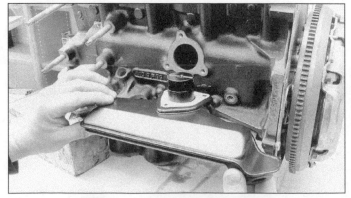

30.13b . . . and pull off the separator and the old gasket (2004 and earlier engine shown)

30.15 Before installing the crankcase vent oil separator on a four-cylinder model, be sure to install a new gasket

30.19a Detach the hose and remove the upper bolts . . .

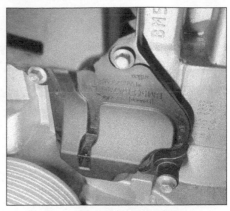

30.19b . . . and lower bolts for the oil separator (2013 and later 1.6L model shown)

30.21 Disconnect the fresh air inlet hose from the front valve cover and from the air intake duct (2004 and earlier V6 models)

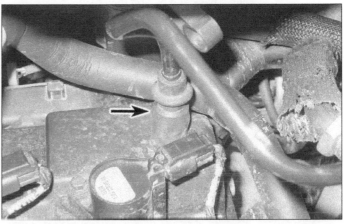

30.23 Disconnect the fresh air inlet hose from the rear valve cover (2004 and earlier V6 models)

arator as opposed to individual components. The oil separator is attached to the front of the engine block with a hose going to the intake manifold.

17   Remove the intake manifold (see Chapter 2A, Section 6).

18   If necessary, detach the hose clip from the oil separator.

19   Remove the bolts attaching the oil sepa-rator to the engine block and remove the oil separator (see illustrations).

20   Installation is reverse of removal.

V6 models

Fresh air inlet hose

21   Disconnect the fresh air inlet hose from the front valve cover (see illustration).

22   Disconnect the fresh air inlet hose from the air intake duct.

23   Disconnect the fresh air inlet hose from the rear valve cover (see illustration).

24   Installation is the reverse of removal.

PCV hose (crankcase ventilation hose)

25   Disconnect the PCV hose from the PCV valve (see illustrations).

30.25a Disconnect the PCV hose from the PCV valve, which is located in the left end of the valley between the cylinder heads (V6 models)

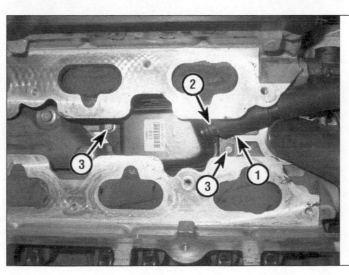

30.25b PCV hose, PCV valve and crankcase vent oil separator installation details (intake manifold removed for clarity) (2004 and earlier V6 models)

1   PCV hose
2   PCV valve
3   Crankcase vent oil separator mounting bolts

30.26 To disconnect the PCV hose (crankcase ventilation hose) from the intake manifold, loosen this hose clamp, slide it back and pull off the hose

30.30 Unscrew the PCV valve to remove it from the crankcase vent oil separator (V6 models)

26   Disconnect the PCV hose from the intake manifold (see illustration).
27   Carefully note the routing the PCV hose and then pull it out and remove it.
28   Installation is the reverse of removal. Make sure that you route the hose exactly the same way that it was routed before removal. Make sure that there are no kinks in the hose.

PCV valve

2004 and earlier models
29   Disconnect the PCV hose from the PCV valve (see illustrations 30.25a and 30.25b).
30   Disconnect the PCV valve from the crankcase vent oil separator (see illustration). Unscrew the PCV valve - don't try to just pull it out.
31   Inspect the PCV valve (see Chapter 1, Section 29).
32   Installation is the reverse of removal.

2005 through 2012 models
33   Disconnect the quick-connect coupling from the PCV valve. On 2008 and later mod-

els, disconnect the wiring from the PCV valve.
34   Turn the valve counterclockwise to remove it.
Note: *Discard the PCV valve. Once it has been removed from the valve cover, it must be replaced with a new one.*
35   Installation is the reverse of removal.

Crankcase vent oil separator (2004 and earlier models)
36   Remove the intake manifold (see Chapter 2B, Section 5)
37   Disconnect the PCV hose from the PCV valve (see illustration 30.25b).
38   Remove the PCV valve (see illustration 30.33).
39   Remove the front cylinder head (see Chapter 2B, Section 11). (You can't remove the crank-case vent oil separator without removing the front cylinder head.)
40   Remove the crankcase vent oil separator mounting bolts (see illustration 30.25b).
41   Remove the crankcase vent oil separator. Remove the old oil separator gasket.
42   Wash out the oil separator with a suit-

able solvent. Blow it out with compressed air (be sure to wear safety goggles during this procedure!) and make sure that the passages are clear.
43   Installation is the reverse of removal. Be sure to use a new gasket and tighten the crankcase vent oil separator bolts to the torque listed in this Chapter's Specifications.

31   Accelerator Pedal Position (APP) sensor - replacement

Note: *The APP sensor is an integral part of the accelerator pedal.*
1   Remove the driver's knee trim panel from below the steering column (see Chapter 11, Section 22).
2   Disconnect the electrical connector from the upper end of the APP sensor (see illustrations).
3   Remove the accelerator pedal/APP sensor mounting fasteners and remove the assembly.
4   Installation is the reverse of removal.

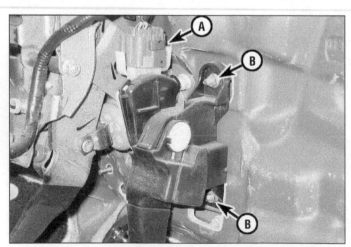

31.2a APP sensor electrical connector (A) and mounting fasteners (B, one not visible) - instrument panel removed for clarity (2009 through 2012 models shown, others similar)

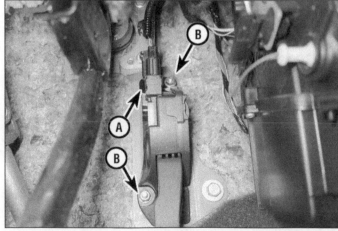

31.2b APP sensor electrical connector (A) and mounting fasteners (B) (2013 and later mode shown)

Chapter 7 Part A
Manual transaxle

Contents

Specifications

General

Transaxle lubricant type .... See Chapter 1
Transaxle lubricant capacity .... See Chapter 1

Torque specifications

Note: *One foot-pound (ft-lb) of torque is equivalent to 12 inch-pounds (in-lbs) of torque. Torque values below approximately 15 ft-lbs are expressed in inch-pounds, since most foot-pound torque wrenches are not accurate at these smaller values.*

	Ft-lbs (unless otherwise indicated)	Nm
Transaxle-to-engine mounting bolts	35	47
Crossmember-to-subframe bolts	85	115
Longitudinal crossmember-to-subframe bolts	66	90

1 General Information

1 The vehicles covered by this manual are equipped with a 5-speed manual transaxle or a 4-speed or 6-speed automatic transaxle. This Part of Chapter 7 contains information on the manual transaxle. Service procedures for the automatic transaxle are contained in Part B. Information on the transfer case used on 4WD models can be found in Part C.

2 The transaxle is contained in a cast-aluminum alloy casing bolted to the engine's left-hand end, and consists of the gearbox and final drive differential - often called a transaxle. The transaxle unit type is stamped on a plate attached to the transaxle.

Transaxle overhaul

3 Because of the complexity of the assembly, possible unavailability of replacement parts and special tools necessary, internal repair procedures for the transaxle are not recommended for the home mechanic. The bulk of the information in this Chapter is devoted to removal and installation procedures.

2 Shift lever - removal and installation

2004 and earlier models
2WD models

1 Apply the parking brake. Place the shift lever in Neutral. Unscrew and remove the shift lever knob.

2 Remove the center console trim panel and rubber shift lever boot.

3 Use a pair of snap-ring pliers to remove the steel snap-ring. Remove the plastic clip from under the steel snap-ring.

4 Remove the rubber O-ring.

5 Raise the vehicle and support it securely on jackstands.

6 Remove the nut and through-bolt securing the shift linkage rod to the shift lever.

7 Lift the gearshift lever from the housing.

8 Installation is the reverse of removal.

4WD models

9 Apply the parking brake. Place the shift lever in Neutral. Unscrew the shift lever knob.

10 Remove the center console trim panel and rubber shift lever boot.

11 Disconnect the shift cables from the shift lever by prying them off of the ballstuds.

12 Using a pair of pliers, remove the spring clips securing the cables to the shifter base.

13 Pull the shift cables forward to release them from the base, then position them out of the way.

14 Remove the four bolts securing the shifter base to the floor, then remove the shifter.

15 Installation is the reverse of removal.

2005 through 2007 models

16   Refer to Chapter 11, Section 21 and remove the center console.
17   Remove the shifter cable retaining clips, then detach the cables from the shift lever.
18   Remove the shift lever assembly mounting fasteners and remove the shifter.
19   Installation is the reverse of removal.

2008 through 2012 models

20   Carefully pull the shift lever boot from the center console. Pull it over the top of the shift knob.
21   Cut off the shift lever crimp ring, then remove the knob and boot assembly.
22   Refer to Chapter 11 and remove the top trim panel of the center console.
23   Pull out the two shift cable retaining clips.
24   Detach the shift cables from the shifter.
25   Remove the shift lever assembly mounting bolts and remove the shifter.
26   Installation is the reverse of removal. Install a new shift knob crimp ring.

3   Shift linkage (2004 and earlier 2WD models) - removal and installation

1   Raise the vehicle and support it securely on jackstands.
2   Remove the nut and through-bolt securing the shift linkage rod to the transaxle shift lever.
3   At the shift lever, remove the nut and through-bolt connecting the shift lever to the shift linkage rod.
4   Remove the shift linkage rod from the vehicle.
5   Installation is the reverse of removal.

4   Shift cables - removal and installation

1   Remove the shift lever (see Section 2).
2   Remove the center console (see Chapter 11, Section 21).
3   Trace the cables to the floor and unbolt the interior grommet.
4   Working inside the engine compartment, remove the shift cable retaining clips then remove the shift cables from the transaxle levers.
5   Disconnect the cable retainer from the retainer bracket by pulling on the black pin then lifting the cable from the bracket.
6   Raise the front of the vehicle and support it securely on jackstands.
7   Remove the nuts securing the shift cable retaining bracket.
8   Inside the engine compartment, carefully remove the shift cables one at a time from the vehicle.
9   Installation is the reverse of removal.

5   Driveaxle oil seals - replacement

1   Oil leaks frequently occur due to wear of the driveaxle oil seals. Replacement of these seals is relatively easy, since the repair can be performed without removing the transaxle from the vehicle.
2   Driveaxle oil seals are located at the sides of the transaxle, where the driveaxles are attached. If leakage at the seal is suspected, raise the vehicle and support it securely on jackstands. If the seal is leaking, lubricant will be found on the sides of the transaxle, below the seals.
3   Refer to Chapter 8, Section 8 and remove the driveaxles.
4   On AWD models, the oil seal is located in the transfer case (see Chapter 7C, Section 3).
5   Use a screwdriver or prybar to carefully pry the oil seal out of the transaxle bore.
6   If the oil seal cannot be removed with a screwdriver or prybar, a special oil seal removal tool (available at auto parts stores) will be required.
7   Using a large section of pipe or a large deep socket (slightly smaller than the outside diameter of the seal) as a drift, install the new oil seal. Drive it into the bore squarely and make sure it's completely seated. Coat the seal lip with transaxle lubricant.
8   Install the driveaxle(s). Be careful not to damage the lip of the new seal.

6   Manual transaxle - removal and installation

Note: *Manual transaxles are available on 2012 and earlier models.*

Removal

2004 and earlier models

1   Remove the battery and battery tray (see Chapter 5, Section 3).
2   Remove the air filter housing (see Chapter 4, Section 9).
3   Disconnect the electrical connector for the back-up light switch.
4   If you're working on a 4WD model, disconnect the shift cables from the transaxle shift levers (see Section 4).
5   Remove the transaxle front and rear wire harness brackets from the transaxle.
6   Disconnect the electrical connector for the vehicle speed sensor (see Chapter 6, Section 17).
7   Attach an engine support fixture to the lifting hook at the transaxle end of the engine. If no hook is provided, use a bolt of the proper size and thread pitch to attach the support fixture chain to a hole at the end of the cylinder head.
Note: *Engine support fixtures can be obtained at most equipment rental yards and some auto parts stores.*

8   Disconnect the clutch release cylinder and line from the transaxle (see Chapter 8, Section 4).
9   Remove the left side transaxle mount and bracket.
10   Remove the front transaxle mount and bracket.
11   Remove the rear transaxle mount through-bolt.
12   Remove the starter (see Chapter 5, Section 7).
13   Remove the four upper transaxle-to-engine mounting bolts.
14   Loosen the driveaxle/hub nuts and the wheel lug nuts, raise the front of the vehicle and support it securely on jackstands. Remove the wheels.
15   Remove the driveaxles (see Chapter 8, Section 8).
16   Drain the transaxle lubricant (see Chapter 1, Section 26).
17   Remove the three bolts securing the transaxle support insulator bracket, then remove the bracket.
18   If you're working on a 2WD model, disconnect the shift linkage bar (see Section 3) and support bar.
19   Remove the subframe crossmember.
20   Remove the longitudinal crossmember.
21   Remove the upper and lower left-side splash shields.
22   If you're working on a 4WD model, remove the transfer case (see Chapter 7C, Section 4).
23   Support the transaxle with a jack - preferably a jack made for this purpose (available at most tool rental yards). Safety chains will help steady the transaxle on the jack.
24   Remove the seven remaining bolts securing the transaxle to the engine.
25   Move the transaxle to the rear to disengage it from the engine block dowel pins. Then carefully remove the transaxle.

2005 through 2012 models

26   Raise the vehicle and support it securely on jackstands.
27   Refer to Chapter 5, Section 3 and remove the battery and the battery tray.
28   Remove the air filter housing and ducts (see Chapter 4, Section 9).
29   Remove the wiring harness bracket nut(s).
30   Disconnect all wiring from the transaxle.
31   Disconnect the shift cables from the transaxle, then remove the shift cable bracket.
32   Disconnect the clutch fluid line from the release cylinder. Plug the open ends.
33   Attach a lifting and support fixture to the engine. Adjust it to take the weight off of the engine and transaxle mounts.

2005 through 2007 models

34   Remove the rear transaxle mount.
35   Remove the three uppermost transaxle-to-engine bolts.
36   Remove the side-to-side crossmember and the front-to-rear crossmember.

37   Remove the wheel and the left splash shield.

38   Refer to Chapter 5, Section 7 and remove the starter.

39   On 4WD models, remove the transfer case (see Chapter 7C, Section 4).

40   Remove both driveaxles and the intermediate shaft (see Chapter 8, Section 8).

41   Remove the front transaxle mount.

42   Remove the interfering section(s) of the exhaust system.

43   Support the transaxle with a jack - preferably a jack made specifically for this purpose (available at most tool rental yards). Use chains or straps to secure the transaxle to the jack.

44   Remove the last six transaxle-to-engine mounting bolts.

45   Roll the transaxle away from the engine, then slowly lower it to the ground.

2008 through 2012 models

46   Remove the nuts from the left transaxle mount bracket. Loosen the through-bolt.

47   Remove the bolt and nuts from the rear transaxle mount.

48   Remove the bolt from the right engine mount.

49   Remove the three uppermost transaxle-to-engine bolts.

50   Remove the wheel and the left splash shield.

51   Remove the side-to-side crossmember.

52   Remove the through-bolt from the left transaxle mount.

53   Remove the front-to-rear crossmember and the left transaxle mount.

54   Remove the right transaxle mount

55   Refer to Chapter 5, Section 7 and remove the starter.

56   Detach the both stabilizer bar links (see Chapter 10, Section 4).

57   Remove both driveaxles and the intermediate shaft (see Chapter 8, Section 8).

58   Use the support fixture to raise the front of the engine about one inch. This will lower the transaxle.

59   Remove the two lower transaxle-to-engine bolts.

60   Support the transaxle with a jack - preferably a jack made specifically for this purpose (available at most tool rental yards). Use chains or straps to secure the transaxle to the jack.

61   Remove the last four transaxle-to-engine bolts.

62   Roll the transaxle away from the engine, then slowly lower it to the ground.

Installation

63   Lubricate the input shaft with a light coat of high-temperature grease. With the transaxle secured to the jack, raise it into position behind the engine and carefully slide it forward, engaging the input shaft with the clutch. Do not use excessive force to install the transaxle - if the input shaft won't slide into place, readjust the angle of the transaxle or turn the input shaft so the splines engage properly with the clutch.

64   Once the transaxle is flush with the engine, install the transaxle-to-engine bolts. Tighten the bolts to the torque listed in this Chapter's Specifications. **Caution:** *Don't use the bolts to force the transaxle and engine together.*

65   The remainder of installation is reverse of removal, noting the following points:

a) *Tighten the suspension crossmember mounting bolts to the torque values listed in this Chapter's Specifications.*

b) *Tighten the driveaxle/hub nuts to the torque value listed in the this Chapter's Specifications.*

c) *Tighten the starter mounting bolts to the torque value listed in the Chapter 5 Specifications.*

d) *If installing the transfer case, refer to Chapter 7C, Section 4.*

e) *Tighten the wheel lug nuts to the torque listed in the Chapter 1 Specifications.*

f) *Fill the transaxle with the correct type and amount of transaxle fluid as described in Chapter 1 Specifications.*

---

7   Manual transaxle overhaul - general information

1   Overhauling a manual transaxle is a difficult job for the do-it-yourselfer. It involves the disassembly and reassembly of many small parts. Numerous clearances must be precisely measured and, if necessary, changed with select-fit spacers and snap-rings. As a result, if transaxle problems arise, it can be removed and installed by a competent do-it-yourselfer,

but overhaul should be left to a transmission repair shop. Rebuilt transaxles may be available - check with your dealer parts department and auto parts stores. At any rate, the time and money involved in an overhaul is almost sure to exceed the cost of a rebuilt unit.

2   Nevertheless, it's not impossible for an inexperienced mechanic to rebuild a transaxle if the special tools are available and the job is done in a deliberate step-by-step manner so nothing is overlooked.

3   The tools necessary for an overhaul include internal and external snap-ring pliers, a bearing puller, a slide hammer, a set of pin punches, a dial indicator and possibly a hydraulic press. In addition, a large, sturdy workbench and a vise or transaxle stand will be required.

4   During disassembly of the transaxle, make careful notes of how each piece comes off, where it fits in relation to other pieces and what holds it in place.

5   Before taking the transaxle apart for repair, it will help if you have some idea what area of the transaxle is malfunctioning. Certain problems can be closely tied to specific areas in the transaxle, which can make component examination and replacement easier. Refer to the *Troubleshooting* Section at the front of this manual for information regarding possible sources of trouble.

---

8   Transaxle mount - replacement

1   Insert a large screwdriver or prybar between the mount and the transaxle and pry up.

2   The transaxle should not move excessively away from the mount. If it does, replace the mount.

3   If you're working on a four-cylinder model, remove the battery (see Chapter 5, Section 3).

4   Remove the air filter housing cover and air intake tube (see Chapter 4, Section 9).

5   Support the transaxle with a jack, remove the nuts and bolts and remove the mount. It may be necessary to raise the transaxle slightly to provide enough clearance to remove the mount.

6   Installation is the reverse of removal. **Note:** *Install all of the mount fasteners before tightening any of them.*

Notes

Chapter 7 Part B
Automatic transaxle

Contents

Specifications

General

Fluid type and capacity	See Chapter 1
Transaxle model	
2008 and earlier models	CD4E (four-speed)
2009 and later models	6F35 (six-speed)

Torque specifications

Note: *One foot-pound (ft-lb) of torque is equivalent to 12 inch-pounds (in-lbs) of torque. Torque values below approximately 15 ft-lbs are expressed in inch-pounds, since most foot-pound torque wrenches are not accurate at these smaller values.*

	Ft-lbs (unless otherwise indicated)	Nm
Crossmember mounting bolts	35	48
Longitudinal crossmember mounting bolts	96	130
Longitudinal crossmember dampener	30	40
Oil cooler bolt (2013 and later)	41	55
Torque converter-to-driveplate nuts		
2012 and earlier	30	40
2013 and later		
1.6L engine	35	48
2.0L engine		
Step 1	30	40
Step 2	Tighten an additional 90 degrees	
Transaxle-to-engine mounting bolts		
2012 and earlier	30	40
2013 and later	35	48
Transaxle mount-to-mount bracket bolt (2013 and later)	109	148
Transaxle mount bracket-to-transaxle bolts (2013 and later)	46	63

1   General information

1   All information on the automatic transaxle is included in this Chapter. Information for the manual transaxle can be found in Chapter 7A.

2   Because of the complexity of the automatic transaxles and the specialized equipment necessary to perform most service operations, this Chapter contains only those procedures related to general diagnosis, routine maintenance, adjustment and removal and installation.

3   If the transaxle requires major repair work, it should be left to a dealer service department or an automotive or transmission repair shop. Once properly diagnosed you can, however, remove and install the transaxle yourself and save the expense, even if the repair work is done by a transmission shop.

2   Diagnosis - general

1   Automatic transaxle malfunctions may be caused by five general conditions:

a) *Poor engine performance*
b) *Improper adjustments*
c) *Hydraulic malfunctions*
d) *Mechanical malfunctions*
e) *Malfunctions in the computer or its signal network*

2   Diagnosis of these problems should always begin with a check of the easily repaired items: fluid level and condition (see Chapter 1, Section 4), shift cable adjustment and shift lever installation. Next, perform a road test to determine if the problem has been corrected or if more diagnosis is necessary. If the problem persists after the preliminary tests and corrections are completed, additional diagnosis should be performed by a dealer service department or other qualified transmission repair shop. Refer to the *Troubleshooting* Section at the front of this manual for information on symptoms of transaxle problems.

Preliminary checks

3   Drive the vehicle to warm the transaxle to normal operating temperature.

4   Check the fluid level as described in Chapter 1, Section 4:

a) *If the fluid level is unusually low, add enough fluid to bring the level within the designated area of the dipstick, then check for external leaks (see following).*

b) *If the fluid level is abnormally high, drain off the excess, then check the drained fluid for contamination by coolant. The presence of engine coolant in the automatic transmission fluid indicates that a failure has occurred in the internal radiator oil cooler walls that separate the coolant from the transmission fluid.*

c) *If the fluid is foaming, drain it and refill the transaxle, then check for coolant in the fluid, or a high fluid level.*

2.28a Locate the shift interlock bypass . . .

2.28b . . . and rotate it counterclockwise using a screwdriver

5   Check the engine idle speed.

Note: *If the engine is malfunctioning, do not proceed with the preliminary checks until it has been repaired and runs normally.*

6   Check and adjust the shift cable, if necessary (see Section 5).

7   If hard shifting is experienced, inspect the shift cable under the steering column and at the manual lever on the transaxle (see Section 5).

Fluid leak diagnosis

8   Most fluid leaks are easy to locate visually. Repair usually consists of replacing a seal or gasket. If a leak is difficult to find, the following procedure may help.

9   Identify the fluid. Make sure it's transmission fluid and not engine oil or brake fluid (automatic transmission fluid is a deep red color).

10   Try to pinpoint the source of the leak. Drive the vehicle several miles, then park it over a large sheet of cardboard. After a minute or two, you should be able to locate the leak by determining the source of the fluid dripping onto the cardboard.

11   Make a careful visual inspection of the suspected component and the area immediately around it. Pay particular attention to gasket mating surfaces. A mirror is often helpful for finding leaks in areas that are hard to see.

12   If the leak still cannot be found, clean the suspected area thoroughly with a degreaser or solvent, then dry it thoroughly.

13   Drive the vehicle for several miles at normal operating temperature and varying speeds. After driving the vehicle, visually inspect the suspected component again.

14   Once the leak has been located, the cause must be determined before it can be properly repaired. If a gasket is replaced but the sealing flange is bent, the new gasket will not stop the leak. The bent flange must be straightened.

15   Before attempting to repair a leak, check to make sure that the following conditions are corrected or they may cause another leak.

Note: *Some of the following conditions cannot be fixed without highly specialized tools and expertise. Such problems must be referred to a qualified transmission shop or a dealer service department.*

Gasket leaks

16   Check the pan periodically. Make sure the bolts are tight, no bolts are missing, the gasket is in good condition and the pan is flat (dents in the pan may indicate damage to the valve body inside).

17   If the pan gasket is leaking, the fluid level or the fluid pressure may be too high, the vent may be plugged, the pan bolts may be too tight, the pan sealing flange may be warped, the sealing surface of the transaxle housing may be damaged, the gasket may be damaged or the transaxle casting may be cracked or porous. If sealant instead of gasket material has been used to form a seal between the pan and the transaxle housing, it may be the wrong type of sealant.

Seal leaks

18   If a transaxle seal is leaking, the fluid level or pressure may be too high, the vent may be plugged, the seal bore may be damaged, the seal itself may be damaged or improperly installed, the surface of the shaft protruding through the seal may be damaged or a loose bearing may be causing excessive shaft movement.

19   Make sure the dipstick tube seal is in good condition and the tube is properly seated. Periodically check the area around the sensors for leakage. If transmission fluid is evident, check the seals for damage.

Case leaks

20   If the case itself appears to be leaking, the casting is porous and will have to be repaired or replaced.

21   Make sure the oil cooler hose fittings are tight and in good condition.

Fluid comes out vent pipe or fill tube

22   If this condition occurs the possible causes are, the transaxle is overfilled, there is coolant in the fluid, the case is porous, the dipstick is incorrect, the vent is plugged or the drain-back holes are plugged.

3.2 Pry at the rear of the bezel to release the clips

3.3 Disconnect the shift bezel connector

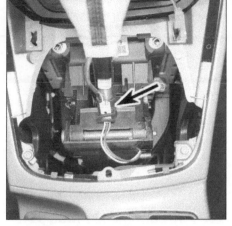

3.4 Disconnect the shift knob connector

Shift interlock bypass

Note: *If the vehicle battery is dead or the shift interlock system is malfunctioning, use this method to shift the transaxle out of park so the vehicle can be moved.*

2012 and earlier models

23   On column shift models, locate the shift interlock bypass cover at the top of the upper steering column cover, near the gauges, and remove it.

24   On console shift models, locate the shift interlock bypass cover at the top right corner of the shifter bezel and remove it.

25   On all models, using a small screwdriver or similar tool, push down on the shift interlock.

26   Press on the brake pedal and select the desired gear position using the shift knob.

2013 and later models

27   Remove the shift bezel (see Section 3).

28   Locate the shift interlock bypass and rotate the tab counterclockwise using a screwdriver or similar tool (see illustrations).

29   Press on the brake pedal and select the desired gear position.

---

3   Shift knob/boot/bezel - replacement

Note: *The following procedure applies to 2013 and later models.*

1   Pull up on the parking brake lever.

2   Using a flat-tipped screwdriver or trim tool, carefully pry upward at the rear of the shifter bezel to release the clips (see illustration).

3   Disconnect the electrical connector for the shift bezel (see illustration).

4   Disconnect the electrical connector at the bottom of the shift knob (see illustration).

5   Remove the two screws attaching the shift knob to the shift lever assembly and remove the shift knob, boot and bezel.

6   Depress the tab on the shift knob to remove the boot and bezel from the shift knob.

7   Installation is reverse of removal.

---

4   Shift lever - replacement

2004 and earlier models

Column-mounted shifter

Warning: *These models are equipped with a Supplemental Restraint System (SRS), more commonly known as airbags. Always disable the airbag system before working in the vicinity of any airbag system component to avoid the possibility of accidental deployment of the airbag(s), which could cause personal injury (see Chapter 12, Section 25).*

Warning: *Do not use a memory saving device to preserve the PCM or radio memory when working on or near airbag system components.*

1   Disconnect the cable from the negative battery terminal (see Chapter 5, Section 1). Wait at least two minutes before proceeding.

2   Remove the steering wheel/airbag module (see Chapter 10, Section 13).

3   Remove the steering column covers (see Chapter 11, Section 23).

4   Remove the clockspring (see Chapter 10, Section 13).

5   Remove the multi-function switch and ignition switch (see Chapter 12, Section 8).

6   Remove the shift cable from the steering column shift lever (see Section 5).

7   Disconnect the electrical connector and wiring harness (see illustration).

8   Remove the three shift lever mounting fasteners, then remove the lever.

9   Installation is the reverse of removal. After you've reconnected the battery, the Powertrain Control Module (PCM) must relearn its idle and fuel mixture trim strategy for optimum drivability and performance (see Chapter 5, Section 1 for this procedure).

Console-mounted shifter

10   Remove the shift lever trim ring by gently prying up on the four trim ring clips.

11   Remove the finish plate, then disconnect the electrical connector.

12   Disconnect the shift cable from the ball-stud, then remove the cable from the shifter assembly.

13   Disconnect any remaining electrical connectors from the shifter.

14   Remove the four bolts, then lift the assembly out of the console.

15   Installation is the reverse of removal.

2005 through 2007 models

16   Disconnect the cable from the negative battery terminal (see Chapter 5, Section 1).

17   Turn the key On, then put the shifter lever in Drive.

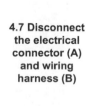

4.7 Disconnect the electrical connector (A) and wiring harness (B)

18   Carefully pry up the oval trim ring around the selector lever using a plastic trim tool or a screwdriver wrapped with tape.

19   Use the same tool to pry up the top trim panel from the console. Disconnect the wiring from the panel.

Note: *Pull up the parking brake lever if necessary.*

20   Detach the end of the cable from the shift lever, then disconnect the cable body from the bracket.

21   Disconnect the wiring from the shifter assembly.

22   Remove the mounting bolts and remove the shifter.

23   Installation is the reverse of removal.

2008 through 2012 models

24   Disconnect the cable from the negative battery terminal (see Chapter 5, Section 1).

25   Use a plastic trim tool or a screwdriver wrapped with tape to carefully pry up the rectangular trim from around the shifter.

26   Remove the interior tray from the center console.

27   Use the trim tool to pry up and remove the top panel from the front section of the console and shifter.

28   Detach the end of the cable from the shift lever, then disconnect the cable body from the bracket.

29   Disconnect the wiring harness and its retainers from the shifter assembly.

30   On 4WD models, pull the 4WD module rearward, then remove it from the rear of the shift lever.

31   Remove the two screws from the sides of the center console.

32   Remove the four shifter mounting bolts and remove the shifter.

33   Installation is the reverse of removal.

2013 and later models

34   Remove the center console (see Chapter 11, Section 21).

35   Pry the shift cable end from the shift lever (see illustration).

36   Remove the retainer securing the shift cable housing to the shift lever assembly and

4.35 Pry the shift cable from the shift lever pivot (2013 and later model shown)

remove the cable (see illustration).

37   Disconnect the shift lever electrical connector.

38   Remove the two nuts and two bolts and the shift lever assembly from the vehicle (see illustration).

39   Installation is reverse of removal. Adjust the shift cable (see Section 5).

---

5   Shift cable - replacement and adjustment

2004 and earlier models
Column-mounted shifter

Warning: *These models are equipped with a Supplemental Restraint System (SRS), more commonly known as airbags. Always disable the airbag system before working in the vicinity of any airbag system component to avoid the possibility of accidental deployment of the airbag(s), which could cause personal injury (see Chapter 12, Section 25).*

Warning: *Do not use a memory saving device to preserve the PCM or radio memory when working on or near airbag system components.*

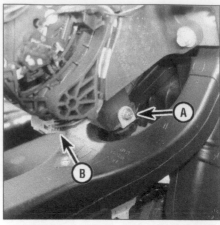

4.36 Remove the shift cable retainer (A) and disconnect the electrical connector (B)

Replacement

1   Disconnect the cable from the negative battery terminal (see Chapter 5, Section 1). Wait at least two minutes before proceeding.

2   Remove the steering column shrouds (see Chapter 11, Section 23).

3   Disconnect the shift cable from the shift lever and steering column bracket (see illustration).

4   Working inside the engine compartment, remove the nuts securing the cable to the firewall (see illustration).

5   Disconnect the shift cable from the shift lever and transaxle mounting bracket (see illustration), then remove the cable from the vehicle.

6   Installation is the reverse of removal. Before installing the steering column shrouds and connecting the battery, be sure to adjust the cable as specified.

Adjustment

7   With the steering column covers removed, place a 0.023 in (0.6mm) feeler gauge between the shift lever and shift lever detent (see illustration).

8   Put the shift lever in the Drive position, then have an assistant hold it firmly in position.

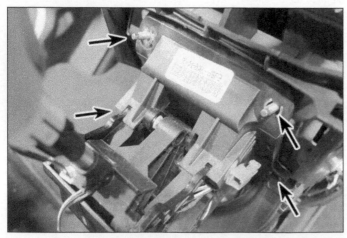

4.38 Remove the two nuts at the top, and two bolts at the bottom of the shifter

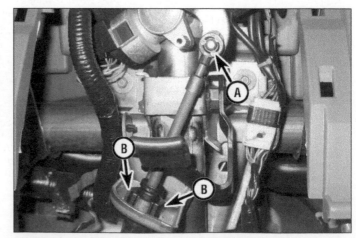

5.3 Remove the C-clip (A) and pull the cable end off the lever, then squeeze the tabs to remove the cable from the bracket (B)

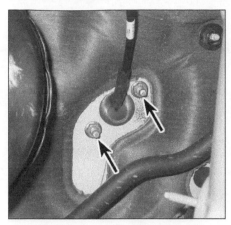

5.4 Inside the engine compartment, remove the two mounting nuts to release the cable from the firewall

5.5 Using a small screwdriver, carefully pry the shift cable from the lever ballstud (A), then squeeze the release tabs under the mounting bracket and slide the cable from the bracket (B)

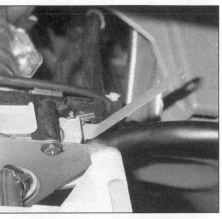

5.7 Insert a feeler gauge between the shift lever and the shift lever detent

9   Inside the engine compartment, remove the cable from the selector ballstud, then loosen the cable adjuster and adjust the cable (see illustration). Be sure the transaxle selector lever is in the Drive position, then connect the cable to the ballstud.

10   Tighten the cable adjuster then remove the feeler gauge and test the shift lever operation in each gear selection position.

11   Replace the steering column shrouds.

12   After you've reconnected the battery, the Powertrain Control Module (PCM) must relearn its idle and fuel mixture trim strategy for optimum drivability and performance (see Chapter 5, Section 1 for this procedure).

Console-mounted shifter

Replacement

13   Disconnect the cable from the negative battery terminal (see Chapter 5, Section 1).

14   Raise the vehicle and securely support it on jackstands

15   Remove the left splash shield.

16   Place the shifter in drive

17   Remove the snow shield from the transaxle.

18   Remove the adjustment bolts from the transaxle shift cable and disconnect the cable from the shifter lever adjustment screw.

19   Disconnect the cable retainers along the length of the cable.

20   Remove the floor console finish panel.

21   Remove the cable body pass through grommet nuts, disconnect the cable from the shifter ball stud and release the cable housing from the shifter.

22   Installation is the reverse of removal. Be certain to adjust the new cable after installation.

Adjustment

23   Disconnect the cable from the negative battery terminal (see Chapter 5, Section 1)

24   Raise the vehicle and securely support it on jackstands

25   Remove the left splash shield.

26   Place the shifter in drive.

27   Loosen the shifter cable adjustment screw and cable bracket nuts enough to allow easy movement of the cable.

28   Align the shifter lever between the two ribs on the transaxle.

29   With the transmission shifter still in Drive, tighten the cable bracket nuts, then tighten the shifter cable adjustment screw.

30   After installing any remaining components, apply the parking brake, lower the vehicle and operate the vehicle in each range to verify the adjustment is correct.

2005 through 2007 models

Replacement

31   Disconnect the cable from the negative battery terminal (see Chapter 5, Section 1).

32   Turn the key On, then shift into Drive.

33   Remove the center console (see Chapter 11).

34   Detach the shift cable from the shift lever and the bracket.

35   Remove the shift cable mounting nuts at the front of the console.

36   In the engine compartment, loosen the mounting bolt of the Z-shaped bracket, then rotate the bracket so you can reach the shift cable bracket.

37   Detach the shift cable from the shift lever of the transaxle. Disconnect the cable from the mounting bracket.

38   Slide the cable behind the air conditioning ducts, then pull it out from under the instrument panel.

39   Installation is the reverse of removal.

Adjustment

40   Remove the bolt and bracket from the shift cable at the transaxle.

41   Slide the adjuster up and depress the two locking tabs.

42   Rotate the adjustment sleeve counterclockwise to unlock.

43   Place the shift lever into D (drive) position.

44   Push the shift lever at the transaxle down until it stops, then back up one click.

45   Rotate the adjustment sleeve clockwise to lock in position.

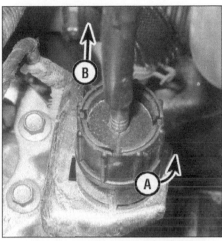

5.9 Turn the adjuster counterclockwise to loosen the adjuster (A); the cable can then be adjusted by moving it up or down (B)

46   Install the bracket and bolt to the shift cable at the transaxle.

2008 through 2012 models

Replacement

47   Disconnect the cable from the negative battery terminal (see Chapter 5, Section 1).

48   Turn the key On, then shift into Drive.

49   Use a plastic trim tool or a screwdriver wrapped with tape to carefully pry up the rectangular ring around the base of the shift lever.

50   Remove the inner tray from the rear of the center console.

51   Use the same tool to pry up and remove the top trim panel from the console.

52   Detach the shift cable from the shift lever and the bracket.

53   Remove the shift cable mounting nuts at the front of the console.

54   Remove the shift cable retainer from the transaxle fill tube.

55   Detach the shift cable from the shift lever of the transaxle.

56   Push in the two tabs on the shift cable, then slide it out of its bracket.

5.72a Remove the cable bolt (1) then disconnect the cable end from the lever (2) . . .

5.72b . . . then disconnect the cable from the bracket (3)

57   Installation is the reverse of removal. Adjust the shift cable.

Adjustment

58   Disconnect the shift cable from the shift lever at the transaxle.
59   Spread the tabs and release the cable adjustment lock.
60   Rotate the shift lever clockwise until it stops, then counter-clockwise on click.
61   Move the shift lever to the D (drive) position.
62   Connect the cable end to the shift lever.
63   Slide the lock downward to lock the shift cable adjustment in place.

2013 and later models
Replacement

64   Raise and support the the vehicle on jack stands.
65   Remove the center console (see Chapter 11, Section 21).
66   Pry the shift cable end from the shift lever (see illustration 4.35).
67   Remove the retainer securing the shift cable housing to the shift lever assembly and remove the cable (see illustration 4.36).
68   Remove the bolts and driver-side floor-to-center console bracket.
69   Remove the vehicle undercover.

70   On FWD models, remove the air filter housing assembly (see Chapter 4, Section 9).
71   On AWD models, remove the battery and battery tray (see Chapter 5, Section 3).
72   On all models, remove the bolt securing the shift cable to the transaxle bracket and disconnect the shift cable from the shift lever and bracket (see illustrations).
73   On AWD models, remove the bolts and secure the exhaust hanger out of the way.
74   On all models, remove the nuts securing the exhaust heat shield to the bottom of the vehicle, and move the heat shield out of the way to access the shift cable.
75   Detach the cable from the cable clip under the vehicle.
76   Working inside of the vehicle, remove the two nuts securing the shift cable grommet to the vehicle's floor (see illustration).
77   Remove the cable from the vehicle by pulling through the passenger compartment.
78   Installation is reverse of removal. Adjust the shift cable.

Adjustment

79   On FWD models, remove the air filter housing (see Chapter 4, Section 9).
80   On AWD models, remove the battery and battery tray (see Chapter 5, Section 3).
81   Disconnect the cable end from the shift

lever at the transaxle (see illustration 5.72a).
82   Unlock the cable adjuster on the transaxle end by sliding the spring-loaded tab forward, and lifting up on the lock (see illustration).
83   Rotate the shift lever clockwise until it stops, then counter-clockwise on click.
84   Move the shift lever to the D (drive) position.
85   Connect the cable end to the shift lever.
86   Slide the lock downward to lock the shift cable adjustment in place.

---

6   Overdrive switch - replacement

Note: *The overdrive switch is part of the shift lever and is replaced as one unit (see Section 4).*

---

7   Transaxle coolers - replacement

Main cooler

Note: *This procedure applies to 2013 and later models only.*
1   Raise and support the vehicle on jack stands.
2   Drain the cooling system (see Chapter 1, Section 22).

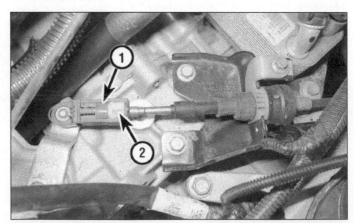

5.76 Remove the two nuts securing the shift cable to the floor (2013 and later model shown)

5.82 Unlock the cable adjuster on the transaxle end by sliding the spring-loaded tab forward (1), and lifting up on the lock (2)

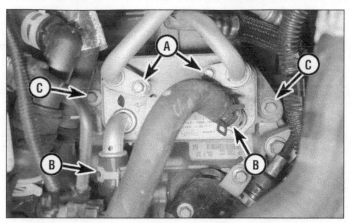

7.4 Disconnect the cooler tubes (A) the coolant hoses (B) and remove the bolts (C) to remove the cooler

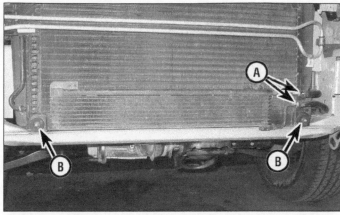

7.10 Using a pair of pliers, disconnect the clamps securing the hoses (A), then remove the two cooler mounting bolts (B) (2007 and earlier models shown)

3   Remove the battery and battery tray (see Chapter 5, Section 3).
4   Remove the bolts and disconnect the transaxle cooler tubes (see illustration). Inspect the O-rings and replace if necessary.
5   Disconnect the coolant hoses from the transaxle cooler.
6   Remove the two bolts attaching the cooler to the transaxle and remove the cooler from the vehicle.
7   Installation is reverse of removal, be sure to top off the coolant and transaxle fluid as described in Chapter 1.

Auxiliary cooler

Note: *This procedure applies only to 2007 and earlier and 2013 and later models with an auxiliary trans fluid cooler. The primary cooler on these vehicles is built into the radiator. On 2008 through 2012 models, the transaxle cooler is part of the air conditioning condenser.*
8   To gain access to the cooler, remove the front bumper cover (see Chapter 11, Section 9).
9   Disconnect the cooler line. On 2013 and later models, the connector covers must be removed first.
10   Remove the bolts securing the cooler (see illustration).
11   Remove the cooler from the vehicle. On 2013 and later models, slide the cooler upwards to disengage the bracket and remove.
12   Installation is the reverse of removal. Be sure to tighten all fasteners securely.

Cooler bypass valve

Note: *2013 and later models with an auxiliary cooler are also equipped with the cooler bypass valve that is temperature controlled.*
13   Raise and support the front of the vehicle on jackstands.
14   Remove the vehicle under cover.
15   Locate the cooler bypass valve under the driver's side of the vehicle. There are four transaxle lines connected to the cooler and the cooler is attached to the transaxle.
Note: *Place a drain pan under the valve to catch*

any fluid when the lines are disconnected.
16   Identify the lines for installation and disconnect the four transaxle cooler lines from the cooler bypass valve.
17   Remove the bolt attaching the bypass valve bracket to the transaxle and remove the valve and bracket.
18   Transfer the bracket from the old part to the new part if necessary.
19   Installation is reverse of removal. Top off the transaxle fluid as necessary (see Chapter 1, Section 25).

Fluid heater control valve

Note: *The trans fluid heater control valve is attached to the engine side of the battery tray/box. Place a drain pan under the valve to catch any fluid when the hoses are disconnected.*
20   Remove the air filter housing (see Chapter 4, Section 9) and battery tray/box (see Chapter 5, Section 3).
21   Loosen the clamps and disconnect the coolant hoses from the valve (see illustration).
Note: *The lower hose can be disconnected using*

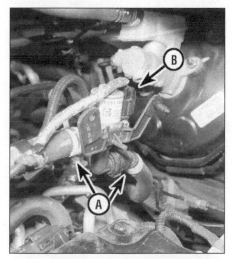

7.21 Disconnect the trans fluid heater control valve coolant hoses (A) then the electrical connector (B)

the quick-disconnect fitting or the hose clamp.
22   Disconnect the electrical connector.
23   Remove the screw and push-in retainer and remove the valve from the battery tray.
24   Installation is reverse of removal. Top off the coolant as necessary (see Chapter 1).

Fluid cooler control valve

Note: *The trans fluid cooler control valve is attached to the fan shroud. Place a drain pan under the valve to catch any fluid when the hoses are disconnected.*
25   Loosen the clamps and disconnect the coolant hoses from the valve, then disconnect the electrical connector (see illustration).
Note: *The lower hose has a quick-connect fitting.*
26   Remove the fan and shroud from the vehicle to access the valve bracket retainers (see Chapter 3, Section 5).
27   Remove the screws from the valve bracket to detach it from the fan shroud.
28   Installation is reverse of removal. Top off the coolant as necessary.

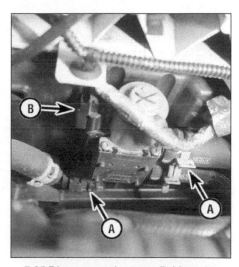

7.25 Disconnect the trans fluid cooler control valve coolant hoses (A) and the electrical connector (B)

8.5 Remove the nut securing the wire harness bracket to the transaxle, then pull the bracket and harness aside

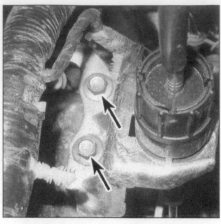

8.6 Disconnect the shift cable and bracket from the transaxle

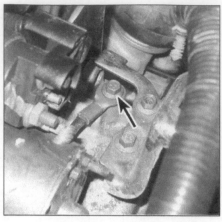

8.8 Remove the bolt securing the battery ground cable

8.10a Remove the upper . . .

8.10b . . . transaxle-to-engine mounting bolts

8 Automatic transaxle - removal and installation

Removal

2004 and earlier models

1   Disconnect the cable from the negative terminal of the battery (see Chapter 5, Section 1).

2   Remove the air filter housing (see Chapter 4, Section 9).

3   Disconnect the electrical connector for the transmission range sensor and the upstream oxygen sensor (see Chapter 6).

4   Disconnect the shift cable from the transaxle shift lever (see Section 5).

5   Remove the transaxle wire harness bracket from the transaxle (see illustration).

6   Remove the shift cable bracket from the transaxle (see illustration).

7   Remove the starter (see Chapter 5, Section 7).

8   Disconnect the battery ground cable at the transaxle (see illustration).

9   Attach an engine support fixture to the lifting hook at the transaxle end of the engine. If no hook is provided, use a bolt of the proper size and thread pitch to attach the support fixture chain to a hole at the end of the cylinder head.

Note: *Engine support fixtures can be obtained at most equipment rental yards and some auto parts stores.*

10   Remove the transaxle-to-engine bolts that are accessible from above (see illustrations).

11   Remove the upper transaxle mount (see illustration).

12   Remove the rear transaxle mount (see illustration).

8.11 Remove the transaxle mount bolt and plate

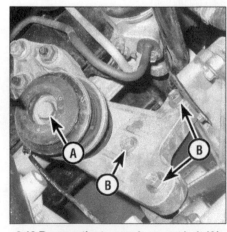

8.12 Remove the transaxle mount bolt (A) and the bracket bolts (B)

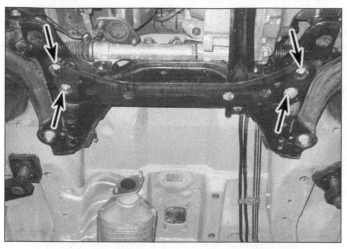

8.18 Remove the subframe crossmember bolts

8.22 Remove the inspection cover for access to the torque converter retaining nuts

13   Remove the upper engine mount bolt from the right side of the engine.
14   Loosen the driveaxle/hub nuts and the wheel lug nuts, raise the front of the vehicle and support it securely on jackstands. Remove the wheels.
15   Remove the under-vehicle splash shield(s).
16   Drain the transaxle lubricant (see Chapter 1, Section 25).
17   Remove the driveaxles (see Chapter 8, Section 8).
18   Remove the subframe crossmember (see illustration).
19   Remove the front section of the exhaust system (see Chapter 4, Section 16).
20   Remove the dampener from the longitudinal crossmember, then remove the crossmember.
21   On 4WD models, remove the transfer case (see Chapter 7C, Section 4).
22   Remove the torque converter inspection cover (see illustration).
23   Mark the relationship of the torque converter to the driveplate (see illustration).
24   Remove the four driveplate-to-torque

8.23 Mark the relationship of the torque converter to the driveplate

converter nuts (see illustration). Turn the crankshaft for access to each nut. Turn the crankshaft in a clockwise direction only (as viewed from the front).
25   Disconnect the fluid cooler hoses from

8.24 Remove the four driveplate-to-torque converter nuts

the transaxle (see illustrations). Be prepared for spillage, and plug the hoses and the lines on the transaxle.
26   Mark and disconnect any electrical connectors accessible from below.

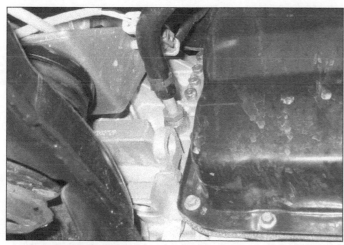

8.25a Disconnect the transaxle cooler lines . . .

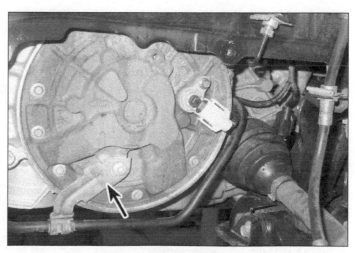

8.25b . . . and mounting bracket

8.27 Remove the transaxle vent tube from the valve body cover

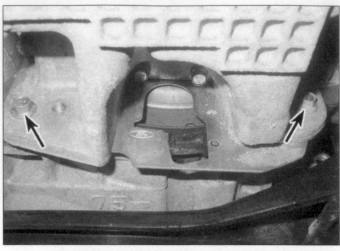

8.29 With the transaxle jack in place, remove the remaining transaxle-to-engine mounting bolts

27 Remove the transaxle vent tube (see illustration).

28 Support the transaxle with a jack - preferably a jack made for this purpose (available at most tool rental yards). Safety chains will help steady the transaxle on the jack.

29 Remove the remaining bolts securing the transaxle to the engine (see illustration).

30 Move the transaxle to the rear to disengage it from the engine block dowel pins and make sure the torque converter is detached from the driveplate. Lower the transaxle with the jack. Clamp a pair of locking pliers on the bellhousing case. The pliers will prevent the torque converter from falling out while you're removing the transaxle.

2005 through 2012 models

31 Remove the battery and the battery tray (see Chapter 5, Section 1). On V6 models, detach the wiring harness from the tray support bracket.

32 Remove the air filter housing (see Chapter 4, Section 9).

33 Unbolt and disconnect the wiring connector from the firewall.

34 Disconnect the shift cable from the transaxle, then disconnect the wiring from the transaxle.

35 Remove the top transaxle mounting bolts.

36 Attach an engine support fixture to the engine. Use the fixture to take the weight off of the transaxle mounts.

Note: *These fixtures can be rented at most rental yards and some auto parts stores.*

37 Raise the vehicle and support it securely on jackstands.

38 Remove the through-bolt from the left transaxle mount.

39 Remove the entire rear transaxle mount.

40 Remove the transverse frame brace. On V6 models, remove the alternator shield and disconnect the oxygen sensor wiring from the oil pan.

41 Remove the plastic splash shields.

42 Remove the through-bolt from the mount, then remove the two cross brace mounting bolts under it. Remove the cross brace bolt from the other side, then remove the cross brace.

43 Refer to Chapter 8 and remove both

driveaxles, the intermediate shaft and the intermediate shaft bracket.

44 On four-cylinder models, detach the exhaust pipe from the exhaust manifold. On V6 models, remove the front section of exhaust pipe, disconnecting it from both exhaust manifolds.

45 On 4WD models, refer to Chapter 7C and remove the transfer case. On 2WD models, remove the vibration damper from the transaxle.

Note: *It may be necessary to raise the engine using the fixture. When finished, lower it to remove the transaxle.*

46 Remove the front mount plate from the transaxle.

47 Disconnect the transaxle fluid cooler lines, then remove the OSS sensor.

48 Remove the starter (see Chapter 5, Section 7).

49 Remove the bellhousing cover, then remove the four torque converter-to-driveplate bolts. Rotate the crankshaft for access to each bolt.

50 Use the fixture to lower the transaxle slightly, then support the transaxle with a

8.57 Detach the harness from the bracket (2013 and later 1.6L model shown)

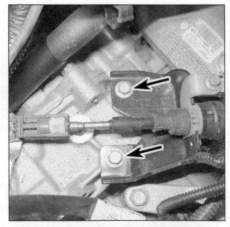

8.58 Remove the bolts attaching the shift cable bracket to the transaxle (2013 and later 1.6L model shown)

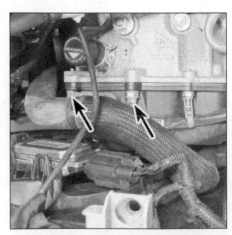

8.59 Remove the nuts attaching the coolant hose bracket to the transaxle (2013 and later 1.6L model shown)

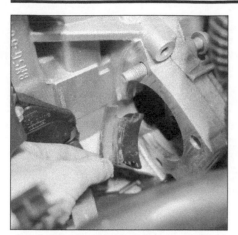

8.63 Remove the starter opening insulator (2013 and later 1.6L model shown)

8.64 Mark a torque converter stud to the driveplate for reference during installation (2013 and later 1.6L model shown)

8.67 Remove the roll restrictor from the transaxle (2013 and later 1.6L model shown)

jack - preferably a jack made for this purpose (available at most equipment rental yards). Safety chains or straps will help steady the transaxle on the jack.

51   Make sure that all components are disconnected from the transaxle.

52   Remove the remaining transaxle mounting bolts, pull the transaxle from the engine and slowly lower the transaxle.

Caution: *Don't let the torque converter fall out of the front of the transaxle. It's heavy when full of fluid.*

2013 and later models

53   Raise and support the vehicle on jackstands and place the transaxle in Neutral.

54   Remove the underbody cover (if equipped).

Note: *If the transaxle is to be replaced, drain the transaxle fluid (see Chapter 1, Section 25).*

55   Remove the battery and battery tray (see Chapter 5, Section 3).

56   On 1.6L models, remove the air filter outlet pipe (see Chapter 4, Section 9).

57   Remove the four nuts for the battery support bracket, detach the harness from the the bracket (see illustration), and remove the bracket from the vehicle (see Section 10).

Note: *These bolts also attach the transaxle mount to the vehicles body.*

58   Disconnect the shift cable from the shift lever and remove the bracket bolts (see illustration).

59   Remove the coolant hose bracket nuts from the transaxle and position the hoses and bracket to the side (see illustrations).

1.5L and 1.6L models

60   Remove the two upper transaxle-to-engine bolts.

61   On 1.6L models, remove the fluid fill cap and vent tube (if equipped).

62   On 1.6L models, place a drain pan under the engine oil filter housing and remove the oil filter. Then remove the oil filter bolt from the center of the cooler (that the oil filter threads onto) and secure the oil cooler out of the way. Inspect the oil cooler seal and replace if damaged.

63   On all models, remove the starter (see Chapter 5, Section 7). Remove the insulator from the starter opening (see illustration).

64   Mark the relationship of the torque converter to the driveplate using a paint pen (see illustration). Remove the torque converter nuts.

65   Disconnect the transaxle harness clips and connectors and secure out of the way. Disconnect any ground cable connections from the transaxle.

66   Remove the driveaxles (see Chapter 8, Section 8).

67   Remove the roll restrictor and bracket from under the transaxle (see illustration).

68   Remove the catalytic converter support bracket (see illustration).

69   On 1.6L FWD models, remove the charge air cooler pipe from below the engine (see illustration).

70   On all models, disconnect the passenger's (right) side tie rod end from the steering knuckle.

71   Disconnect the passenger's (right) side lower balljoint from the steering knuckle.

72   Disconnect the passenger's (right) brake hose bracket from the strut.

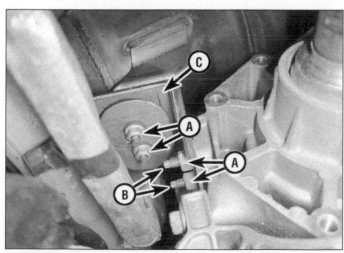

8.68 Remove the nuts (A) and studs (B) and the catalytic converter bracket (C) (2013 and later 1.6L model shown)

8.69 Remove the charge air cooler pipe (2013 and later 1.6L FWD model shown)

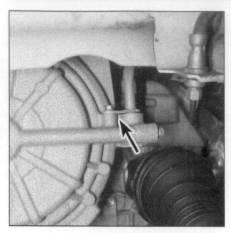

8.76a Remove the transmission cooler line bolts . . .

8.76b . . . and pull the lines from the transaxle (2013 and later 1.6L model shown)

73   On AWD models, remove the transfer case (see Chapter 7C, Section 4).

74   Attach an engine support fixture to the lifting hook at the transaxle end of the engine. If no hook is provided, use a bolt of the proper size and thread pitch to attach the support fixture chain to a hole at the end of the cylinder head.

75   Remove the transaxle mount-to-mount bracket bolt and mount bracket.

76   Disconnect the transaxle cooler lines (see illustrations). If the seals are stuck in the transaxle, remove them carefully. Check the seals for damage. Replace if damaged. Disconnect any cooler line brackets from the transaxle.

77   Remove the transaxle cooler bolts and secure the cooler out of the way (see Section 7).

78   Remove the left-side inner fender splash shield shield (see Chapter 11, Section 10).

79   Remove the four bolts and remove the support bar on the left (driver's) side between the subframe and radiator lower support (see illustration).

80   On 1.5L models with auto start/stop, remove the nuts attaching the heater coolant pump to the transaxle, disconnect the electri-

cal connector, and secure the pump out of the way.

81   On all models, support the transaxle with a jack - preferably a jack made for this purpose (available at most tool rental yards). Safety chains will help steady the transaxle on the jack.

82   Remove the six lower engine-to-transaxle bolts (see illustration).

83   Remove the single remaining transaxle-to-engine bolt on the front side of the engine/transaxle (see illustration).

84   Pull the transaxle from the engine and slowly lower the jack.

Caution: *Don't let the torque converter fall out of the front of the transaxle. It's heavy - especially when full of fluid.*

2.0L and 2.5L models

85   Detach the harness from the front of the transaxle.

86   Remove the fluid fill cap and vent tube (if equipped).

87   On 2017 and later 2.0L models, remove the charge air cooler pipe from under the vehicle.

88   Remove the roll restrictor and bracket from under the transaxle.

89   Attach an engine support fixture to the lifting hook at the transaxle end of the engine. If no hook is provided, use a bolt of the proper size and thread pitch to attach the support fixture chain to a hole at the end of the cylinder head.

90   Remove the transaxle mount-to-mount bracket bolt and mount bracket.

91   Disconnect the transaxle cooler lines. If the seals are stuck in the transaxle, remove them carefully. Check the seals for damage. Replace if damaged. Disconnect any trans cooler line brackets from the transaxle.

92   Remove the transaxle cooler bolts and secure the cooler out of the way.

93   Remove the three upper transaxle-to-engine bolts. One bolt secures the ground cable.

94   On 2016 and earlier 2.0L models, loosen the clamp and disconnect the charge air cooler pipe from the throttle body. Secure the pipe out of the way.

95   Place a drain pan under the oil filter housing and remove the oil filter. Then disconnect the electrical connectors and remove the oil filter housing bolts. Inspect the seal and replace if damaged.

96   Remove the starter (see Chapter 5, Section 7). Remove the insulator from the starter opening.

97   Mark the relationship of the torque converter to the driveplate using a paint pen. Remove the torque converter nuts.

98   Disconnect the transaxle harness clips and connectors and secure out of the way.

99   On 2.5L models with active warm up option, remove the bolt attaching the cooler bypass valve bracket to the transaxle.

100 On AWD models, disconnect the crankcase vent tube and remove the transfer case (see Chapter 7C, Section 4).

101 On all models, remove the left and right driveaxles (see Chapter 8, Section 8).

102 Disconnect the passenger's (right) side tie rod end from the steering knuckle (see Chapter 10).

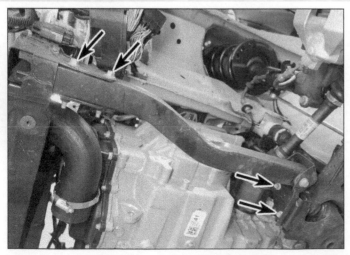

8.79 Remove the support bar (2013 and later 1.6L model shown)

8.82 Remove six lower engine-to-transaxle bolts (five are visible in this picture)

103 Disconnect the passenger's (right) side lower balljoint from the steering knuckle (see Chapter 10).

104 Disconnect the passenger's (right) brake hose bracket from the strut.

105 Remove the catalytic converter support bracket.

106 On 2.0L models, remove the driver's (left) inner fender splash shield (see Chapter 11).

107 On all models, remove the four bolts and remove the support bar on the driver's (left) side between the subframe and vehicle body.

108 On 2.5L models, remove the bolts and the exhaust manifold support bracket from the rear of the engine.

109 On all models, support the transaxle with a jack - preferably a jack made for this purpose (available at most tool rental yards). Safety chains will help steady the transaxle on the jack.

110 Remove the eight lower engine-to-transaxle bolts.

111 Pull the transaxle from the engine and slowly lower the jack.

Caution: *Don't let the torque converter fall out of the front of the transaxle. It's heavy - especially when full of fluid.*

Installation

112 Installation is the reverse of removal, noting the following points:

a) *As the torque converter is reinstalled, ensure that the drive tangs at the center of the torque converter hub engage with the recesses in the automatic transaxle fluid pump inner gear. This can be confirmed by turning the torque converter while pushing it towards the transaxle. If it isn't fully engaged, it will "clunk" into place.*

b) *Lightly coat the torque converter hub with grease.*

c) *Verify the dowel pins are installed in the engine block and not stuck in the transaxle.*

d) *When installing the transaxle, make sure the matchmarks you made on the torque converter and driveplate line up.*

e) *Install all of the driveplate-to-torque converter nuts before tightening any of them.*

f) *Tighten the driveplate-to-torque converter nuts to the torque listed in this Chapter's Specifications.*

g) *On 2013 and later models, use thread locker on the roll restrictor bolts.*

h) *Tighten the transaxle mounting bolts to the torque listed in this Chapter's Specifications.*

i) *Tighten the suspension crossmember mounting bolts to the torque values listed in this Chapter's Specifications.*

j) *Tighten the driveaxle/hub nuts to the torque value listed in the Chapter 8 Specifications.*

k) *Tighten the wheel lug nuts to the torque listed in the Chapter 1 Specifications.*

l) *Fill the transaxle with the correct type and amount of automatic transmission fluid as described in Chapter 1, Section 25.*

m) *On completion, adjust the shift cable as described in Section 5.*

n) *On 2012 and earlier models, after you've reconnected the battery, the Powertrain Control Module (PCM) must relearn its idle and fuel mixture trim strategy for optimum driveability and performance (see Chapter 5, Section 1 for this procedure).*

9 Automatic transaxle overhaul - general information

1 In the event of a problem occurring, it will be necessary to establish whether the fault is electrical, mechanical or hydraulic in nature, before repair work can be contemplated. Diagnosis requires detailed knowledge of the transaxle's operation and construction, as well as access to specialized test equipment, and so is deemed to be beyond the scope of this manual. It is therefore essential that problems with the automatic transaxle are referred to a dealer service department or other qualified repair facility for assessment.

2 Note that a faulty transaxle should not be removed before the vehicle has been diagnosed by a knowledgeable technician equipped with the proper tools, as troubleshooting must be performed with the transaxle installed in the vehicle.

10 Transaxle mounts - replacement

Left mount

2012 and earlier models

1 Remove the battery and battery tray (see Chapter 5, Section 3).

2 Remove the air filter housing (see Chapter 4, Section 9).

3 Place a floor jack under the transaxle and jack up enough to support the transaxle's weight. Jack up slightly to raise the transaxle mount.

Note: *Place a block of wood between the floor jack and the transaxle.*

4 Remove the mount through-bolt.

5 Remove the bolts/nuts attaching the mount to the vehicle's body and remove the mount.

6 Installation is reverse of removal.

2013 and later models

7 Remove the battery and battery tray (see Chapter 5, Section 3).

8 Remove the four nuts for the battery support bracket, detach the harness from the the bracket, and remove the bracket from the vehicle (see illustration). These nuts also attach the mount to the body.

8.83 Remove the remaining transaxle-to-engine bolt

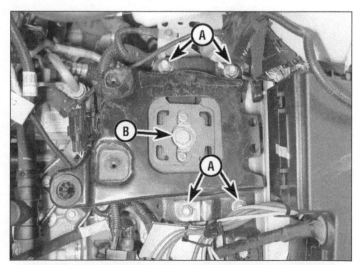

10.8 Remove the four nuts (A), support the transaxle and remove the bolt (B) (2013 and later model shown)

9   Place a floor jack under the transaxle and jack up enough to support the transaxle's weight. Jack up slightly to raise the transaxle mount.
Note: *Place a block of wood between the floor jack and the transaxle.*

10   Remove the bolt securing the transaxle mount to the mount bracket on the transaxle and remove the mount (see illustration 10.8).

11   Installation is reverse of removal. Tighten the fasteners to the torque listed in this Chapter's Specifications.

Front mount

Note: *Only 2012 and earlier models are equipped with a front transaxle mount. On 2004 and earlier models, the mount is attached to the transaxle with a bolt that goes up and through the mount. On 2008 through 2012 models, the mount attaches to a bracket on the transaxle with a through-bolt.*

12   Raise the front vehicle and support it securely on jackstands.

13   Remove the left splash shield.

14   Remove the four bolts and the rear cross brace.

15   On 2004 and earlier models, remove the bolt that goes up through the mount.

16   On 2005 through 2012 models, remove the thru-bolt that attaches the mount to the bracket on the transaxle.

17   On all models, remove the two front and one rear cross brace bolts and remove the cross brace with the mount attached.

18   On 2004 and earlier models, remove the mount from the cross brace.

19   On 2005 through 2012 models, unbolt the mount from the cross brace.

20   Installation is reverse of removal.

Rear mount

Note: *Only 2012 and earlier models are equipped with a rear transaxle mount.*

21   Remove the battery and battery tray (see Chapter 5, Section 3).

22   Remove the air filter housing (see Chapter 4, Section 9).

23   Remove the clamps for the power steering lines as necessary to access the rear mount bolts.

24   Place a floor jack under the transaxle and jack up enough to support the transaxle's weight. Jack up slightly to raise the transaxle mount.

25   Remove the through-bolt attaching the mount to the transaxle.

26   Remove the three bolts attaching the mount to the vehicle and remove the mount from the vehicle.

27   Installation is reverse of removal.

Chapter 7 Part C
Transfer case

Contents

Specifications

General

Transfer case fluid type .... See Chapter 1

Torque specifications   Ft-lbs (unless otherwise indicated)   Nm

Note: One foot-pound (ft-lb) of torque is equivalent to 12 inch-pounds (in-lbs) of torque. Torque values below approximately 15 ft-lbs are expressed in inch-pounds, since most foot-pound torque wrenches are not accurate at these smaller values.

	Ft-lbs (unless otherwise indicated)	Nm
Crossmember mounting bolts	30	40
Transfer case-to-transaxle bolts/nuts	33	45
Transfer case bracket-to-engine block mounting bolts	30	40
Transfer case output flange nut (2013 and later models)	173	235

1   General information

1   Due to the complexity of the transfer case covered in this manual and the need for specialized equipment to perform most service operations, this chapter contains only routine maintenance and removal and installation procedures.

2   4WD or AWD mode may be disabled if:

a) The system detects an overheat condition in the rear drive unit, the active torque coupling is disabled and switches to FWD mode.

b) The system detects an overheat condition in the transfer case, the active torque coupling is disabled and switches to FWD mode.

c) On 2013 and later models, if the spare tire or a tire of a different size than the rest is installed, the AWD system may disable automatically and switch to FWD mode.

3   4WD indicator light (2005 through 2007 models)

a) On steady - 4WD system is currently locked in 4WD mode due to heat protection mode. Stop the vehicle and let cool for a minimum of five minutes until the indicator turns off.

b) Blinks continuously - 4WD system is disabled due to heat protection mode. Stop the vehicle and let cool for a minimum of five minutes until the indicator turns off.

c) Blinks 3, 6, 8, or 10 times every minute - the 4WD system requires service (DTCs may be stored).

4   AWD messages (2013 and later models)

a) Check AWD - The AWD system has detected an error. This message may appear if the spare is on the vehicle.

b) AWD off - The AWD system has detected an error and has disabled itself to protect system components.

c) AWD temporarily disabled - The AWD system is disabled due to heat protection mode. Stop the vehicle and let cool for a minimum of 10 minutes.

d) AWD restored - The AWD system has cooled down from heat protection mode and AWD operation is available.

5   If the transfer case requires major repair work, it should be taken to a dealer service department or an automotive or transmission repair shop. You can, however, remove and install the transfer case yourself and save the expense of that labor, even if the repair work is done by a transmission shop.

AWD drive cycle (2013 and later models)

6   The AWD drive cycle must be performed after any of the following service procedures:

a) PCM is replaced.

b) Rear Drive Unit (RDU) is replaced. The RDU contains the Active Torque Coupling (ATC).

7   When the PCM and/or the RDU are replaced, the 4-digit ATC bar code must be entered into the PCM using a scan tool by a

qualified technician or repair facility. After entering the information into the PCM, perform the following AWD drive cycle procedure.

8 First, accelerate the vehicle from 0 to 30 MPH in a straight line. Repeat this three times at low, medium and full acceleration speeds. Ensure the front wheels do not spin.

9 Next, on a dry paved surface, drive the vehicle at 5 MPH with the wheels in fully locked position (left or right turn). Ensure there is no driveline bind.

2 Transfer case rear output shaft oil seal - removal and installation

1 Raise the vehicle and support it securely on jackstands.

2 Drain the transfer case lubricant (see Chapter 1, Section 28).

3 Remove the driveshaft (see Chapter 8, Section 9).

4 Mark the relative positions of the pinion, nut and flange.

5 On 2012 and earlier models, use a beam- or dial-type inch-pound torque wrench to determine the torque required to rotate the pinion. Record it for use later. Count the number of threads visible between the end of the nut and the end of the pinion shaft and record it for use later.

6 On all models, remove the flange mounting nut using a chain wrench to hold the pinion flange while loosening the locknut.

7 Remove the companion flange; a small puller may be required for removal.

8 Pry out the seal with a screwdriver or a seal removal tool. Don't damage the seal bore.

9 Lubricate the lips of the new seal with multi-purpose grease and tap it evenly into position with a seal installation tool or a large socket. Make sure it enters the housing squarely and is tapped into its full depth.

10 Align the mating marks made before disassembly and install the companion flange. If necessary, tighten the pinion nut to draw the flange into place.

11 On 2012 and earlier models, tighten the nut carefully until the original number of threads are exposed and the marks are aligned. Measure the torque required to rotate the pinion and tighten the nut in small increments until it matches the figure recorded earlier.

12 On 2013 and later models, tighten the flange nut to the torque listed in this Chapter's Specifications.

13 On all models, connect the driveshaft, add the specified lubricant to the transfer case (see Chapter 1, Section 28) and lower the vehicle.

3 Transfer case driveaxle oil seal (right side) - removal and installation

1 Remove the wheel cover or hub cap. Break the hub nut loose with a socket and large breaker bar.

2 Loosen the wheel lug nuts, raise the vehicle and support it securely on jackstands. Remove the wheel.

3 Remove the right driveaxle and intermediate shaft (see Chapter 8, Section 8).

4 Remove the exhaust crossover pipe (see Chapter 4, Section 16).

5 Remove the three bolts securing the heat shield then remove the heat shield.

6 Remove the dust shield.

7 Carefully pry out the oil seal with a seal removal tool or a large screwdriver; make sure you don't scratch the seal bore.

8 Using a seal installer or a large deep socket as a drift, install the new oil seal. Drive it into the bore squarely and make sure it's completely seated.

9 Lubricate the lip of the new seal with multi-purpose grease, then install a new dust shield.

10 Install the intermediate shaft and driveaxle (see Chapter 8, Section 8).

11 The remainder of installation is the reverse of removal. Check the transfer case lubricant level and add some, if necessary, to bring it to the appropriate level (see Chapter 1, Section 28).

4 Transfer case - removal and installation

2004 and earlier models

Four-cylinder models

1 Loosen the right front wheel lug nuts. Raise the front of the vehicle and support it securely on jackstands. Remove the wheel.

2 Unbolt the front portion of the driveshaft (see Chapter 8, Section 9) and suspend it from a piece of wire (don't let it hang from the center support bearing). Remove the right driveaxle and intermediate shaft (see Chapter 8, Section 8).

3 Remove the four bolts securing the crossmember and remove the crossmember.

4 Working on the right side of the transfer case, remove the upper two mounting nuts and two lower mounting bolts. Remove the transfer case from the vehicle.

5 Installation is the reverse of removal, noting the following points:
Install a new O-ring seal to the case.

a) *Tighten the driveshaft fasteners to the torque listed in the Chapter 8 Specifcations.*

b) *Tighten the transfer case and crossmember fasteners to the torque listed in this Chapter's Specifications.*

c) *Refill the transfer case with the proper type and amount of fluid (see Chapter 1, Section 28).*

d) *Tighten the wheel lug nuts to the torque listed in the Chapter 1 Specifications.*

V6 models

6 Remove the wheel cover or hub cap. Break the hub nut loose with a socket and large breaker bar. Loosen the wheel lug nuts.

7 Raise the front of the vehicle and support it securely on jackstands. Remove the wheel.

8 Unbolt the front portion of the driveshaft (see Chapter 8, Section 9), suspend it from a piece of wire (don't let it hang from the center support bearing).

9 Remove the right driveaxle and intermediate shaft (see Chapter 8, Section 8). Then remove the exhaust system crossover pipe and rear exhaust manifold (see Chapter 4, Section 16).

10 Remove the four bolts securing the crossmember and remove the crossmember.

11 Remove the alternator (see Chapter 5, Section 6).

12 Remove the six bolts securing the transfer case bracket to the engine block.

13 Working on the right side of the transfer case, remove the three transfer case mounting bolts.

14 Working on the left side, remove the final mounting bolt then remove the transfer case from the vehicle.

2005 through 2008 models

15 Raise the vehicle and support it securely on jackstands.

16 Drain the transfer case fluid (see Chapter 1, Section 28).

17 Remove the driveshaft (see Chapter 8, Section 9).

18 Remove the crossmember brace.

Manual transaxle models

19 Remove the two transaxle mounting bolts and two mounting nuts.

20 After removing the two nuts, tighten both nuts together on the front stud, then use them to remove the front stud.

Note: *You may have to clean the threads of the front stud using a die or a thread chaser.*

21 Pull the transfer case forward and off of the rear stud and remove it. Discard the large O-ring.

Automatic transaxle models

22 Refer to Chapter 8, Section 8 and remove the right driveaxle and the intermediate shaft.

23 On V6 models, refer to Chapter 2B and remove the interfering exhaust manifold.

24 Remove the transfer case heat shield.

25 Remove the transfer case mounting bracket. Detach the transfer case vent tube.

26 Remove the transfer case mounting bolts, then remove the transfer case.

2009 and later models

27 Raise the vehicle and support it securely on jackstands. Remove the under-vehicle splash shield.

28 Drain the transfer case fluid (see Chapter 1, Section 28).

29 Remove the driveshaft (see Chapter 8, Section 9).

30 On 2012 and earlier models, remove the crossmember brace.

31 Refer to Chapter 8, Section 8 and remove the right driveaxle and the intermediate shaft.

5.1 Pull the knob straight off from the mode select switch

32 On V6 models, refer to Chapter 2B and remove the interfering exhaust manifold.
33 On turbocharged models, remove the turbocharger outlet pipe (see Chapter 4).
34 Remove the transfer case heat shield.
35 On 2012 and earlier models, remove the two exhaust bracket nuts.
36 Remove the transfer case support bracket.
37 Detach the transfer case vent tube.
38 Remove the transfer case mounting bolts, then remove the transfer case.

All models

39 Installation is the reverse of removal, noting the following points:

a) *Install a new O-ring seal to the case.*
b) *Tighten the exhaust system fasteners to the torque listed in the Chapter 4 Specifications.*
c) *Tighten the driveshaft fasteners to the torque listed in Chapter 8 Specifications.*
d) *Tighten the transfer case and crossmember fasteners to the torque listed in this Chapter's Specifications.*
e) *Refill the transfer case with the proper type and of lubricant (see Chapter 1, Section 28).*
f) *Tighten the wheel lug nuts to the torque listed in the Chapter 1 Specifications.*

5 Mode select switch - replacement

Note: *2004 and earlier models have the ability to select the 4WD operation mode. On 2005 and later models, when a vehicle is equipped with 4WD or AWD, the vehicle is always in 4WD or AWD mode.*

1 Pull the switch knob straight off (see illustration).
2 Remove the instrument center trim panel (see Chapter 11).
3 Disconnect the switch electrical connector.
4 Remove the two mounting screws, then remove the switch.
5 Installation is the reverse of removal.

6 4X4 module (2012 and earlier models) - replacement

1 Remove the center console (see Chapter 11, Section 21).
2 On 2007 and earlier models, locate the 4X4 module attached to the center console bracket and disconnect the electrical connector.
3 Remove the two nuts and remove the module from the vehicle.
4 On 2008 through 2012 models, locate the 4X4 module attached to the shifter assembly base and disconnect the electrical connector.
5 Firmly pull on the module to disengage the three clips and remove the module from the vehicle.
6 On all models, installation is reverse of removal.

7 All-Wheel Drive (AWD) module (2013 and later models) - replacement

1 Remove the passenger's side trim panel between the rear quarter window and rear hatch opening (see Chapter 11).
2 Disconnect the AWD module electrical connector.
3 Remove the nuts and remove the AWD module from the vehicle.
4 Installation is reverse of removal.

Notes

Chapter 8
Clutch and driveaxles

Contents

Specifications

Clutch fluid type .... See Chapter 1

Torque specifications

Ft-lbs (unless otherwise indicated)   Nm

Note: *One foot-pound (ft-lb) of torque is equivalent to 12 inch-pounds (in-lbs) of torque. Torque values below approximately 15 ft-lbs are expressed in inch-pounds, since most foot-pound torque wrenches are not accurate at these smaller values.*

Clutch

	Ft-lbs	Nm
Clutch master cylinder mounting nuts	17	23
Clutch pressure plate-to-flywheel bolts	21	29
Clutch release cylinder mounting bolts		
2004 and earlier models	15	20
2005 through 2012 models	106 in-lbs	12

Driveaxles

	Ft-lbs	Nm
Driveaxle/hub nut*		
Front		
2005 and earlier models	214	290
2006 through 2012 models	221	300
2013 and later models		
Step 1	59	80
Step 2	Tighten an additional 90 degrees	
Rear		
2012 and earlier models	214	290
2013 and later models	98	133
Intermediate shaft bearing bracket nuts		
2012 and earlier models	20	27
2013 and later models		
Step 1 (lower nut)	44 in-lbs	5
Step 2 (upper nut)	18	25
Step 3 (lower nut)	18	25

Driveshaft

	Ft-lbs	Nm
Center bearing mounting fasteners		
2012 and earlier (nuts)	35	47
2013 and later (bolts)	18	25
Universal joint cap bolts	17	23
Driveshaft-to-differential pinion flange bolts		
2012 and earlier models	52	70
2013 and later models	26	35
Driveshaft-to-transfer case flange bolts		
2012 and earlier models	30	40
2013 and later models	26	35

*Use new fasteners

Torque specifications (continued)   Ft-lbs (unless otherwise indicated)   Nm

Note: *One foot-pound (ft-lb) of torque is equivalent to 12 inch-pounds (in-lbs) of torque. Torque values below approximately 15 ft-lbs are expressed in inch-pounds, since most foot-pound torque wrenches are not accurate at these smaller values.*

Rear differential and Active Torque Coupling (ATC)

	Ft-lbs	Nm
ATC housing-to-differential bolts		
2012 and earlier models	19	26
2013 and later models	30	40
ATC driveshaft flange nut (2012 and earlier models)	180	244
ATC driveshaft flange bolt (2013 and later models)	40	54
ATC stub shaft bolts	43	58
Differential mass damper bolts	40	54
Differential bracket-to-frame bolts (2004 and earlier models)	85	115
Differential-to-bracket bolts (2004 and earlier models)	59	80
Differential-to-front mount bolts (2005 and later models)	66	90
Front mount-to-frame bolts		
2005 through 2007 models	111	150
2008 and later models	66	90
Differential-to-rear bracket bolts (2005 and later models)	66	90

1   General information

1   The information in this Chapter deals with the components from the rear of the engine to the drive wheels, except for the transaxle, which is dealt with in the previous Chapter.
2   Since nearly all the procedures covered in this Chapter involve working under the vehicle, make sure it's securely supported on sturdy jackstands or on a hoist where the vehicle can be easily raised and lowered.

2   Clutch - description and check

1   All vehicles with a manual transaxle have a single dry plate, diaphragm spring-type clutch. The clutch disc has a splined hub which allows it to slide along the splines of the transaxle input shaft. The clutch and pressure plate are held in contact by spring pressure exerted by the diaphragm in the pressure plate.
2   The clutch release system is operated by hydraulic pressure. The hydraulic release system consists of the clutch pedal, a master cylinder and a shared common reservoir with the brake master cylinder, a release (or slave) cylinder and the hydraulic line connecting the two components.
3   When the clutch pedal is depressed, a pushrod pushes against brake fluid inside the master cylinder, applying hydraulic pressure to the release cylinder, which pushes the release bearing against the diaphragm fingers of the clutch pressure plate.
4   Terminology can be a problem when discussing the clutch components because common names are in some cases different from those used by the manufacturer. For example, the driven plate is also called the clutch plate or disc, the clutch release bearing is sometimes called a throwout bearing, the release cylinder is sometimes called the slave cylinder.
5   Unless you're replacing components with obvious damage, do these preliminary checks to diagnose clutch problems:

a) *The first check should be of the fluid level in the clutch master cylinder. If the fluid level is low, add fluid as necessary and inspect the hydraulic system for leaks. If the master cylinder reservoir is dry, bleed the system as described in Section 5 and recheck the clutch operation.*
b) *To check "clutch spin-down time," run the engine at normal idle speed with the transaxle in Neutral (clutch pedal up - engaged). Disengage the clutch (pedal down), wait several seconds and shift the transaxle into Reverse. No grinding noise should be heard. A grinding noise would most likely indicate a bad pressure plate or clutch disc.*
c) *To check for complete clutch release, run the engine (with the parking brake applied to prevent vehicle movement) and hold the clutch pedal approximately 1/2-inch from the floor. Shift the transaxle between 1st gear and Reverse several times. If the shift is rough, component failure is indicated.*
d) *Visually inspect the pivot bushing at the top of the clutch pedal to make sure there's no binding or excessive play.*

3   Clutch master cylinder - removal and installation

2004 and earlier models

Removal

1   Working under the dashboard, disconnect the clutch master cylinder pushrod from the pedal and unscrew the clutch master cylinder bracket nut.
2   Clamp a pair of locking pliers onto the clutch fluid feed hose, a couple of inches downstream of the brake fluid reservoir (the clutch master cylinder is supplied with fluid from the brake fluid reservoir). The pliers should be just tight enough to prevent fluid flow when the hose is disconnected. Disconnect the reservoir hose from the clutch master cylinder.
3   Using a flare-nut wrench, disconnect the hydraulic line fitting at the cylinder. Have rags handy, as some fluid will be lost as the line is removed. Cap or plug the ends of the line to prevent fluid leakage and the entry of contaminants.
4   Remove the remaining mounting nut and detach the cylinder from the firewall.
Caution: *Don't allow brake fluid to come into contact with the paint, as it will damage the finish.*

Installation

5   Place the master cylinder in position on the firewall and install the mounting nut finger-tight.
6   Connect the hydraulic line fitting to the clutch master cylinder and tighten it finger-tight (since the cylinder is still a bit loose, it'll be easier to start the threads into the cylinder).
7   Tighten the mounting nut to the torque listed in this Chapter's Specifications, then tighten the hydraulic line fitting securely.
8   Attach the fluid feed hose from the reservoir to the clutch master cylinder and tighten the hose clamp. Remove the locking pliers.
9   Working under the dash, install the remaining mounting nut and tighten it to the torque listed in this Chapter's Specifications. Connect the pushrod to the clutch pedal.
10   Fill the reservoir with the recommended brake fluid (see Chapter 1) and bleed the clutch system as outlined in Section 5.

2005 through 2012 models

11   Disconnect the clutch hose from the brake master cylinder, then plug the hose and the port on the clutch master cylinder.
12   Remove the clutch master cylinder mounting nuts.
13   Detach the pushrod from the master cylinder.
14   Place rags under the clutch master cylinder to absorb spillage.

15 Using a flare-nut wrench, disconnect the fluid line, then remove the clutch master cylinder. Plug the openings to prevent contamination.

16 Installation is the reverse of removal. Fill the reservoir with the recommended brake fluid (see Chapter 1), then bleed the system as outlined in Section 5.

4 Clutch release cylinder - removal and installation

2004 and earlier models

Removal

1 Disconnect the hydraulic line at the release cylinder using a flare-nut wrench. Have a small can and rags handy, as some fluid will be spilled as the line is removed. Plug the line to prevent excessive fluid loss and contamination.

2 Remove the release cylinder mounting bolts.

3 Remove the release cylinder.

Installation

4 Connect the hydraulic line fitting to the release cylinder, using your fingers only at this time (since the cylinder is still a bit loose, it'll be easier to start the threads into the cylinder).

5 Tighten the mounting bolts to the torque listed in this Chapter's Specifications.

6 Tighten the hydraulic fitting securely, using a flare-nut wrench.

7 Check the fluid level in the brake fluid reservoir, adding the recommended fluid (see Chapter 1) until the level is correct.

8 Bleed the system as described in Section 5, then recheck the brake fluid level.

2005 through 2012 models

9 Refer to Chapter 7A and remove the transaxle from the vehicle.

10 Detach the fluid line from the clutch release cylinder.

11 Remove the mounting bolts and the release cylinder.

12 Installation is the reverse of removal. Fill the reservoir with the recommended brake fluid (see Chapter 1), then bleed the system as outlined in Section 5.

5 Clutch hydraulic system - bleeding

1 Bleed the hydraulic system whenever any part of the system has been removed or the fluid level has fallen so low that air has been drawn into the master cylinder. The bleeding procedure is very similar to bleeding a brake system.

2 Fill the brake master cylinder reservoir with new brake fluid (see Chapter 1).

Caution: *Do not re-use any of the fluid coming from the system during the bleeding operation or use fluid which has been inside an open container for an ex-tended period of time.*

3 Have an assistant depress the clutch pedal and hold it. Open the bleeder valve on the release cylinder, allowing fluid and any air to escape. Close the bleeder valve when the flow of fluid (and bubbles) ceases. Once closed, have your assistant release the pedal.

4 Continue this process until all air is evacuated from the system, indicated by a solid stream of fluid being ejected from the bleeder valve each time with no air bubbles. Keep a close watch on the fluid level inside the brake master cylinder reservoir - if the level drops too far, air will get into the system and you'll have to start all over again.

Note: *Wash the area with water to remove any excess brake fluid.*

5 Check the brake fluid level again, and add some, if necessary, to bring it to the appropriate level. Check carefully for proper operation before placing the vehicle into normal service.

6 Clutch components - removal, inspection and installation

Warning: *Dust produced by clutch wear is hazardous to your health. Do not blow it out with compressed air and do not inhale it. Do not use gasoline or petroleum-based solvents to remove the dust. Brake system cleaner should be used to flush the dust into a drain pan. After the clutch components are wiped clean with a rag, dispose of the contaminated rags and cleaner in a covered, marked container.*

Removal

1 Access to the clutch components is normally accomplished by removing the transaxle, leaving the engine in the vehicle. If the engine is being removed for major overhaul, check the clutch for wear and replace worn components as necessary. However, the relatively low cost of the clutch components compared to the time and trouble spent gaining access to them warrants their replacement anytime the engine or transaxle is removed, unless they are new or in near-perfect condition. The following procedures are based on the assumption the engine will stay in place.

2 Remove the transaxle from the vehicle (see Chapter 7A). Support the engine while the transaxle is out. Preferably, an engine support fixture or a hoist should be used to support it from above.

3 The clutch fork and release bearing can remain attached to the transaxle housing for the time being.

4 To support the clutch disc during removal, install a clutch alignment tool through the clutch disc hub.

5 Carefully inspect the flywheel and pressure plate for indexing marks. The marks are usually an X, an O or a white letter. If they cannot be found, scribe or paint marks yourself so the pressure plate and the flywheel will be in the same alignment during installation (see illustration).

6 Turning each bolt a little at a time, loosen the pressure plate-to-flywheel bolts. Work

in a criss-cross pattern until all spring pressure is relieved. Then hold the pressure plate securely and completely remove the bolts, followed by the pressure plate and clutch disc.

Inspection

7 Ordinarily, when a problem occurs in the clutch, it can be attributed to wear of the clutch driven plate assembly (clutch disc). However, all components should be inspected at this time.

8 Inspect the flywheel for cracks, heat checking, grooves and other obvious defects. If the imperfections are slight, a machine shop can machine the surface flat and smooth, which is highly recommended regardless of the surface appearance. Refer to Chapter 2A for the flywheel removal and installation procedure.

9 Inspect the lining on the clutch disc. There should be at least 1/16-inch of lining above the rivet heads. Check for loose rivets, distortion, cracks, broken springs and other obvious damage. As mentioned above, ordinarily the clutch disc is routinely replaced, so if in doubt about the condition, replace it with a new one.

10 The release bearing should also be replaced along with the clutch disc (see Section 7).

11 Check the machined surfaces and the diaphragm spring fingers of the pressure plate. If the surface is grooved or otherwise damaged, replace the pressure plate. Also check for obvious damage, distortion, cracking, etc. Light glazing can be removed with emery cloth or sandpaper. If a new pressure plate is required, new and re-manufactured units are available.

12 Check the pilot bearing in the end of the crankshaft for excessive wear, scoring, dryness, roughness and any other obvious damage. If any of these conditions are noted, replace the bearing.

13 Removal can be accomplished with a slide hammer and puller attachment, which are available at most auto parts stores or tool rental yards. Refer to Chapter 2A for the flywheel removal procedure (it must be removed before the pilot bearing is removed).

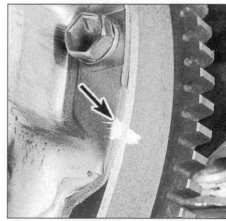

6.5 Mark the relationship of the pressure plate to the flywheel (if you're planning to re-use the old pressure plate)

Installation

14   To install a new pilot bearing, lightly lubricate the outside surface with grease, then drive it into the recess with a bearing driver or a socket. Install the flywheel (see Chapter 2A).

15   Before installation, clean the flywheel and pressure plate machined surfaces with brake cleaner. It's important that no oil or grease is on these surfaces or the lining of the clutch disc. Handle the parts only with clean hands.

16   Position the clutch disc and pressure plate against the flywheel with the clutch held in place with an alignment tool (see illustration). Make sure the disc is installed properly (most replacement clutch discs will be marked "flywheel side" or something similar - if not marked, install the clutch disc with the damper springs toward the transaxle).

17   Tighten the pressure plate-to-flywheel bolts only finger-tight, working around the pressure plate.

18   Center the clutch disc by ensuring the alignment tool extends through the splined hub and into the pilot bearing in the crankshaft. Wiggle the tool up, down or side-to-side as needed to center the disc. Tighten the pressure plate-to-flywheel bolts a little at a time, working in a criss-cross pattern to prevent distorting the cover. After all of the bolts are snug, tighten them to the torque listed in this Chapter's Specifications. Remove the alignment tool.

19   Using high-temperature grease, lubricate the inner groove of the release bearing (see Section 7). Also place a small amount of grease on the release lever contact areas and the transaxle input shaft bearing retainer.

20   Install the clutch release bearing (see Section 7).

21   Install the transaxle and all components removed previously.

---

7   Clutch release bearing and lever - removal, inspection and installation

Warning: *Dust produced by clutch wear is hazardous to your health. Do not blow it out with compressed air and do not inhale it. Do not use gasoline or petroleum-based solvents to remove the dust. Brake system cleaner should be used to flush the dust into a drain pan. After the clutch components are wiped clean with a rag, dispose of the contaminated rags and cleaner in a covered, marked container.*

Note: *This procedure applies only to early models that use a release lever and an externally mounted release cylinder.*

Removal

1   Remove the transaxle (see Chapter 7A).

2   Pull the clutch release fork off the ballstud and slide the release bearing off the input shaft along with the release fork.

Inspection

3   Wipe off the bearing with a clean rag and inspect it for damage, wear and cracks. Don't

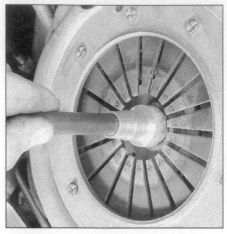

6.16 Center the clutch disc in the pressure plate with a clutch alignment tool

immerse the bearing in solvent - it's sealed for life and immersion in solvent will ruin it.

4   Hold the center of the bearing and rotate the outer portion while applying pressure (see illustration). If the bearing doesn't turn smoothly or if it's noisy or rough, replace it.

Note: *Considering the difficulty involved with replacing the release bearing, we recommend replacing the release bearing whenever the clutch components are replaced.*

Installation

5   Lightly lubricate the friction surfaces of the release bearing, ballstud and the input shaft bearing retainer with high-temperature grease.

6   Install the release lever and bearing onto the input shaft.

7   The remainder of installation is the reverse of removal.

---

8   Driveaxles - removal and installation

Front

Removal

1   Loosen the wheel lug nuts, raise the vehicle and support it securely on jackstands. Remove the wheel.

2   Remove the under-vehicle splash shield, if equipped.

3   Insert a punch into the brake disc cooling vanes and allow it to rest against the caliper mounting bracket, then remove the driveaxle/hub nut and discard it (see illustration).

4   If equipped with antilock brakes, remove the wheel speed sensor and set it aside.

5   Disconnect the brake hose from the strut assembly and position aside.

6   Separate the lower control arm from the steering knuckle (see Chapter 10).

7   Swing the knuckle/hub assembly out (away from the vehicle) until the end of the driveaxle is free of the hub. If the driveaxle splines stick in the hub, tap on the end of

7.4 To check the bearing, hold it by the outer race and rotate the inner race while applying pressure; if the bearing doesn't turn smoothly or if it is noisy, replace the bearing

the driveaxle with a brass drift and hammer (see illustration). Support the outer end of the driveaxle with a piece of wire to avoid unnecessary strain on the inner CV joint.

8   On 2012 and earlier models, if you're removing the right driveaxle, carefully pry the inner CV joint off the intermediate shaft using a large screwdriver or prybar positioned between the CV joint housing and the intermediate shaft bearing support.

9   On 2013 and later 2WD models, if you're removing the right driveaxle, remove the intermediate shaft bearing support strap (the intermediate shaft and driveaxle are one unit) (see illustration).

10   If you're removing the left driveaxle, pry the inner CV joint out of the transaxle using a large screwdriver or prybar positioned between the transaxle and the CV joint housing (see illustration). Be careful not to damage the differential seal.

11   Support the CV joints and carefully remove the driveaxle from the vehicle.

8.3 Remove the driveaxle/hub nut (2013 and later model shown)

8.7 Use a brass drift and hammer to drive the axle from the hub if necessary (2013 and later model shown)

8.9 Remove the nuts and intermediate shaft bearing support strap (2013 and later models)

8.10 Carefully pry the inner end of the driveaxle from the transaxle

Installation

12   Pry the old spring clip from the inner end of the driveaxle (left side) or outer end of the intermediate shaft (right side) and install a new one. Lubricate the differential or intermediate shaft seal with multi-purpose grease and raise the driveaxle into position while supporting the CV joints.

Note: *Position the spring clip with the opening facing down; this will ease insertion of the driveaxle and prevent damage to the clip.*

13   Push the splined end of the inner CV joint into the differential side gear (left side) or onto the intermediate shaft (right side) and make sure the spring clip locks in its groove.

14   Apply a light coat of multi-purpose grease to the outer CV joint splines, pull out on the steering knuckle assembly and install the stub axle into the hub.

15   On 2013 and later models, ensure the intermediate shaft support bearing strap is installed. Tighten the nuts to the torque listed in this Chapter's Specifications.

16   On all models, insert the balljoint stud into the steering knuckle and tighten the pinch bolt to the torque listed in the Chapter 10 Specifications.

Note: *On 2005 and later models, the manufacturer recommends replacing the pinch bolt and nut with new ones.*

17   Install a new driveaxle/hub nut. Tighten the driveaxle/hub nut to the torque listed in this Chapter's Specifications.

Caution: *The manufacturer recommends that the driveaxle/hub nut be tightened before the vehicle is lowered in order to keep the vehicle's weight from damaging the wheel bearing. This can be done by having an assistant use the brake to lock the hub while you tighten the nut, or by bracing a long prybar across two of the wheel studs.*

18   Grasp the inner CV joint housing (not the driveaxle) and pull out to make sure the driveaxle has seated securely in the transaxle or on the intermediate shaft.

19   Install the wheel and lug nuts, then lower the vehicle. Tighten the lug nuts to the torque listed in the Chapter 1 Specifications.

Intermediate shaft

Note: *On 2013 and later models, the intermediate shaft and driveaxle are one unit.*

Removal

20   Remove the right side wheel cover or hub cap. Break the hub nut loose with a socket and large breaker bar.

21   Loosen the wheel lug nuts, raise the vehicle and support it securely on jackstands. Remove the wheel.

22   Separate the lower control arm from the steering knuckle (see Chapter 10).

23   Remove the driveaxle/hub nut from the axle and discard it.

24   Swing the knuckle/hub assembly out (away from the vehicle) until the end of the driveaxle is free of the hub. Support the outer end of the driveaxle with a piece of wire to avoid unnecessary strain on the inner CV joint.

Note: *If the driveaxle splines stick in the hub, tap on the end of the driveaxle with a plastic hammer.*

25   Remove the driveaxle (see Step 6).

26   Remove the bearing support nuts (see illustration) and slide the intermediate shaft out of the transaxle. Be careful not to damage the differential seal when pulling the shaft out.

27   Check the support bearing for smooth operation by turning the shaft while holding the bearing. If you feel any roughness, take the intermediate shaft to an automotive machine shop or other qualified repair facility to have a new bearing installed.

Note: *On 4WD models, always replace the transfer case seal any time the intermediate shaft is removed from it.*

Installation

28   Lubricate the lips of the transaxle seal with multi-purpose grease. Carefully guide the intermediate shaft into the transaxle side gear

then install the mounting nuts for the bearing support. Tighten the nuts to the torque listed in this Chapter's Specifications.

29   The remainder of installation is the reverse of removal.

Rear (4WD models)

2004 and earlier models

Removal

30   Block the front wheels to prevent the vehicle from rolling. Loosen the wheel lug nuts, raise the rear of the vehicle and support it securely on jackstands. Remove the wheel.

31   Remove the driveaxle/hub nut and discard it.

32   Remove the coil spring (see Chapter 10).

33   Remove the nut and detach the trailing arm from the lower suspension arm balljoint (see Chapter 10). Pry the trailing arm outward and remove the outer end of the driveaxle from the hub.

Caution: *Don't let the driveaxle hang by the inner CV joint.*

8.26 Remove the nuts securing the intermediate shaft bearing support (2012 and earlier models)

8.34 Carefully pry the inner end of the driveaxle from the differential (2012 and earlier model shown)

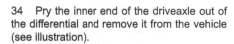

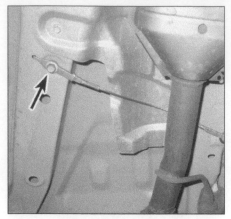

9.3 Remove the ground strap mounting bolt (2012 and earlier models)

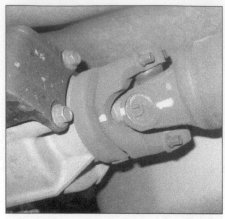

9.4 Mark the relationship of the driveshaft to the differential pinion and/or transfer case yoke (2012 and earlier models shown)

34   Pry the inner end of the driveaxle out of the differential and remove it from the vehicle (see illustration).

Installation
35   Pry the old spring clip from the inner end of the driveaxle and install a new one.
36   Apply a light film of grease to the area on the inner CV joint stub shaft where the seal rides, then insert the splined end of the inner CV joint into the differential. Make sure the spring clip locks in its groove.
37   Apply a light film of grease to the outer CV joint splines, pry the trailing arm outward and insert the outer end of the driveaxle into the hub.
38   Install the coil spring (see Chapter 10).
39   Connect the trailing arm to the lower suspension arm and install the nut. Using a floor jack, raise the trailing arm to simulate normal ride height, then tighten the nut to the torque listed in the Chapter 10 Specifications.
40   Install a new driveaxle/hub nut. Tighten the driveaxle/hub nut to the torque listed in this Chapter's Specifications.
Caution: *The manufacturer recommends that the driveaxle/hub nut be tightened before the vehicle is lowered in order to keep the vehicle's weight from damaging the wheel bearing. This can be done by having an assistant use the brake to lock the hub while you tighten the nut, or by bracing a long prybar across two of the wheel studs.*
41   Install the wheel and lug nuts, then lower the vehicle. Tighten the lug nuts to the torque listed in the Chapter 1 Specifications.

2005 through 2012 models
42   Block the front wheels to keep the vehicle from rolling. Loosen the rear wheel lug nuts, raise the vehicle and support it securely on jackstands, then remove the wheel. Remove the rear wheel driveaxle hub nut and discard it.
43   Refer to Chapter 10 and remove the coil spring.
44   Remove the wheel speed sensor wiring harness mounting bolt.
45   Remove the four bolts from the stabilizer

bar bracket. Discard the bolts (the manufacturer recommends using new bolts).
46   Place a floor jack under the knuckle and support it.
47   Remove the bolt from the lower balljoint, then separate the balljoint from the knuckle.
48   Separate the inner and outer ends of the driveaxle from the differential and the hub.
Caution: *Work carefully to avoid damaging the CV joints and boots.*
49   Installation is the reverse of removal. Tighten the driveaxle/hub nut to the torque listed in this Chapter's Specifications. Tighten all suspension fasteners to the torque values listed in the Chapter 10 Specifications.
Caution: *The manufacturer recommends that the driveaxle/hub nut be tightened before the vehicle is lowered in order to keep the vehicle's weight from damaging the wheel bearing. This can be done by having an assistant use the brake to lock the hub while you tighten the nut, or by bracing a long prybar across two of the wheel studs.*

2013 and later models
50   Block the front wheels to keep the vehicle from rolling. Loosen the rear wheel lug nuts,

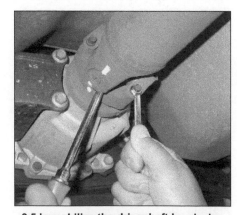

9.5 Immobilize the driveshaft by placing a screwdriver into the universal joint while loosening the bolts (2012 and earlier models shown)

raise the vehicle and support it securely on jackstands, then remove the wheel. Remove the rear wheel driveaxle hub nut and discard it.
51   Remove the rear hub and bearing assembly (see Chapter 10, Section 12).
52   Remove the bolt and remove the rear wheel speed sensor from the spindle.
53   Pry the inner end of the driveaxle out of the differential and remove it from the vehicle through the spindle.
54   Installation is the reverse of removal. Tighten the driveaxle/hub nut to the torque listed in this Chapter's Specifications. Tighten the suspension fasteners to the torque listed in the Chapter 10 Specifications.

---

9   Driveshaft (4WD models) - removal and installation

Removal
Note: *The manufacturer recommends replacing driveshaft fasteners with new ones when installing the driveshaft.*
1   Raise the vehicle and support it securely on jackstands, place the selector lever in Neutral.

9.6 Remove the bolts securing the driveshaft to the transaxle/transfer case (2012 and earlier models shown)

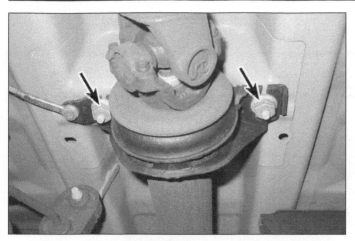

9.7 Remove the nuts securing the center support bearing (2012 and earlier models shown)

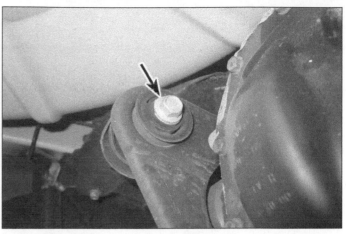

11.4 Differential mounting bracket-to-subframe mounting bolt

2   Remove the underbody covers (if equipped).
3   On 2012 and earlier models, remove the bolt securing the ground strap (see illustration).
4   On all models, use chalk or a scribe to mark the relationship of the driveshaft to the differential and/or transfer case pinion yoke. This ensures correct alignment when the driveshaft is reinstalled (see illustration).
5   Remove the bolts securing the universal joint clamps or flange to the differential pinion yoke (see illustration).
6   Working at the front of the drive shaft, remove the bolts securing the driveshaft to the transaxle/transfer case (see illustration).
7   Remove the nuts/bolts securing the center support bearing (see illustration).
8   With the help from an assistant carefully remove the driveshaft from the vehicle.

Installation

9   Installation is the reverse of removal. Make sure the universal joint caps are properly placed in the flange seat. Tighten the fasteners to the torque listed in this Chapter's Specifications.

10 Rear driveaxle oil seals (4WD models)

1   Raise the rear of the vehicle and support it securely on jackstands. Place the transaxle in Neutral with the parking brake off. Block the front wheels to prevent the vehicle from rolling.
2   Remove the driveaxle(s) (see Section 8) or driveshaft (input shaft seal) (see Section 9).
3   If replacing the input shaft seal, remove the driveshaft flange nut or bolt, place an index mark on the flange and input shaft and remove the flange. The input shaft seal may be part of the active torque coupling.
4   Carefully pry out the oil seal with a seal removal tool or a large screwdriver. Be careful not to damage or scratch the seal bore.
5   Using a seal installer or a large deep socket as a drift, install the new oil seal. Drive it into the bore squarely and make sure it's completely seated.
6   Lubricate the lip of the new seal with multi-purpose grease, then install the drive-axle or driveshaft flange. Be careful not to damage the lip of the new seal.

7   Check the differential lubricant level and add some, if necessary, to bring it to the appropriate level (see Chapter 1).

11 Differential (4WD models) - removal and installation

2004 and earlier models

1   Raise the rear of the vehicle and support it securely on jackstands. Block the front wheels to prevent the vehicle from rolling.
2   Remove the driveshaft (see Section 9).
3   Remove the driveaxles (see Section 8).
4   Remove the bolt securing the differential bracket to the subframe (see illustration).
5   Remove the three bolts securing the differential mass damper (see illustration).
6   At the top of the differential, disconnect the electrical connector.
7   Support the rear of the differential with a floor jack. Remove the four bolts securing the differential to the mounting brackets (see illustration).

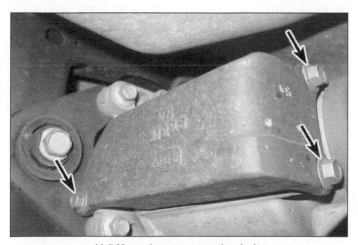

11.5 Mass damper mounting bolts

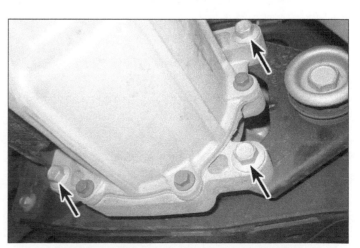

11.7 Remove the four bolts securing the differential to its mounting brackets (one bolt not visible in photo)

8   Loosen the bolts securing the brackets-to-subframe, rotate the brackets and carefully remove the differential.

9   Installation is the reverse of removal, noting the following points:

a) *Install all the mass damper bolts finger-tight first, then tighten them in the proper sequence (see illustration).*

b) *Tighten all fasteners to the torque listed in this Chapter's Specifications.*

2005 through 2012 models

10   Raise the vehicle and support it securely on jackstands.

11   Remove the spare tire.

12   Refer to Section 8 and remove the rear driveaxles.

13   Remove the driveshaft (see Section 9).

14   Place a floor jack under the differential. Tie the differential securely to the jack using tie-downs or similar straps.

15   Disconnect the wiring from the differential.

16   Remove the differential front mounting brackets.

17   Have an assistant stabilize the differential while you remove the three rear mounting bolts. Carefully lower the differential and remove it

18   Installation is the reverse of removal. Tighten the fasteners to the torque values listed in this Chapter's Specifications and the Chapter 10 Specifications.

2013 and later models

19   Raise the vehicle and support it securely on jackstands.

20   Remove the driveaxles (see Section 8).

21   Remove the rear subframe with the differential attached (see Chapter 10).

22   Remove the four mounting bolts at the front of the differential assembly.

23   Remove the two mounting bolts at the rear of the differential assembly.

24   Remove the three bolts for each left and right mounting bracket.

25   Remove the differential from the subframe.

26   Installation is reverse of removal. Tighten all fasteners to the torque listed in this Chapter's Specifications and the Chapter 10 Specifications.

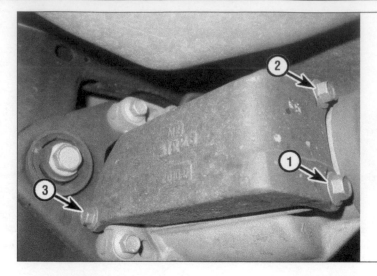

11.9 In the sequence shown, torque the bolts to the specification listed in this Chapter

12   Active Torque Coupling (ATC) - replacement

Note: *The ATC is attached to the front of the rear differential. The following procedure describes basic removal and installation.*

1   Raise and support the vehicle in jackstands.

2   Note the 4-digit ATC number on the barcode. The 4-digit number is also etched on the ATC itself, inside of the housing.

3   Remove the driveshaft (see Section 9).

4   Disconnect the ATC electrical connector.

5   Remove the four Torx bolts attaching the ATC to the rear differential.

Caution: *The inside of the differential between the ATC housing and differential should be dry. If it is not, this can be signs of an overheated ATC. Ensure the housing is clean and dry before installation.*

6   Remove the ATC from the differential. Use a prybar as necessary. On 2012 and earlier models, use care not to damage the ATC coil harness during removal of the housing.

7   If the ATC or the seal for the input flange is to be replaced, hold the flange and remove the bolt or nut securing the flange to the input shaft. Before removing the flange from the shaft, place an index mark on the flange and shaft for installation.

8   If replacing the ATC, remove the four Torx screws attaching the ATC stub shaft to the ATC and remove the stub shaft.

9   If a new ATC is installed in the housing, record the 4-digit code before installing into the housing. If the code no longer matches the housing label, cross out the old code and write the new code on the label.

10   On 2012 and earlier models, the ATC coil is attached to the rear of the ATC. To remove the coil and yoke, remove the snap ring and use a special puller to remove the coil and yoke from the ATC.

11   On 2013 and later models, to remove the ATC coil from the differential housing, remove the snap ring, carefully pull the coil and connector from the housing, then remove the three Torx screws and remove the coil yoke.

Note: *The coil and coil yoke must be replaced together.*

12   Installation is reverse of removal.

13   Place a bead of sealant around the housing mating surface before installing onto the differential.

14   Tighten the fasteners to the torque listed in this Chapter's Specifications.

Chapter 9
Brakes

Contents

Specifications

General

Brake fluid type	See Chapter 1

Disc brakes

Brake pad minimum thickness	See Chapter 1
Disc lateral runout limit	0.004 inch
Disc minimum thickness	Cast into disc
Minimum pad lining thickness	See Chapter 1

Drum brakes

Maximum drum diameter	Cast into drum
Shoe lining minimum thickness	See Chapter 1

Torque specifications

	Ft-lbs (unless otherwise indicated)	Nm

Note: *One foot-pound (ft-lb) of torque is equivalent to 12 inch-pounds (in-lbs) of torque. Torque values below approximately 15 ft-lbs are expressed in inch-pounds, since most foot-pound torque wrenches are not accurate at these smaller values.*

	Ft-lbs (unless otherwise indicated)	Nm
Caliper mounting bolts		
Front		
2004 and earlier models	26	35
2005 through 2007 models		
Models with rear drum brakes	26	35
Models with rear disc brakes	33	45
2008 and 2009 models	37	50
2010 through 2012 models	18	25
2013 and later models	21	28
Rear	21	28
Caliper mounting bracket bolts		
2005 and earlier models	111	150
2006 and later models	129	175
Wheel cylinder mounting bolts		
2004 and earlier models	120 in-lbs	14
2005 and 2006 models	108 in-lbs	12
2007 through 2012 models	97 in-lbs	11
Backing plate mounting bolts (drum brake models)	63	85
Flexible brake hose mounting bracket bolts	133 in-lbs	15
Master cylinder mounting nuts		
2005 and earlier models	18	25
2006 through 2008 models	22	30
2009 and later models	18	25
Power brake booster mounting nuts	17	23
Parking brake lever mounting bolts	22	30
Parking brake cable bracket to body mounting bolts	97 in-lbs	11
Brake hose-to-caliper banjo bolt (rear, 2005 through 2008 models)	26	35
Vacuum pump mounting bolts	89 in-lbs	10
Wheel lug nuts	See Chapter 1	

2.2 ABS control unit

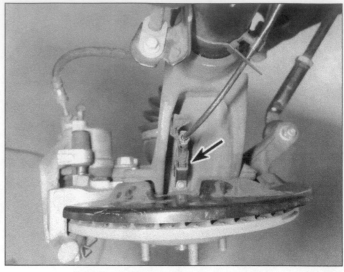

2.9a Front wheel speed sensor location

1   General Information

1   The vehicles covered by this manual are equipped with hydraulically operated front and rear brake systems. The front brakes are disc type and the rear brakes are disc or drum type. Both the front and rear brakes are self adjusting. The disc brakes automatically compensate for pad wear, while the drum brakes incorporate an adjustment mechanism that is activated as the parking brake is applied.

Hydraulic system

2   The hydraulic system consists of two separate circuits. The master cylinder has separate reservoir chambers for the two circuits, and, in the event of a leak or failure in one hydraulic circuit, the other circuit will remain operative. A dynamic proportioning valve, integral with the ABS hydraulic unit, provides brake balance to each individual wheel.

Power brake booster and vacuum pump

3   The power brake booster, utilizing engine manifold vacuum, and atmospheric pressure to provide assistance to the hydraulically operated brakes, is mounted on the firewall in the engine compartment. An auxiliary vacuum pump, mounted below the power brake booster in the left side of the engine compartment, provides additional vacuum to the booster under certain operating conditions.

Parking brake

4   The parking brake operates the rear brakes only. On 2016 and earlier models, the parking brake lever in the center console engages the rear brakes through cable actuation. On 2017 models, an Electric Parking Brake (EPB) system is used, which operates the rear brake calipers with an electric motor (actuator) mounted on each rear caliper

Service

5   After completing any operation involving disassembly of any part of the brake system, always test drive the vehicle to check for proper braking performance before resuming normal driving. When testing the brakes, perform the tests on a clean, dry, flat surface. Conditions other than these can lead to inaccurate test results.

6   Test the brakes at various speeds with both light and heavy pedal pressure. The vehicle should stop evenly without pulling to one side or the other. Avoid locking the brakes, because this slides the tires and diminishes braking efficiency and control of the vehicle.

7   Tires, vehicle load and wheel alignment are factors which also affect braking performance.

Precautions

8   There are some general cautions and warnings involving the brake system on this vehicle:
- a) *Use only brake fluid conforming to DOT 3 specifications.*
- b) *The brake pads and linings contain fibers that are hazardous to your health if inhaled. Whenever you work on brake system components, clean all parts with brake system cleaner. Do not allow the fine dust to become airborne. Also, wear an approved filtering mask.*
- c) *Safety should be paramount whenever any servicing of the brake components is performed. Do not use parts or fasteners that are not in perfect condition, and be sure that all clearances and torque specifications are adhered to. If you are at all unsure about a certain procedure, seek professional advice. Upon completion of any brake system work, test the brakes carefully in a con-*

trolled area before putting the vehicle into normal service. If a problem is suspected in the brake system, don't drive the vehicle until it's fixed.
- d) *Used brake fluid is considered a hazardous waste and it must be disposed of in accordance with federal, state and local laws. DO NOT pour it down the sink, into septic tanks or storm drains, or on the ground. Clean up any spilled brake fluid immediately and then wash the area with large amounts of water. This is especially true for any finished or painted surfaces.*

2   Anti-lock Brake System (ABS) - general information

General information

1   The anti-lock brake system is designed to maintain vehicle steerability, directional stability and optimum deceleration under severe braking conditions on most road surfaces. It does so by monitoring the rotational speed of each wheel and controlling the brake line pressure to each wheel during braking. This prevents the wheels from locking up.

2   The ABS system has three main components - the wheel speed sensors, the electronic control unit (ECU) and the hydraulic unit (see illustration). Four wheel speed sensors - one at each wheel - send a variable voltage signal to the control unit, which monitors these signals, compares them to its program and determines whether a wheel is about to lock up. When a wheel is about to lock up, the control unit signals the hydraulic unit to reduce hydraulic pressure (or not increase it further) at that wheel's brake caliper. Pressure modulation is handled by electrically-operated solenoid valves.

2.9b Rear wheel speed sensor location (2WD model shown)

3   If a problem develops within the system, an "ABS" warning light will glow on the dashboard. Sometimes, a visual inspection of the ABS system can help you locate the problem. Carefully inspect the ABS wiring harness. Pay particularly close attention to the harness and connections near each wheel. Look for signs of chafing and other damage caused by incorrectly routed wires. If a wheel sensor harness is damaged, the sensor must be replaced.

Warning: *Do not try to repair an ABS wiring harness. The ABS system is sensitive to even the smallest changes in resistance. Repairing the harness could alter resistance values and cause the system to malfunction. If the ABS wiring harness is damaged in any way, it must be replaced.*

Caution: *Make sure the ignition is turned off before unplugging or reattaching any electrical connections.*

Diagnosis and repair

4   If a dashboard warning light comes on and stays on while the vehicle is in operation, the ABS system requires attention. Although special electronic ABS diagnostic testing tools are necessary to properly diagnose the system, you can perform a few preliminary checks before taking the vehicle to a dealer service department.

   a) Check the brake fluid level in the reservoir.
   b) Verify that the computer electrical connectors are securely connected.
   c) Check the electrical connectors at the hydraulic control unit.
   d) Check the fuses.
   e) Follow the wiring harness to each wheel and verify that all connections are secure and that the wiring is undamaged.

5   If the above preliminary checks do not rectify the problem, the vehicle should be diagnosed by a dealer service department or other qualified repair shop. Due to the complex nature of this system, all actual repair work must be done by a qualified automotive technician.

Wheel speed sensor - removal and installation

6   Loosen the wheel lug nuts, raise the vehicle and support it securely on jackstands. Remove the wheel.
7   Make sure the ignition key is turned to the Off position.
8   Trace the wiring back from the sensor, detaching all brackets and clips while noting its correct routing, then disconnect the electrical connector.
9   Remove the mounting bolt and carefully pull the sensor out from the knuckle or rear hub and bearing assembly, as applicable (see illustrations).
10   Installation is the reverse of the removal procedure. Tighten the mounting bolt securely.
11   Install the wheel and lug nuts, tightening them securely. Lower the vehicle and tighten the lug nuts to the torque listed in the Chapter 1 Specifications.

3   Disc brake pads (front) - replacement

Warning: *Disc brake pads must be replaced on both front or both rear wheels at the same time - never replace the pads on only one wheel. Also, the dust created by the brake system is harmful to your health. Never blow it out with compressed air and don't inhale any of it. An approved filtering mask should be worn when working on the brakes. Do not, under any circumstances, use petroleum-based solvents to clean brake parts. Use brake system cleaner only!*

1   Remove the cap from the brake fluid reservoir.
2   Loosen the wheel lug nuts, raise the vehicle and support it securely on jackstands. Block the wheels at the opposite end.
3   Remove the wheels. Work on one brake assembly at a time, using the assembled brake for reference if necessary.
4   Inspect the brake disc carefully as outlined in Section 6. If machining is necessary, follow the information in that Section to remove the disc, at which time the pads can be removed as well.
5   Push the piston back into its bore to provide room for the new brake pads. A C-clamp can be used to accomplish this (see Illustration). As the piston is depressed to the bottom of the caliper bore, the fluid in the master cylinder will rise. Make sure that it doesn't overflow. If necessary, siphon off some of the fluid.

2001 through 2007 models/ 2013 and later models

6   Follow the accompanying photos (illustrations 3.6a through 3.6m), for the actual pad replacement procedure. Be sure to stay in order and read the caption under each illustration.

Note: *Remember to work on one caliper at a time so that the other caliper can be referred to if necessary.*

3.5 Before removing the caliper, be sure to depress the piston into the bottom of its bore in the caliper with a large C-clamp to make room for the new pads

3.6a Always wash the brakes with brake cleaner before disassembling anything

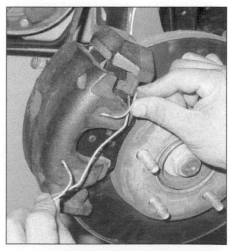

3.6b On 2004 and earlier models and 2013 and later models, remove the wire-type brake caliper anti-rattle clip

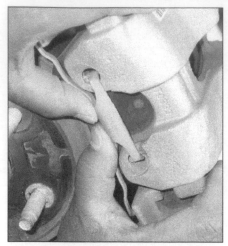

3.6c On 2005 through 2007 models, push the clip up to disengage the tangs from the holes in the caliper

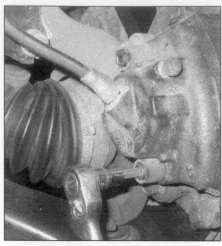

3.6d Pry the covers off, then remove the bolts

3.6e Remove the caliper and use a piece of wire to tie it to the coil spring

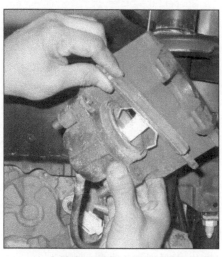

3.6f Unclip the inner pad from the piston . . .

3.6g . . . then remove the outer pad from the caliper mounting bracket

3.6h Pull out the mounting bolts and clean them off, then apply a coat of high-temperature grease to them

3.6i Apply anti-squeal compound to the back of the pads

3.6j Clip the inner pad into the piston

3.6k Install the outer pad

3.6l Install the caliper and tighten the caliper mounting bolts to the torque listed in this Chapter's Specifications, then install covers

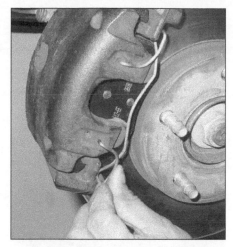

3.6m Install the anti-rattle clip, then depress the brake pedal a few times to bring the pads into contact with the disc

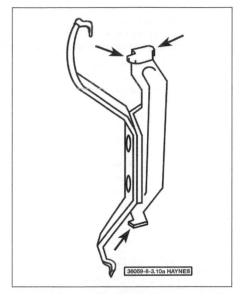

3.7a The top anchor housing spring tab has a protrusion on each side. The bottom tab has only one protrusion - left side shown

2008 through 2012 models

7   Follow the photo sequence for 2007 and earlier models/2013 and later models (see illustrations 3.6a through 3.6m), with the following differences:

a)   *On these models there is a long spring (anchor housing spring) attached on the outside of each caliper.*

Warning: *These anchor housing springs have different tabs on the top and bottom and must be reinstalled in the exact same position as they were originally (see illustration).*

b)   *To release the anchor housing spring on the left front (driver's side) caliper, push the center of the spring inward (toward the rear) while pulling the bottom tab out*

of the hole in the caliper (see illustration). Swivel and rotate the spring so that the top tab can be guided out of its hole in the caliper (see illustration).

c)   *To release the spring on the right front (passenger's side) caliper, push the center of the spring inward (toward the rear) while pulling the top tab out of the hole in the caliper. Swivel and rotate the spring so that the bottom tab can be guided out of its hole in the caliper.*

d)   *Swivel the brake caliper and remove the pads from the caliper mounting bracket.*

e)   *Apply a small amount of lubricant to the four contact points on the brake caliper mounting bracket.*

f)   *After installing the brake pads, install the anchor housing spring to its original position. It is installed in the exact reverse of removal. If the anchor housing spring is damaged, replace it with a new one.*

8   When reinstalling the caliper, be sure to tighten the mounting bolts to the torque listed (see this Chapter's Specifications). After the job has been completed, firmly depress the brake pedal a few times to bring the pads into contact with the disc. Check the level of the brake fluid, adding some if necessary. Check the operation of the brakes carefully before placing the vehicle into normal service.

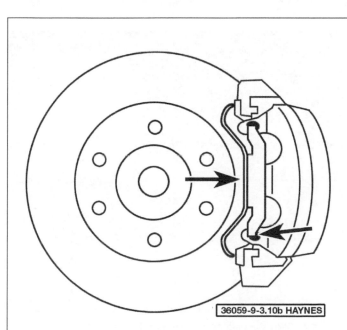

3.7b Push the center of the spring inward (to the rear) while simultaneously pulling the lower tab from the hole in the caliper (left side caliper)

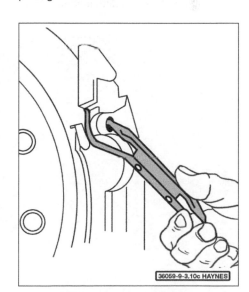

3.7c Swivel and rotate the spring to release the upper tab from the hole in the caliper

4   Disc brake pads (rear) - replacement

Warning: *Disc brake pads must be replaced on both front or both rear wheels at the same time - never replace the pads on only one wheel. Also, the dust created by the brake system is harmful to your health. Never blow it out with compressed air and don't inhale any of it. An approved filtering mask should be worn when working on the brakes. Do not, under any circumstances, use petroleum-based solvents to clean brake parts. Use brake system cleaner only!*

1   Loosen the wheel lug nuts, raise the vehicle and support it securely on jackstands. Block the wheels at the opposite end.

2   Remove the rear wheels.

3   On 2008 and earlier models, push the piston back into its bore to provide room for the new brake pads. A C-clamp can be used to accomplish this (see illustration 3.5). As the piston is depressed to the bottom of the caliper bore, the fluid in the master cylinder will rise. Make sure that it doesn't overflow. If necessary, siphon off some of the fluid.

Caution: *2013 and later rear calipers have screw-type parking brake actuators - do not attempt to use a C-clamp to depress the piston on these models.*

Note: *If you're working on a 2017 model, deactivate the Electric Parking Brake (EPB) before performing this procedure (see Section 15). Once the calipers have been reinstalled, reactivate the EPB system.*

4   Follow the accompanying photos (illustrations 4.4a through 4.4m), for the actual pad replacement procedure. Be sure to stay in order and read the caption under each illustration.

5   After the job has been completed, firmly depress the brake pedal a few times to bring the pads into contact with the disc. Check the level of the brake fluid, adding some if necessary. Check the operation of the brakes carefully before placing the vehicle into normal service.

4.4a Wash the brakes with brake cleaner before disassembling anything

4.4b Remove the caliper spring

4.4c Remove the dust caps . . .

4.4d . . . and unscrew the caliper mounting bolts

4.4e Slide the caliper off of the brake disc and support it with wire - don't let it hang by the hose

4.4f Remove the pads from the caliper mounting bracket. Note that the inner pad is the one with the anti-rattle spring

4.4g On 2013 and later models, use a caliper piston retracting tool to turn the caliper piston back into the caliper body. The cutouts on the piston face must be perpendicular to the plane of the mounting bolt holes so they will align with the tab on the inner brake pad

4.4h Install the pads into the mounting bracket

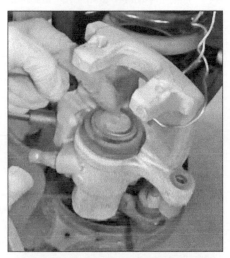

4.4i On 2013 and later models, be sure the cutouts on the caliper piston are positioned like this (to align with the tab on the inner brake pad)

4.4j Lower the caliper back into place

4.4k Apply a light film of of high-temperature brake grease to the sliding surface of the caliper mounting bolts

4.4l Tighten the caliper mounting bolts to the torque listed in this Chapter's Specifications, then reinstall the dust caps

4.4m Reinstall the caliper spring

5　Disc brake caliper - removal and installation

Warning: *Dust created by the brake system is harmful to your health. Never blow it out with compressed air and don't inhale any of it. An approved filtering mask should be worn when working on the brakes. Do not, under any circumstances, use petroleum-based solvents to clean brake parts. Use brake system cleaner only.*

Note: *If replacement is indicated (usually because of fluid leakage), it is recommended that the calipers be replaced, not overhauled. New and factory rebuilt units are available on an exchange basis, which makes this job quite easy. Always replace the calipers in pairs - never replace just one of them.*

Note: *If you're working on a 2017 model, deactivate the Electric Parking Brake (EPB) before performing this procedure (see Section 15). Once the calipers have been reinstalled, reactivate the EPB system.*

Removal

1　Loosen the front wheel lug nuts, raise the front of the vehicle and place it securely on jackstands. Remove the wheel.

2　All except 2005 through 2008 rear calipers and 2013 and later front calipers: Break loose the brake hose fitting (but don't try to unscrew it yet), then remove the caliper mounting bolts (see Section 3). Lift the caliper off its mounting bracket, then unscrew it from the brake hose. Plug the brake hose to keep contaminants out of the brake system and to prevent losing any more brake fluid than is necessary.

3　2005 through 2008 rear calipers: Unscrew the brake hose banjo bolt, detach the hose from the caliper and retrieve the sealing washers. Remove the caliper mounting bolts and lift the caliper from its mounting bracket.

4　2013 and later front calipers: Unscrew the brake hose fitting from the caliper. Use a flare-nut wrench, if available, to avoid rounding-off the fitting.

5　2013 and later rear calipers: Detach the parking brake cable from the caliper.
Note: *If the caliper is being removed for access to another component, don't disconnect the hose. Be sure to support it with a length of wire (see illustration 3.6g).*

6　Remove the inner brake pad from the caliper, where applicable.

Installation

7　Install the caliper by reversing the removal procedure. Make sure the brake hose isn't twisted. On 2005 through 2008 rear calipers, use new sealing washers on either side of the brake hose fitting, then tighten the banjo bolt to the torque listed in this Chapter's Specifications.

8　If the hose was disconnected, bleed the brake circuit according to the procedure in Section 11. Make sure there are no leaks from the hose connections.

9　Install the wheel, then lower the vehicle to the ground. Tighten the lug nuts to the torque listed in the Chapter 1 Specifications. Depress the brake pedal a few times to bring the brake pads into contact with the disc. Check the operation of the brakes carefully.

6　Brake disc - inspection, removal and installation

Inspection

1　Loosen the wheel lug nuts, raise the vehicle and support it securely on jackstands. Remove the wheel and install the lug nuts to hold the disc in place against the hub flange.
Note: *If the lug nuts don't contact the disc when screwed on all the way, install washers under them.*

2　Remove the brake caliper as outlined in Section 5. It isn't necessary to disconnect the brake hose. After removing the caliper bolts, suspend the caliper out of the way with a piece of wire. If you're working on the

6.2 The caliper mounting bracket is retained by two bolts

front brakes (or rear brakes on 2013 and later models), remove the two caliper mounting bracket-to-knuckle bolts (see illustration) and detach the mounting bracket.

3　Visually inspect the disc surface for score marks and other damage. Light scratches and shallow grooves are normal after use and may not always be detrimental to brake operation, but deep scoring requires disc removal and refinishing by an automotive machine shop. Be sure to check both sides of the disc (see illustration). If pulsating has been noticed during application of the brakes, suspect disc runout.

4　To check disc runout, place a dial indicator at a point about 1/2-inch from the outer edge of the disc (see illustration). Set the indicator to zero and turn the disc. The indicator reading should not exceed the specified allowable runout limit. If it does, the disc should be refinished by an automotive machine shop.
Note: *The discs should be resurfaced regardless of the dial indicator reading, as this will impart a smooth finish and ensure a perfectly flat surface, eliminating any brake pedal pulsation or other undesirable symptoms related to questionable discs. At the very least, if you elect not*

6.3 The brake pads on this vehicle were obviously neglected, as they wore down completely and cut deep grooves into the disc - wear this severe means the disc must be replaced

6.4a To check disc runout, mount a dial indicator as shown and rotate the disc

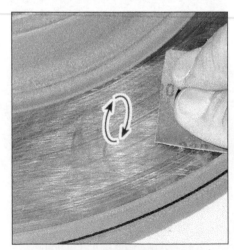

6.4b Using a swirling motion, remove the glaze from the disc surface with sandpaper or emery cloth

6.5a The minimum thickness dimension is cast into the back side of the disc

6.5b Use a micrometer to measure disc thickness

to have the discs resurfaced, remove the glaze from the surface with emery cloth or sandpaper, using a swirling motion (see illustration).

5   It's absolutely critical that the disc not be machined to a thickness under the specified minimum thickness. The minimum (or discard) thickness is cast or stamped into the inside of the disc (see illustration). The disc thickness can be checked with a micrometer (see illustration).

Removal

6   Remove the lug nuts which were installed to hold the disc in place and remove the disc from the hub.

Installation

7   Place the disc in position over the threaded studs.

8   Install the caliper mounting bracket and caliper, tightening the bolts to the torque values listed in this Chapter's Specifications.

9   Install the wheel, then lower the vehicle to the ground. Tighten the lug nuts to the torque listed (see Chapter 1 Specifications). Depress the brake pedal a few times

to bring the brake pads into contact with the disc. Bleeding won't be necessary unless the brake hose was disconnected from the caliper. Check the operation of the brakes carefully before driving the vehicle.

---

7   Drum brake shoes - replacement

Warning: *Drum brake shoes must be replaced on both wheels at the same time - never replace the shoes on only one wheel. Also, the dust created by the brake system is harmful to your health. Never blow it out with compressed air and don't inhale any of it. An approved filtering mask should be worn when working on the brakes. Do not, under any circumstances, use petroleum-based solvents to clean brake parts. Use brake system cleaner only!*

Caution: *Whenever the brake shoes are replaced, the return and hold-down springs should also be replaced. Due to the continuous heating/cooling cycle the springs are subjected to, they can lose tension over a period of time and may allow the shoes to drag on the drum and wear at a much faster rate than normal.*

1   Loosen the wheel lug nuts, raise the

rear of the vehicle and support it securely on jackstands. Block the front wheels to keep the vehicle from rolling.

2   Release the parking brake.

3   Remove the wheel.

Note: *All four rear brake shoes must be replaced at the same time, but to avoid mixing up parts, work on only one brake assembly at a time.*

4   Remove the brake drum. If the brake drum cannot be easily pulled off the axle and shoe assembly, make sure the parking brake is completely released. If the drum still cannot be pulled off, the brake shoes will have to be retracted. This is done by first removing the plug from the brake drum (see illustration). With the plug removed, pull the lever off the adjuster star wheel with a hooked tool while turning the adjuster wheel with another screwdriver, moving the shoes away from the drum. The drum should now come off.

5   Clean the brake shoe assembly with brake system cleaner before beginning work.

2007 and earlier models

6   Follow illustrations 7.6a through 7.6w for the shoe replacement procedure, then proceed to Step 8.

7.4 Remove this plug for access to the adjuster star wheel

7.6a Unhook the return spring from its hole in the front shoe

7.6b Unhook the front brake shoe hold-down clip

7.6c Remove the front shoe and disengage the anchor spring

7.6d Disengage the cable from the parking brake lever

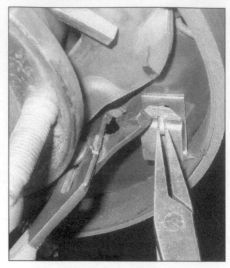

7.6e Unhook the rear brake shoe hold-down clip

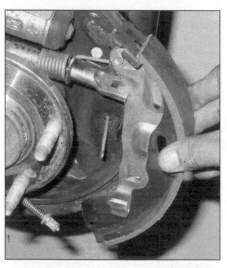

7.6f Remove the rear shoe assembly

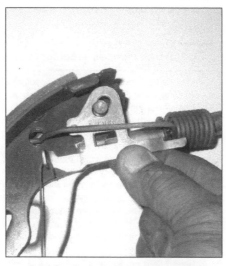

7.6g On the rear shoe, remove the self-adjust lever . . .

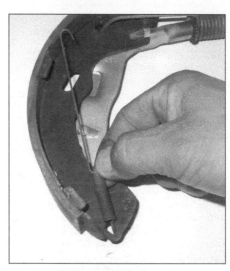

7.6h . . . and spring . . .

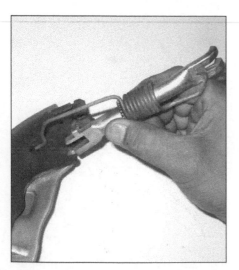

7.6i . . . then remove the adjuster screw and return spring

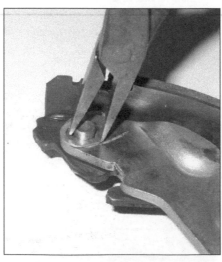

7.6j Pry open the C-clip and detach the parking brake lever from the rear shoe

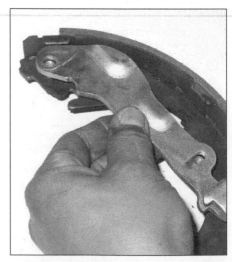

7.6k Place the parking brake lever on the new rear shoe . . .

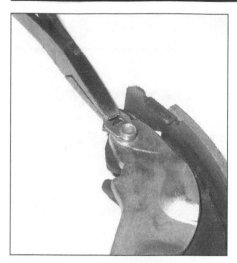

7.6l . . . and secure it in place with the C-clip

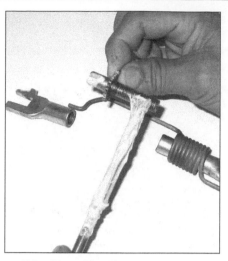

7.6m Clean the adjuster screw, then lubricate the threads and ends with high-temperature grease

7.6n Insert the adjuster screw into its slot on the rear shoe and attach the return spring

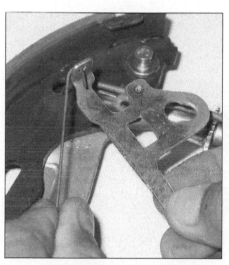

7.6o Attach the self-adjust lever and spring

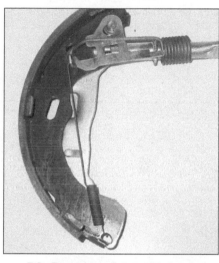

7.6p Rear view of assembled rear shoe/adjuster assembly

7.6q Apply high-temperature grease to the friction points of the backing plate

7.6r Attach the rear brake shoe hold-down clip

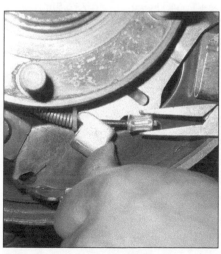

7.6s Attach the parking brake cable to the parking brake lever

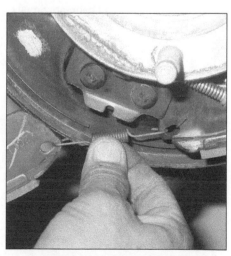

7.6t Attach the anchor spring to the bottom of each shoe

7.6u Install the adjuster screw, making sure it engages the slot in the front shoe

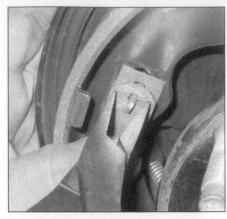

7.6v Secure the front shoe with the hold-down clip

7.6w Attach the upper return spring to the hole in the front shoe

2008 and later models

7   Follow illustrations 7.7a through 7.7s for the shoe replacement procedure, then proceed to Step 8.

8   Before reinstalling the drum, it should be checked for cracks, score marks, deep scratches and hard spots, which will appear as small discolored areas. If the hard spots cannot be removed with fine emery cloth or if any of the other conditions listed above exist, the drum must be taken to an automotive machine shop to have it resurfaced.

Note: *Professionals recommend resurfacing the drums each time a brake job is done. Resurfacing will eliminate the possibility of out-of-round drums. If the drums are worn so much that they can't be resurfaced without exceeding the maximum allowable diameter (stamped or cast into the drum), then new ones will be required. At the very least, if you elect not to have the drums resurfaced, remove the glaze from the surface with emery cloth using a swirling motion.*

7.7a If the drums haven't been previously removed, you'll have to cut off these retaining clips; they don't have to be reinstalled

7.7b Spray all components thoroughly with brake cleaner, then let the assembly air dry; work on one side at a time so the complete side can be used for reference

7.7c Remove this upper spring . . .

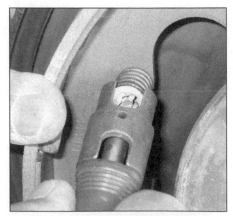

7.7d . . . then use this special tool to remove the shoe hold-down springs

7.7e Remove one shoe and the lower spring

7.7f Unhook the spring and remove
the adjuster

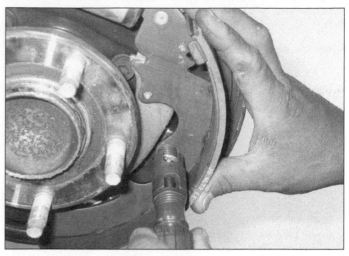

7.7g Remove the remaining shoe by releasing the hold-down
spring, then disconnect the parking brake cable
from the lever

7.7h Pull back the dust boots of the wheel
cylinder to check for leakage; if any fluid
is present, the wheel cylinder
should be replaced

7.7i Clean the backing plates, then apply
high-temperature brake grease to the
areas where the shoes slide

7.7j Transfer the adjuster lever to the new
leading shoe and secure it with a
new retaining clip

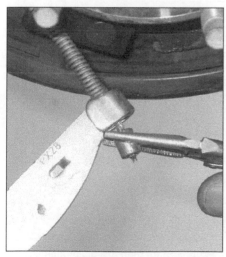

7.7k Engage the parking brake cable
with the lever . . .

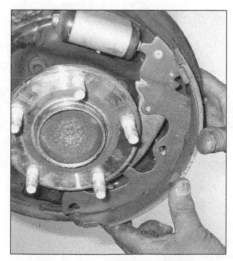

7.7l . . . then install the rear shoe

7.7m Install the hold-down spring and use
the special tool to install it over the pin

7.7n Lubricate the threads of the adjuster with high-temperature brake grease

7.7o Install the adjuster, connect the spring to the trailing shoe, then pull the other end of the spring over the forward end of the adjuster until it seats in the notch

7.7p Install the lower return spring and the front shoe . . .

7.7q . . . making sure the adjuster lever properly engages with the adjuster wheel and the shoe engages with the wheel cylinder and adjuster

7.7r Secure the front shoe with the hold-down spring

7.7s Install the upper return spring, then make a preliminary adjustment before reinstalling the brake drum

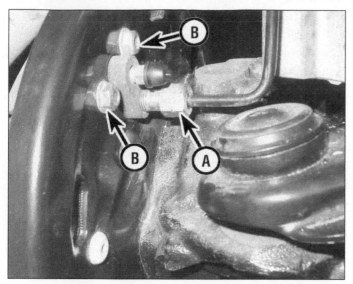

8.4 Disconnect the brake line (A) then remove the wheel cylinder mounting bolts (B)

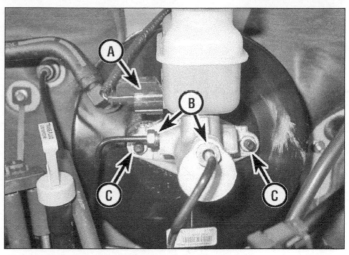

9.5a Master cylinder mounting details (2012 and earlier models)

A   Fluid level warning switch connector
B   Brake line fittings
C   Mounting nuts

9   Install the brake drum on the axle flange. Using a screwdriver inserted through the adjusting hole in the brake drum (see illustration 7.4), turn the adjuster star wheel until the brake shoes drag on the drum as the drum is rotated, then back off the star wheel until the shoes don't drag. Reinstall the plug in the drum.

10   Mount the wheel and install the lug nuts. Lower the vehicle and tighten the lug nuts to the torque listed in the Chapter 1 Specifications.

11   Make a number of forward and reverse stops and operate the parking brake to adjust the brakes until satisfactory pedal action is obtained.

12   Check the operation of the brakes carefully before driving the vehicle.

9.5b Master cylinder mounting details (2013 and later models)

A   Fluid level warning switch connector
B   Brake line fittings
C   Mounting nuts
D   Remote hose fitting

8   Wheel cylinder - removal and installation

Note: *If replacement is indicated (usually because of fluid leakage or sticky operation), it is recommended that the wheel cylinders be replaced, not overhauled. Always replace the wheel cylinders in pairs - never replace just one of them.*

Removal

1   Raise the rear of the vehicle and support it securely on jackstands. Block the front wheels to keep the vehicle from rolling.

2   Remove the brake shoe assembly (see Section 7).

3   Remove all dirt and foreign material from around the wheel cylinder.

4   Disconnect the brake line (see illustration). Don't pull the brake line away from the wheel cylinder.

5   Remove the wheel cylinder mounting bolts.

6   Detach the wheel cylinder from the brake backing plate and immediately plug the brake line to prevent fluid loss and contamination.

Installation

7   Place the wheel cylinder in position and install the bolts finger-tight. Connect the brake line to the cylinder, being careful not to cross thread the fitting. Tighten the wheel cylinder mounting bolts to the torque listed in this Chapter's Specifications. Now tighten the brake line fitting securely.

8   Install the brake shoe assembly (see Section 7).

9   Bleed the brakes (see Section 11).

10   Check the operation of the brakes carefully before driving the vehicle.

9   Master cylinder - removal and installation

Removal

1   The master cylinder is located in the engine compartment, mounted to the power brake booster. Remove as much fluid as you can from the reservoir with a syringe, such as

an old turkey baster.

Warning: *If a baster is used, never again use it for the preparation of food.*

2   Place rags under the fluid fittings and prepare caps or plastic bags to cover the ends of the lines once they are disconnected.

Caution: *Brake fluid will damage paint. Cover all body parts and be careful not to spill fluid during this procedure.*

3   If you're working on a 2013 or later model:

a) *Remove the battery and battery tray (see Chapter 5).*

b) *Disconnect the brake fluid remote reservoir hose from the plastic reservoir on the master cylinder (see illustration 9.5b).*

4   On manual transaxle models, disconnect the clutch master cylinder supply hose from the brake fluid reservoir.

5   Loosen the fittings at the ends of the brake lines where they enter the master cylinder (see illustrations). To prevent rounding off the corners on these nuts, the use of a flare-nut wrench, which wraps around the nut,

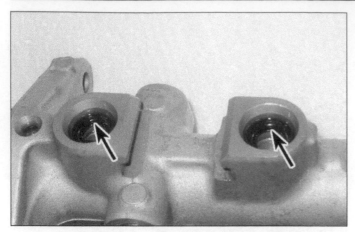

9.7 After the reservoir has been removed, replace the O-rings with new ones

9.9 The best way to bleed air from the master cylinder before installing it on the vehicle is with a pair of bleeder tubes that direct brake fluid into the reservoir during bleeding

9.14 Install a new O-ring onto the master cylinder sleeve

is preferred. Pull the brake lines slightly away from the master cylinder and plug the ends to prevent contamination.

6   Disconnect the electrical connector at the brake fluid level switch on the master cylinder reservoir, then remove the nuts attaching the master cylinder to the power booster. Pull the master cylinder off the studs and out of the engine compartment. Again, be careful not to spill the fluid as this is done.

7   If a new master cylinder is being installed, remove the reservoir from the master cylinder, install new O-rings in the master cylinder (see illustration), and transfer the reservoir to the new master cylinder.

a) *On 2012 and earlier models, the reservoir is secured by retaining clips.*

b) *On 2013 and later models, the reservoir is secured by a retaining pin.*

Installation

8   Bench bleed the new master cylinder before installing it. Mount the master cylinder in a vise, with the jaws of the vise clamping on the mounting flange.

9   Attach a pair of master cylinder bleeder tubes to the outlet ports of the master cylinder (see illustration).

10   Fill the reservoir with brake fluid of the recommended type (see Chapter 1).

11   Slowly push the pistons into the master cylinder (a large Phillips screwdriver can be used for this) - air will be expelled from the pressure chambers and into the reservoir. Because the tubes are submerged in fluid, air can't be drawn back into the master cylinder when you release the pistons.

12   Repeat the procedure until no more air bubbles are present.

13   Remove the bleed tubes, one at a time, and install plugs in the open ports to prevent fluid leakage and air from entering. Install the reservoir cap.

14   Install the master cylinder over the studs on the power brake booster and tighten the attaching nuts only finger-tight at this time. **Note:** *Be sure to install a new O-ring into the sleeve of the master cylinder (see illustration).*

15   Thread the brake line fittings into the master cylinder. Since the master cylinder is still a bit loose, it can be moved slightly in order for the fittings to thread in easily. Do not strip the threads as the fittings are tightened.

16   Fully tighten the mounting nuts, then the brake line fittings. Tighten the nuts to the torque listed (see Chapter 1 Specifications).

17   Fill the master cylinder reservoir with fluid, then bleed the master cylinder and the brake system as described in Section 11. To bleed the cylinder on the vehicle, have an assistant

depress the brake pedal and hold the pedal to the floor. Loosen the fitting to allow air and fluid to escape. Repeat this procedure on both fittings until the fluid is clear of air bubbles. **Caution:** *Have plenty of rags on hand to catch the fluid - brake fluid will ruin painted surfaces. After the bleeding procedure is completed, rinse the area under the master cylinder with clean water.*

18   Test the operation of the brake system carefully before placing the vehicle into normal service. **Warning:** *Do not operate the vehicle if you are in doubt about the effectiveness of the brake system. It is possible for air to become trapped in the anti-lock brake system hydraulic control unit, so, if the pedal continues to feel spongy after repeated bleedings or the BRAKE or ANTI-LOCK light stays on, have the vehicle towed to a dealer service department or other qualified shop to be bled with the aid of a scan tool.*

10   Brake hoses and lines - inspection and replacement

1   About every six months, with the vehicle raised and placed securely on jackstands, the flexible hoses which connect the steel brake lines with the front and rear brake assemblies should be inspected for cracks, chafing of the outer cover, leaks, blisters and other damage. These are important and vulnerable parts of the brake system and inspection should be complete. A light and mirror will be needed for a thorough check. If a hose exhibits any of the above defects, replace it with a new one.

Flexible hoses

2   Clean all dirt away from the ends of the hose.

3   To disconnect a brake hose from the brake line, unscrew the metal tube nut with a flare nut wrench, then remove the U-clip from the female fitting at the bracket and remove the hose from the bracket (see illustrations).

4   Disconnect the hose from the caliper.

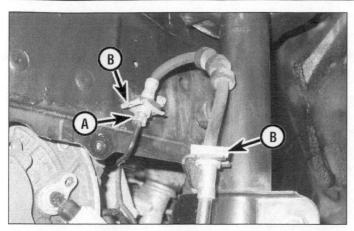

10.3a Unscrew the brake line threaded fitting with a flare-nut wrench to protect the fitting corners from being rounded off (A), then pull off the U-clips (B) with a pair of pliers (front shown, rear similar)

10.3b On 2013 and later front calipers, unbolt the brake hose from the bracket on the strut

If you're replacing a rear hose on a 2005 through 2008 model, discard the sealing washers on either side of the fitting.

5   On 2005 through 2008 rear brake hoses, attach the new brake hose to the caliper or wheel cylinder using nwe sealing washers. Tighten the banjo fitting bolt to the torque listed in this Chapter's Specifications.

6   On all except 2005 through 2008 rear calipers, thread the hose fitting into the caliper body and tighten it securely.

7   To reattach a brake hose to the metal line, insert the end of the hose through the frame bracket, make sure the hose isn't twisted, then attach the metal line by tightening the tube nut fitting securely. Install the U-clip at the frame bracket. On 2013 and later front hoses, connect the hose to the strut bracket and tighten the bolt securely.

8   Carefully check to make sure the suspension or steering components don't make contact with the hose. Have an assistant push down on the vehicle and also turn the steering wheel lock-to-lock during inspection.

Caution: *The weight of the vehicle must be on the suspension, so the vehicle should not be raised while positioning the hose.*

9   Bleed the brake system (see Section 11).

Metal brake lines

10   When replacing brake lines, be sure to use the correct parts. Don't use copper tubing for any brake system components. Purchase steel brake lines from a dealer parts department or auto parts store.

11   Prefabricated brake line, with the tube ends already flared and fittings installed, is available at auto parts stores and dealer parts departments. These lines can be bent to the proper shapes using a tubing bender.

12   When installing the new line make sure it's well supported in the brackets and has plenty of clearance between moving or hot components.

13   After installation, check the master cylinder fluid level and add fluid as necessary. Bleed the brake system as outlined in Section 11 and test the brakes carefully before placing the vehicle into normal operation.

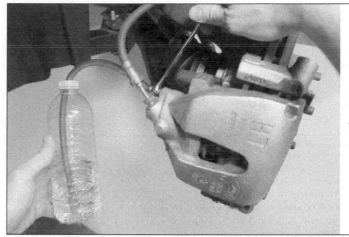

11.8 When bleeding the brakes, a hose is connected to the bleed screw at the caliper and submerged in brake fluid - air will be seen as bubbles in the tube and container (all air must be expelled before moving to the next wheel)

11   Brake hydraulic system - bleeding

Warning: *If air has found its way into the hydraulic control unit, the system must be bled with the use of a scan tool. If the brake pedal feels "spongy" even after bleeding the brakes, or the ABS light on the instrument panel does not go off, or if you have any doubts whatsoever about the effectiveness of the brake system, have the vehicle towed to a dealer service department or other repair shop equipped with the necessary tools for bleeding the system.*

Warning: *Wear eye protection when bleeding the brake system. If the fluid comes in contact with your eyes, immediately rinse them with water and seek medical attention.*

Note: *Bleeding the brake system is necessary to remove any air that's trapped in the system when it's opened during removal and installation of a hose, line, caliper, wheel cylinder or master cylinder.*

1   It will probably be necessary to bleed the system at all four brakes if air has entered the system due to low fluid level, or if the brake lines have been disconnected at the master cylinder.

2   If a brake line was disconnected only at a wheel, then only that caliper or wheel cylinder must be bled.

3   If a brake line is disconnected at a fitting located between the master cylinder and any of the brakes, that part of the system served by the disconnected line must be bled.

4   Remove any residual vacuum (or hydraulic pressure) from the brake power booster by applying the brake several times with the engine off.

5   Remove the master cylinder reservoir cap and fill the reservoir with brake fluid. Reinstall the cap.

Note: *Check the fluid level often during the bleeding operation and add fluid as necessary to prevent the fluid level from falling low enough to allow air bubbles into the master cylinder.*

6   Have an assistant on hand, as well as a supply of new brake fluid, an empty clear plastic container, a length of plastic, rubber or vinyl tubing to fit over the bleeder valve and a wrench to open and close the bleeder valve.

7   Beginning at the right rear wheel, loosen the bleeder screw slightly, then tighten it to a point where it's snug but can still be loosened quickly and easily.

8   Place one end of the tubing over the bleeder screw fitting and submerge the other end in brake fluid in the container (see illustration).

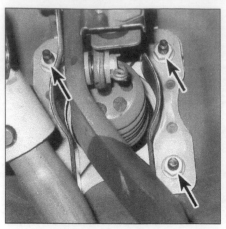

12.9 To detach the power brake booster from the firewall, remove these four nuts (one nut not visible in picture)

13.5 Turn the adjusting nut until the desired handle travel is obtained (center console removed for clarity)

13.16 Parking brake cable adjuster nut (console removed for clarity)

9   Have the assistant slowly depress the brake pedal and hold it in the depressed position.

10   While the pedal is held depressed, open the bleeder screw just enough to allow a flow of fluid to leave the valve. Watch for air bubbles to exit the submerged end of the tube. When the fluid flow slows after a couple of seconds, tighten the screw and have your assistant release the pedal.

11   Repeat Steps 9 and 10 until no more air is seen leaving the tube, then tighten the bleeder screw and proceed to the left rear wheel, the right front wheel and the left front wheel, in that order, and perform the same procedure. Be sure to check the fluid in the master cylinder reservoir frequently.

12   Never use old brake fluid. It contains moisture which can boil, rendering the brake system inoperative.

13   Refill the master cylinder with fluid at the end of the operation.

14   Check the operation of the brakes. The pedal should feel solid when depressed, with no sponginess. If necessary, repeat the entire process.

Warning: *Do not operate the vehicle if you are in doubt about the effectiveness of the brake system. It is possible for air to become trapped in the anti-lock brake system hydraulic control unit, so, if the pedal continues to feel spongy after repeated bleedings or the BRAKE or ANTI-LOCK light stays on, have the vehicle towed to a dealer service department or other qualified shop to be bled with the aid of a scan tool.*

12   Power brake booster - check, removal and installation

Operating check

1   Depress the brake pedal several times with the engine off and make sure that there is no change in the pedal reserve distance.

2   Depress the pedal and start the engine. If the pedal goes down slightly, operation is normal.

Airtightness check

3   Start the engine and turn it off after one or two minutes. Depress the brake pedal several times slowly. If the pedal goes down farther the first time but gradually rises after the second or third depression, the booster is airtight.

4   Depress the brake pedal while the engine is running, then stop the engine with the pedal depressed. If there is no change in the pedal reserve travel after holding the pedal for 30 seconds, the booster is airtight.

Removal and installation

5   2012 and earlier models: Remove the nuts attaching the master cylinder to the booster and carefully pull the master cylinder forward until it clears the mounting studs (see Section 9). Be careful not to bend or kink the brake lines. On V6 models it may be necessary to disconnect the EVAP canister purge valve.

6   2013 and later models: Remove the master cylinder (see Section 9).

Note: *It might also be necessary to remove the hydraulic lines from the master cylinder to the ABS Hydraulic Control Unit.*

7   Disconnect the vacuum hose where it attaches to the power brake booster.

8   Remove the brake light switch (see Section 15), then detach the pushrod from the top of the brake pedal.

Caution: *On 2008 and later models, the clevis locking pin must be discarded and replaced with a new one.*

9   Remove the nuts attaching the booster to the firewall (see illustration).

10   Carefully lift the booster unit away from the firewall and out of the engine compartment.

11   To install the booster, place it into position and tighten the retaining nuts to the torque settings listed (see this Chapter's Specifications). Connect the pushrod to the brake pedal.

12   Install the master cylinder. Reconnect the vacuum hose.

13   On 2013 and later models, bleed the brake system (see Section 11).

14   Carefully test the operation of the brakes before placing the vehicle in normal service.

13   Parking brake - adjustment

1   The parking brake systems have gone through several changes over the years. 2012 and older models with rear drum brakes that utilizes a cable to operate a lever inside the drum, which expands the brake shoes. 2008 and earlier disc brake models utilize a drum-in-hat design disc brake configuration for the rear with a set of brake shoes inside the drum portion of the disc for the parking brake. 2013 and later models use actuators in the rear caliper pistons that operate rear brakes through cable actuation..

2   The manually operated parking brake system works by applying tension to a series of cables that are attached to the parking brake lever and to the rear brake shoes, parking brake shoes, or brake calipers. An indicator light on the dash will illuminate when the parking brake has been applied.

Parking brake adjustment check

3   The parking brake lever, when properly adjusted, should travel three to five clicks, when a moderate pulling force is applied. If it travels less than the specified minimum number of clicks, there's a chance the parking brake might not be releasing completely and might be dragging on the drum. If the lever can be pulled up more than the specified maximum number of clicks, the parking brake may not hold adequately on an incline, allowing the car to roll.

2007 and earlier models

4   Release the parking brake lever.

5   Remove the parking brake lever boot by squeezing inward on the bottom of the boot, then lift the boot to access the adjustment nut (see illustration).

6   Adjust the nut so when the lever is pulled, you hear or feel three to five clicks (drum brake models) or five clicks (disc brake models) of the lever's ratchet mechanism.

7   Verify the parking brake is actually holding the vehicle when applied.

2008 through 2012 models

Adjustment

8   These models feature an automatic adjuster incorporated into the parking brake pedal; manual cable adjustment is not required.

Cable tension release

9   Tension on the parking brake cables must be released if a cable or the parking brake pedal is to be replaced.

10   To remove tension on the brake cables, have an assistant pull down on the cable union between the front and rear parking brake cables, while you watch the brake cable as it winds around the parking brake lever cam on the parking brake pedal hub. When the lever rotates to its highest position look for the 4 mm hole at about the 1 o'clock position and insert a pin of the appropriate size to keep the pedal from applying tension to the cable.

11   To reload the tension back onto the cable, perform the same procedure in reverse.

2013 and later models

12   The following procedure will require two people. One inside the vehicle and one underneath the vehicle. Raise the vehicle and support it securely on jackstands. Place the shift lever in Neutral.

13   Remove the center console trim (see Chapter 11).

14   If equipped, remove any of the factory media hub located in the center console.

15   Release the parking brake completely.

16   Remove the parking brake locking nut (see illustration).

17   Loosen the parking brake adjustment nut.

18   Use a feeler gauge between the brake caliper abutment and the parking brake lever. (Where the parking brake lever comes into contact with the brake caliper.) A measurement of 0.027 inch (0.7 mm) should be obtained from both sides of the vehicle.

19   Adjust the parking brake cable nut until the desired distance has been obtained.

20   Tighten the locking nut securely.

21   The remainder of the installation is the reverse of removal.

14   Parking brake shoes (2005 through 2008 models with rear disc brakes) - replacement

1   Release the parking brake.

2   Loosen the rear wheel lug nuts. Raise the vehicle and support it securely on jackstands. Remove the wheels. Be sure to block the front tires.

3   Remove the brake calipers (see Section 5) and the brake discs (see Section 6).

4   Clean the parking brake assembly with brake system cleaner before beginning work.

5   Remove the brake shoe retracting spring.

6   Remove the adjusting screw spring, then remove the adjusting screw.

7   Remove the hold-down springs from the shoes.

8   Remove the parking brake shoes.

9   Clean the brake disc/parking brake drum and check it for score marks, deep grooves, hard spots (which will appear as small discolored areas) and cracks. If the disc/drum is worn, scored or out-of-round, it can be resurfaced by an automotive machine shop.

10   Install the shoes by reversing the removal procedure. Turn the adjusting screw so the disc just fits over the new shoes. When the disc is installed, the shoes should not rub as the disc is turned. If you have a brake shoe adjusting gauge, adjust the diameter of the shoes to 0.042-inch less than that of the drum surface of the rear brake disc.

11   Repeat this procedure for the other parking brake assembly.

12   Install the brake discs (see Section 6) and the brake calipers (see Section 5).

13   Remove the rubber plug from the brake backing plate and, using a screwdriver or brake adjusting tool, turn the adjusting screw star wheel until the parking brake shoes start to drag as the disc is turned, then back off the start wheel until the shoes don't drag.

14   Install the rear wheels and lug nuts, lower the vehicle and tighten the lug nuts to the torque listed (see this Chapter's Specifications).

15   Electric Parking Brake (EPB) - general information, deactivation/activation and actuator replacement

Warning: *Dust created by the brake system is harmful to your health. Never blow it out with compressed air and don't inhale any of it. An approved filtering mask should be worn when working on the brakes. Do not, under any circumstances, use petroleum-based solvents to clean brake parts. Use brake system cleaner only.*

Caution: *Do NOT perform any repairs to the rear braking system on a vehicle equipped with the Electric Parking Brake (EPB) unless you've deactivated the system. Follow the procedure outlined in this section.*

Note: *2017 models are equipped with the Electric Parking Brake (EPB) system.*

General information

1   The Electric Parking Brake (EPB) system utilizes an actuator mounted on each rear brake caliper. The EPB control module monitors the functions of the actuators for the engagement and release and is integrated into the ABS control module. The control module also stores any trouble codes related to the EPB system. If a malfunction occurs in the EPB system, the controller will turn on the parking brake warning light on the instrument panel.

2   The red brake warning light is for hydraulic issues; the yellow brake warning light is for issues with the EPB system. In some cases, both the red and yellow warning lights can come on informing you of a problem with the EPB system. The yellow parking brake warning light will also come on when the EPB system is in the service mode.

EPB Deactivation procedure

a) *Turn the ignition to ON. Depress the accelerator pedal and push the Electric Parking Brake (EPB) switch down to the brake-released position.*

b) *Continue to hold the accelerator pedal and EPB switch.*

c) *Turn the ignition to OFF, then within 5 seconds, turn the ignition to ON.*

d) *Turn the ignition OFF and release the EPB switch.*

e) *The EPB is now in the "inactive" or "service mode" and is safe to perform the repairs needed.*

EPB Activation procedure

a) *After completing the service work, you'll need to reactivate the EPB system.*

b) *Turn the ignition on (don't start the engine) and hold the accelerator pedal to the floor as you move the EPB switch in the activate position (upwards).*

c) *Turn the ignition switch to the OFF position without releasing the accelerator pedal or the EPB switch.*

d) *Within 5 seconds turn the ignition switch to the ON position.*

e) *Release the accelerator pedal and EPB switch.*

f) *The EPB is now in "active" mode and is ready for use.*

Actuator replacement

3   Loosen the rear wheel lug nuts. Raise the vehicle and support it securely on jackstands. Remove the wheels. Be sure to block the front tires.

4   Disconnect the electrical connector from the actuator and unclip the wire retainer.

5   Remove the two bolts securing the actuator to the caliper. Discard the bolts and obtain new ones for installation.

6   Work the actuator off of the caliper by slightly twisting from left to right as you pull it outward.

7   Discard the O-ring.

8   To install, replace the O-ring and fasteners. Use the Activation procedure to initiate the EPB system.

16.1 Disconnect the electrical connector (A) then rotate the switch clockwise to remove it - 2006 and earlier modes

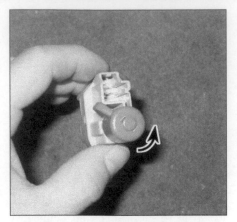

16.3 Rotate the knob counterclockwise until it stops to unlock the switch - 2006 and earlier modes

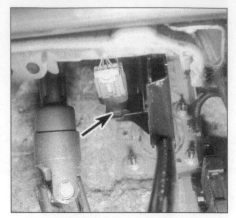

16.8 Rotate the switch 1/8 of a turn to remove it

16   Brake light switch - replacement

Caution: *The switch must be removed and installed with the pedal in its at-rest position or damage to the switch will result.*

2006 and earlier models
Removal
1   Disconnect the wiring from the switch (see illustration).
2   Turn the switch clockwise to remove it.

Installation
3   Turn the knob on the switch fully counterclockwise until it stops (see illustration).
4   Start the engine, then depress the brake pedal and hold it.
5   Place the switch into its bracket, then turn it counterclockwise.
6   Connect the wiring to the switch.

2007 and later models
7   If you're working on a 2013 or later model, remove the driver's-side knee bolster (see Chapter 11, Section 22).
8   Turn the switch clockwise 1/8-turn, then remove it (see illustration).
9   Disconnect the wiring from the switch.
10   Installation is the reverse of removal.

17   Brake Vacuum Pump (2013 and later models) - description, removal and installation

Description
1   With the smaller displacement engines in today's vehicles, but higher demands on the various systems, new and innovative ways of accomplishing those needed tasks have been incorporated into the modern car. A brake vacuum pump is one of these systems that have been added. Vacuum pumps have been on vehicles for decades, and with all the new systems, they are being reintroduced into the modern vehicle as a standard component.

2   Today's braking systems are one of those systems that have become more and more complex as the requirements for occupant safety and ABS controls keep advancing. Because of the smaller displacement engines with less available vacuum as well as the increased requirements placed on the braking systems a vacuum pump driven by the cam has been added.

The reasons for the vacuum pump comes down to the available vacuum from the engine based on the braking system requirements. Here are the new brake control systems that have been incorporated into the 2013 and later models.

TCS - Traction Control Systems - This system controls wheel spin during acceleration.

ECS - Electronic Stability Control - This system helps to maintain vehicle control during lateral movements such as during a skid or aggressive turning speeds in a curve. This system regulates brake pressure to the wheels and adjusts engine torque to a safe level until the event is over.

RSC - Roll Stability Control - Like the ESC, the function of the roll stability system is to evaluate the pitch or roll of the vehicle and adjust the brake engagement to each wheel by modulating the brake pressure.

Trailer Sway Control - As the name implies, to maintain stability while pulling a trailer.

Hill Start Assist - This system allows the vehicle to maintain its position on a steep incline without rolling backwards. It is also referred to as the "Hill Hold Assist System" For the manual transmission vehicles this becomes very apparent when stopped at a traffic light on a hilly road and you begin to take off, the vehicle will not have the tendency to roll backward but will maintain its position when you remove your foot from the brake pedal and apply the accelerator.

Active Cruise Control System - Even though the cruise control system is meant to allow the driver to set a given speed, the active cruise control system also can maintain a safe distance from the car in front of you. This distance is maintained by reducing the vehicle speed but also by applying braking force to the wheels as needed.

ABS - Anti-Lock Braking System - None of this would be possible without the anti-lock braking system. The ABS was once just a way to control a car on a slippery or icy road and allow the driver to regain control. Now, that same system can not only prevent that from ever happening but can avoid the possibility of it happening by monitoring the other various systems on the car such as the forward and rearward facing

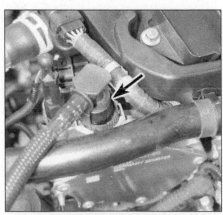

17.5 Squeeze the fitting retainer and detach the vacuum hose

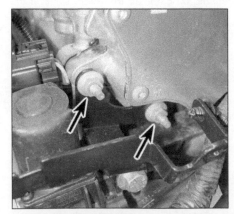

17.6 EVAP purge valve bracket nuts (1.6L models)

17.7 Remove the nut and bolt and detach the wiring harness bracket (1.6L models)

17.8 Vacuum pump mounting bolts (1.6L model shown, 2.0L similar)

cameras and infrared sensors. The modern car can think, make decisions, and if needed, stop an accident from occurring especially when the driver's attention has been adverted.

Vacuum pump - removal and installation

3   Remove the engine compartment appearance cover by pulling it free from the rubber connectors.

4   Remove the air filter outlet duct (see Chapter 4).

5   Disconnect the vacuum tube on the vacuum pump by squeezing the sides together (see illustration).

6   On 1.6L models, remove the EVAP purge valve and bracket (see illustration).

7   On 1.6L models, remove the fasteners and detach the engine wiring harness bracket (see illustration).

8   Remove the bolts and detach the vacuum pump from the cylinder head (see illustration).

9   Clean the mating surfaces before installing.

10   Install a new gasket (1.6L engine) or O-ring gasket (2.0L engine) and install the pump to the cylinder head.

Caution: *Align the vacuum pump and the cam coupling teeth before installing it onto the engine. Failure to do so can result in damage to the pump and/or the camshaft.*

11   Install the bolts and tighten them to the torque listed in this Chapter's Specifications.

12   The remainder of installation is the reverse of removal.

18 Troubleshooting

PROBABLE CAUSE	CORRECTIVE ACTION

No brakes - pedal travels to floor

PROBABLE CAUSE	CORRECTIVE ACTION
1 Low fluid level 2 Air in system	1 and 2 Low fluid level and air in the system are symptoms of another problem a leak somewhere in the hydraulic system. Locate and repair the leak
3 Defective seals in master cylinder	3 Replace master cylinder
4 Fluid overheated and vaporized due to heavy braking	4 Bleed hydraulic system (temporary fix). Replace brake fluid (proper fix)

Brake pedal slowly travels to floor under braking or at a stop

PROBABLE CAUSE	CORRECTIVE ACTION
1 Defective seals in master cylinder	1 Replace master cylinder
2 Leak in a hose, line, caliper or wheel cylinder	2 Locate and repair leak
3 Air in hydraulic system	3 Bleed the system, inspect system for a leak

Brake pedal feels spongy when depressed

PROBABLE CAUSE	CORRECTIVE ACTION
1 Air in hydraulic system	1 Bleed the system, inspect system for a leak
2 Master cylinder or power booster loose	2 Tighten fasteners
3 Brake fluid overheated (beginning to boil)	3 Bleed the system (temporary fix). Replace the brake fluid (proper fix)
4 Deteriorated brake hoses (ballooning under pressure)	4 Inspect hoses, replace as necessary (it's a good idea to replace all of them if one hose shows signs of deterioration)

Troubleshooting (continued)

PROBABLE CAUSE	CORRECTIVE ACTION

Brake pedal feels hard when depressed and/or excessive effort required to stop vehicle

PROBABLE CAUSE	CORRECTIVE ACTION
1 Power booster faulty	1 Replace booster
2 Engine not producing sufficient vacuum, or hose to booster clogged, collapsed or cracked	2 Check vacuum to booster with a vacuum gauge. Replace hose if cracked or clogged, repair engine if vacuum is extremely low
3 Brake linings contaminated by grease or brake fluid	3 Locate and repair source of contamination, replace brake pads or shoes
4 Brake linings glazed	4 Replace brake pads or shoes, check discs and drums for glazing, service as necessary
5 Caliper piston(s) or wheel cylinder(s) binding or frozen	5 Replace calipers or wheel cylinders
6 Brakes wet	6 Apply pedal to boil-off water (this should only be a momentary problem)
7 Kinked, clogged or internally split brake hose or line	7 Inspect lines and hoses, replace as necessary

Excessive brake pedal travel (but will pump up)

PROBABLE CAUSE	CORRECTIVE ACTION
1 Drum brakes out of adjustment	1 Adjust brakes
2 Air in hydraulic system	2 Bleed system, inspect system for a leak

Excessive brake pedal travel (but will not pump up)

PROBABLE CAUSE	CORRECTIVE ACTION
1 Master cylinder pushrod misadjusted	1 Adjust pushrod
2 Master cylinder seals defective	2 Replace master cylinder
3 Brake linings worn out	3 Inspect brakes, replace pads and/or shoes
4 Hydraulic system leak	4 Locate and repair leak

Brake pedal doesn't return

PROBABLE CAUSE	CORRECTIVE ACTION
1 Brake pedal binding	1 Inspect pivot bushing and pushrod, repair or lubricate
2 Defective master cylinder	2 Replace master cylinder

Brake pedal pulsates during brake application

PROBABLE CAUSE	CORRECTIVE ACTION
1 Brake drums out-of-round	1 Have drums machined by an automotive machine shop
2 Excessive brake disc runout or disc surfaces out-of-parallel	2 Have discs machined by an automotive machine shop
3 Loose or worn wheel bearings	3 Adjust or replace wheel bearings
4 Loose lug nuts	4 Tighten lug nuts

Brakes slow to release

PROBABLE CAUSE	CORRECTIVE ACTION
1 Malfunctioning power booster	1 Replace booster
2 Pedal linkage binding	2 Inspect pedal pivot bushing and pushrod, repair/lubricate
3 Malfunctioning proportioning valve	3 Replace proportioning valve
4 Sticking caliper or wheel cylinder	4 Repair or replace calipers or wheel cylinders
5 Kinked or internally split brake hose	5 Locate and replace faulty brake hose

Brakes grab (one or more wheels)

PROBABLE CAUSE	CORRECTIVE ACTION
1 Grease or brake fluid on brake lining	1 Locate and repair cause of contamination, replace lining
2 Brake lining glazed	2 Replace lining, deglaze disc or drum

PROBABLE CAUSE	CORRECTIVE ACTION

Vehicle pulls to one side during braking

PROBABLE CAUSE	CORRECTIVE ACTION
1 Grease or brake fluid on brake lining	1 Locate and repair cause of contamination, replace lining
2 Brake lining glazed	2 Deglaze or replace lining, deglaze disc or drum
3 Restricted brake line or hose	3 Repair line or replace hose
4 Tire pressures incorrect	4 Adjust tire pressures
5 Caliper or wheel cylinder sticking	5 Repair or replace calipers or wheel cylinders
6 Wheels out of alignment	6 Have wheels aligned
7 Weak suspension spring	7 Replace springs
8 Weak or broken shock absorber	8 Replace shock absorbers

Brakes drag (indicated by sluggish engine performance or wheels being very hot after driving)

PROBABLE CAUSE	CORRECTIVE ACTION
1 Brake pedal pushrod incorrectly adjusted	1 Adjust pushrod
2 Master cylinder pushrod (between booster and master cylinder)	2 Adjust pushrod incorrectly adjusted
3 Obstructed compensating port in master cylinder	3 Replace master cylinder
4 Master cylinder piston seized in bore	4 Replace master cylinder
5 Contaminated fluid causing swollen seals throughout system	5 Flush system, replace all hydraulic components
6 Clogged brake lines or internally split brake hose(s)	6 Flush hydraulic system, replace defective hose(s)
7 Sticking caliper(s) or wheel cylinder(s)	7 Replace calipers or wheel cylinders
8 Parking brake not releasing	8 Inspect parking brake linkage and parking brake mechanism, repair as required
9 Improper shoe-to-drum clearance	9 Adjust brake shoes
10 Faulty proportioning valve	10 Replace proportioning valve

Brakes fade (due to excessive heat)

PROBABLE CAUSE	CORRECTIVE ACTION
1 Brake linings excessively worn or glazed	1 Deglaze or replace brake pads and/or shoes
2 Excessive use of brakes	2 Downshift into a lower gear, maintain a constant slower speed (going down hills)
3 Vehicle overloaded	3 Reduce load
4 Brake drums or discs worn too thin	4 Measure drum diameter and disc thickness, replace drums or discs as required
5 Contaminated brake fluid	5 Flush system, replace fluid
6 Brakes drag	6 Repair cause of dragging brakes
7 Driver resting left foot on brake pedal	7 Don't ride the brakes

Brakes noisy (high-pitched squeal)

PROBABLE CAUSE	CORRECTIVE ACTION
1 Glazed lining	1 Deglaze or replace lining
2 Contaminated lining (brake fluid, grease, etc.)	2 Repair source of contamination, replace linings
3 Weak or broken brake shoe hold-down or return spring	3 Replace springs
4 Rivets securing lining to shoe or backing plate loose	4 Replace shoes or pads
5 Excessive dust buildup on brake linings	5 Wash brakes off with brake system cleaner
6 Brake drums worn too thin	6 Measure diameter of drums, replace if necessary
7 Wear indicator on disc brake pads contacting disc	7 Replace brake pads
8 Anti-squeal shims missing or installed improperly	8 Install shims correctly

Troubleshooting (continued)

PROBABLE CAUSE   CORRECTIVE ACTION

Brakes noisy (scraping sound)

1 Brake pads or shoes worn out; rivets, backing plate or brake | 1 Replace linings, have discs and/or drums machined (or replace) shoe metal contacting disc or drum

Brakes chatter

PROBABLE CAUSE	CORRECTIVE ACTION
1 Worn brake lining	1 Inspect brakes, replace shoes or pads as necessary
2 Glazed or scored discs or drums	2 Deglaze discs or drums with sandpaper (if glazing is severe, machining will be required)
3 Drums or discs heat checked	3 Check discs and/or drums for hard spots, heat checking, etc. Have discs/drums machined or replace them
4 Disc runout or drum out-of-round excessive	4 Measure disc runout and/or drum out-of-round, have discs or drums machined or replace them
5 Loose or worn wheel bearings	5 Adjust or replace wheel bearings
6 Loose or bent brake backing plate (drum brakes)	6 Tighten or replace backing plate
7 Grooves worn in discs or drums	7 Have discs or drums machined, if within limits (if not, replace them)
8 Brake linings contaminated (brake fluid, grease, etc.)	8 Locate and repair source of contamination, replace pads or shoes
9 Excessive dust buildup on linings	9 Wash brakes with brake system cleaner
10 Surface finish on discs or drums too rough after machining	10 Have discs or drums properly machined (especially on vehicles with sliding calipers)
11 Brake pads or shoes glazed	11 Deglaze or replace brake pads or shoes

Brake pads or shoes click

PROBABLE CAUSE	CORRECTIVE ACTION
1 Shoe support pads on brake backing plate grooved or	1 Replace brake backing plate excessively worn
2 Brake pads loose in caliper	2 Loose pad retainers or anti-rattle clips
3 Also see items listed under Brakes chatter	

Brakes make groaning noise at end of stop

PROBABLE CAUSE	CORRECTIVE ACTION
1 Brake pads and/or shoes worn out	1 Replace pads and/or shoes
2 Brake linings contaminated (brake fluid, grease, etc.)	2 Locate and repair cause of contamination, replace brake pads or shoes
3 Brake linings glazed	3 Deglaze or replace brake pads or shoes
4 Excessive dust buildup on linings	4 Wash brakes with brake system cleaner
5 Scored or heat-checked discs or drums	5 Inspect discs/drums, have machined if within limits (if not, replace discs or drums)
6 Broken or missing brake shoe attaching hardware	6 Inspect drum brakes, replace missing hardware

Rear brakes lock up under light brake application

PROBABLE CAUSE	CORRECTIVE ACTION
1 Tire pressures too high	1 Adjust tire pressures
2 Tires excessively worn	2 Replace tires
3 Defective proportioning valve	3 Replace proportioning valve

PROBABLE CAUSE	CORRECTIVE ACTION

Brake warning light on instrument panel comes on (or stays on)

1 Low fluid level in master cylinder reservoir (reservoirs with fluid level sensor)	1 Add fluid, inspect system for leak, check the thickness of the brake pads and shoes
2 Failure in one half of the hydraulic system	2 Inspect hydraulic system for a leak
3 Piston in pressure differential warning valve not centered	3 Center piston by bleeding one circuit or the other (close bleeder valve as soon as the light goes out)
4 Defective pressure differential valve or warning switch	4 Replace valve or switch
5 Air in the hydraulic system	5 Bleed the system, check for leaks
6 Brake pads worn out (vehicles with electric wear sensors - small	6 Replace brake pads (and sensors) probes that fit into the brake pads and ground out on the disc when the pads get thin)

Brakes do not self adjust

Disc brakes

1 Defective caliper piston seals	1 Replace calipers. Also, possible contaminated fluid causing soft or swollen seals (flush system and fill with new fluid if in doubt)
2 Corroded caliper piston(s)	2 Same as above

Drum brakes

1 Adjuster screw frozen	1 Remove adjuster, disassemble, clean and lubricate with high-temperature grease
2 Adjuster lever does not contact star wheel or is binding	2 Inspect drum brakes, assemble correctly or clean or replace parts as required
3 Adjusters mixed up (installed on wrong wheels after brake job)	3 Reassemble correctly
4 Adjuster cable broken or installed incorrectly (cable-type adjusters)	4 Install new cable or assemble correctly

Electric Parking Brake (EPB)

1 Red warning light is on, never comes on, or is flashing	1 Low brake fluid level, wiring/connections, faulty brake fluid level sensor, parking brake is applied, brake booster vacuum sensor, EPB, IPC, or ABS concerns
2 Yellow warning light is on or never comes on	2 Faulty parking brake switch, connections/wiring, EPB module problem
3 Message center displays parking brake malfunction or "Service Parking Brake System Now"	3 Have system checked for codes
4 Message center displays "Maintenance Mode"	4 System is in the service mode
5 Rear brakes are dragging	5 Service the rear brake system
6 Parking brake system does not release or fails to engage	6 Check for codes and have the system checked for network communication problems

Rapid brake lining wear

1 Driver resting left foot on brake pedal	1 Don't ride the brakes
2 Surface finish on discs or drums too rough	2 Have discs or drums properly machined
3 Also see Brakes drag	

Notes

Chapter 10
Suspension and steering systems

Contents

Specifications

Torque specifications

Ft-lbs (unless otherwise indicated)   Nm

Note: *One foot-pound (ft-lb) of torque is equivalent to 12 inch-pounds (in-lbs) of torque. Torque values below approximately 15 ft-lbs are expressed in inch-pounds, because most foot-pound torque wrenches are not accurate at these smaller values.*

Warning: *The manufacturer recommends that all suspension fasteners be replaced with new fasteners whenever removed.*

Front suspension

	Ft-lbs	Nm
Balljoint-to-steering knuckle pinch bolt		
2005 and earlier models	52	70
2006 models		
Vehicles built up to 9/2005	52	70
Vehicles built after 9/2005	46	63
2007 through 2012 models	46	63
2013 and later models	61	83
Control arm		
Front pivot bolt		
2012 and earlier models	85	115
2013 and later models		
Step 1	59	80
Step 2	Tighten an additional 360 degrees (one turn)	
Rear mounting bolt(s)		
2005 and earlier models	148	200
2006 models	129	175
2007 through 2012 models	110	150
2013 and later models		
Step 1	85	115
Step 2	Tighten an additional 90 degrees (1/4 turn)	
Stabilizer bar		
Bracket bolts/nuts		
2012 and earlier	52	70
2013 and later		
Step 1	85	115
Step 2	Tighten an additional 90 degrees (1/4 turn)	
Link nuts		
2004 and earlier models	35	47
2005 and 2006 models		
Upper nut	46	63
Lower nut	41	55
2007 and later models	46	63

Front suspension (continued)

Strut

Damper shaft nut

2012 and earlier models	76	103
2013 and later models	41	55

Strut-to-steering knuckle bolt(s)

2012 and earlier models	85	115

2013 and later models

Step 1	59	80
Step 2	Tighten an additional 180 degrees (1/4 turn)	

Strut upper mounting nuts

2004 and earlier models	41	55
2005	30	40
2006 (vehicles built up to 9/2005)	30	40
2006 (vehicles built after 9/2005)	35	47
2007 through 2012 models	35	47
2013 and later models	26	35

Subframe

2005 and earlier models

Bolts	148	200
Nuts	111	150

2006 models

Bolts

Vehicles built up to 9/2005	148	200
Vehicles built after 9/2006	129	175
Nuts	111	150

2007 through 2012 models

Bolts	129	175
Nuts	111	150
Lateral support crossmember bolts	85	115

2013 and later models

Front bolts	85	115

Rear bolts

Step 1	103	140
Step 2	Tighten an additional 180 degrees (1/2 turn)	

Support bracket bolts

Inner	35	47
Outer	46	62

Load path bar

To subframe

Bar-to-bracket bolt/nut	35	48
Bracket-to-subframe bolts	18	25
Bar-to-radiator lower support	18	25

Rear suspension

Shock absorber

Upper mounting nut

2005 and earlier models	156 in-lbs	18
2006 through 2012 models	30	40
2013 and later models	18	25

Lower mounting bolt/nut

2004 and earlier models	85	115
2005 through 2012 models	129	175
2013 and later models	85	115

Stabilizer bar

Bracket bolts

2009 through 2012 models

Upper	66	90
Lower	85	115
2013 and later models	44	60

Link nuts

2012 and earlier models	41	55

2013 and later models

Upper	115 in-lbs	13
Lower	52	70

Suspension arms

Upper arm inner pivot bolt

2008 and earlier models	85	115
2009 through 2012 models	129	175
2013 and later models	85	115
Upper arm-to-knuckle bolt (2013 and later models)	85	115

Rear suspension (continued)

Lower arm inner pivot bolt
    2008 and earlier models 85   115
    2009 through 2012 models 129   175
    2013 and later models
        Front 85   115
        Rear 66   90
Lower arm-to-rear knuckle (2013 and later models) 85   115
Trailing arm-to-body bolt
    2004 and earlier models 85   115
    2005 through 2008 models 111   150
    2009 through 2012 models 92   125
    2013 and later models 85   115
Balljoint nuts
    2004 and earlier models 76   103
    2005 through 2012 models 46   63
Hub and bearing assembly mounting bolts (2013 and later models) 81   110
Subframe mounting bolts
    2005 and earlier models 85   115
    2006 and 2007 models
        Front 76   103
        Rear 85   115
    2008 and later models 85   115

Steering

Steering shaft pinch bolts (2005 and earlier models)
    Upper (to steering column shaft) 17   23
    Lower (to steering gear) 30   40
Steering shaft coupler pinch bolts (2006 through 2008 models)
    Upper 18   25
    Lower 22   30
Steering shaft coupler pinch bolts (2008 through 2012 models)
    Upper 52   70
    Lower 46   63
Steering shaft pinch bolt (2009 and later models)
    2009 through 2012 models 46   63
    2013 and later models 21   28
Steering shaft -to-steering column pinch bolt (2012 and earlier models) 17   23
Steering gear mounting bolts
    2004 and earlier models 93   126
    2005 through 2007 models 85   115
    2008 through 2012 models 92   125
    2012 and later models (in sequence: driver's side, passenger's side, center rear)
        Step 1 81   110
        Step 2 .... Loosen one full turn
        Step 3 41   56
        Step 4 .... Tighten an additional 180 degrees (1/2 turn)
Steering column mounting bolts
    2004 and earlier models .... 156 in-lbs   18
    2005 through 2007 models 18   25
    2008 and later models 21   28
Steering column through-bolt (2009 through 2012 models) 18   25
Steering wheel
    2007 and earlier models (pinion shaft bolt) .... 108 in-lbs   12
    2008 through 2012 models (bolt) 30   40
    2013 and later models (bolt) 35   48
Power steering pump (2007 and earlier models)
    Four-cylinder models 18   25
    V6 models
        2004 and earlier models
            Bolts 18   25
            Stud bolts .... 89 in-lbs   10
        2005 through 2007 models 18   25
        Pulley bolt 36   49
Tie-rod end-to-steering knuckle nut
    2004 and earlier models 37   50
    2005 through 2007 models 41   55
    2008 and later models 59   80

1 General information

Front suspension

1   The front suspension (see illustrations) is a MacPherson strut design. The upper end of each strut/coil spring assembly is attached to the vehicle's body strut support. The lower end of the strut assembly is connected to the upper end of the steering knuckle. The steering knuckle is attached to a balljoint mounted on the outer end of the suspension control arm. A stabilizer bar reduces body roll.

Rear suspension – 2012 and earlier models

2   The rear suspension (see illustration) employs a trailing arm, two lateral suspension arms, a coil spring and shock absorber per side.

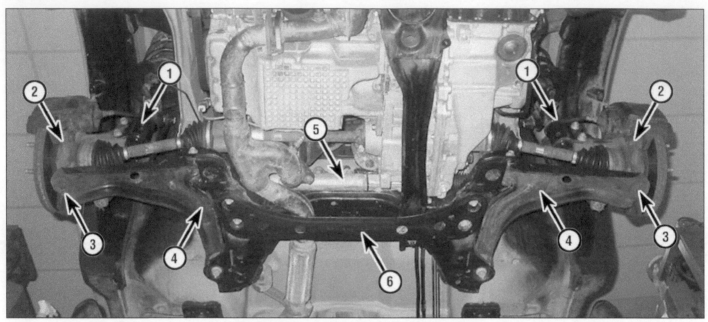

1.1a Front suspension and steering components

1 Strut/coil spring assembly   3 Balljoint   5 Steering gear
2 Steering knuckle   4 Control arm   6 Subframe

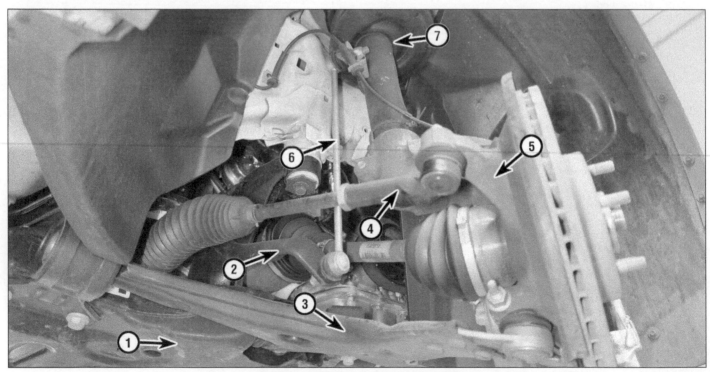

1.1b Front suspension and steering components (right side shown)

1 Subframe   4 Tie-rod end   7 Strut/coil spring assembly
2 Stabilizer bar   5 Steering knuckle
3 Control arm   6 Stabilizer bar link

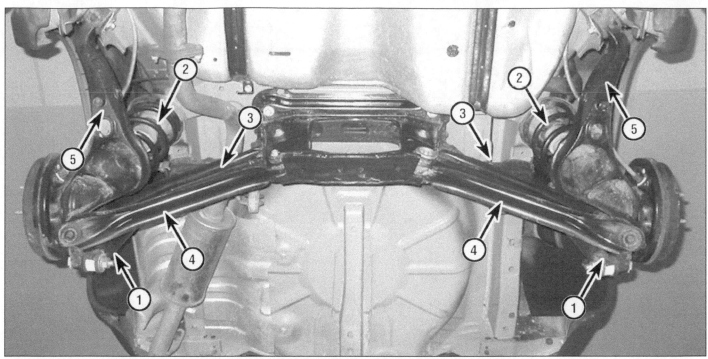

1.2 Rear suspension components (2012 and earlier models)

1 Shock absorber
2 Coil spring

3 Upper suspension arm
4 Lower suspension arm

5 Trailing arm

Rear suspension - 2013 and later models

3   The suspension on the 2013 and later models is quite different from the previous

models. The rear suspension on 2013 and later models consists of a shock absorber, coil spring, a stabilizer bar, a front and rear lower control arm, an upper control arm and a trailing arm/knuckle (see illustrations).

Steering

4   The rack-and-pinion steering gear is located behind the engine/transaxle assembly on the front subframe and actuates the tie rods, which are attached to the steering

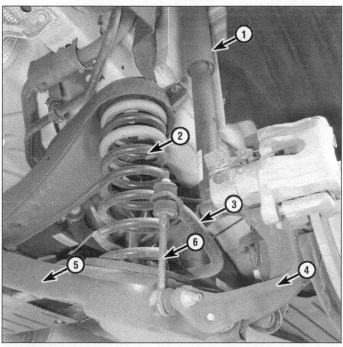

1.3a Rear suspension components - rear view (2013 and later models)

1 Shock absorber
2 Coil Spring
3 Stabilizer bar

4 Knuckle
5 Lower arm (rear)
6 Stabilizer bar link

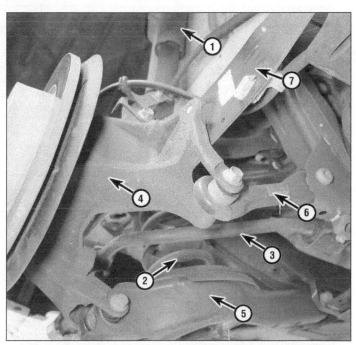

1.3b Rear suspension components - front view (2013 and later models)

1 Shock absorber
2 Coil spring
3 Stabilizer bar
4 Knuckle

5 Lower arm (rear)
6 Lower arm (front)
7 Trailing arm

2.3 Disconnect the stabilizer bar link

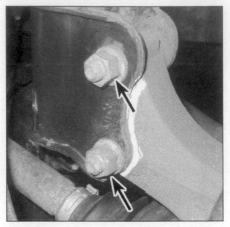

2.4 To detach the strut from the steering knuckle, remove the two nuts, then knock out the bolts with a hammer and punch

2.7 Mark the relationship of the studs to the bearing plate (2012 and earlier models only), then remove the upper mounting nuts or bolts

knuckles. The inner ends of the tie-rods are protected by rubber boots which should be inspected periodically for secure attachment, tears and leaking lubricant (which would indicate a failed rack seal).

5   2007 and earlier models use a hydraulic power assist system. This consists of a belt-driven pump and associated lines and hoses. The fluid level in the pump reservoir should be checked periodically (see Chapter 1).

6   2008 and later models have an electric power assist system. These vehicles do not have a pump or hoses, so no maintenance is necessary. The steering wheel operates the steering shaft, which actuates the steering gear through universal joints. Looseness in the steering can be caused by wear in the steering shaft universal joints, the steering gear, the tie-rod ends and loose retaining bolts.

Precautions

7   Frequently, when working on the suspension or steering system components, you may come across fasteners which seem impossible to loosen. These fasteners on the underside of the vehicle are continually subjected to water, road grime, mud, etc., and can become rusted or "frozen," making them extremely difficult to remove. In order to unscrew these stubborn fasteners without damaging them (or other components), be sure to use lots of penetrating oil and allow it to soak in for a while. Using a wire brush to clean exposed threads will also ease removal of the nut or bolt and prevent damage to the threads. Sometimes a sharp blow with a hammer and punch will break the bond between a nut and bolt threads, but care must be taken to prevent the punch from slipping off the fastener and ruining the threads. Heating the stuck fastener and surrounding area with a torch sometimes helps too, but isn't recommended because of the obvious dangers associated with fire. Long breaker bars and extension, or "cheater," pipes will increase leverage, but never use an extension pipe on a ratchet - the ratcheting mechanism could be damaged.

Sometimes tightening the nut or bolt first will help to break it loose. Fasteners that require drastic measures to remove should always be replaced with new ones.

8   Since most of the procedures dealt with in this Chapter involve jacking up the vehicle and working underneath it, a good pair of jackstands will be needed. A hydraulic floor jack is the preferred type of jack to lift the vehicle, and it can also be used to support certain components during various operations. **Warning:** *Never, under any circumstances, rely on a jack to support the vehicle while working on it. Whenever any of the suspension or steering fasteners are loosened or removed they must be inspected and, if necessary, replaced with new ones of the same part number or of original equipment quality and design. Torque specifications must be followed for proper reassembly and component retention. Never attempt to heat or straighten any suspension or steering components. Instead, replace any bent or damaged part with a new one.*

---

2   Strut assembly (front) - removal, inspection and installation

Warning: *Suspension fasteners are a critical part of the driveability and safe operation of any vehicle. New hardware should be used. Do not reuse the original fasteners.*

Removal

1   Loosen the wheel lug nuts, raise the vehicle and support it securely on jackstands. Remove the wheel.

2   Detach the brake hose bracket from the strut. If the vehicle is equipped with ABS, detach the speed sensor wiring harness from the strut.

3   Detach the stabilizer bar link from the bracket on the strut (see illustration). On later models, remove the brake jounce hose clip, then move the hose out of the way.

4   2012 and earlier models - Remove the strut-to-knuckle nuts (see illustration) and knock the bolts out with a hammer and punch.

5   2013 and later models - Remove the steering knuckle (see Section 6), and the cowl cover (see Chapter 11).

6   Separate the strut from the steering knuckle. Be careful not to overextend the inner CV joint. Also, don't let the steering knuckle fall outward and strain the brake hose.

7   On 2012 and earlier models, mark the relationship of one of the outer upper mounting studs to the strut tower (see illustration). On all models, support the strut and spring assembly with one hand and remove the strut-to-strut tower nuts or bolts. Remove the assembly out from the fenderwell.

Inspection

8   Check the strut body for leaking fluid, dents, cracks and other obvious damage which would warrant repair or replacement.

9   Check the coil spring for chips or cracks in the spring coating (this will cause premature spring failure due to corrosion). Inspect the spring seat for cuts, hardness and general deterioration.

10   If any undesirable conditions exist, proceed to the strut disassembly procedure (see Section 2).

Installation

11   Guide the strut assembly up into the fenderwell. On 2012 and earlier models, align the previously made matchmarks and insert the upper mounting studs through the holes in the body. Once the studs protrude, install the nuts so the strut won't fall back through. On 2013 and later models, align the bolt holes and install the bolts.

Note: *This is most easily accomplished with the help of an assistant, as the strut is quite heavy and awkward.*

3.3 Make sure the spring compressor tool is on securely

3.4 Loosen and remove the retaining nut

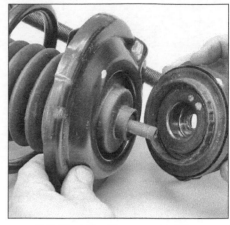

3.5a Remove the upper bearing and spring seat . . .

12   2012 and earlier models - Slide the steering knuckle into the strut flange and insert the two bolts. Install the nuts, align the previously made matchmarks and tighten them to the torque listed in this Chapter's Specifications.

13   2013 and later models - Reinstall the steering knuckle and cowl cover. Be sure to tighten the fasteners to the torque listed in this Chapter's Specifications. Tighten the caliper mounting bolts to the torque listed in the Chapter 9 Specifications. Tighten the drive-axle/hub nut to the torque listed in the Chapter 8 Specifications.

14   Connect the brake hose to the strut bracket. If the vehicle is equipped with ABS, connect the speed sensor wiring harness to its bracket.

15   Connect the stabilizer bar link to the strut bracket. Tighten the nut to the torque listed in this Chapter's Specifications.

16   Install the wheel and lug nuts, then lower the vehicle and tighten the lug nuts to their proper torque specifications.

17   Tighten the upper mounting nuts to the torque listed in this Chapter's Specifications.

18   Have the front end alignment checked, and if necessary, adjusted.

3   Strut/spring assembly - replacement

Warning: *Before attempting to disassemble the front suspension strut, a tool to hold the coil spring in compression must be obtained. Do not attempt to use makeshift methods. Uncontrolled release of the spring could cause damage and personal injury. Use a high-quality spring compressor, and carefully follow the tool manufacturer's instructions provided with it. After removing the coil spring with the compressor still installed, place it in a safe, isolated area.*

Warning: *Suspension fasteners are a critical part of the driveability and safe operation of any vehicle. New hardware should be used. Do not reuse the original fasteners.*

1   If the front suspension struts exhibit signs of wear (leaking fluid, loss of damping capability, sagging or cracked coil springs) then they should be disassembled and overhauled as necessary. The struts themselves cannot be serviced, and should be replaced if faulty; the springs and related components can be replaced individually.

To maintain balanced characteristics on both sides of the vehicle, the components on both sides should be replaced at the same time.

2   With the strut removed from the vehicle (see Section 3), clean away all external dirt, then mount it in a vise.

3   Install the coil spring compressor tools (ensuring that they are fully engaged), and compress the spring until all tension is relieved from the upper mount (see illustration).

4   Hold the strut piston rod with an Allen key and unscrew the thrust bearing retaining nut with a box-end wrench (see illustration).

5   Withdraw the top mounting, thrust bearing, upper spring seat and spring, followed by the boot and the bump stop (see illustrations).

6   If a new spring is to be installed, the original spring must now be carefully released from the compressor. If it is to be re-used, the spring can be left in compression.

7   With the strut assembly now completely disassembled, examine all the components for wear and damage, and check the bearing for smoothness of operation. Replace components as necessary.

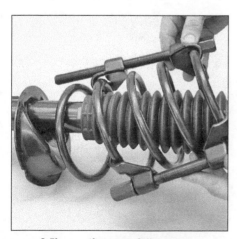

3.5b . . . then carefully remove the spring . . .

3.5c . . . followed by the boot . . .

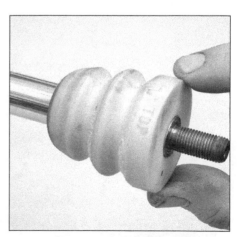

3.5d . . . and the bump stop

3.9 When installing the spring, make sure the ends fit into the recessed portion of the seats

4.2 Remove the lower nut and detach the link from the bar

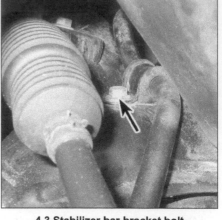

4.3 Stabilizer bar bracket bolt (other bolt not visible in photo)

8   Examine the strut for signs of fluid leakage. Check the strut piston rod for signs of pitting along its entire length, and check the strut body for signs of damage. Test the operation of the strut, while holding it in an upright position, by moving the piston through a full stroke, and then through short strokes of 2 to 4 inches. In both cases, the resistance felt should be smooth and continuous. If the resistance is jerky, uneven, or if there is any visible sign of wear or damage to the strut, replacement is necessary.

9   Reassembly is a reversal of dismantling, noting the following points:

a) *The coil springs must be installed with the paint mark at the bottom.*

b) *Make sure that the coil spring ends are correctly located in the upper and lower seats before releasing the compressor (see illustration).*

c) *Check that the bearing is correctly installed on the piston rod seat.*

d) *Tighten the thrust bearing retaining nut to the specified torque.*

4   Stabilizer bar and bushings - removal and installation

Warning: *Suspension fasteners are a critical part of the driveability and safe operation of any vehicle. New hardware should be used. Do not reuse the original fasteners.*

Front stabilizer bar

2012 and earlier models

1   Loosen the front wheel lug nuts. Raise the front of the vehicle and support it securely on jackstands. Apply the parking brake and block the rear wheels to keep the vehicle from rolling off the stands. Remove the front wheels.

2   Detach the stabilizer bar links from the bar (see illustration). If the ballstud turns with the nut, use an Allen wrench to hold the stud.

3   Unbolt the stabilizer bar bushing clamps (see illustration). Guide the stabilizer bar out from between the subframe and the body.

4   While the stabilizer bar is off the vehicle, slide off the retainer bushings and inspect

them. If they're cracked, worn or deteriorated, replace them.

5   Clean the bushing area of the stabilizer bar with a stiff wire brush to remove any rust or dirt.

6   Lubricate the inside and outside of the new bushing with vegetable oil (used in cooking) to simplify reassembly.

Note: *The slits of the bushings must face the rear of the vehicle.*

Caution: *Don't use petroleum or mineral-based lubricants or brake fluid - they will lead to deterioration of the bushings.*

7   Installation is the reverse of removal. Tighten the fasteners to the torque values listed in this Chapter's Specifications.

2013 and later models

8   Disconnect the negative battery cable (see Chapter 5). Lock the steering wheel from turning. Loosen the lug nuts to the front wheels, then raise the vehicle and support it securely on jackstands. Remove the wheels.

9   Follow the procedures listed in the *Subframe - removal and installation* (see Section 21) to unbolt and lower the subframe,

4.12 Stabilizer bar link nuts (2013 and later models shown)

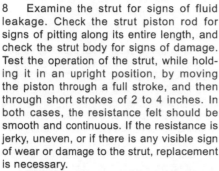

4.13 Stabilizer bar bracket bolts

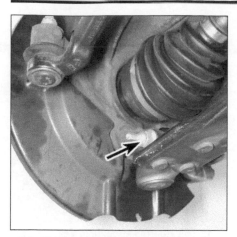

5.2a To detach the control arm from the steering knuckle balljoint, remove this nut and bolt . . .

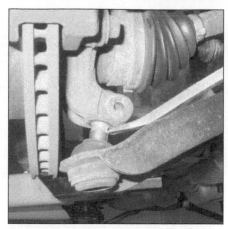

5.2b . . . and pry the balljoint out of the steering knuckle with a prybar or large screwdriver

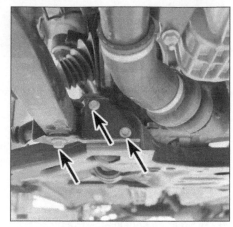

5.2c On 2013 and later models, remove the load path bar-to-bracket bolts and bracket to subframe bolts

unbolt the stabilizer bar brackets and remove the bar. When installing, tighten all fasteners to the torque listed in this Chapter's Specifications.

Note: *The subframe only has to be lowered about four inches.*

10   The stabilizer bar brackets are secured by the control arm rear mounting bolts/nuts.

Rear stabilizer bar

11   Loosen the rear wheel lug nuts. Raise the rear of the vehicle and support it securely on jackstands. Block the front wheels to keep the vehicle from rolling. Remove the rear wheels. (On models so equipped, remove the rear splash shield.

12   Remove the stabilizer bar link nuts from the stabilizer bar (see illustration).

13   Unscrew the bolts and remove the stabilizer bar brackets (see illustration).

14   While the stabilizer bar is off of the vehicle, slide off the the bushings and inspect for cracks, worn areas, and deterioration from age. Replace the bushings if they show any signs of damage.

15   Clean any debris off with a stiff wire brush and remove any rust from the mounting areas.

16   Replace all hardware fasteners with new fasteners.

17   Installation is the reverse of removal.

---

5   Control arm (front) - removal, inspection and installation

---

Warning: *Suspension fasteners are a critical part of the driveability and safe operation of any vehicle. New hardware should be used. Do not reuse the original fasteners.*

Removal

1   Loosen the wheel lug nuts on the side to be dismantled, raise the front of the vehicle, support it securely on jackstands and remove the wheel.

2   Remove the pinch bolt securing the balljoint to the control arm. Use a prybar to disconnect the control arm from the steering knuckle (see illustrations).

3   Remove the control arm front pivot bolt and rear mounting bolt(s) (see illustrations).

4   Remove the control arm.

Inspection

5   Check the control arm for distortion and the bushings for wear, replacing parts as necessary. Do not attempt to straighten a bent control arm.

Note: *Before tightening the pivot bolt and the bolt on the rear of the control arm, raise the outer end of the control arm with a floor jack to simulate normal ride height.*

Installation

6   Installation is the reverse of removal. Tighten all of the fasteners to the torque values listed in this this Chapter's Specifications.

Caution: *Before tightening the front pivot bolt, raise the outer end of the control arm with a floor jack to simulate normal ride height.*

7   Install the wheel and lug nuts, lower the vehicle and tighten the lug nuts to the torque listed in the Chapter 1 Specifications.

8   It's a good idea to have the front wheel alignment checked and, if necessary, adjusted after this job has been performed.

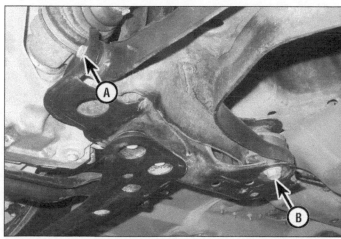

5.3a 2012 and earlier models - to detach the front of the control arm from the subframe, remove the pivot bolt (A) and the rear mounting bolt (B)

5.3b 2013 and later models shown - to detach the front of the control arm from the subframe, remove the pivot bolt (A) and the rear mounting bolts (B)

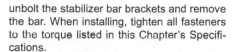

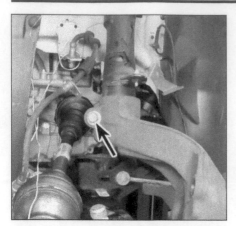

6.11a Steering knuckle-to-strut pinch bolt (2013 and later models)

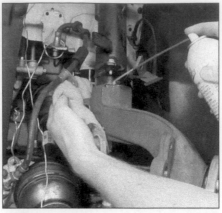

6.11b Spray some penetrating oil around the strut-to-steering knuckle joint and allow it to soak in awhile (2013 and later models)

6.12 The steering knuckle is a tight fit on the strut - you might have to use a little force to get it off (2013 and later models)

6   Steering knuckle and hub - removal and installation

Warning: *Dust created by the brake system is harmful to your health. Never blow it out with compressed air and don't inhale any of it. Do not, under any circumstances, use petroleum-based solvents to clean brake parts. Use brake system cleaner only.*

Warning: *Suspension fasteners are a critical part of the driveability and safe operation of any vehicle. New hardware should be used. Do not reuse the original fasteners.*

Removal

1   Loosen the wheel lug nuts, raise the vehicle and support it securely on jackstands, then remove the wheel. Loosen the driveaxle hub nut (see Chapter 8).

2   Remove the brake caliper (don't disconnect the hose) and the brake disc (see Chapter 9), and disconnect the brake hose from the strut. Hang the caliper from the coil spring with a piece of wire - don't let it hang by the brake hose.

3   If the vehicle is equipped with ABS, remove the wheel speed sensor (see Chapter 9).

4   2012 and earlier models: Loosen, but don't remove the strut-to-steering knuckle nuts and bolts (see Section 2).

5   Separate the tie-rod end from the steering knuckle arm (see Section 15).

6   Remove the balljoint-to-steering knuckle pinch bolt (see Section 5).

7   2012 and earlier models: Remove the driveaxle/hub nut and push the driveaxle from the hub as described in Chapter 8. Support the end of the driveaxle with a piece of wire.

8   Separate the balljoint from the steering knuckle (see Section 5).

9   Remove the brake disc shield.

10   2012 and earlier models: The strut-to-knuckle bolts can now be removed.

11   2013 and later models - Remove the steering knuckle-to-strut pinch bolt. Apply some penetrating oil where the strut meets the knuckle (see illustrations).

12   Separate the steering knuckle from the strut. On 2013 and later models, it might be necessary to tap the knuckle from the strut with a mallet (see illustration). On 2013 and later models, guide the knuckle out of the hub, then support it with wire.

Caution: *On 2013 and later models, do not use a chisel to pry between the pinch joint gap to try to remove the knuckle from the strut.*

Installation

13   Guide the knuckle and hub assembly into position, inserting the driveaxle into the hub.

14   2012 and earlier models: Push the knuckle into the strut flange and install the bolts and nuts, but don't tighten them yet.

15   2013 and later models: Push the knuckle onto the strut, aligning the blade on the strut with the pinch gap in the knuckle.

16   Connect the balljoint to the knuckle and tighten the pinch bolt/nut to the torque listed in this Chapter's Specifications.

17   Attach the tie-rod to the steering knuckle arm (see Section 15). Tighten the strut bolt nuts (2012 and earlier models) or the knuckle-to-strut pinch bolt (2013 and later models) and the tie-rod nut to the torque values listed in this Chapter's Specifications.

18   Install the brake disc shield, disc and caliper (see Chapter 9).

19   Install the driveaxle/hub nut and tighten it to the torque listed in the Chapter 8 Specifications.

20   Install the wheel and lug nuts. Lower the vehicle and tighten the lug nuts to the torque listed in the Chapter 1 Specifications.

21   Tighten the driveaxle/hub nut to the torque listed in the Chapter 8 Specifications.

22   Install the wheel and lug nuts, then lower the vehicle and tighten the lug nuts to the torque listed in the Chapter 1 Specifications.

23   Have the front-end alignment checked and, if necessary, adjusted.

7   Hub and bearing assembly (front) - replacement

Warning: *Suspension fasteners are a critical part of the driveability and safe operation of any vehicle. New hardware should be used. Do not reuse the original fasteners.*

Note: *The hub and bearing assembly is only removed from the steering knuckle by using a press. The steering knuckle must be removed in order to remove the hub and bearing assembly. A special fixture and press adapters are available through the manufacturer's special tool department, but suitable alternative equipment may be available. If you don't have access to a press and the proper adapters, take the steering knuckle to a repair shop or an automotive machine shop to have the bearing pressed out of the steering knuckle and the new one pressed in.*

1   Remove the steering knuckle (see Section 6).

Disassembly

2   Mount the steering knuckle in the press (preferably using the recommended fixture) with the hub flange facing down. Using the proper adapter, press the hub and flange out of the wheel bearing and knuckle.

3   If the bearing inner race did not come out with the hub, press it off the hub using a bearing splitter and the proper adapter.

4   Remove the knuckle from the press, then remove the snap-ring.

5   Place the steering knuckle back in the fixture (if used) and press the bearing out towards the outside (the same direction as the hub and flange was removed) (2012 and earlier models) or towards the inside (2013 and later models).

6   Thoroughly clean the hub and flange, especially the spindle area where the bearing rides. Also clean the knuckle bore.

Reassembly

7   Place the knuckle on the press. Press the bearing into place with the proper-size adapter (the adapter must only contact the outer race of the bearing). On 2012 and earlier models, the bearing is pressed into the knuckle from the outside. On 2013 and later models, the bearing is pressed into the knuckle from the inside.

8   Remove the steering knuckle from the press and install a new snap-ring.

8.3a Remove the fasteners securing the rear trim cover . . .

8.3b . . . followed by the trim cover . . .

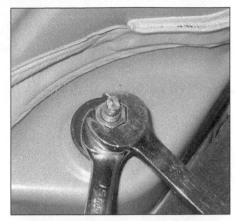

8.3c . . . then unscrew the two upper shock mounting nuts

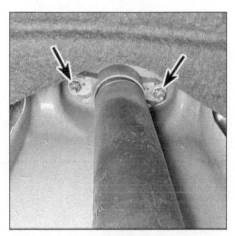

8.4 Shock absorber upper mounting bolts (2013 and later models)

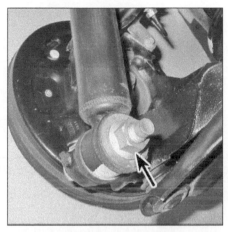

8.5 Shock absorber lower mounting nut (2012 and earlier models)

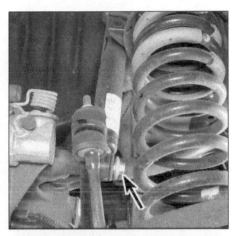

8.6 Shock absorber lower mounting bolt (2013 and later models)

9   Place the knuckle back on the press, outer side facing up, and supported by the properly sized support cup (the inside diameter of the support cup must be large enough to accept the hub spindle).

10   Set the hub spindle squarely into the wheel bearing and, using a driver that contacts the center of the hub flange, press the hub fully into the bearing.

11   Remove the knuckle and hub assembly from the press and check to see that the hub rotates freely.

12   Install the steering knuckle (see Section 6).

---

8   Shock absorber (rear) - removal, inspection and installation

---

Warning: *Suspension fasteners are a critical part of the driveability and safe operation of any vehicle. New hardware should be used. Do not reuse the original fasteners.*

1   Loosen the wheel lug nuts, raise the vehicle and support it securely on jackstands. Block the front wheels to prevent the vehicle from rolling. Remove the wheel.

2   Support the outer end of the lower arm

with a floor jack. Raise the jack slightly to take the spring pressure off the shock absorber lower mount.

Warning: *The jack must remain in this position throughout the entire procedure.*

3   2012 and earlier models - Working inside the vehicle, remove the rear quarter trim panel for access to the shock absorber upper mounting nut. Remove the nuts, upper retainer, bushing and lower retainer (see illustrations).

4   2013 and later models - The upper shock absorber bolts are accessed from inside the fenderwell (see illustration). Use an extension and socket to reach the bolts and remove them.

5   2012 and earlier models: Unscrew the shock absorber lower mounting nut, then pull the shock off the mounting pin (see illustration). Note how the retainers are positioned.

6   2013 and later models: Remove the shock absorber lower mounting bolt and detach the shock from the rear knuckle (see illustration).

7   2013 and later models: Remove the nut from the damper rod and slide the mount off (see illustration). Transfer the mount to the new shock absorber and tighten the mounting nut securely.

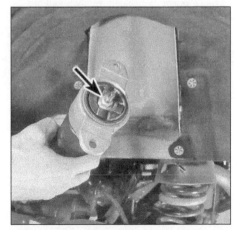

8.7 Remove this nut and slide the upper mount off the damper rod

8   Installation is the reverse of the removal procedure. Tighten the mounting fasteners to the torque listed in this Chapter's Specifications.

Caution: *Before tightening the lower mounting nut or bolt, raise the rear suspension with the floor jack to simulate normal ride height.*

9.2 Remove the nut from the upper ballstud

9.3 Remove the upper pivot bolt from the inner end of the upper suspension arm

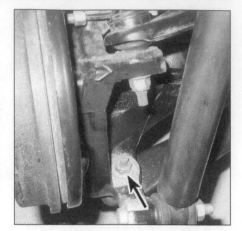

9.5 Remove the nut from the lower ballstud

9 Suspension arms (rear, 2012 and earlier models) - removal and installation

Warning: *Suspension fasteners are a critical part of the driveability and safe operation of any vehicle. New hardware should be used. Do not reuse the original fasteners.*

1 Loosen the wheel lug nuts, raise the vehicle and support it securely on jackstands. Block the front wheels to prevent the vehicle from rolling. Remove the wheel.

Upper arm

2 Remove the nut from the upper ballstud at the outer end of the arm (see illustration), then separate the arm from the boss on the trailing arm. If the ballstud sticks in its boss, a two-jaw puller or balljoint removal tool can be used to force it out.
3 Remove the pivot bolt from the inner end of the arm (see illustration). Remove the arm from the vehicle.
4 Installation is the reverse of removal. Tighten the fasteners to the torque listed in this Chapter's Specifications.

Caution: *Before tightening the inner pivot bolt, raise the rear suspension with the floor jack to simulate normal ride height.*

Lower arm

5 Remove the nut from the lower ballstud at the outer end of the arm (see illustration), then separate the arm from the boss on the trailing arm. If the ballstud sticks in its boss, use a balljoint removal tool to force it out.
6 Remove the pivot bolt from the inner end of the arm (see illustration). Remove the arm from the vehicle.
7 Installation is the reverse of removal. Tighten the fasteners to the torque listed in this Chapter's Specifications.

Caution: *Before tightening the inner pivot bolt, raise the rear suspension with the floor jack to simulate normal ride height.*

Trailing arm

8 Remove the coil spring (see Section 11).
9 Remove the brake shoe assembly or the caliper and brake disc, as applicable (see Chapter 9).
10 Unbolt the brake hose and brake line brackets from the trailing arm. Unbolt the park-

ing brake cable brackets from the trailing arm.
11 If equipped, remove the ABS wheel speed sensor and unbolt the harness brackets from the trailing arm.
12 On 4WD/AWD models, remove the driveaxle from the hub (see Chapter 8).
13 Mark the relationship of the toe adjusting cam to the trailing arm mounting bracket (see illustration).
14 Detach the lower arm from the knuckle (see Step 9).
15 Unscrew the pivot bolt and separate the trailing arm from the vehicle.
16 Inspect the trailing arm pivot bushing for signs of deterioration. If it is in need of replacement, take the trailing arm to an automotive machine shop to have the bushing replaced.
17 Installation is the reverse of removal, noting the following points:

a) *Align the matchmarks on the toe adjusting cam and the mounting bracket.*
b) *Before fully tightening the trailing arm pivot bolt, raise the rear end of the trailing arm with a floor jack to simulate normal ride height.*
c) *Tighten all fasteners to the proper torque specifications.*

9.6 Remove the lower pivot bolt from the inner end of the lower suspension arm

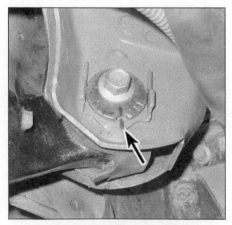

9.13 Mark the relationship of the toe adjusting cam to the trailing arm mounting bracket

10.2 Front lower arm pivot fasteners

10.14a Mark the position of the cam to the subframe at the front . . .

10.14b . . . and rear of the inner pivot bolt

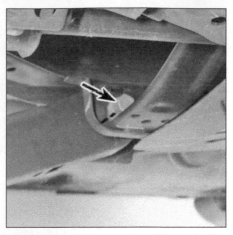

10.15 Rear lower arm-to-subframe bolt

d) It won't be necessary to bleed the brakes unless a hydraulic fitting was loosened.

e) Have the rear wheel alignment checked and, if necessary, adjusted.

10 Suspension arms (rear, 2013 and later models) - removal and installation

Warning: Suspension fasteners are a critical part of the driveability and safe operation of any vehicle. New hardware should be used. Do not reuse the original fasteners.

1   Loosen the wheel lug nuts, raise the vehicle and support it securely on jackstands. Block the front wheels to prevent the vehicle from rolling. Remove the wheels.

Lower arm, front

Left side

2   Support the rear lower arm with a floor jack positioned under the coil spring pocket, then raise the jack slightly. Remove the bolts securing the arm to the subframe and the knuckle (see illustration).

3   Detach the arm from the knuckle and subframe and remove it.

4   Installation is the reverse of removal. Tighten the pivot fasteners to the torque listed in this Chapter's Specifications.

Caution: Raise the suspension with the floor jack to simulate normal ride height before tightening the fasteners.

Right side

5   Remove the shock absorber lower mounting bolts from each side of the vehicle (see Section 8).

6   Remove the splash shields from the underbody.

7   Remove the two bolts securing the right and left trailing arms to the chassis (see illustration 10.32).

8   Support the subframe with a floor jack and block of wood or a transmission jack placed under the center of the crossmember.

Remove the subframe mounting bolts and lower the rear suspension about 2.5 to 3 inches (see Section 21). This will give you the needed room to remove the inner pivot bolt.

Note: The fuel tank impedes removal of the bolt if the subframe isn't lowered.

9   Remove the bolts securing the arm to the subframe and the knuckle (see illustration 10.2).

10   Installation is the reverse of removal. Tighten the suspension fasteners to the torque listed in this Chapter's Specifications.

11   Install the arm and fasteners, raise the subframe and install the subframe mounting bolts, tightening them first. Then connect the shock absorbers to the knuckles and tighten the lower arm fasteners and shock absorber lower mounting fasteners.

Caution: Raise the suspension with the floor jack to simulate normal ride height before tightening the pivot fasteners and shock absorber lower mounting bolts.

Lower arm, rear

12   Remove the coil spring (see Section 11).

13   All Wheel Drive models: Remove the differential mounting crossmember bracket bolts, then set the bracket out of the way.

14   Mark the positions of the toe-adjusting cams on the inner pivot bolt (see illustrations).

15   Remove the rear lower arm pivot bolts (see illustration).

16   Detach the arm from the subframe.

17   Installation is the reverse of removal. Tighten the pivot bolts to the torque listed in this Chapter's Specifications.

Caution: Raise the suspension with the floor jack to simulate normal ride height before tightening the fasteners.

Upper arm

18   Support the rear lower arm with a floor jack positioned under the coil spring pocket, then raise the jack slightly.

19   Remove the upper arm to subframe fasteners and the fasteners that secure the upper arm to the wheel knuckle (see illustration).

20   Work the upper control arm out of the brackets and remove it from the vehicle.

21   Installation is the reverse of removal. Tighten the pivot bolts to the torque listed in this Chapter's Specifications.

Caution: Raise the suspension with the floor jack to simulate normal ride height before tightening the fasteners.

Trailing arm and knuckle

22   Loosen the wheel lug nuts, raise the vehicle and support it securely on jackstands. Remove the rear wheels.

23   Remove the hub and bearing assembly (see Section 12).

Caution: Support the driveaxle with a length of wire - don't let it hang by the inner CV joint.

24   Remove the wheel speed sensor from the knuckle and disconnect the attached bracket (see Chapter 9). Set the sensor off to the side, out of the way of the knuckle.

25   Disconnect the stabilizer bar link from the knuckle (see Section 4).

26   Using a floor jack placed underneath the coil spring pocket, raise the rear suspension to normal ride height.

27   Remove the shock absorber lower mounting bolt (see Section 8).

10.19 Upper arm fasteners

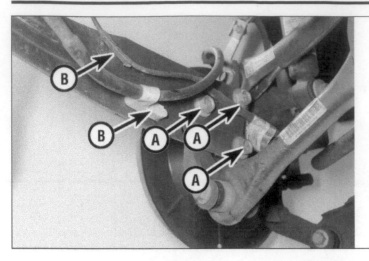

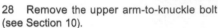

10.30 (A) Do not remove these bolts. (B) Remove the parking brake cable and wire harness only

10.32 Trailing arm pivot shaft bolts

28   Remove the upper arm-to-knuckle bolt (see Section 10).

29   Remove the front and rear lower arm-to-knuckle bolts.

30   Unbolt the parking brake cable and ABS wheel speed sensor wire harness from the trailing arm. Do not remove the trailing arm-to-knuckle bolts. The knuckle and the brace are to be removed as a unit (see illustration).

31   Remove the under body protective shield bolts to expose the trailing arm pivot shaft bolts.

32   Remove the trailing arm pivot shaft bolts (see illustration).

33   Detach the trailing arm from the suspension arms and remove it.

34   Installation is the reverse of removal. Tighten all fasteners to the torque values listed in this Chapter's Specifications.

Caution: *Raise the suspension with the floor jack to simulate normal ride height before tightening the fasteners.*

11   Coil spring (rear) - removal and installation

Warning: *Always replace the springs as a set - never replace just one of them.*

Warning: *Suspension fasteners are a critical part of the driveability and safe operation of any vehicle. New hardware should be used.*

Do not reuse the original fasteners.

1   Loosen the wheel lug nuts, raise the vehicle and support it securely on jackstands. Block the front wheels to prevent the vehicle from rolling. Remove the wheel.

2012 and earlier models

Note: *On models with rear disc brakes, remove the caliper and set it aside.*

2   Unclip the brake line from the body and trailing arm mounting brackets (see illustration).

3   Support the trailing arm with a floor jack (see illustration).

4   On models with rear drum brakes, it may be necessary to disconnect the brake line from the wheel cylinder and move the brake line and bracket out of the way.

5   If equipped, remove the ABS wheel speed sensor and unbolt the harness bracket from the trailing arm.

6   Remove the upper suspension arm (see Section 9 or Section 10). Loosen the lower arm inner bolt.

7   Detach the lower end of the shock absorber from the trailing arm (see Section 8).

8   Mark the position of the spring to the spring insulator (see illustration).

9   Slowly lower the floor jack, pull the trailing arm down and remove the coil spring.

10   Check the spring for cracks and chips,

replacing the springs as a set if any defects are found. Also check the upper insulator for damage and deterioration, replacing it if necessary.

11   Installation is the reverse of removal. Be sure to position the lower end of the coil spring in the depressed area of the trailing arm. Tighten all fasteners to the torque values listed in this Chapter's Specifications.

2013 and later models

12   Support the rear lower arm with a floor jack positioned under the coil spring pocket. Raise the jack slightly.

Note: *If the same coil spring is to be installed, mark the position of the spring to the subframe so it can be installed in the same orientation.*

13   Loosen the rear lower arm inner pivot bolt (see Section 10).

14   Remove the rear lower arm-to-knuckle bolt (see illustration).

15   Check the spring for cracks and chips, replacing the springs as a set if any defects are found. Also check the upper insulator for damage and deterioration, replacing it if necessary.

16   Installation is the reverse of the removal procedure. Tighten the lower arm bolts to the torque listed in this Chapter's Specifications.

Caution: *Raise the suspension with the floor jack to simulate normal ride height before tightening the fasteners.*

11.2 Remove the clip securing the brake line

11.3 Using a floor jack, support the trailing arm

11.8 Mark the relationship of the spring to the spring insulator

11.14 Slowly lower the jack and remove the spring

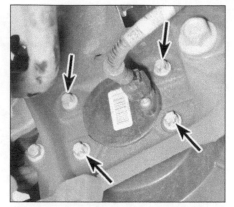

12.7 The manufacturer requires that all four torx bolts should be replaced with new bolts

12 Hub and bearing assembly (rear) - removal and installation

Warning: *Suspension fasteners are a critical part of the driveability and safe operation of any vehicle. New hardware should be used. Do not reuse the original fasteners.*

2012 and earlier models

1 Due to the special tools and expertise required to press the hub and bearing from the steering knuckle, this job should be left to a professional mechanic. However, the trailing arm may be removed and the assembly taken to an automotive machine shop or other qualified repair facility equipped with the necessary tools. See Section 9 for the trailing arm removal procedure.

2013 and later models

2 Loosen the wheel lug nuts, raise the vehicle and support it securely on jackstands. Block the front wheels to prevent the vehicle from rolling. Remove the wheel.
3 If you're working on an AWD model, remove the driveaxle/hub nut (see Chapter 8).
4 Remove brake caliper and disc (see Chapter 9).
5 If you're working on an AWD model, push the driveaxle into the hub using a flange

puller. Push it in just far enough to create enough room to get a socket on the hub and bearing assembly bolts.
6 On FWD models, disconnect the ABS wheel speed sensor electrical connector from the hub (see Chapter 9).
7 Remove the four bolts that secure the hub and bearing assembly to the rear knuckle (see illustration). Discard the original bolts. Remove the hub and bearing assembly from the knuckle.
Caution: *Always use new fasteners on any suspension parts or components.*
Note: *Occasionally the hub can be rusted into place. Penetrating oil and a few hammer blows will generally free the hub from the knuckle.*
8 Clean the mating surfaces of both the replacement hub and the knuckle where the two will touch.
9 Align the bolt holes of the hub assembly with the bolt holes in the knuckle. Be sure the hub has fully seated into the knuckle before tightening the replacement bolts.
10 Tighten the bolts to the torque listed in this Chapter's Specifications.
11 If you're working on an AWD model, pull the driveaxle into the hub with a tubular-type puller that threads onto the end of the drive-axle stub shaft.
12 Install the brake disc, caliper mounting bracket and caliper. Tighten the fas-

teners to the torque listed in the Chapter 9 Specifications.
13 If you're working on an AWD model, install a new hub nut and tighten it to the torque listed in the Chapter 8 Specifications.
14 The remander of the installation is the reverse of removal.

13 Steering wheel - removal and installation

Warning: *These models are equipped with a Supplemental Restraint System (SRS), more commonly known as airbags. Always disable the airbag system before working in the vicinity of any airbag system component to avoid the possibility of accidental deployment of the airbag(s), which could cause personal injury (see Chapter 12).*
Warning: *Do not use a memory saving device to preserve the PCM or radio memory when working on or near airbag system components.*
Warning: *Steering fasteners are a critical part of the driveability and safe operation of any vehicle. New hardware should be used. Do not reuse the original fasteners.*

2007 and earlier models
Note: *On these models, the steering wheel and airbag are serviced as an assembly.*

Steering wheel/airbag removal
1 Turn the ignition key to Off, then disconnect the cable from the negative terminal of the battery (see Chapter 5). Wait at least two minutes before proceeding.
2 Turn the steering wheel so the wheels are pointing straight ahead. Open the cover on the bottom of the steering wheel (see illustration).
3 Disengage the connector lock and disconnect the airbag module electrical connector (see illustration).
Note: *It may take quite a few turns before the steering wheel will come off, but don't loosen the bolt any more than necessary.*
4 Inside the cover, loosen the pinion shaft until the steering wheel can be slid off the steering column (see illustration).

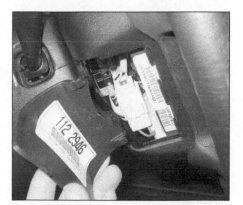

13.2 At the bottom of the steering wheel, open the cover to gain access to the airbag harness

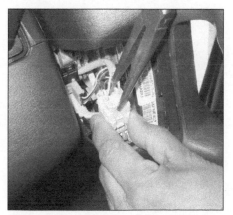

13.3 With a pair of pliers, disengage the airbag connector

13.4 Using a socket, loosen the steering wheel pinion bolt

13.5 Lift the steering wheel off the steering shaft then unplug the electrical connector

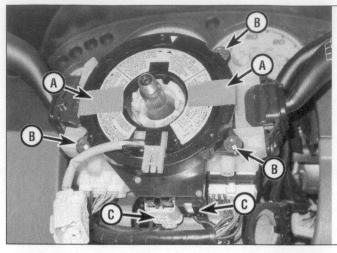

13.7 Apply two pieces of tape to the clockspring to prevent it from rotating (A), then remove the three airbag clockspring mounting screws (B) and disconnect the electrical connectors (C)

5   Pull off the steering wheel and unplug the electrical connector for the horn (see illustration). Set the steering wheel/airbag module in a safe, isolated area.
Warning: *Carry the steering wheel/airbag module with the trim side facing away from you, and set the steering wheel/airbag module down with the trim side facing up. Don't place anything on top of the steering wheel/airbag module.*

Clockspring removal
6   If it is necessary to remove the clockspring, remove the steering column covers (see Chapter 11).
7   Unplug the clockspring electrical connector, then remove the three screws and detach it from the combination switch (see illustration).

Clockspring installation
8   When installing the clockspring, make absolutely sure that the clockspring is centered with the arrow on the clockspring pointing up. This shouldn't be a problem as long as you taped the clockspring and you have not turned the steering shaft while the wheel was removed. If for some reason the hub was turned, center the clockspring by turning the

hub clockwise by hand until it stops (but don't force it, as the cable could break). Rotate the hub counterclockwise about 2-3/4 turns and align the two pointers (see illustration).

Steering wheel installation
9   If removed, reinstall the steering column covers (see Chapter 11).
10   To install the wheel, plug in the horn connector and place the steering wheel on the steering shaft.
11   Inside the cover, tighten the pinion shaft to the torque listed in this Chapter's Specifications.
12   Plug in the electrical connector for the airbag module and close the locking device.
13   After you've reconnected the battery, the Powertrain Control Module (PCM) must relearn its idle and fuel mixture trim strategy for optimum driveability and performance (see Chapter 5, Section 1 for this procedure).

2008 and later models
Driver's airbag removal
14   Point the front wheels straight ahead. Disconnect the negative battery cable (see Chapter 5).

13.8 The clockspring must be centered with the arrow pointing up

15   Insert a small Allen wrench or a similar tool into the airbag release hole on the bottom of the steering wheel (see illustration). On 2013 and later models, use a flat screwdriver and come in from the back of the steering wheel (see illustrations).

13.15a 2008 through 2012 models: Insert a small stiff tool into this opening and push it against the spring inside to release the airbag module

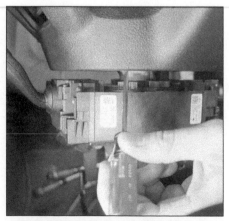

13.15b 2013 and later models: Use a flat screwdriver at about a 50-degree angle from the steering column shaft. Disengage the clips by rotating the screwdriver

13.15c 2013 and later models: The best method is to have the steering wheel at a 90-degree angle while twisting the screwdriver at the same time lifting the edge of the airbag off of the steering wheel. This will keep the airbag from reattaching while you're trying to remove the opposite side

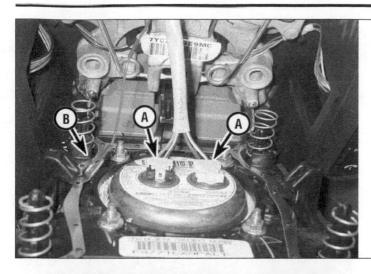

13.17a Release the connector locks, then disconnect the airbag wiring (A) and the horn wire (B)

13.17b 2013 and later models: Before removing the airbag, look for this connector . . .

13.17c . . . pull the connector out of the pocket it's in and disconnect it . . .

13.17d . . . then disconnect the main and safety clips to the airbag itself

the trim side facing away from you, and set it down with the trim side facing up. Don't place anything on top of the airbag module.

Steering wheel removal

Caution: *Return the steering wheel to the centered position before removing it.*

19   Disconnect the wiring from the clock-spring (see illustration), then remove the steering wheel bolt.

20   On 2008 through 2012 models, attach a conventional steering wheel puller using the two tapped holes in the steering wheel as pulling points. Tighten the center bolt until the steering wheel is free, then remove it.

Note: *On 2013 and later models, no puller is necessary. The steering wheel should lift off easily with the bolt removed.*

21   Use tape to secure the center rotor of the clockspring to the column cover so that it can't rotate while the steering wheel is off (see illustration).

16   Push the tool straight inward to disengage the wire clip from the three airbag module hooks, then lift off the airbag module.

17   Disconnect the two airbag electrical connectors and the horn connector (see illustrations).

18   Place the airbag module in a safe place with the trim side up.

Warning: *Carry the airbag module with*

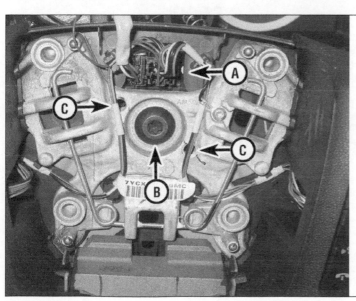

13.19 Disconnect the electrical connectors (A) and remove the bolt (B) before installing the puller in the tapped holes (C, 2008 through 2012 only)

13.21 Tape the clockspring in place so that it does not change position

13.22 Clockspring alignment details (2013 and later models)

1   *Electrical connector at 12 o'clock*
2   *Wiring harness visible in sight glass*
3   *Arrows in alignment*

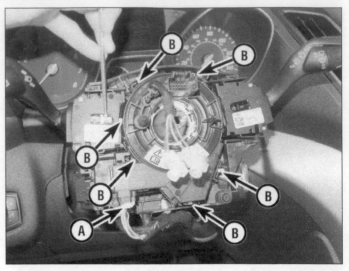

13.25 Clockspring mounting details (2013 and later models shown)

A   *Electrical connector*
B   *Retaining tabs*

Installation

22   Make sure that the clockspring is centered. If it has become uncentered, rotate the inner rotor counterclockwise until a little resistance can be felt:

a)   *2008 through 2012 models: Turn the inner rotor clockwise about four turns. The window should be at the four o'clock position, the yellow area should be visible in the window, the arrow should be aligned at the six o'clock position and the wiring and connector should be at twelve o'clock.*

b)   *2013 and later models: Turn the inner rotor clockwise until the electrical connector is in the 12 o'clock position, then turn it clockwise three turns and return the electrical connector to the 12 o'clock position. The wiring harness should be visible through the sight glass and the two arrows at the bottom should be aligned (see illustration).*

23   The remainder of installation is the reverse of removal. Tighten the steering wheel bolt to the torque listed in this Chapter's Specifications.
Note: *There will be clicks heard when the airbag module snaps into place properly.*

Clockspring - removal and installation

24   Remove steering wheel, if not already done.
25   On 2008 to 2012 models, the clockspring is held in place with two screws. On 2013 and later models, disconnect the clockspring electrical connector and release the six locking tabs. Then pull the clockspring off of the steering column (see illustration).
26   Installation is the reverse of removal. Be sure to center the clockspring as described earlier.

14   Steering column - removal and installation

Warning: *These models are equipped with a Supplemental Restraint System (SRS), more commonly known as airbags. Always disable the airbag system before working in the vicinity of any airbag system component to avoid the possibility of accidental deployment of the airbag(s), which could cause personal injury (see Chapter 12).*
Warning: *Do not use a memory saving device to preserve the PCM or radio memory when working on or near airbag system components.*
Warning: *Steering fasteners are a critical part of the driveability and safe operation of any vehicle. New hardware should be used. Do not reuse the original fasteners.*

Removal

1   Park the vehicle with the wheels pointing straight ahead. Disconnect the negative battery cable (see Chapter 5).
2   Remove the steering wheel (see Section 13), then turn the ignition key to the Lock position to prevent the steering shaft from turning.
Caution: *If this is not done, the airbag clockspring could be damaged.*
3   Remove the steering column covers (see Chapter 11).
4   Remove the clockspring (see Section 13).
5   Disconnect the electrical connectors for the multi-function switch (see Chapter 12).
6   If equipped, remove the shift cable from the shift lever and cable mounting bracket (see Chapter 11).
7   Mark the relationship of the U-joint to the intermediate shaft, then remove and discard the pinch bolt and nut (see illustration).

8   Remove the steering column mounting fasteners (see illustrations), lower the column and pull it to the rear, making sure nothing is still connected, then remove the column.

Installation

9   Guide the steering column into position, then install the steering column mounting fasteners and tighten them to the torque listed in this Chapter's Specifications.
10   Connect the U-joint to the intermediate shaft. Install a new intermediate shaft pinch bolt and nut, tightening it to the torque listed in this Chapter's Specifications.
11   The remainder of installation is the reverse of removal. After you've reconnected the battery, the Powertrain Control Module (PCM) must relearn its idle and fuel mixture trim strategy for optimum driveability and performance (see Chapter 5, Section 1 for this procedure).

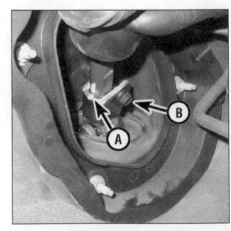

14.7 Mark the U-joint-to-intermediate shaft (A), then remove the pinch bolt and nut (B)

14.8a On 2008 through 2012 models, remove the through-bolt at the top of the steering column . . .

14.8b . . . and the two bolts securing the underside

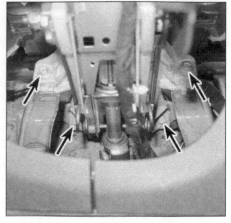

14.8c Steering column mounting bolts - 2013 and later models

15 Tie-rod ends - removal and installation

Warning: *Steering fasteners are a critical part of the driveability and safe operation of any vehicle. On models with self-locking nuts, new hardware should be used. Do not reuse the original fasteners.*

Removal

1 Loosen the front wheel lug nuts. Apply the parking brake, raise the front of the vehicle and support it securely on jackstands. Remove the front wheel.

2 Remove the cotter pin (on models so equipped) (see illustration) and loosen the nut on the tie-rod end stud.

Note: *Not all models use cotter pins - others have self-locking nuts.*

3 Hold the tie-rod with a pair of locking pliers or wrench and loosen the jam nut enough to mark the position of the tie-rod end in relation to the threads (see illustrations).

4 Disconnect the tie-rod end from the steering knuckle arm with a puller (see illustration).

Remove the nut and detach the tie-rod end.

5 Unscrew the tie-rod end from the tie-rod.

Installation

6 Thread the tie-rod end on to the marked position and insert the stud into the steering knuckle arm. Tighten the jam nut securely.

7 Install the nut on the stud and tighten it to the torque listed in this Chapter's Specifications. On models that use a castle nut, install a new cotter pin. If the hole for the cotter pin doesn't line up with one of the slots in the nut, tighten the nut an additional amount until it does.

8 Install the wheel and lug nuts. Lower the vehicle and tighten the lug nuts to the torque listed in the Chapter 1 Specifications.

9 Have the alignment checked and, if necessary, adjusted.

16 Steering gear boots - replacement

Warning: *Steering fasteners are a critical part of the driveability and safe operation of any vehicle. On models with self-locking nuts, new*

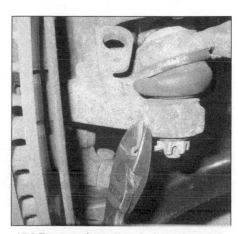

15.2 Remove the cotter pin from the castle nut and loosen - but don't remove - the nut

hardware should be used. Do not reuse the original fasteners.

1 Loosen the lug nuts, raise the vehicle and support it securely on jackstands. Remove the wheel.

2 Remove the tie-rod end and jam nut (see Section 15).

15.3a Loosen the jam nut . . .

15.3b . . . then mark the position of the tie-rod end in relation to the threads

15.4 Disconnect the tie-rod end from the steering knuckle arm with a puller

16.3a The outer ends of the steering gear boot are secured by band-type clamps; they're easily released with a pair of pliers

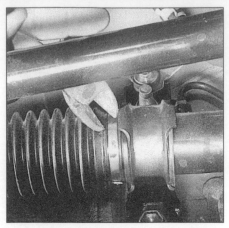

16.3b The inner ends of the steering gear boots are retained by boot clamps which must be cut off and discarded

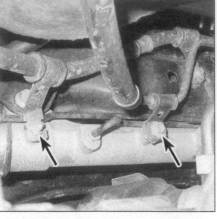

17.3 Remove the fasteners securing the pressure and return line brackets

3   Remove the outer steering gear boot clamp with a pair of pliers (see illustration). Cut off the inner boot clamp with a pair of diagonal cutters (see illustration). Slide off the boot.

4   Before installing the new boot, wrap the threads and serrations on the end of the steering rod with a layer of tape so the small end of the new boot isn't damaged.

5   Slide the new boot into position on the steering gear until it seats in the groove in the steering rod and install new clamps.

6   Remove the tape and install the tie-rod end (see Section 15).

7   Install the wheel and lug nuts. Lower the vehicle and tighten the lug nuts to the torque listed in the Chapter 1 Specifications.

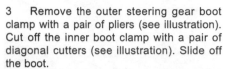

17   Steering gear - removal and installation

Warning: *These models are equipped with airbags. Always disable the airbag system*

before working in the vicinity of airbag system components (see Chapter 12). Make sure the steering column shaft is not turned while the steering gear is removed or you could damage the airbag system clockspring. To prevent the shaft from turning, turn the ignition key to the lock position before beginning work, and run the seat belt through the steering wheel and clip it into its latch.

Warning: *Steering fasteners are a critical part of the driveability and safe operation of any vehicle. New hardware should be used. Do not reuse the original fasteners.*

2007 and earlier models
Removal
Note: *On some vehicles equipped with the 3.0L V6 engine, it may be necessary to remove the EGR valve (see Chapter 2B).*

1   Disconnect the cable from the negative terminal of the battery (see Chapter 1).

2   Remove the rear transaxle mount and bracket.

3   Detach the power steering pressure and

return line bracket bolts (see illustration).

4   From inside the vehicle under the dashboard, slide the steering shaft boot up from the firewall and remove and discard the intermediate shaft pinch bolt and nut (see Section 14).

5   Drain the power steering fluid from the remote power steering reservoir. This can be accomplished with a suction gun or large syringe, or by disconnecting the fluid hose and draining the fluid into a container.

6   Loosen the front wheel lug nuts. Raise the vehicle and place it securely on jackstands. Remove both front wheels.

7   Detach the tie-rod ends from the steering knuckles (see Section 15).

8   Mark the U-joint and the steering gear input shaft with alignment markings for later installation, then separate the U-joint from the steering gear input shaft (see illustration).

9   Place a drain pan under the steering gear and detach the power steering pressure and return lines (see illustration). Cap the ends to prevent excessive fluid loss and contamination.

10   Remove the two steering gear mounting bolts (see illustration).

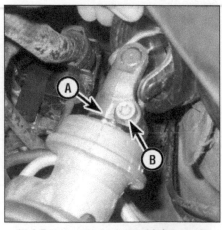

17.8 Remove the universal joint cover, then mark the relationship of the universal joint to the steering gear input shaft (A) and remove the U-joint pinch bolt (B)

17.9 Remove the bolt securing the power steering pressure and return lines

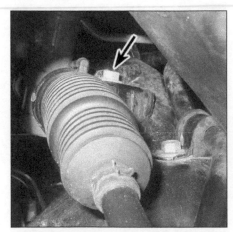

17.10 Steering gear mounting bolt (left side shown, right side similar)

18.6 Disconnect the supply hose from the pump

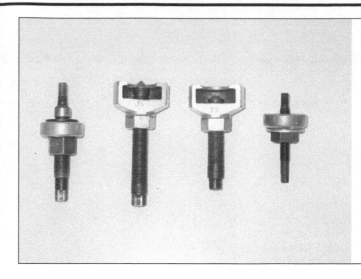

18.7 A pulley removal kit is needed to remove the power steering pulley from the four-cylinder engine pump

11 Remove the steering gear from the vehicle.

Installation
12 Installation is the reverse of removal, noting the following points:

  a) *When connecting the steering gear input shaft to the intermediate shaft U-joint, be sure to align the matchmarks, and install a new pinch bolt and nut*

  b) *Tighten all fasteners to the torque values listed in this Chapter's Specifications.*

  c) *Tighten all fasteners for the rear transaxle mount and mounting bracket to the torque listed in the Chapter 7A Specifications.*

  d) *Tighten the lug nuts to the torque listed in the Chapter 1 Specifications.*

  e) *After you've reconnected the battery, the Powertrain Control Module (PCM) must relearn its idle and fuel mixture trim strategy for optimum driveability and perform any relearn procedures.*

  f) *Fill the power steering reservoir with the recommended fluid (See Chapter 1).*

  g) *Bleed the power steering hydraulic system as described in Section 20.*

  h) *Have the front end alignment checked and, if necessary, adjusted.*

2008 through 2012 models
Removal
13 Disconnect the cable from the negative terminal of the battery (see Chapter 1).
14 Loosen the front wheel lug nuts. Raise the vehicle and support it securely on jackstands. Point the front wheels straight ahead. Remove the front wheels.
15 Unbolt the steering gear from the steering shaft.
16 Loosen the two steering gear mounting bolts.
17 Remove the steering gear shield, if so equipped.
18 Separate the tie-rods from the steering knuckles (see Section 15).
19 On 4WD models, remove the through-bolt from the rear transaxle mount.

20 On four-cylinder models with an automatic transaxle, remove the transmission vibration damper.
21 Remove the steering gear mounting bolts.
Note: *It may be necessary for the rear of the front subframe to be lowered slightly in order to remove these bolts.*
22 Remove the steering gear through the left side.
Note: *On 4WD models, pull the driveshaft down to get sufficient clearance.*

Installation
23 Installation is the reverse of removal, noting the following points:

  a) *Center the steering gear in its travel before installing it.*

  b) *Tighten all fasteners to the torque values listed in this Chapter's Specifications.*

  c) *Have the front end alignment checked and, if necessary, adjusted.*

2013 and later models
24 Remove the subframe (see Section 21).
25 Remove the boot from the steering gear input shaft.
26 Remove the steering gear bolts, rotate the stabilizer bar upward and detach the steering gear from the subframe.
27 Installation is the reverse of removal. Tighten the steering gear mounting bolts, in sequence, to the torque listed in this Chapter's Specifications.
28 Install the subframe (see Section 21).

---

18 Power steering pump (2007 and earlier models) - removal and installation

---

Note: *2008 and later models are equipped with EPS (Electronic Power Steering) and do not have a power steering pump.*
1 Disconnect the cable from the negative battery terminal (see Chapter 1).
2 Using a large syringe or suction gun, suck as much fluid out of the power steering

fluid reservoir as possible. Place a drain pan under the vehicle to catch any fluid that spills out when the hoses are disconnected.

Removal
Four-cylinder engine
3 Remove the drivebelt.
4 Disconnect the electrical connector at the power steering pump.
5 Unscrew and remove the bolt securing the high pressure fluid line support bracket. Using a flare-nut wrench, unscrew the pressure line fitting from the pump.
6 Disconnect the supply hose from the power steering pump (see illustration).
7 Unscrew and remove the four mounting bolts, and withdraw the power steering pump from its bracket. The pulley can be removed on the work bench, using a pulley removal tool (see illustration).

V6 engine
8 Remove the ground wire (see illustration).
9 Loosen the right front wheel lug nuts. Raise the front of the vehicle and support it securely on jackstands, then remove the wheel. Remove the drivebelt.

18.8 Remove the nut securing the ground strap

18.12 With the pulley secured, remove the pulley mounting nut

18.14 Unplug the electrical connector (A), then remove the bolt securing the pressure line bracket (B)

10   Install the wheel and lower the vehicle to the ground.
11   Support the engine with a floor jack and block of wood.
12   Remove the pulley from the power steering pump (see illustration).
13   Disconnect the fluid return hose from the pump, allowing the fluid to drain into a suitable container.
14   Unplug the electrical connector from the power steering fluid pressure switch and remove the bolt securing the pressure line bracket (see illustration).
15   Disconnect the power steering pressure hose from the pump, allowing the fluid to drain into a suitable container. Use two wrenches; a backup wrench to hold the fitting on the pump tube, the other to loosen the fitting.
16   Under the power steering reservoir, remove the refrigerant line bracket-to-body nut.
17   Remove the upper engine mount.
18   Unscrew the power steering mounting bolts, then remove the bracket and the pump (see illustrations).

Installation

19   Installation is the reverse of removal, noting the following points:
a)   *Tighten the bolts and unions securely.*
b)   *The O-ring on the high-pressure outlet should be replaced. Use a special tool or a tapered tube to slide the new O-ring onto the pipe union (see illustrations).*
c)   *On 3.0L V6 models, when installing the engine mounts, refer to Chapter 2B.*
d)   *Install the drivebelt.*
e)   *After you've reconnected the battery, the Powertrain Control Module (PCM) must relearn its idle and fuel mixture trim strategy for optimum driveability and performance (see Chapter 5, Section 1 for this procedure).*
f)   *Fill the power steering reservoir with the recommended fluid. Bleed the power steering hydraulic system as described in Section 19.*

19   Power steering system (2007 and earlier models) - bleeding

1   The power steering system must be bled whenever a line is disconnected. Bubbles can be seen in power steering fluid that has air in it and the fluid will often have a tan or milky appearance. Low fluid level can cause air to mix with the fluid, resulting in a noisy pump as well as foaming of the fluid.
2   Open the hood and check the fluid level in the reservoir, adding the specified fluid necessary to bring it up to the proper level.
3   Start the engine and slowly turn the steering wheel several times from left-to-right and back again. Do not turn the wheel completely from lock-to-lock. Check the fluid level, topping it up as necessary until it remains steady and no more bubbles are visible.

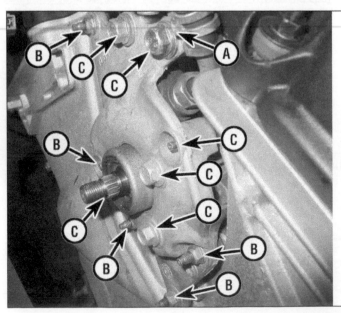

18.18a Remove the screw securing the pressure line bracket (A) followed by the five mounting nuts (B), then remove the six bolts securing the mounting bracket (C) . . .

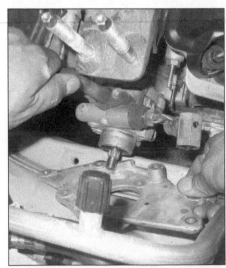

18.18b . . . and carefully remove the bracket and pump

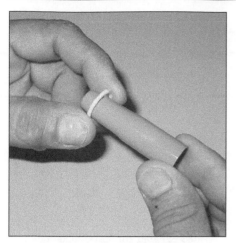

18.19a Slide the teflon seal onto the seal tool . . .

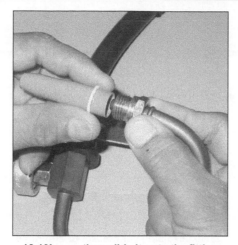

18.19b . . . then slide it onto the fitting

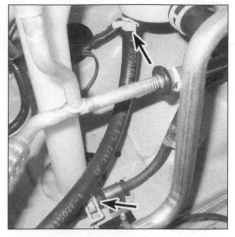

20.2 Using a pair of pliers, release the clamps securing the hoses

20 Power steering fluid cooler (2007 and earlier models) - removal and installation

Removal

1   Using a large syringe or suction gun, suck as much fluid out of the power steering fluid reservoir as possible. Place a drain pan under the vehicle to catch any fluid that spills out when the hoses are disconnected.
2   Disconnect the cooler hoses (see illustration).
3   Remove the front bumper cover (see Chapter 11).
4   Unscrew the fluid cooler mounting nuts and remove the cooler from the vehicle (see illustration).

Installation

5   Installation is the reverse of removal. Fill the power steering reservoir with the recommended fluid. Bleed the power steering hydraulic system as described in Section 19.

21 Subframe - removal and installation

Warning: *These models are equipped with airbags. Always disable the airbag system before working in the vicinity of airbag system components (see Chapter 12). Make sure the steering column shaft is not turned while the steering gear is removed or you could damage the airbag system clockspring. To prevent the shaft from turning, turn the ignition key to the lock position before beginning work, and run the seat belt through the steering wheel and clip it into its latch.*
Warning: *Suspension and steering fasteners are a critical part of the driveability and safe operation of any vehicle. New hardware should be used. Do not reuse the original fasteners.*

1   Disconnect the negative battery cable (see Chapter 5).

20.4 Remove the two fasteners securing the cooler

Front subframe

2007 and earlier models

2   Loosen the front wheel lug nuts. Raise the front of the vehicle and support it securely on jackstands. Apply the parking brake and block the rear wheels to keep the vehicle from rolling off the stands. Remove the front wheels.
3   Turn the wheels to the straight ahead position and lock the steering wheel in place.
4   Install an engine support brace across the fenders to carry the weight of the engine and transmission.
5   Remove the rear transaxle mount.
6   Locate and remove the power steering hose bracket bolt.
7   Remove the lateral support crossmember.
8   Remove the splash shields.
9   Remove the engine support crossmember.
10   Remove the balljoint-to-steering knuckle pinch bolts from each side of the vehicle (see Section 5).
11   Disconnect the stabilizer bar links from the stabilizer bar (see Section 4).
12   Remove the steering gear bolts.

13   Lift the steering gear out of the sub-frame mounting area.
14   On four-cylinder models, remove the flexible exhaust pipe. On the V6 models, remove the dual-converter Y pipe section.
15   On 4WD models, remove the driveshaft (see Chapter 8).
16   On manual transaxle models, disconnect the shift linkage rods from the transaxle.
17   Support the subframe with a pair of floor jacks.
18   Loosen completely, but do not remove, the two subframe rear mounting bolts.
19   Remove the subframe front mounting nuts.
20   Remove the subframe rear mounting bolts.
21   Carefully lower the subframe with the help of an assistant. Be sure to check for any obstructions, wiring, or components that will need to be moved or guided out of the way as you lower the subframe.
22   Installation is the reverse of removal. Tighten the fasteners to the torque listed in this Chapter's Specifications.
Caution: *When installing the rear bolts, be sure the bolts are fully engaged in their caged nuts before tightening them.*
23   Have the wheel alignment checked and, if necessary, adjusted.

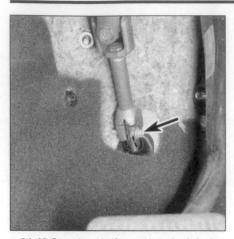

21.46 Steering shaft coupler pinch bolt

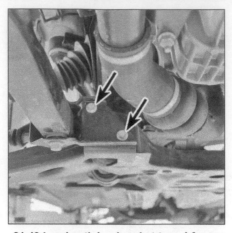

21.49 Load path bar bracket-to-subframe bolts (right side shown)

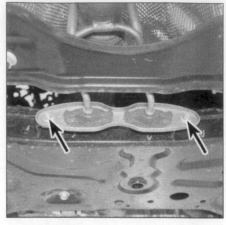

21.53 Exhaust support bracket bolts

2008 through 2012 models

24   Loosen the front wheel lug nuts. Raise the front of the vehicle and support it securely on jackstands. Apply the parking brake and block the rear wheels to keep the vehicle from rolling off the stands. Remove the front wheels.

25   Turn the wheels to the straight ahead position and lock the steering wheel in place.

26   Remove the front driveaxles (see Chapter 8).

27   If you're working on a 4WD/AWD model, remove the front portion of the driveshaft (see Chapter 8).

28   Remove the lateral support crossmember.

29   Remove the exhaust. On four-cylinder models, remove the flexible exhaust pipe. On V6 models, remove the dual-converter Y pipe section.

30   Remove the left-side splash shield.

31   Remove the front insulator bolt from the engine support crossmember.

32   Remove the two engine support crossmember bolts.

33   Remove the two lower bolts from the transmission mount.

34   Remove the engine support crossmember.

35   Remove the rear transaxle mount through-bolt.

36   Remove the steering coupler-to-steering gear pinch bolt.

Note: *Mark the relationship of the coupler to the steering gear shaft.*

37   Detach the tie-rod ends from the steering knuckles (see Section 15).

38   Detach the stabilizer bar links from the stabilizer bar (see Section 4).

39   Remove the pinch bolts and detach the balljoints from the steering knuckles (see Section 5).

40   Support the subframe with a pair of floor jacks.

41   Loosen, but do not remove, the two subframe rear mounting bolts.

42   Remove the subframe mounting nuts.

43   Remove the subframe mounting bolts.

44   Carefully lower the subframe with the help of an assistant. Be sure nothing is still attached or in the way.

a) *Installation is the reverse of removal. Tighten all fasteners to the torque values listed in this Chapter's Specifications.*

b) *Have the wheel alignment checked and, if necessary, adjusted.*

2013 and later models

45   Turn the wheels to the straight ahead position and lock the steering wheel in place.

46   Working under the dash, remove the pinch bolt and disconnect the steering column shaft coupler from the steering gear input shaft (see illustration).

47   Loosen the front wheel lug nuts. Raise the front of the vehicle and support it securely on jackstands. Apply the parking brake and block the rear wheels to keep the vehicle from rolling off the stands. Remove the front wheels.

48   Remove the under-vehicle splash shields.

49   Remove the load path bar bracket bolts (see illustration), if equipped.

50   Disconnect the stabilizer bar links from the stabilizer bar (see Section 4).

51   Separate the tie-rod ends from the steering knuckles (see Section 15).

52   Remove the pinch bolts and disconnect the balljoints from the steering knuckles (see Section 5).

53   Remove the exhaust support bracket bolts and brace the flexible exhaust pipe section from moving too far out of its original position (see illustration).

Note: *Securing the flexible exhaust pipe section from dangling or hanging by its own weight will prevent it from internally tearing the flexible membrane.*

54   Remove the roll restrictor through-bolt (see illustration).

55   Remove the clips securing the EPAS (Electronic Power Assisted Steering) wiring harness to the subframe.

Note: *At this point it's a good idea to match mark the subframe location to the body section so that it can be easily reattached to its original position.*

56   Support the subframe with a pair of floor jacks or a transmission jack.

Note: *Transmission jack head adapters are available that will fit in place of most floor jack heads.*

57   Double-check for any obstructions or wiring that needs to be moved or disconnected, then lower the subframe following the next set of sequences.

a) *Remove the innermost bolts securing the lower stiffener brackets from each side of the chassis to the subframe.*

b) *Remove the two small bolts from each of the lower stiffener brackets.*

c) *Remove the larger bolt from each of the lower stiffener brackets.*

d) *Remove the upper main bolts securing the sub-frame to the body, located above the stabilizer to end link attachment points.*

58   Lower the subframe by about four inches and remove the heat shield from the top of the steering gear. The ground strap is attached to the heat shield too. Position the ground strap out of the way. Then reach up and disconnect the electrical connections to the steering gear.

59   Carefully lower the subframe the remainder of the way.

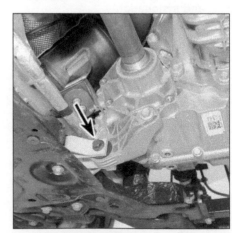

21.54 Remove the roll restrictor through-bolt

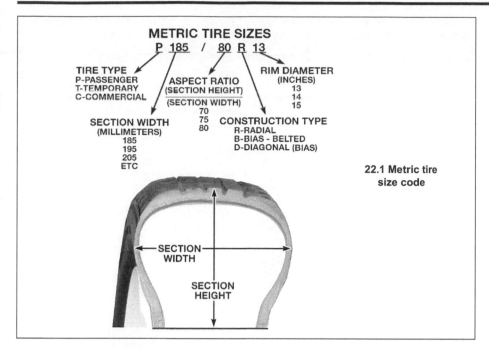

22.1 Metric tire size code

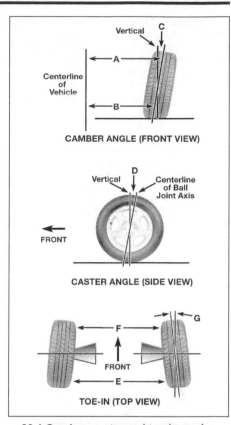

23.1 Camber, caster and toe-in angles

A minus B = C (degrees camber)
D = degrees caster
E minus F = toe-in (measured in inches)
G = toe-in (expressed in degrees)

60   Installation is the reverse of removal. Tighten the fasteners to the torque listed in this Chapter's Specifications.
61   Have the wheel alignment checked and, if necessary, adjusted.

Rear subframe

2012 and earlier models

62   Loosen the rear wheel lug nuts. Raise the vehicle and support it securely on jackstands. Place the vehicle in neutral so that the wheels can be rotated once it has been lifted off the ground. Remove the rear wheels.
63   If you're working on a 4WD model, remove the rear differential (see Chapter 8).
64   Remove the rear stabilizer bar link upper nuts and detach the links (see Section 4).
65   Remove the four rear control arm-to-subframe bolts (see Section 9).
66   Remove any of the exhaust hangers attached to the subframe area.
67   Support the center of the subframe with a floor jack, then remove the four subframe mounting bolts.
68   With the help of an assistant, carefully lower the subframe.
69   Installation is the reverse of removal. Tighten all fasteners to the torque values listed in this Chapter's Specifications. Have the wheel alignment checked and, if necessary, adjusted.

2013 and later models

70   Loosen the rear wheel lug nuts. Raise the vehicle and support it securely on jackstands. Place the vehicle in neutral so that the wheels can be rotated once it has been lifted off the ground. Remove the rear wheels.
71   Remove the coil springs (see Section 11).
72   Detach the upper arms from the knuckles (see Section 10).
73   Detach the front lower arms from the knuckle (see Section 10).

74   Remove any exhaust hangers that are attached to the subframe.
75   If you're working on an AWD model:
   a) *Remove the rear driveaxles and the driveshaft (see Chapter 8).*
   b) *Disconnect the rear differential vent hose retaining clips.*
   c) *Unplug the rear differential electrical connector, then unclip the harness and set it out of the way.*
76   Remove the shield rfom the bottom of the subframe.
77   Support the subframe (or differential/subframe on AWD models), then carefully lower the subframe while checking for any obstructions, wires or components that may hinder the removal.
78   Installation is the reverse of removal. Tighten all suspension fasteners to the torque values listed in this Chapter's Specifications. On AWD models, tighten the driveshaft and driveaxle fasteners to the torque values listed in the Chapter 8 Specifications.
Caution: *Before tightening the suspension fasteners, raise the outer ends of the rear lower arms with a floor jack to simulate normal ride height.*
79   Have the wheel alignment checked and, if necessary, adjusted.

22   Wheels and tires - general information

1   All vehicles covered by this manual are equipped with metric-sized fiberglass or steel belted radial tires (see illustration). Use of other size or type of tires may affect the ride and handling of the vehicle. Don't mix different types of tires, such as radials and bias belted, on the same vehicle as handling may be seriously affected. It's recommended that tires be replaced in pairs on the same axle, but if only

one tire is being replaced, be sure it's the same size, structure and tread design as the other.
2   Because tire pressure has a substantial effect on handling and wear, the pressure on all tires should be checked at least once a month or before any extended trips.
3   Wheels must be replaced if they are bent, dented, leak air, have elongated bolt holes, are heavily rusted, out of vertical symmetry or if the lug nuts won't stay tight. Wheel repairs that use welding or peening are not recommended.
4   Tire and wheel balance is important in the overall handling, braking and performance of the vehicle. Unbalanced wheels can adversely affect handling and ride characteristics as well as tire life. Whenever a tire is installed on a wheel, the tire and wheel should be balanced by a shop with the proper equipment.

23   Wheel alignment - general information

1   A wheel alignment refers to the adjustments made to the wheels so they are in proper angular relationship to the suspension and the ground. Wheels that are out of proper alignment not only affect vehicle control, but also increase tire wear. The front end angles normally measured are camber, caster and toe-in (see illustration). Front toe-in and cam-

ber are adjustable; if the caster is not correct, check for bent components. Rear toe-in is also adjustable, and rear camber is adjustable on 2013 and later models.

2   Getting the proper wheel alignment is a very exacting process, one in which complicated and expensive machines are necessary to perform the job properly. Because of this, you should have a technician with the proper equipment perform these tasks. We will, however, use this space to give you a basic idea of what is involved with a wheel alignment so you can better understand the process and deal intelligently with the shop that does the work.

3   Toe-in is the turning in of the wheels. The purpose of a toe specification is to ensure parallel rolling of the wheels. In a vehicle with zero toe-in, the distance between the front edges of the wheels will be the same as the distance between the rear edges of the wheels. The actual amount of toe-in is normally only a fraction of an inch. On the front end, toe-in is controlled by the tie-rod end position on the tie-rod. On the rear end, it's controlled by a cam at the front of the suspension trailing arm (2012 and earlier models) or on the inner pivot bolt of the front lower arm (2013 and later models). Incorrect toe-in will cause the tires to wear improperly by making them scrub against the road surface.

4   Camber is the tilting of the wheels from vertical when viewed from one end of the vehicle. When the wheels tilt out at the top, the camber is said to be positive (+). When the wheels tilt in at the top the camber is negative (-). The amount of tilt is measured in degrees from vertical and this measurement is called the camber angle. This angle affects the amount of tire tread which contacts the road and compensates for changes in the suspension geometry when the vehicle is cornering or traveling over an undulating surface. On the front end it is adjusted by altering the position of the strut upper mount. on the rear end of 2013 and later models, it's adjusted by replacing the upper suspension arm with a replacement of slightly different length.

5   Caster is the tilting of the front steering axis from the vertical. A tilt toward the rear is positive caster and a tilt toward the front is negative caster.

Chapter 11
Body

Contents

1   General Information

Warning: *The models covered by this manual are equipped with Supplemental Restraint Systems (SRS), more commonly known as airbags. Always disable the air-bag system before working in the vicinity of any airbag system components to avoid the possibility of accidental deployment of the airbags, which could cause personal injury (see Chapter 12).*

1   Certain body components are particu-larly vulnerable to accident damage and can be unbolted and repaired or replaced. Among these parts are the hood, doors, tailgate, lift-gate, bumpers and front fenders.

2   Only general body maintenance prac-tices and body panel repair procedures within the scope of the do-it-yourselfer are included in this Chapter.

2.1 Make sure the damaged area is perfectly clean and rust free. If the touch-up kit has a wire brush, use it to clean the scratch or chip. Or use fine steel wool wrapped around the end of a pencil. Clean the scratched or chipped surface only, not the good paint surrounding it. Rinse the area with water and allow it to dry thoroughly

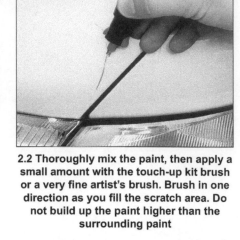

2.2 Thoroughly mix the paint, then apply a small amount with the touch-up kit brush or a very fine artist's brush. Brush in one direction as you fill the scratch area. Do not build up the paint higher than the surrounding paint

2   Repairing minor paint scratches

1   No matter how hard you try to keep your vehicle looking like new, it will inevitably be scratched, chipped or dented at some point. If the metal is actually dented, seek the advice of a professional. But you can fix minor scratches and chips yourself. Buy a touch-up paint kit from a dealer service department or an auto parts store. To ensure that you get the right color, you'll need to have the specific make, model and year of your vehicle and, ideally, the paint code, which is located on a special metal plate under the hood or in the door jamb.

3   Body repair - minor damage

Plastic body panels

1   The following repair procedures are for minor scratches and gouges. Repair of more serious damage should be left to a dealer service department or qualified auto body shop. Below is a list of the equipment and materials necessary to perform the following repair procedures on plastic body panels.

Wax, grease and silicone removing solvent
Cloth-backed body tape
Sanding discs
Drill motor with three-inch disc holder
Hand sanding block
Rubber squeegees
Sandpaper
Non-porous mixing palette
Wood paddle or putty knife
Wood paddle or putty knife
Curved-tooth body file
Flexible parts repair material

Flexible panels (bumper trim)

2   Remove the damaged panel, if necessary or desirable. In most cases, repairs can be carried out with the panel installed.
3   Clean the area(s) to be repaired with a

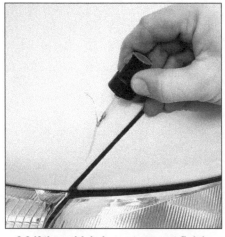

2.3 If the vehicle has a two-coat finish, apply the clear coat after the color coat has dried

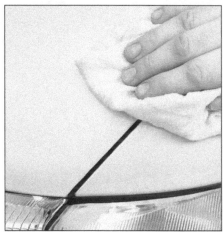

2.4 Wait a few days for the paint to dry thoroughly, then rub out the repainted area with a polishing compound to blend the new paint with the surrounding area. When you're happy with your work, wash and polish the area

wax, grease and silicone removing solvent applied with a water-dampened cloth.
4   If the damage is structural, that is, if it extends through the panel, clean the backside of the panel area to be repaired as well. Wipe dry.
5   Sand the rear surface about 1-1/2 inches beyond the break.
6   Cut two pieces of fiberglass cloth large enough to overlap the break by about 1-1/2 inches. Cut only to the required length.
7   Mix the adhesive from the repair kit according to the instructions included with the kit, and apply a layer of the mixture approximately 1/8-inch thick on the backside of the panel. Overlap the break by at least 1-1/2 inches.
8   Apply one piece of fiberglass cloth to the adhesive and cover the cloth with additional adhesive. Apply a second piece of fiberglass cloth to the adhesive and immediately cover the cloth with additional adhesive in sufficient quantity to fill the weave.
9   Allow the repair to cure for 20 to 30 min-

utes at 60-degrees to 80-degrees F.
10   If necessary, trim the excess repair material at the edge.
11   Remove all of the paint film over and around the area(s) to be repaired. The repair material should not overlap the painted surface.
12   With a drill motor and a sanding disc (or a rotary file), cut a "V" along the break line approximately 1/2-inch wide. Remove all dust and loose particles from the repair area.
13   Mix and apply the repair material. Apply a light coat first over the damaged area; then continue applying material until it reaches a level slightly higher than the surrounding finish.
14   Cure the mixture for 20 to 30 minutes at 60-degrees to 80-degrees F.
15   Roughly establish the contour of the area being repaired with a body file. If low areas or

pits remain, mix and apply additional adhesive.

16   Block sand the damaged area with sandpaper to establish the actual contour of the surrounding surface.

17   If desired, the repaired area can be temporarily protected with several light coats of primer. Because of the special paints and techniques required for flexible body panels, it is recommended that the vehicle be taken to a paint shop for completion of the body repair.

Steel body panels

Repairing simple dents

18   When repairing dents, the first job is to pull the dent out until the affected area is as close as possible to its original shape. There is no point in trying to restore the original shape completely as the metal in the damaged area will have stretched on impact and cannot be restored to its original contours. It is better to bring the level of the dent up to a point that is about 1/8-inch below the level of the surrounding metal. In cases where the dent is very shallow, it is not worth trying to pull it out at all.

19   If the backside of the dent is accessible, it can be hammered out gently from behind using a soft-face hammer. While doing this, hold a block of wood firmly against the opposite side of the metal to absorb the hammer blows and prevent the metal from being stretched.

20   If the dent is in a section of the body which has double layers, or some other factor makes it inaccessible from behind, a different technique is required. Drill several small holes through the metal inside the damaged area, particularly in the deeper sections. Screw long, self-tapping screws into the holes just enough for them to get a good grip in the metal. Now pulling on the protruding heads of the screws with locking pliers can pull out the dent.

21   The next stage of repair is the removal of paint from the damaged area and from an inch or so of the surrounding metal. This is easily done with a wire brush or sanding disk in a drill motor, although it can be done just as effectively by hand with sandpaper. To complete the preparation for filling, score the surface of the bare metal with a screwdriver or the tang of a file or drill small holes in the affected area. This will provide a good grip for the filler material. To complete the repair, see the Section on filling and painting.

Repair of rust holes or gashes

22   Remove all paint from the affected area and from an inch or so of the surrounding metal using a sanding disk or wire brush mounted in a drill motor. If these are not available, a few sheets of sandpaper will do the job just as effectively.

23   With the paint removed, you will be able to determine the severity of the corrosion and decide whether to replace the whole panel, if possible, or repair the affected area. New body panels are not as expensive as most people think and it is often quicker to install a new panel than to repair large areas of rust.

24   Remove all trim pieces from the affected area except those which will act as a guide to the original shape of the damaged body, such as headlight shells, etc. Using metal snips or a hacksaw blade, remove all loose metal and any other metal that is badly affected by rust. Hammer the edges of the hole in to create a slight depression for the filler material.

25   Wire-brush the affected area to remove the powdery rust from the surface of the metal. If the back of the rusted area is accessible, treat it with rust inhibiting paint.

26   Before filling is done, block the hole in some way. This can be done with sheet metal riveted or screwed into place, or by stuffing the hole with wire mesh.

27   Once the hole is blocked off, the affected area can be filled and painted. See the following subsection on filling and painting.

Filling and painting

28   Many types of body fillers are available, but generally speaking, body repair kits which contain filler paste and a tube of resin hardener are best for this type of repair work. A wide, flexible plastic or nylon applicator will be necessary for imparting a smooth and contoured finish to the surface of the filler material. Mix up a small amount of filler on a clean piece of wood or cardboard (use the hardener sparingly). Follow the manufacturer's instructions on the package, otherwise the filler will set incorrectly.

29   Using the applicator, apply the filler paste to the prepared area. Draw the applicator across the surface of the filler to achieve the desired contour and to level the filler surface. As soon as a contour that approximates the original one is achieved, stop working the paste. If you continue, the paste will begin to stick to the applicator. Continue to add thin layers of paste at 20-minute intervals until the level of the filler is just above the surrounding metal.

30   Once the filler has hardened, the excess can be removed with a body file. From then on, progressively finer grades of sandpaper should be used, starting with a 180-grit paper and finishing with 600-grit wet-or-dry paper. Always wrap the sandpaper around a flat rubber or wooden block, otherwise the surface of the filler will not be completely flat. During the sanding of the filler surface, the wet-or-dry paper should be periodically rinsed in water. This will ensure that a very smooth finish is produced in the final stage.

31   At this point, the repair area should be surrounded by a ring of bare metal, which in turn should be encircled by the finely feathered edge of good paint. Rinse the repair area with clean water until all of the dust produced by the sanding operation is gone.

32   Spray the entire area with a light coat of primer. This will reveal any imperfections in the surface of the filler. Repair the imperfections with fresh filler paste or glaze filler and once more smooth the surface with sandpaper. Repeat this spray-and-repair procedure until you are satisfied that the surface of the filler and the feathered edge of the paint are perfect. Rinse the area with clean water and allow it to dry completely.

33   The repair area is now ready for painting. Spray painting must be carried out in a warm, dry, windless and dust free atmosphere. These conditions can be created if you have access to a large indoor work area, but if you are forced to work in the open, you will have to pick the day very carefully. If you are working indoors, dousing the floor in the work area with water will help settle the dust that would otherwise be in the air. If the repair area is confined to one body panel, mask off the surrounding panels. This will help minimize the effects of a slight mismatch in paint color. Trim pieces such as chrome strips, door handles, etc., will also need to be masked off or removed. Use masking tape and several thickness of newspaper for the masking operations.

34   Before spraying, shake the paint can thoroughly, then spray a test area until the spray painting technique is mastered. Cover the repair area with a thick coat of primer. The thickness should be built up using several thin layers of primer rather than one thick one. Using 600-grit wet-or-dry sandpaper, rub down the surface of the primer until it is very smooth. While doing this, the work area should be thoroughly rinsed with water and the wet-or-dry sandpaper periodically rinsed as well. Allow the primer to dry before spraying additional coats.

35   Spray on the top coat, again building up the thickness by using several thin layers of paint. Begin spraying in the center of the repair area and then, using a circular motion, work out until the whole repair area and about two inches of the surrounding original paint is covered. Remove all masking material 10 to 15 minutes after spraying on the final coat of paint. Allow the new paint at least two weeks to harden, then use a very fine rubbing compound to blend the edges of the new paint into the existing paint. Finally, apply a coat of wax

---

4   Body repair - major damage

1   Major damage must be repaired by an auto body shop specifically equipped to perform body and frame repairs. These shops have the specialized equipment required to do the job properly.

2   If the damage is extensive, the frame must be checked for proper alignment or the vehicle's handling characteristics may be adversely affected and other components may wear at an accelerated rate.

3   Due to the fact that all of the major body components (hood, fenders, etc.) are separate and replaceable units, any seriously damaged components should be replaced rather than repaired. Sometimes the components can be found in a wrecking yard that specializes in used vehicle components, often at considerable savings over the cost of new parts.

These photos illustrate a method of repairing simple dents. They are intended to supplement *Body repair - minor damage* in this Chapter and should not be used as the sole instructions for body repair on these vehicles.

1 If you can't access the backside of the body panel to hammer out the dent, pull it out with a slide-hammer-type dent puller. In the deepest portion of the dent or along the crease line, drill or punch hole(s) at least one inch apart . . .

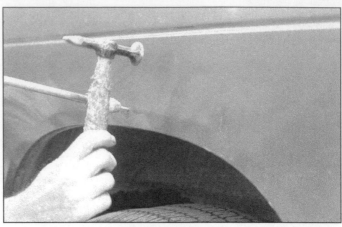

2 . . . then screw the slide-hammer into the hole and operate it. Tap with a hammer near the edge of the dent to help 'pop' the metal back to its original shape. When you're finished, the dent area should be close to its original contour and about 1/8-inch below the surface of the surrounding metal

3 Using coarse-grit sandpaper, remove the paint down to the bare metal. Hand sanding works fine, but the disc sander shown here makes the job faster. Use finer (about 320-grit) sandpaper to feather-edge the paint at least one inch around the dent area

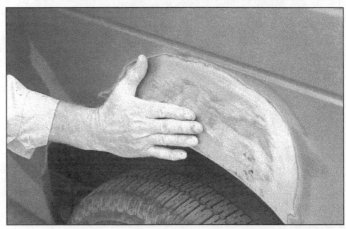

4 When the paint is removed, touch will probably be more helpful than sight for telling if the metal is straight. Hammer down the high spots or raise the low spots as necessary. Clean the repair area with wax/silicone remover

5 Following label instructions, mix up a batch of plastic filler and hardener. The ratio of filler to hardener is critical, and, if you mix it incorrectly, it will either not cure properly or cure too quickly (you won't have time to file and sand it into shape)

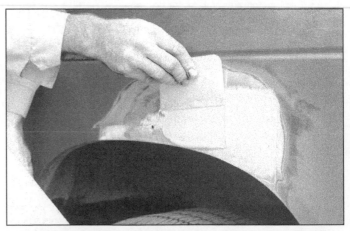

6 Working quickly so the filler doesn't harden, use a plastic applicator to press the body filler firmly into the metal, assuring it bonds completely. Work the filler until it matches the original contour and is slightly above the surrounding metal

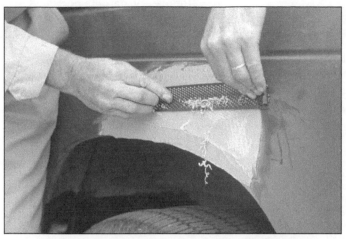

7 Let the filler harden until you can just dent it with your fingernail. Use a body file or Surform tool (shown here) to rough-shape the filler

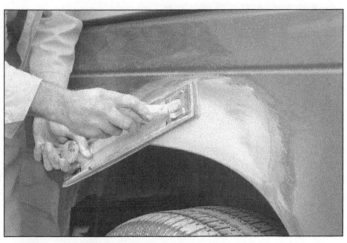

8 Use coarse-grit sandpaper and a sanding board or block to work the filler down until it's smooth and even. Work down to finer grits of sandpaper - always using a board or block - ending up with 360 or 400 grit

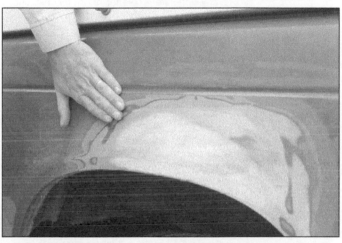

9 You shouldn't be able to feel any ridge at the transition from the filler to the bare metal or from the bare metal to the old paint. As soon as the repair is flat and uniform, remove the dust and mask off the adjacent panels or trim pieces

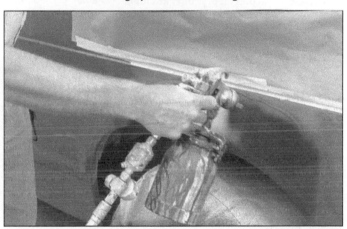

10 Apply several layers of primer to the area. Don't spray the primer on too heavy, so it sags or runs, and make sure each coat is dry before you spray on the next one. A professional-type spray gun is being used here, but aerosol spray primer is available inexpensively from auto parts stores

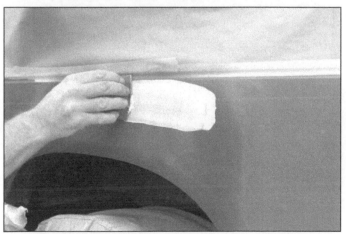

11 The primer will help reveal imperfections or scratches. Fill these with glazing compound. Follow the label instructions and sand it with 360 or 400-grit sandpaper until it's smooth. Repeat the glazing, sanding and respraying until the primer reveals a perfectly smooth surface

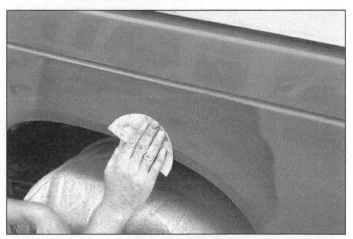

12 Finish sand the primer with very fine sandpaper (400 or 600-grit) to remove the primer overspray. Clean the area with water and allow it to dry. Use a tack rag to remove any dust, then apply the finish coat. Don't attempt to rub out or wax the repair area until the paint has dried completely (at least two weeks)

5   Fastener and trim removal

1   There is a variety of plastic fasteners used to hold trim panels, splash shields and other parts in place in addition to typical screws, nuts and bolts. Once you are familiar with them, they can usually be removed with-

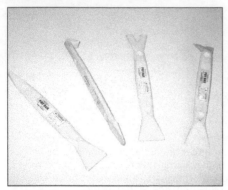

5.4 These small plastic pry tools are ideal for prying off trim panels

out too much difficulty.

2   The proper tools and approach can prevent added time and expense to a project by minimizing the number of broken fasteners and/or parts.

3   The following illustration shows various types of fasteners that are typically used on most vehicles and how to remove and install them (see illustration). Replacement fasteners are commonly found at most auto parts stores, if necessary.

4   Trim panels are typically made of plastic and their flexibility can help during removal. The key to their removal is to use a tool to pry the panel near its retainers to release it without damaging surrounding areas or breaking-off any retainers. The retainers will usually snap out of their designated slot or hole after force is applied to them. Stiff plastic tools designed for prying on trim panels are available at most auto parts stores (see illustration). Tools that are tapered and wrapped in protective tape, such as a screwdriver or small pry tool, are also very effective when used with care.

6   Upholstery, carpets and vinyl trim - maintenance

Upholstery and carpets

1   Every three months remove the floormats and clean the interior of the vehicle (more frequently if necessary). Use a stiff whiskbroom to brush the carpeting and loosen dirt and dust, then vacuum the upholstery and carpets thoroughly, especially along seams and crevices.

2   Dirt and stains can be removed from carpeting with basic household or automotive carpet shampoos available in spray cans. Follow the directions and vacuum again, then use a stiff brush to bring back the "nap" of the carpet.

3   Most interiors have cloth or vinyl upholstery, either of which can be cleaned and maintained with a number of material-specific cleaners or shampoos available in auto supply stores. Follow the directions on the product for usage, and always spot-test any upholstery cleaner on an inconspicuous area (bottom edge of a backseat cushion) to

Fasteners

This tool is designed to remove special fasteners. A small pry tool used for removing nails will also work well in place of this tool

A Phillips head screwdriver can be used to release the center portion, but light pressure must be used because the plastic is easily damaged. Once the center is up, the fastener can easily be pried from its hole

Here is a view with the center portion fully released. Install the fastener as shown, then press the center in to set it

This fastener is used for exterior panels and shields. The center portion must be pried up to release the fastener. Install the fastener with the center up, then press the center in to set it

This type of fastener is used commonly for interior panels. Use a small blunt tool to press the small pin at the center in to release it . . .

. . . the pin will stay with the fastener in the released position

Reset the fastener for installation by moving the pin out. Install the fastener, then press the pin flush with the fastener to set it

This fastener is used for exterior and interior panels. It has no moving parts. Simply pry the fastener from its hole like the claw of a hammer removes a nail. Without a tool that can get under the top of the fastener, it can be very difficult to remove

ensure that it doesn't cause a color shift in the material.

4   After cleaning, vinyl upholstery should be treated with a protectant.

Note: *Make sure the protectant container indicates the product can be used on seats - some products may make a seat too slippery.*

Caution: *Do not use protectant on vinyl-covered steering wheels.*

5   Leather upholstery requires special care. It should be cleaned regularly with saddle-soap or leather cleaner. Never use alcohol, gasoline, nail polish remover or thinner to clean leather upholstery.

6   After cleaning, regularly treat leather upholstery with a leather conditioner, rubbed in with a soft cotton cloth. Never use car wax on leather upholstery.

7   In areas where the interior of the vehicle is subject to bright sunlight, cover leather seating areas of the seats with a sheet if the vehicle is to be left out for any length of time.

Vinyl trim

8   Don't clean vinyl trim with detergents, caustic soap or petroleum-based cleaners. Plain soap and water works just fine, with a soft brush to clean dirt that may be ingrained.

Wash the vinyl as frequently as the rest of the vehicle.

9   After cleaning, application of a high-quality rubber and vinyl protectant will help prevent oxidation and cracks. The protectant can also be applied to weather-stripping, vacuum lines and rubber hoses, which often fail as a result of chemical degradation, and to the tires.

7   Hood - removal, installation and adjustment

Note: *The hood is somewhat heavy and awkward to remove and install - at least two people should perform this procedure.*

Removal

Note: *On models that lack a washer hose disconnection point for hood removal - if the hose is made of flexible rubber, the manufacturer recommends that the hose be cut near the hood hinge, and fitted with a line coupler. If the hose is made of solid corrugated plastic, follow the illustration sequence mentioned below for removal.*

1   Open the hood, then place blankets or pads over the fenders and cowl area of the body. This will protect the body and paint as

the hood is lifted off.

2   Disconnect any cables or wires that will interfere with removal. On early models, disconnect the windshield washer tubing from the fitting on the bottom side of the hood (see illustration).

3   On later models without a washer line disconnection fitting, follow the photo sequence (7.3a through 7.3d) below to detach the line from the hood:

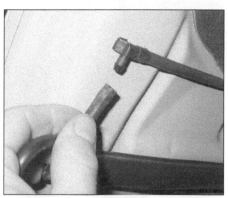

7.2 Disconnect the windshield washer fluid line at the connector on the edge of the hood (early models shown)

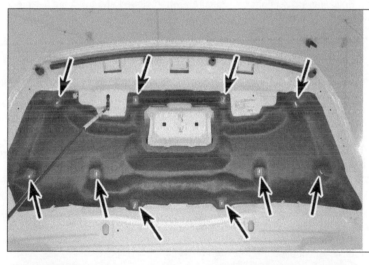

7.3a Remove the hood insulator panel fasteners from the underside of the hood

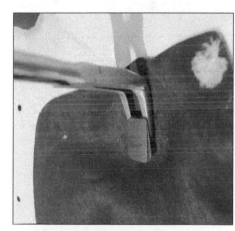

7.3b To remove the plastic fastener, squeeze the ends of the clip then pull it out

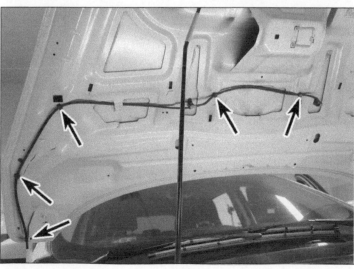

7.3c Detach the hose retaining clips from the points along the hood

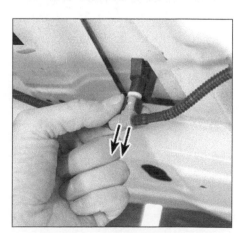

7.3d To disconnect the washer connection at the nozzle, rotate the sleeve and pull off the fitting

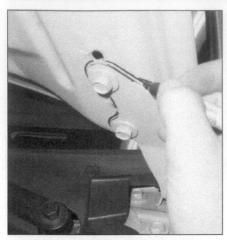

7.4 Draw alignment marks around the hood hinges to ensure proper alignment of the hood when it's reinstalled

7.10 To adjust the hood latch horizontally or vertically, draw a reference mark and loosen the latch bolts

7.11a To adjust the vertical height of the leading edge of the hood so that it's flush with the fenders, turn each edge cushion clockwise (to lower the hood) or counterclockwise (to raise the hood) - early models shown

4　Make marks around the hood hinge to ensure proper alignment during installation (see illustration).

5　Have an assistant support the weight of the hood and remove the hinge-to-hood bolts and lift off the hood.

6　Installation is the reverse of removal. Align the hinge bolts with the marks made in Step 4.

Adjustment

7　Fore-and-aft and side-to-side adjustment of the hood is done by moving the hinge plate slot after loosening the bolts or nuts.

8　Mark around the entire hinge plate so you can determine the amount of movement.

9　Loosen the bolts and move the hood into correct alignment. Move it only a little at a time. Tighten the hinge bolts and carefully lower the hood to check the position.

Note: *On later models, the position of the hood latch cannot be adjusted*

10　If necessary after installation, the entire hood latch assembly can be adjusted up-and-down as well as from side-to-side on the

radiator support so the hood closes securely and flush with the fenders. Scribe a line or mark around the hood latch mounting bolts to provide a reference point, then loosen them and reposition the latch assembly, as necessary (see illustration). Following adjustment, retighten the mounting bolts.

11　Finally, adjust the hood bumpers on the radiator support so the hood, when closed, is flush with the fenders (see illustrations).

12　The hood latch assembly, as well as the hinges, should be periodically lubricated with white, lithium-based grease to prevent binding and wear.

8　Hood latch and release cable - removal and installation

Hood latch

1　Mark the position of the latch to aid in alignment on installation.

2　Remove the hood latch mounting bolts.

3　Release the cable outer sleeve, then

unhook the end of the cable (see illustrations). Disconnect the hood latch electrical connector on later models. Remove the latch.

4　Installation is the reverse of removal. Adjust the latch so that the hood engages securely when closed and the hood bumpers are slightly compressed.

Cable

5　Working in the passenger's compartment-on older models the release lever is mounted on the lower dash panel. On later models it is located on the kick panel. Follow the illustrations for details on removal of the release handle (see illustrations). On later models, the kick panel removal can be found in Section 24.

6　Attach a piece of thin wire or string to the end of the cable (interior).

7　Working in the engine compartment, disconnect the hood release cable from the latch assembly as described in Steps 1 and 2. Unclip all the cable retaining clips on the radiator support and the inner fenderwell.

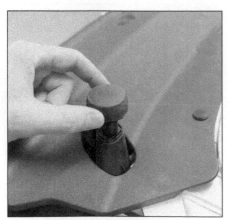

7.11b This same principle applies to the later model design

8.3a Depress the locking tabs on the cable housing to release the cable from the hood latch

8.3b Remove the cable end from the slotted portion of the latch mechanism

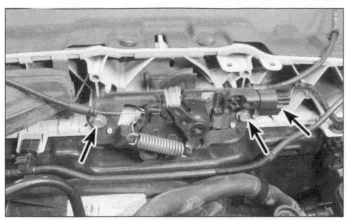

8.3c Later versions of the hood latch may have a different configuration, but are removed basically the same way

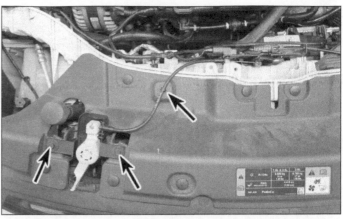

8.3d The newer version safety latch has been moved - remove the bolts and detach the cable retaining clip

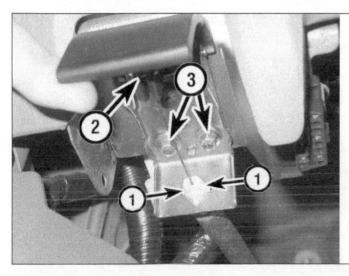

8.5a On older models, lift upward on the handle and depress the tabs on the backside of the cable housing using needle nose pliers (1) to release it from the base of the handle, then detach the cable end from the lever (2). To remove the release handle, remove the mounting bolts (3).

8.5b On later models, the inside release handle is a lever type. The access cover needs to be removed first. Use a pocket screwdriver or trim tool to gently pry the cover off . . .

8   Pull the cable forward into the engine compartment until you can see the wire or string, then remove the wire or string from the old cable and fasten it to the new cable. On some models, remove the battery tray

(see Chapter 5) and unbolt the battery junction box from the fenderwell area to gain access to the hood release cable.

9   With the new cable attached to the wire or string, pull the wire or string back through

the firewall until the new cable reaches the inside handle.

10   Check that the cable grommet is properly seated in the engine compartment firewall (see illustration).

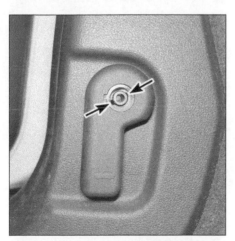

8.5c . . . to remove the lever, insert two thin flathead screwdrivers into the recess to depress the tabs, then slide off the lever. Once the lever is removed, remove the kick panel trim . . .

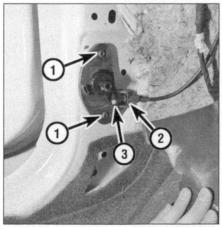

8.5d . . . remove the two fasteners that secure the cable bracket to the kick panel area (1) - the cable can then be removed by sliding the cable housing out of the clip (2), then detaching the cable end (3)

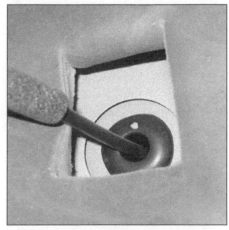

8.10 Be sure the grommet is completely sealing the opening

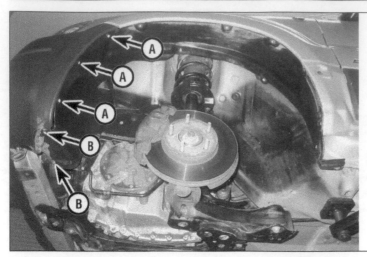

9.3 Remove the pin retainers (A) and the bolts (B) from the lower section of the inner fender splash shield from each side of the vehicle

9.5 Remove the cover-to-fender retainers from each side of the vehicle

11   Working in the passenger's compartment, reinstall the new cable into the hood release lever, making sure the cable housing fits snugly into the notch in the handle bracket.
Note: *Pull on the cable from the passenger's compartment until the cable stop seats in the grommet on the firewall correctly.*
12   The remainder of installation is the reverse of removal.

9   Bumpers and bumper covers - removal and installation

1   Disconnect the cable from the negative battery terminal (see Chapter 5).
2   Apply the parking brake, raise the vehicle and support it securely on jackstands.

2007 and earlier models
Front bumper cover and bumper
3   Remove the pin retainers from the inner fender splash shield (see illustration).
4   Pull the splash shield away from the front of the bumper area.
5   Remove the front bumper cover push-pin type retainers that retain the cover to the

9.7 Remove the bumper cover lower mounting bolts

9.8 Remove the retainers from the front of the bumper cover

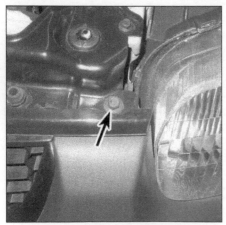

9.9 Remove the upper bumper cover mounting bolts - one on each side of the radiator support

9.10a Remove the front bumper mounting nuts from the left side . . .

9.10b . . . and the right side of the vehicle

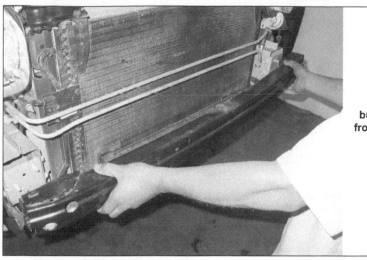

9.11 Lift the bumper from the front of the vehicle

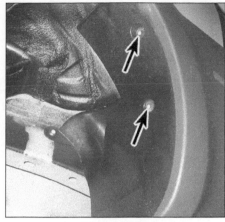

9.13 Remove the retainers from the inner fender splash shield

fender (see illustration).

6   Disconnect the fog lamp connectors, if equipped.

7   Remove the bumper cover lower bolts (see illustration).

8   Remove the pin-type retainers from the front of the bumper cover (see illustration).

9   Remove the two upper bumper cover mounting bolts (see illustration), then remove the cover.

10   Remove the front bumper mounting nuts (see illustrations).

11   Remove the front bumper (see illustration).

12   Installation is the reverse of removal.

Rear bumper cover and bumper

13   Remove the pin retainers from the inner fender splash shield (see illustration). Pull the splash shield away from the rear bumper cover.

14   Remove the rear bumper cover mounting screws from the quarter panel (see illustration).

15   Remove the pin-type retainers from below (see illustration).

16   Remove the rear bumper cover mounting bolts (see illustration), then remove the cover.

17   Remove the rear bumper mounting nuts (see illustrations).

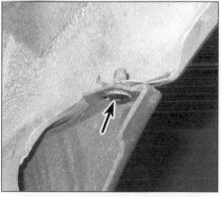

9.14 Remove the screws from the rear bumper cover and fender - one on each side of the vehicle

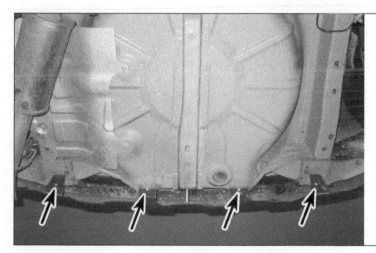

9.15 Remove the retainers from the underside of the rear bumper cover

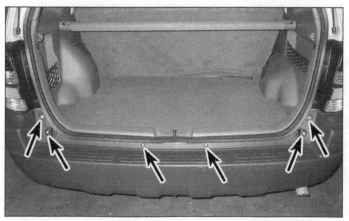

9.16 Remove the rear bumper cover mounting bolts

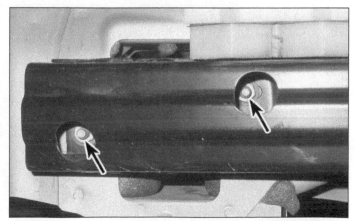

9.17a Remove the rear bumper mounting nuts from the left side . . .

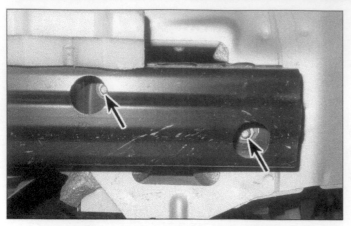

9.17b . . . and the right side of the vehicle

9.18 Remove the bumper from the vehicle

9.22 Grille
and bumper
cover fastener
locations

9.23 Reach behind the bumper cover to
disconnect the wiring from the fog lights

18   Remove the rear bumper (see illustration).

19   Installation is the reverse of removal.

2008 through 2012 models

Front bumper cover

20   Remove the front wheels or alternate turning them for access to the splash shield fasteners.

21   Remove the three pushpins and two bolts that secure the front bumper cover to the inner fender splash shields.

Note: *It's easiest to pull the front of the splash shields out of the way for access to the bumper cover clips.*

22   Remove the air dam fasteners, then remove the two pushpins from the center of the bumper (see illustration).

23   Disconnect the wiring from the parking sensor (if equipped) and the fog lights (see illustration).

24   With the hood open, remove the fasteners at each top corner of the grille (see illustration).

25   Unclip each end of the bumper from the

fender. On 2013 and later models, there are three bolts that will need to be removed as well (see illustrations).

26   Carefully remove the bumper cover from the vehicle, having an assistant help if necessary.

27   Installation is the reverse of removal.

Note: *To remove the front bumper- disengage the plastic clips from each side of the bumper, remove the ambient air temperature sensor from the bumper (if equipped), then remove the front bumper mounting nuts on each side.*

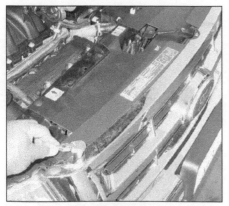

9.24 Remove the two screws from the top
corners of the grille

9.25a The ends of the bumper cover are
attached to the fenders with these
claw-type tabs

9.25b Push in the protrusions on the tabs
and detach the bumper from the fenders

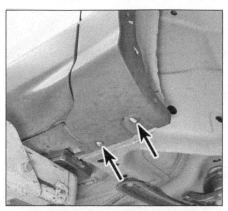

9.39a Remove the outer fender trim rear fasteners . . .

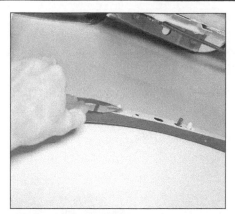

9.39b . . . disengage the inner plastic clips along the inside with needle nose pliers . . .

9.39c . . . use a plastic trim tool to release the end clips, then remove the fender outer trim.

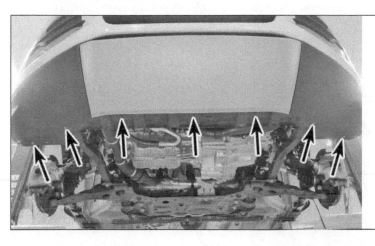

9.40 Bumper cover lower fastener locations

9.41 Disconnect the front bumper cover electrical connector

Rear bumper cover

Note: *On these model years, the replacement procedure for the rear bumper itself is very similar to the procedure mentioned in Steps 65 through 67 (2013 and later models).*

28   Remove the rear wheels for access to the rear wheelwell splash shield fasteners, if necessary.

29   Remove the four screws from the center section of the bumper cover.

30   Remove the rear wheelwell splash shields.

31   Remove the bumper cover screw from each side.

32   Remove the liftgate alignment bumpers.

33   Disconnect any interfering wiring.

34   Carefully lift the bumper cover off.

35   Installation is the reverse of removal.

2013 and later models

Front bumper cover

36   Remove the front wheels.

37   Remove the engine front under cover (see Chapter 1).

38   Remove the front fenderwell splash shields (see Section 10).

39   Remove the fender outer trim (see illustrations).

40   Remove the front bumper cover lower retaining bolts (see illustration).

41   Working from the driver's side fenderwell, disconnect the front bumper cover electrical connector (see illustration).

42   Remove the hood safety release latch (see Section 8).

43   Remove the upper cover fasteners (see illustration), then remove the cover.

44   Remove the front bumper cover upper fasteners (see illustration).

9.43 Upper trim cover plastic fastener locations

9.44 Bumper cover upper fastener locations

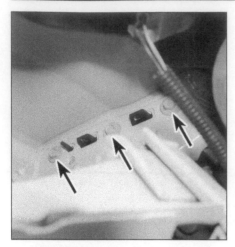

9.45a Remove the bolts securing the bumper cover to the fender (both sides) . . .

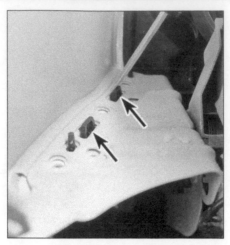

9.45b . . . then release the retaining tabs to free the bumper cover ends.

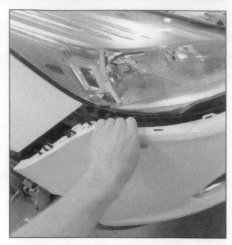

9.46 Pull sharply on the bumper cover ends to disengage them from the fenders and headlight housing clips

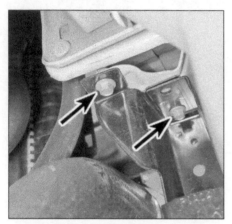

9.51 Bumper-to-subframe bolt locations

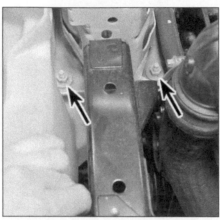

9.52 Bumper lower mounting nuts

48   Remove the active grille shutter assembly (see Chapter 3).
49   Remove the single bolt at each side; one bolt retains the PCM plastic casing (frontmost fastener), the other retains the windshield washer reservoir (also the frontmost fastener).
50   Detach any wiring harness clips that may be attached to the bumper.
51   Working from below, remove the bumper-to-subframe bolts on both sides (see illustration).
52   Remove the front bumper lower mounting nuts on both sides (see illustration).
53   Remove the bumper adjusting bolt assemblies (see illustration).
54   Remove the bumper upper mounting nuts on both sides (see illustration), then remove the front bumper.
55   Installation is the reverse of removal.

Rear bumper cover

56   On a level surface, raise and support the rear of the vehicle on jackstands.
57   Remove the taillight housings (see Chapter 12, Section 19) to gain access to the hidden fasteners.

45   Remove the front bumper cover side fasteners at each end of the fenderwells, then use a screwdriver to release each retaining tab (see illustrations).
46   Pull the bumper cover ends free from the fenders and headlight housing clips (see illus-tration), then remove the bumper cover from the vehicle. This step may be easiest with the help of an assistant.

Front bumper

47   Remove the front bumper cover (see previous Steps).

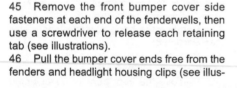

9.53 Bumper adjuster bolts

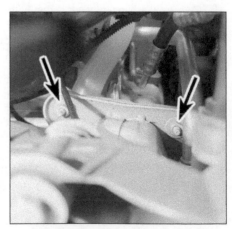

9.54 Bumper upper mounting nuts

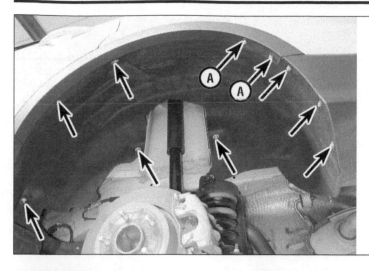

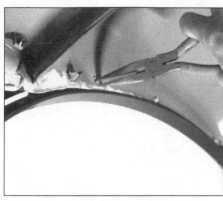

9.58 Remove the rear fender-well outer trim moulding fasteners (A) and the remaining splash shield fasteners

9.59a Disengage the inner plastic clips retaining the trim moulding with needle-nose pliers . . .

9.59b . . . then, using a plastic trim tool, work the remainder of the wheel trim moulding off the fender.

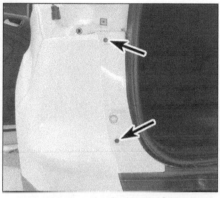

9.60a Remove the fasteners from the taillight housing opening . . .

9.60b . . . then remove the retaining bolt on each side.

58   Remove the fenderwell splash shield (see illustration).
59   Remove the fenderwell trim molding (see illustrations).
60   Remove the fasteners from the removed-taillight housing area and from the exposed moulding trim section (see illustrations).
61   Disconnect the rear bumper electrical connector(s) (see illustration).
62   Remove the fasteners securing the bum-

per to the vehicle from the underside (see illustration).
63   Gently pry and remove the rear bumper from the vehicle, using an assistant's help if necessary.
64   Installation is the reverse of removal.

Rear bumper
Note: *On models with a trailer tow package: remove the retaining bolts at the sides of*

the bumper, then remove the rearmost fasteners. Disconnect the electrical connector (if equipped), then slide out the bumper/ hitch for removal.
65   Remove the rear bumper cover (see previous Steps).
66   Remove the mounting fasteners, then remove the rear bumper.
67   Installation is the reverse of removal.

9.61 Disconnect the bumper cover electrical connector(s) from underneath

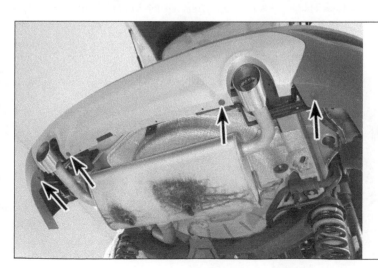

9.62 Remove the bumper cover lower fasteners

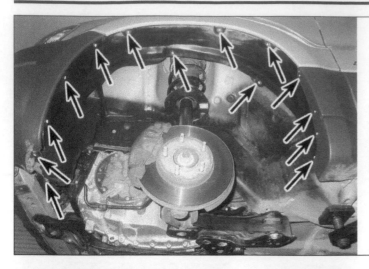

10.3a Remove the inner fender splash shield pin-type retainers and mounting screws

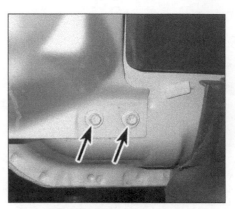

10.3b On pin-type retainers, loosen the center screw . . .

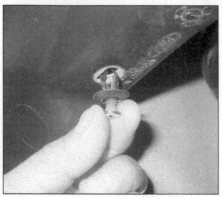

10.3c . . . then pop the retainer from the fenderwell

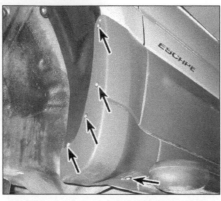

10.6 Remove the fender molding screws and detach the molding

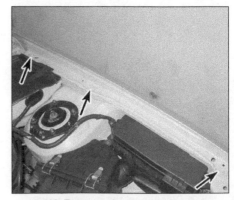

10.8 Location of the lower fender-to-body bolts

10   Front fender - removal and installation

2007 and earlier models

1   Loosen the front wheel lug nuts. Raise the vehicle, support it securely on jackstands and remove the front wheel(s).
2   Remove the headlight housing (see Chapter 12).
3   Remove the inner fender splash shield (see illustrations).

Note: *The fender splash shield is fastened using either mounting screws or pin-type fasteners.*
4   Remove the sidemarker lights, if equipped (see Chapter 12).
5   If you're removing the passenger's side fender, remove the radio antenna (see Chapter 12). If you're removing the driver's side fender, remove the hood prop rod from the fender and prop the hood open.

4WD models

6   Remove the fender molding (see illustration).

7   Remove the rocker panel molding.
Note: *The rocker panel molding must be released from inside the door panel. Remove the door trim panel (see Section 13). The molding retainer clips are not reusable. Install new clips on reassembly.*

All 2007 and earlier models

8   Remove the lower fender-to-body bolts (see illustration).
9   Open the front door and remove the upper fender-to-body bolt (see illustration).
10   Remove the remaining fender mounting

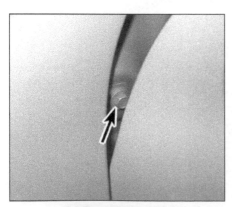

10.9 Remove the upper fender-to-body bolt

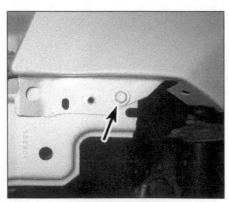

10.10a Location of the fender corner bolt

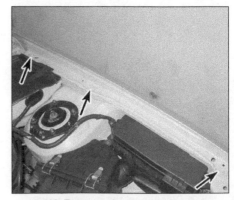

10.10b Remove the three bolts along the top of the fender

10.25a Fenderwell inner splash shield fastener locations

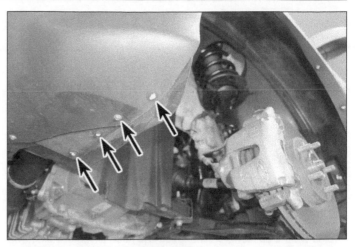

10.25b Fenderwell underside splash shield fastener locations

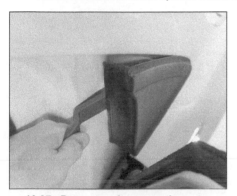

10.27a Pry-out and remove the outer
A-pillar trim panel . . .

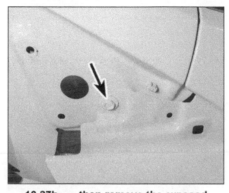

10.27b . . . then remove the exposed
fender bolt.

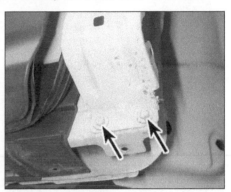

10.28 The fender lower bolts are hidden by
the rocker panel trim

bolts (see illustrations).

11   Lift off the fender. It's a good idea to have an assistant support the fender while it's being moved away from the vehicle to prevent damage to the surrounding body panels.

12   Installation is the reverse of removal. Check the alignment of the fender to the hood and front edge of the door before final tightening of the fender fasteners.

2008 through 2012 models

13   Refer to Section 9 and remove the front bumper cover.

14   Remove the inner fender splash shield.

15   On vehicles with rocker panel moldings, remove the screws from the front fenderwells, then remove the running boards, if equipped. **Note:** *The rocker molding clips must be replaced with new ones on installation.*

16   Use a plastic trim tool or a screwdriver wrapped with tape to carefully pry off the rocker molding, starting from one end and working down its length. Remove the rocker molding.

17   If you're working on the right fender, remove the antenna mast and the grommet at its base.

18   Remove the two bolts at the lower rear edge of the fender (see illustration 10.8).

19   Open the door, then remove the bolt at the top rear (see illustration 10.9).

20   Remove the two bolts behind the headlight.

21   Remove the three bolts along the top edge from the engine compartment (see illustration 10.10b)

22   Carefully lift off the fender. The fender must be lifted over the antenna base.

23   Installation is the reverse of removal. Check the alignment of the fender to the hood and the front edge of the door before final tightening of the bolts.

2013 and later models

24   Remove the cowl panel cover (see Section 12).

25   Remove the fenderwell splash shield fasteners (see illustrations), then fold the splash shield as necessary to allow for removal.

26   Refer to Section 9 and remove the front bumper cover (and rocker panel trim).

27   Open the door, pry-off and remove the body trim panel with a plastic trim tool, then remove the exposed bolt (see illustrations).

28   Remove the two bolts at the lower rear edge of the fender (see illustration).

29   From inside the fender, remove the plastic baffle and the inner bolt securing the fender to the vehicle (see illustrations).

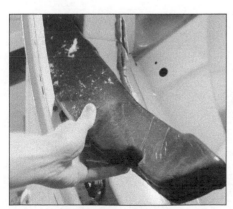

10.29a Remove the inner
fender baffle, then . . .

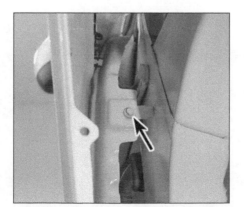

10.29b . . . remove the fender bolt hidden
behind the baffle

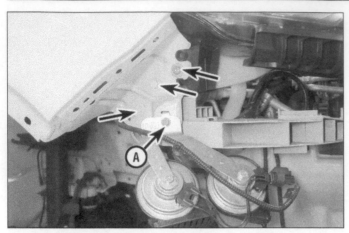

10.30 On the passenger's side, you will need to remove the horn bracket (A) and the bolts securing the fender

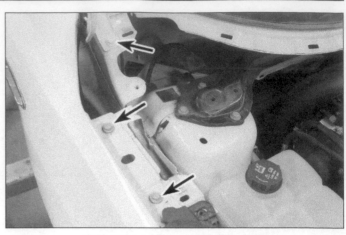

10.31 Remove the upper fender bolts

30   Remove the bolts located underneath the headlight housing, also removing the horn bracket assembly (see illustration).
31   Remove the three bolts along the top edge from the engine compartment (see illustration).
32   Carefully lift off the fender.
33   Installation is the reverse of removal. Check the alignment of the fender to the hood and the front edge of the door before final tightening of the bolts.

11.2 Location of the grille mounting nut - one on each side (2007 and earlier models)

11   Radiator grille - removal and installation

2007 and earlier models

1   Remove the front bumper cover (see Section 9).
2   Remove the radiator grille mounting nuts, if equipped (see illustration).
3   Release the radiator grille clips along the

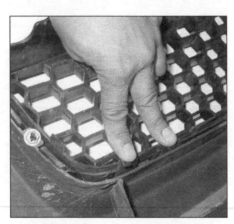

11.3 Press the radiator grille clips to release (2007 and earlier models)

perimeter of the grille (see illustration). The retaining clips can be disengaged by simply pressing down on the tabs.
4   Pull the grille out and remove it.
5   Installation is the reverse of removal.

2008 through 2012 models

6   Refer to Section 9 and remove the front bumper cover.

Mariner

7   Push each of the twelve grille clips outward, then remove the grille from the bumper cover.
8   Installation is the reverse of removal.

Escape/Tribute

9   Remove the two screws, the pinch nuts and the retaining clips, then separate the grille from the bumper cover.
10   Installation is the reverse of removal.

2013 and later models

Upper grille

11   Open the hood and remove the bumper cover upper air deflector cover only (along with the hood release safety latch) (see Section 9).
12   Remove the upper radiator grille support bracket.
13   Unscrew the retaining nuts and remove the upper radiator grille.
14   Installation is the reverse of removal.

12.2a Location of the right cowl cover fasteners

12.2b Location of the left cowl cover fasteners

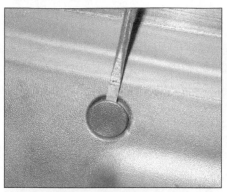

12.2c Remove the screw covers to access the upper fasteners

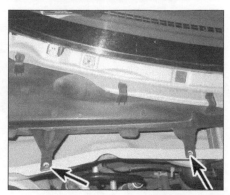

12.3 Location of the vent tray mounting screws

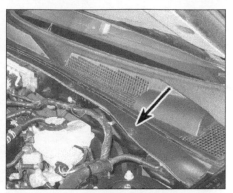

12.6 Remove all of the plastic fasteners that secure the cowl

12.7 Remove the cowl end trim covers

12.11 Pry the clips off of the edge of the cowling cover - the clips can be reused upon installation

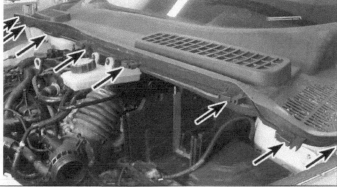

Lower grille (license plate grille)

15   Remove the bottom pushpin securing fastener from the grille.

16   Reaching on the inside of the grille, disengage the retaining tabs along the edges and remove the lower radiator grille.

17   Installation is the reverse of removal.

---

12   Cowl cover and vent tray - removal and installation

2007 and earlier models

1   Remove the wiper arms (see Chapter 12).

2   Remove the push pin fasteners and mounting screws securing the cowl cover (see illustrations).

Note: *Use a small screwdriver to pop the screw covers up to access the cowl cover mounting screws.*

3   If the vent tray needs to be removed, first remove the wiper motor linkage assembly as described in Chapter 12, then remove the vent tray mounting bolts (see illustration).

4   Installation is the reverse of removal.

2008 to 2012 models

5   Refer to Chapter 12 and remove the wiper arms.

6   Remove the pushpin clips from the grille (see illustration).

7   Remove the end caps from the cowl grille (see illustration).

8   Pull the grille from the retaining clips and remove it.

9   Installation is the reverse of removal.

2013 and later models

10   Remove windshield washer arms (see Chapter 12).

11   Remove the eight metal clips securing the cowl cover to the cowling (see illustration).

12   Lift the cowl cover upwards, noting the locations of the inner plastic pressure clips, then disengage the clips with a trim tool (see illustration). Remove the cowl cover.

13   To remove the lower cowl cover, remove these bolts at each side (see illustration) then lift-off the lower cowl cover.

14   Installation is the reverse of removal.

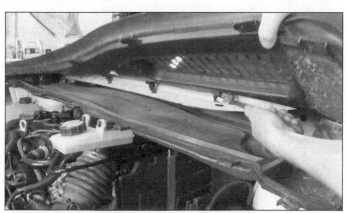

12.12 Lift up the cowl cover enough to free the inner clips with a trim tool

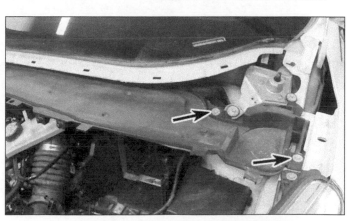

12.13 Lower cowl cover mounting bolts

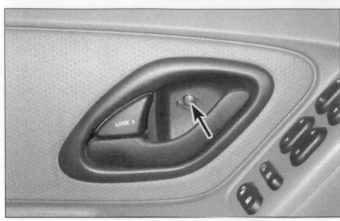

13.1a Remove the screw from the bezel . . .

13.1b . . . then remove the bezel from the handle

13 Door trim panels - removal and installation

Warning: *The models covered by this manual are equipped with a Supplemental Restraint System (SRS), more commonly known as airbags. Always disarm the airbag system before working in the vicinity of any airbag system component to avoid the possibility of acci-*dental deployment of the airbag, which could cause personal injury (see Chapter 12).*

2007 and earlier models
Removal

Front and rear doors

1   Remove the bezel from the door latch release handle (see illustrations).
2   Remove the screw from the door pull cup (see illustration).

3   On the rear door, remove the sail panel (see illustration).
4   Remove the door trim panel retaining screws (see illustrations), then carefully pry the panel out until the clips disengage (see illustration). Work slowly and carefully around the outer edge of the trim panel until it's free.

Note: *The front doors use two screws to secure the door trim panels, one near the outer*

13.2 Remove the cup screw

13.3 Carefully pry the sail panel from the door

13.4a Carefully pry the screw cover off the door panel and remove the door panel mounting screw

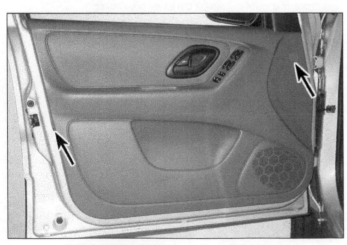

13.4b Location of the front door panel screws

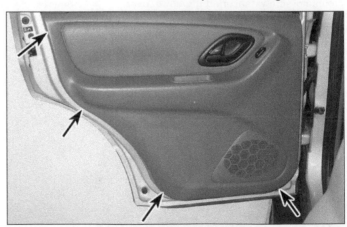

13.4c The rear door panel is not secured with mounting screws - carefully pry the door panel using a flat-bladed screwdriver or door panel tool until the door clips release

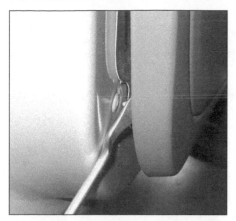

13.5 Using a panel tool, pry the door panel where the clips are located

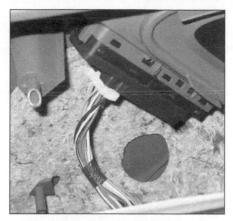

13.6a Unplug the door panel window and lock switch connector - front door shown

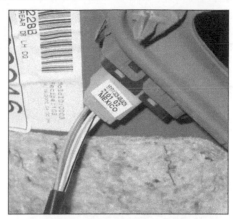

13.6b Unplug the rear door window and lock switch connector

edge and one along the inside panel. The rear door panel is not secured with screws.

5   Using a screwdriver or trim panel tool, carefully pry around the perimeter of the panel to disengage the retaining clips (see illustration).
6   Unplug any wiring harness connectors (see illustrations) and remove the panel.
7   For access to the door outside handle or the door window regulator inside the door, raise the window fully, then carefully peel back the plastic watershield (see illustrations).

Liftgate
8   Remove the liftgate panel mounting screws (see illustration).
9   Using a screwdriver or trim panel tool, pry out the clips and remove the trim panel from the liftgate (see illustration).
10   For access to other components inside the door, carefully peel back the plastic watershield.

Installation
11   Prior to installation of the door trim

panel, be sure to reinstall any clips in the panel which may have come out when you removed the panel.
12   Position the wire harness connectors for the power door lock switch and the power window switch (if equipped) on the back of the panel, then place the panel in position in the door. Press the door panel into place until the clips are seated.
13   The remainder of the installation is the reverse of removal.

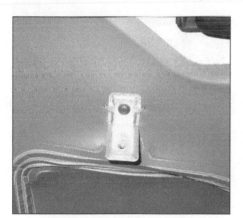

13.7a Remove the door panel bracket screw

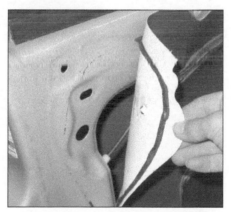

13.7b Peel back the plastic watershield

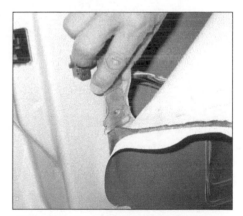

13.7c Use a scraper or razor blade to cut through the old sealant

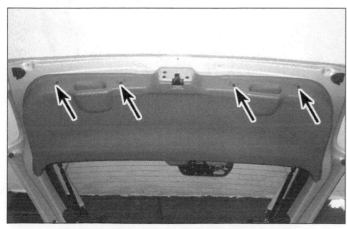

13.8 Remove the screws from the liftgate panel

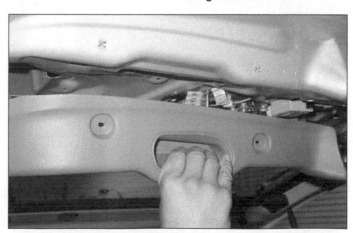

13.9 Lower the panel and separate it from the upper door rim

2008 - 2012 models
Removal

Front and rear doors

14 Remove the mirror control panel (see illustration), then disconnect the electrical connector.

15 Pull the door release handle partially open, then remove the trim panel under it (see illustrations). Remove the screw under the panel.

16 Pull up the cover from the door pull handle, then remove the screw under it (see illustration). Pull up the switch panel and dis-connect the wiring (see illustrations).

17 Remove the screws from the perimeter of the door panel (see illustration).

18 Pull the door panel out to release the clips, then up to lift it free (see illustration). Disconnect the wiring from the rear of the panel.

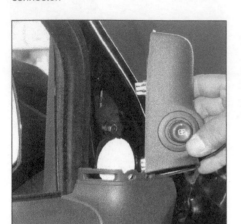

13.14 Pull off the mirror control panel; if necessary, carefully pry it off with a plastic trim tool

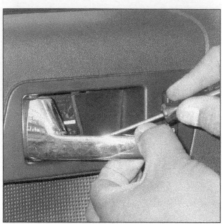

13.15a Peel back the cover under the door release handle . . .

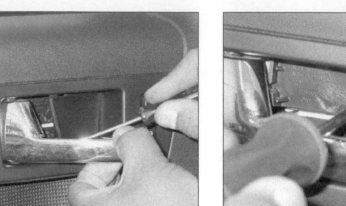

13.15b . . . then remove the screw under it

13.16a Peel back the cover at the bottom of the door pull handle, then remove the screw under it

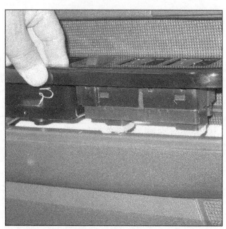

13.16b Pry the switch panel up . . .

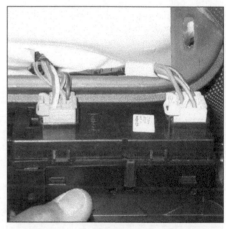

13.16c . . . and disconnect the wiring

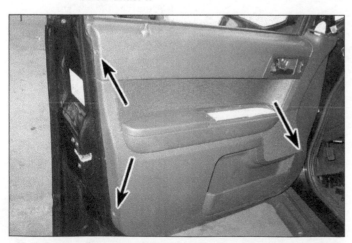

13.17 Remove the screws from the perimeter of the door trim panel

13.18 Carefully pry the door panel loose, then lift it to disengage it from the top of the door

19   Push in the locking tabs, then press the door handle from the panel. Pull the handle back through the panel to remove it.
20   Installation is the reverse of removal.

Liftgate

21   Open the liftgate, then remove the four screws from the bottom edge of the panel.
22   Open the rear window.

23   Pull out on the panel to release the eight inner retaining clips.
24   Installation is the reverse of removal.

Installation

25   Put all of the retaining clips onto the liftgate trim panel before installing it.
26   Press the panel back into place, making sure that all of the retaining clips are engaged.

27   The remainder of installation is the reverse of removal.

2013 and later models

Door trim panel

28   To remove the door trim panel follow the steps, in sequence, listed in the photos below (see illustrations).

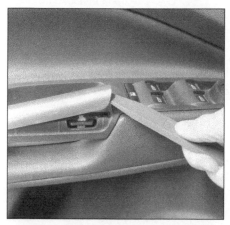
13.28a Remove the trim cover from the inside pull handle

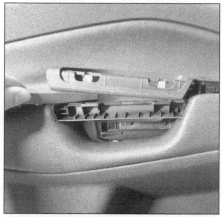
13.28b Pry the window switch assembly up and off of the door

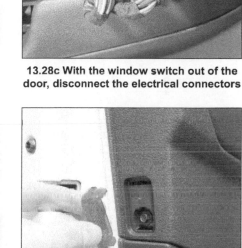
13.28c With the window switch out of the door, disconnect the electrical connectors

13.28d Pry out the trim behind the inside release lever

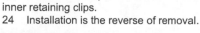
13.28e Remove the screws securing the door panel that were hidden by the trim panels

13.28f Remove the reflector on the edge of the door to expose the screw; remove the screw

13.28g Remove the lower screw securing the door panel

13.28h Starting at the lower corner and working your way around the perimeter, use a trim tool to pry the door panel free of the pressure clips

13.28i Pull the door panel up and off the window sill, then disconnect the electrical connector on the inside of the panel

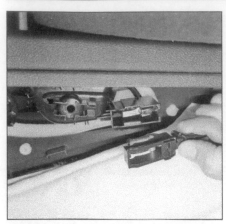

13.28j Remove the door latch cable by depressing the tab and pulling the cable bracket straight out

13.28k If you need further access to the inside of the door, gently peel back the moisture barrier plastic. The plastic barrier can be reused upon installation. Use a razor blade to aid in cutting/removing the rubber sealant, if necessary. Preserve as much of the rubber sealant in its original shape on the plastic as possible.

29   Installation is the reverse of removal.

Liftgate trim panel

30   To remove the liftgate trim panel- follow the steps, in sequence, listed in the photos below (see illustrations).

14   Door - removal, installation and adjustment

Warning: *The models covered by this manual are equipped with a Supplemental Restraint System (SRS), more commonly known as air-* *bags. Always disarm the airbag system before working in the vicinity of any airbag system component to avoid the possibility of accidental deployment of the airbag, which could cause personal injury (see Chapter 12).*
Note: *The door is heavy and somewhat awkward to remove and install - at least two peo-*

13.30a Pry the center trim off, using a plastic trim tool

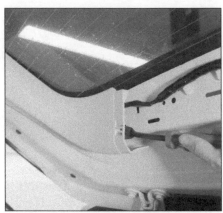

13.30b Using a more rigid trim tool, release the clips and pry the upper corner trim off . . .

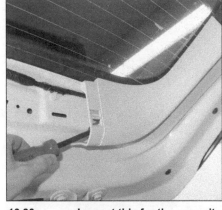

13.30c . . . and repeat this for the opposite upper corner trim

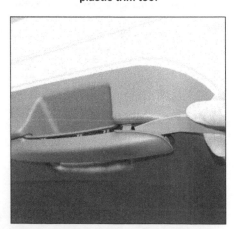

13.30d Pry the trim off of the grab handle with a flat plastic trim tool

13.30e With the grab handle trim removed, remove the two small liftgate panel securing torx bolts

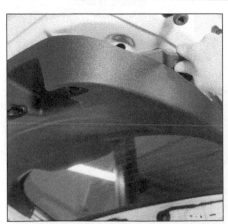

13.30f Starting from the corner, work your way around the panel and release the liftgate pressure clips, then remove the liftgate panel

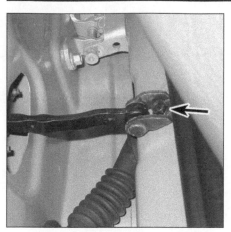

14.6 Remove the bolt from the door stop strut

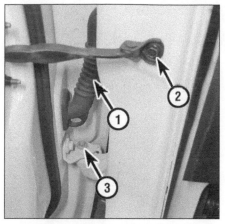

14.10 Disconnect the front door electrical connector (1), remove the door stop bolt (2), then remove the door lower mounting bolt (3)

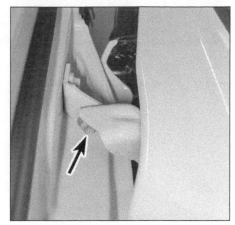

14.11 Remove the front door upper mounting bolt

ple should perform this procedure.
Note: *The rear doors are replaced and adjusted in the same way as the front doors*

Removal and installation
1   Raise the window completely in the door. Open the door all the way and support it from the ground on jacks or blocks of wood covered with rags to prevent damaging the paint.
2   Disconnect the cable from the negative battery terminal.

2007 and earlier models
3   Remove the door trim panel and watershield as described in Section 13.
4   Disconnect all electrical connections, ground wires and harness retaining clips from the door.
5   From the door side, detach the rubber conduit between the body and the door. Then, pull the wiring harness through the conduit hole and remove it from the door.
Note: *It is a good idea to label all connections to aid in the reassembly process.*
6   Remove the door stop strut bolt(s) (see illustration).

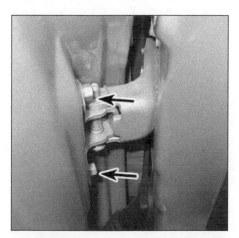

14.13 Remove the door hinge bolts with the door supported (bottom hinge similar)

2008 through 2012 models
7   Remove the weatherstrip and trim panel from the lower front windshield pillar.
8   Disconnect the front door wiring inside the pillar.
9   Remove the door stop bolt.

2013 and later models
10   While the door is being supported, disconnect the door wiring harness electrical connector, remove the door stop bolt, then remove the door lower mounting bolt (see illustration).
11   Remove the door upper mounting bolt (see illustration) and remove the door.

All models
12   On 2012 and earlier models, mark around the door hinges with a pen or a scribe to facilitate realignment during installation.
13   On 2012 and earlier models, with an assistant holding the door, remove the hinge-to-door bolts (see illustration) and lift the door off.
14   Installation is the reverse of removal.
15   Reconnect the battery. After you're

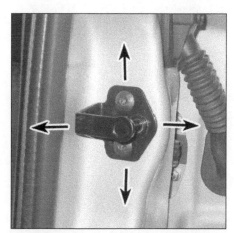

14.19 Adjust the door lock striker by loosening the mounting screws and gently tapping the striker in the desired direction

done, the Powertrain Control Module (PCM) must relearn its idle and fuel trim strategy for optimum driveability and performance (see Chapter 5).

Adjustment
16   Having proper door-to-body alignment is a critical part of a well-functioning door assembly. First check the door hinge pins for excessive play. Fully open the door and lift up and down on the door without lifting the body. If a door has 1/16-inch or more play, the hinges should be replaced.
17   Door-to-body alignment adjustments are made by loosening the hinge-to-body bolts or hinge-to-door bolts and moving the door. Proper body alignment is achieved when the top of the doors are parallel with the roof section, the front door is flush with the fender, the rear door is flush with the rear quarter panel and the bottom of the doors are aligned with the lower rocker panel. If these goals can't be reached by adjusting the hinge-to-body or hinge-to-door bolts, body alignment shims may have to be purchased and inserted behind the hinges to achieve correct alignment.
18   To adjust the door-closed position, scribe a line or mark around the striker plate to provide a reference point, then check that the door latch is contacting the center of the latch striker. If it is not, adjust the up and down position first.
19   Finally adjust the latch striker sideways position, so that the door panel is flush when closed and provides positive engagement with the latch mechanism (see illustration).

---
15   Door latch, lock cylinder and handles - removal and installation

---

Caution: *Wear gloves when working inside the door openings to protect against cuts from sharp metal edges.*
Caution: *Disconnect the cable from the negative battery terminal*

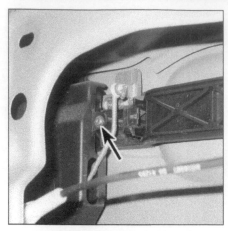

15.2 Loosen this door latch screw

15.3 Remove the door handle cover

15.4a Slide the door handle toward the
rear of the vehicle to remove it . . .

2007 and earlier models
Front handles and door
latch assembly

1   Raise the window, then remove the door trim panel and watershield (see Section 13).
Caution: *Take care not to scratch the paint on the outside of the door. Wide masking tape applied around the handle opening before beginning the procedure can help avoid scratches.*

2   Working inside the door, remove the screw that secures the door latch assembly (see illustration).

3   Remove the door handle cover from the exterior of the door (see illustration).

4   Remove the door handle by sliding it toward the rear of the vehicle (see illustrations).

5   Working inside the door panel, remove the screw (see illustration) and loosen the remote control assembly.

6   Release the clip and position the interior door handle to the side (see illustration).

7   Remove the door latch screws (see illustration). The door latch screws must not be reused. Always install new door latch screws on reassembly.

8   Disconnect the door latch electrical connector (see illustration).

15.4b . . . and remove the rubber
insulators

9   Slide the door latch remote control assembly toward the front of the vehicle.

10   Remove the anti-theft guard from the latch assembly (see illustration).

11   Release the clip and and disconnect the exterior door handle actuating rod (see illustrations).

12   Installation is the reverse of removal.

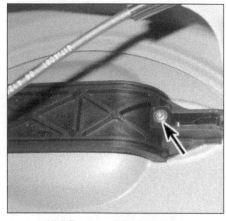

15.5 Remove this door latch
assembly screw

Rear handles and door latch
assembly

13   Raise the window, then remove the door trim panel and watershield (see Section 13).

14   Remove the electrical connector from the rear door latch (see illustration).

15   Remove the bolts that retain the door glass top run (see illustration).

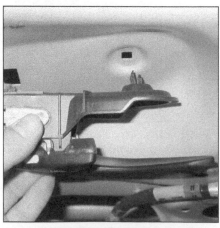

15.6 Pull the door handle from the door

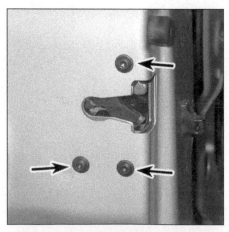

15.7 Remove the door latch screws (install
new ones when reinstalling the latch)

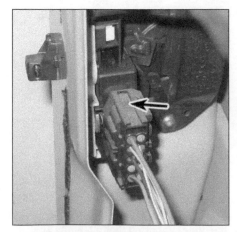

15.8 Release the clip and disconnect the
electrical connector

15.10 Remove the anti-theft guard assembly from the latch

15.11a Release the clip . . .

15.11b . . . then remove the actuating rod from the door handle

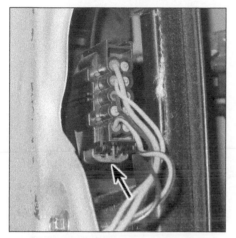

15.14 Release the clip for the electrical connector and unplug it from the door latch assembly

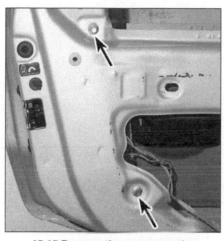

15.15 Remove the screws on the door glass top run

15.16 Lift the door glass top run from the door

16   Separate the door glass top run from the door (see illustration).
17   Remove the door handle screws from inside the door panel.
Caution: *Take care not to scratch the paint on the outside of the door. Wide masking*

tape applied around the handle opening before beginning the procedure can help avoid scratches.
18   Remove the cover from the outside of the door (see illustration).
19   Remove the door handle by sliding

it toward the rear of the vehicle (see illustration).
20   Working inside the door panel, remove the screw and loosen the remote control assembly.
21   Disconnect the door handle actuating rod (see illustration).

15.18 Remove the cover from the rear door handle

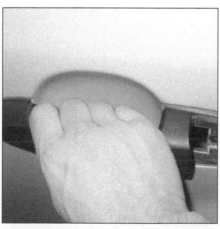

15.19 Slide the door handle toward the rear of the vehicle

15.21 Release the clip from the actuating rod

15.22 Remove the door latch screws but do not reuse them

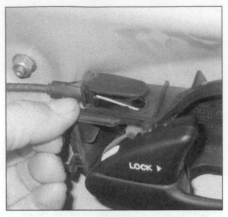

15.24a Release the door handle cable conduit by lifting . . .

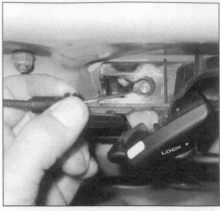

15.24b . . . then sliding the cable end through the slotted portion of the handle casting

15.25a Pry the door handle clips . . .

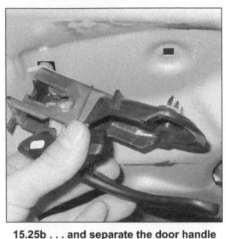

15.25b . . . and separate the door handle from the door

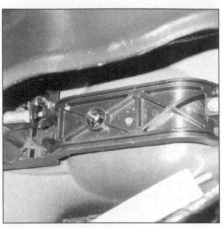

15.28 The door handle reinforcement plate

22   Remove the door latch screws (see illustration). The door latch screws must not be reused. Always install new door latch screws on reassembly.
23   Remove the rear door latch.
24   Release the cable conduit and disconnect the interior door handle cable (see illustrations).
25   Remove the rear door handle assembly (see illustrations).

26   Installation is the reverse of removal.

2008 through 2012 models
Door handles
27   Refer to Section 13 and remove the door trim panel and watershield.
28   Working inside the door, remove the screw from the handle reinforcement plate (see illustration).
29   Pull on the exterior handle, then rotate the

rear handle cover to remove it (see illustration).
30   With the handle still pulled, slide the handle to the rear and lift it out (see illustration).
31   Installation is the reverse of removal.

Front door latch
Note: *The rear door latch is replaced in a similar manner as the front door latch*
32   Remove the front door handle (see Steps 27 through 30).

15.29 With the handle pulled out, the rear button can be turned and removed

15.30 Slide the handle to remove it, then remove the gaskets

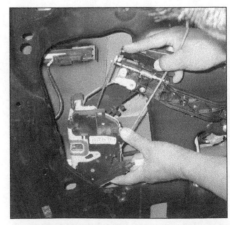

15.36 Remove the latch assembly through the access hole

15.44 Remove the door handle and lock cylinder retaining screw

15.45 Remove the door lock cylinder by pulling it straight out

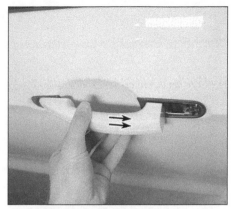

15.46a Slide the door handle rearwards first . . .

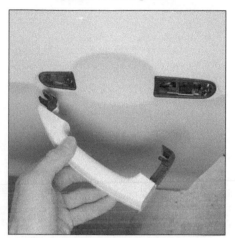

15.46b . . . then pull it away from the door to remove it

15.49 Remove the door handle reinforcement bracket screw and gaskets

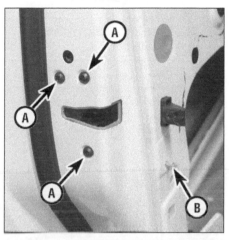

15.50 Remove the latch mounting bolts (A) and release the latch retaining clip (B)

33   Loosen the reinforcement plate screw enough to allow it to slide off.
34   Remove the clip from the interior door handle. Disconnect the cable from the handle and remove it.
35   Remove the latch screws from the door edge and discard them.
36   Disconnect the wiring from the latch, then remove the latch (see illustration).
37   Installation is the reverse of removal. Replace the latch screws with new ones.

Lock cylinder
38   Remove the front door latch (see Steps 32 through 36).
39   Remove the theft guard plate.
40   Disconnect the exterior door rod.
41   Detach the lock cylinder rod and separate the reinforcement from the latch.
42   Release the two retaining clips, then remove the lock cylinder.
43   Installation is the reverse of removal.

2013 and later models
Front door handles and lock cylinder
Note: *The procedure for the rear door handles is similar to the front door handle replacement procedure*
44   Open the door and pry-out the access

cover, then remove the door lock cylinder and handle retaining screw (see illustration).
45   Pull the lock cylinder and cover straight out (see illustration).
Note: *On keyless entry models, it will be necessary to gently pull and guide out the attached door handle electrical connector from the door to be disconnected - a small screwdriver or hook tool may be helpful in this process*
46   Slide the door handle rearwards, then pull it out of the door (see illustrations).
47   Installation is the reverse of removal.

Front door latch
Note: *The procedure for the rear door latch is similar to the front door latch replacement procedure*
48   Remove the door trim panel (see Section 13) and window regulator assembly (Section 17).
49   Remove the outside door handle reinforcement bracket screw and rubber gaskets (see illustration).
50   Remove the latch mounting bolts and release the latch retaining clip (see illustration).
51   Disconnect the latch electrical connector(s), release the wiring harness clip, then push the grommet and cable through the door and remove the latch assembly (see illustration).

52   Once removed from the door, unbolt the front door latch from the bracket assembly and release/disconnect the necessary cables.
53   Installation is the reverse of removal.

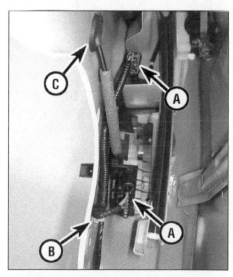

15.51 Disconnect the latch electrical connectors (A), release the wiring harness clip (B), then push the grommet and cable through the door (C)

16.3 Remove the weatherstrip from the front door glass channel

16.4 Access the window retaining bolts through the holes in the door frame

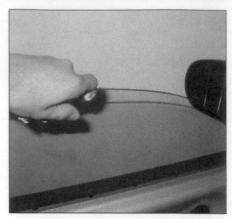

16.6 Lift the window glass from the door

16   Door window glass - removal and installation

Caution: *Wear gloves when working inside the door openings to protect against cuts from sharp metal edges.*
Caution: *Disconnect the cable from the negative terminal of the battery (except while raising/lowering the window position ONLY) while performing the procedure*

Front door window glass - 2012 and earlier models

1   Remove the door trim panel and the plastic watershield (see Section 13).
2   Lower the window glass all the way down into the door.
3   Remove the weatherstrip from the front door interior glass channel (see illustration).
4   Raise the window just enough to access

the window retaining bolts through the holes in the door frame (see illustration).
5   Place a rag over the glass to help prevent scratching the glass and remove the two glass mounting bolts.
6   Remove the glass by pulling it up and out (see illustration).
7   Installation is the reverse of removal.

Front and rear door window glass - 2013 and later models

Note: *The replacement procedure for the rear door window glass is very similar to the front, noting the following differences: remove the rear door window regulator first (see Section 17), remove the window glass-to-regulator mounting bolts (instead of releasing tabs) and when removing the actual glass, carefully rotate the glass 90-degrees to allow for removal.*

8   Remove the door trim panel (see Section 13).
9   Remove the door speaker (see Chapter 12).
10   Remove the door sail trim panel (see Section 18).
11   For the remaining front door window glass removal steps, follow the photos below in sequence (see illustrations).
12   Installation is the reverse of removal.

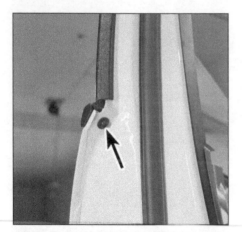

16.11a Remove the retaining screw on the inside of the door . . .

16.11b . . . then, using a flat trim tool, pry the outside door trim up and off of the door

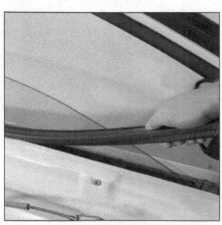

16.11c Pull off the inside window sill trim

16.11d Remove the retaining screw for the plastic trim piece . . .

16.11e . . . then remove the plastic trim from the window frame

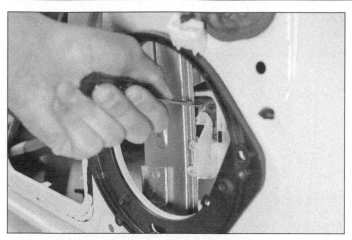

16.11f Reinstall the window switch and raise/lower the window so that the attachment clips on both sides are visible, then depress the plastic attachment clip with a screwdriver as you raise the glass out of the track to disengage it

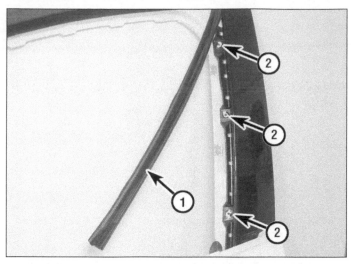

16.11g Working on the outside of the door, peel away the weatherstripping (1), then remove the trim panel torx screws (2)

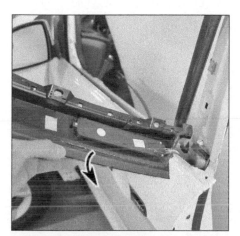

16.11h Position and temporarily tape aside the touch-entry code access panel

16.11i Peel back the remainder of the upper window frame weatherstripping

16.11j Carefully guide the window glass out of the door

Rear door window glass - 2012 and earlier models

13   Remove the door panel and watershield (see Section 13).

14   Lower the window glass all the way down into the door.
15   Remove the interior sail panel and the exterior sail panel (see illustration).

16   Remove the weatherstrip from the rear door glass channel (see illustration).
17   Remove the door belt line molding (see illustration).

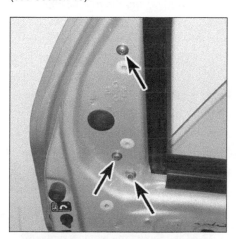

16.15 Remove the exterior sail panel mounting screws

16.16 Remove the weatherstrip from the rear door glass channel

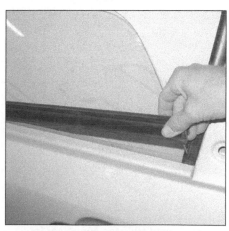

16.17 Remove the door belt line molding piece

16.18 Remove the bolts that retain the door glass top run

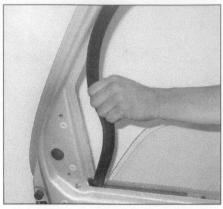

16.20 Lift the door glass top run from the door interior

16.23 Remove the rubber access cover and slide the harness from the slotted portion of the access hole

18   Remove the door glass top run mounting bolts (see illustration).
19   Position the door glass down and forward, tilting it slightly.
20   Remove the door glass top run (see illustration).
21   Remove the glass by pulling it up and out.
22   Installation is the reverse of the removal procedure.

Liftgate glass - 2012 and earlier models

Note: *The liftgate window glass on 2013 and later models is fixed, and is not meant to be easily replaced by the home mechanic - on these model years, the liftgate glass replacement procedure is better left to a window replacement professional*

23   Open the liftgate and remove the two rubber access covers (see illustration).

24   Disconnect the liftgate electrical connector (see illustration).
25   Disconnect the heated grid wire electrical connector (see illustration).
26   Have an assistant support the liftgate and remove the two liftgate support struts (see illustrations).
27   Remove the two liftgate window hinge nuts (see illustration).
28   Installation is the reverse of removal.

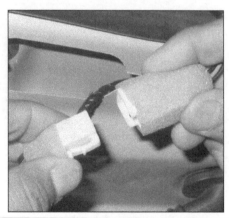

16.24 Disconnect the liftgate electrical connector

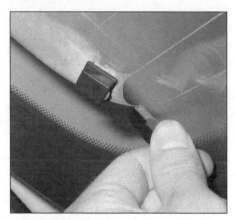

16.25 Disconnect the heated grid wire

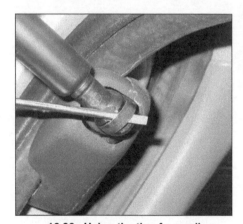

16.26a Using the tip of a small screwdriver, pry the release clip . . .

16.26b . . . and slide the connector forward and off the ballstud

16.27 Location of the liftgate window hinge nut on the left side of the vehicle

17.6 Disconnect the window regulator motor connector

17.7 Window regulator mounting bolts (2008 to 2012 models shown, earlier models similar)

17.13 Window regulator motor electrical connector

17 Door window regulator assembly - removal and installation

Caution: *Wear gloves when working inside the door openings to protect against cuts from sharp metal edges.*
Caution: *Disconnect the cable from the negative terminal of the battery (except while raising/lowering the window position ONLY) while performing the procedure*

Front - 2012 and earlier models

1   Remove the door trim panel and the plastic watershield (see Section 13).

17.14 Window regulator assembly mounting bolts

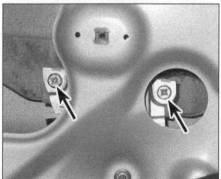

17.21 Location of the rear window glass mounting bolts

2   Lower the window until the window glass-to-regulator mounting bolts are visible.
3   Disconnect the cable from the negative battery terminal (see Chapter 1).
4   Remove the screws that retain the door window glass (see Section 16).
5   Lift the window glass and loop tape over the top of the door to securely support the glass at the top.
6   Disconnect the electrical connector from the window regulator motor (see illustration).
7   Remove the regulator/motor assembly mounting bolts (see illustration).
8   Remove the window regulator/motor assembly from the door.
9   Reconnect the battery. After you're done, the Powertrain Control Module (PCM) must relearn its idle and fuel trim strategy for optimum driveability and performance (see Chapter 5).
10 Installation is the reverse of removal. Lubricate the rollers and wear points on the regulator with white grease before installation.

Front - 2013 and later models

11 Remove the door trim panel and plastic moisture shield (see Section 13) and door speaker (see Chapter 12).
12 Disconnect the the window glass from the regulator (see Section 16), then raise the window glass all the way up by hand and loop tape over the top of the door to securely support the glass at the top.
13 Disconnect the window regulator motor

electrical connector (see illustration).
14 Remove the fasteners securing the window regulator assembly to the door (see illustration). You'll notice some of the holes are slotted to allow those bolts to only be loosened rather than removed. This makes reinstalling the assembly much easier because you can hang the assembly by the mounting bolts while you align and install the remaining bolts.
15 Collapse the two window track guides towards each other, then rotate them counterclockwise to about 90 degrees. Carefully guide the assembly out of the door cavity.
16 Installation is the reverse of removal.

Rear - all models

17 Remove the door trim panel and the plastic watershield (see Section 13).
18 On 2013 and later models, remove the rear door latch (see Section 15)
19 Lower the window until the window glass-to-regulator mounting bolts are visible.
20 Disconnect the cable from the negative battery terminal (see Chapter 1).
21 Remove the screws that retain the door window glass (see illustration).
22 Lift and slide the door glass all the way up, then tape it securely to the top of the door frame.
23 Disconnect the electrical connector from the window regulator motor.
24 Remove the regulator/motor assembly mounting bolts (see illustration).

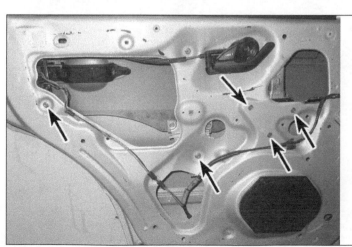

17.24 Location of the power window regulator/motor mounting bolts (2012 and earlier model shown, other models similar)

18.2a Removing the door sail trim panel - 2012 and earlier models

18.2b Removing the door sail trim panel - 2013 and later models

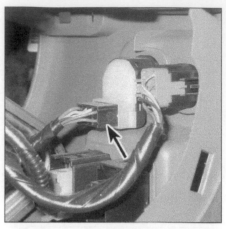

18.3a Power mirror electrical connector - 2012 and earlier models

25   Remove the window regulator/motor assembly from the door.

26   Installation is the reverse of removal. Lubricate the rollers and wear points on the regulator with white grease before installation.

27   Reconnect the battery. After you're done, the Powertrain Control Module (PCM) must relearn its idle and fuel trim strategy for optimum driveability and performance (see Chapter 5).

18   Mirrors - removal and installation

Outside mirror assemblies

1   Remove the front door trim panel (see Section 13).

2   Using a plastic trim tool, remove the front door interior sail panel (see illustrations).

3   Disconnect the electrical connector from the mirror. On 2013 and later models, remove the mounting bolts as well, then remove the mirror assembly (see illustrations).

4   On 2012 and earlier models, remove the retaining bolts and detach the mirror from the vehicle (see illustration).

5   Installation is the reverse of removal.

Mirror glass

6   Position the mirror glass all the way down.

7   Place a suitable rag between the mirror glass and mirror housing to prevent the mirror from cracking or breaking, then use a screwdriver to gently dislodge the mirror glass from the motor retaining tabs. Be careful not to let the mirror glass fall.

8   If equipped, disconnect the electrical connector(s) and remove the mirror glass.

9   Installation is the reverse of removal. Make sure all the clips snap securely in place, and operate the mirror glass controls in every direction to make certain that it functions correctly.

Inside mirror

Note: *The way the interior mirror is mounted may vary, as there are a few different mounting options across the models mentioned in the manual. The most common mirror option for replacement is mentioned below. Some options might include mirror retaining bolts or extra trim covers to be removed first. Others may require the use of factory tools to complete replacement. If in doubt, this procedure should be performed by a dealer service department or other qualified repair shop.*

10   Disconnect the auto-dimmer electrical connector, if equipped. Remove the compass module if so equipped.

11   Pry the tension clip at the base of the mirror stalk with a screwdriver (see illustration). Push the screwdriver out, using leverage to release the mirror from the bracket.

12   To install the mirror, insert the mirror into the bracket, pushing downward until the mirror is secured.

13   If the mount plate itself has come off the windshield, adhesive kits are available at auto parts stores to resecure it. Follow the instructions included with the kit.

19   Liftgate - removal, installation and adjustment

Note: *The liftgate is heavy and somewhat awkward to remove and install - at least two people should perform this procedure.*

Removal and installation
2012 and earlier models

1   Remove both rear pillar trim panels. Also remove the seat belt anchors nearest the liftgate.

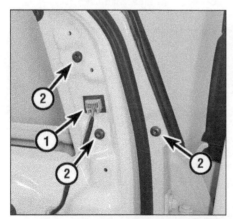

18.3b Power mirror electrical connector (1) and mirror assembly mounting bolts (2) - 2013 and later models

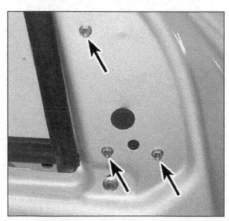

18.4 Remove the mirror assembly mounting fasteners - 2012 and earlier models

18.11 Gently pry under the mirror to aid in sliding it upward

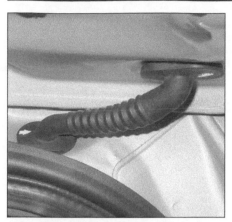

19.6a Detach the rubber conduit from the door cavities to expose the wiring harness

19.6b Remove the rubber conduit from the body and the liftgate to expose the heater grid ground wire

19.7a Release the clip and slide it forward to release the liftgate support strut from the ballstud (lower strut clip shown)

2   Have an assistant hold the liftgate in the open position.
3   Remove the three retainers from the rear of the headliner, then lower the headliner as much as possible without damaging it.
4   Disconnect the washer hose located above the headliner and immediately plug the hose to prevent it from damaging the headliner as it drains. Pull out the washer hose grommet and place it on the outside of the vehicle.
5   Disconnect all electrical connections, ground wires and harness retaining clips from the liftgate.
Note: *It is a good idea to label all connections to aid the reassembly process.*
6   From the liftgate side, detach the rubber conduit between the body and the liftgate (see illustrations). Then pull the wiring harness through the conduit hole and remove it from the liftgate.
7   Remove the liftgate support struts (see illustrations).
8   Mark around the liftgate hinges with a pen or a scribe to facilitate realignment during installation.
Note: *Be sure to draw a reference line around the hinges before removing the bolts.*

9   With an assistant holding the liftgate, remove the hinge-to-liftgate bolts and lift the door off (see illustration).
10   Installation is the reverse of removal.

2013 and later models
11   Remove the liftgate trim paneling (see Section 13).
12   Disconnect the electrical connectors at each side, then release the wiring harness clips from the liftgate (see illustration). Pull the wiring through the holes in the liftgate body.
13   Mark around the liftgate hinges with a pen or a scribe to facilitate realignment during installation.
Note: *Be sure to draw a reference line around the hinges before removing the bolts.*
14   Have an assistant hold the liftgate open, then release the ends of the support struts by prying back the retainers with a small screwdriver. Detach the support struts from the liftgate (see illustrations 19.7a and 19.7b).
15   Remove the hinge bolts and lift off the liftgate (see illustration 19.9).
16   Installation is the reverse of removal.

Adjustment
17   Having proper liftgate-to-body alignment

is a critical part of a well-functioning liftgate assembly. First check the liftgate hinge pins for excessive play. Fully open the liftgate and move it side-to-side. If it has 1/16-inch or more play, the hinges should be replaced.
18   Liftgate-to-body alignment adjustments are made by loosening the hinge-to-body bolts or hinge-to-liftgate bolts and moving the door. Proper body alignment is achieved when the top of the liftgate is parallel with the roof section, the sides of the liftgate are flush with the rear quarter panels and the bottom of the liftgate is aligned with the lower sill panel. If these goals can't be reached by adjusting the hinge-to-body or hinge-to-liftgate bolts, body alignment shims may have to be purchased and inserted behind the hinges to achieve correct alignment.
19   To adjust the liftgate-closed position, scribe a line or mark around the striker plate to provide a reference point, then check that the liftgate latch is contacting the center of the latch striker. If not, adjust the up and down position first.
20   Finally adjust the latch striker sideways position, so that the liftgate panel is flush with the rear quarter panel and provides positive engagement with the latch mechanism.

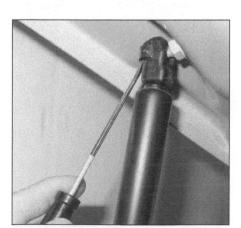

19.7b Release the upper support strut clip and release it from the ballstud

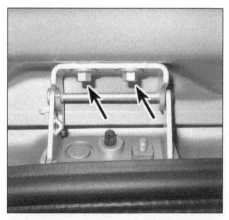

19.9 Location of the hinge-to-liftgate bolts

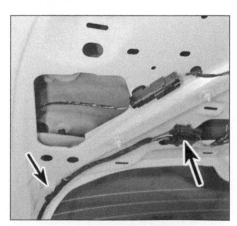

19.12 Disconnect the electrical connectors at each side, then release the wiring harness clips

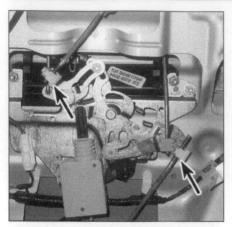

20.2a Disconnect the actuating rods from the liftgate latch remote control

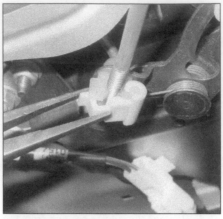

20.2b Use needle-nose pliers to release the locking tab from the clip

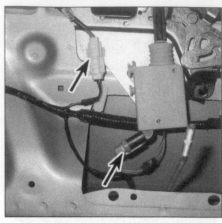

20.3 Remove the liftgate latch remote control connectors

20 Liftgate latch remote control, latch and window latch - removal and installation

Note: *Before replacing any components, check for any trouble codes related to the liftgate system. Verify the concern by properly diagnosing the trouble code, then proceed with any repairs or part replacement that is recommended by the trouble code diagnostics.*

Note: *The liftgate system has taken on many new roles from years past. If equipped, the anti-pinch features, passive entry with hands free activation and the interconnection between the interior light system and the anti-theft system has brought on more complex electronics than previous model generations. If you are experiencing problems with the liftgate system, have the system properly diagnosed before attempting to replace any components.*

2012 and earlier models
Liftgate latch remote control assembly (2007 and earlier models)

1   Open the liftgate and remove the trim panel and watershield as described in Section 13.

Note: *All door lock rods are attached by plastic clips. The plastic clips can be removed by unsnapping the portion engaging the connecting rod and then pulling the rod out of its locating hole.*

2   Disconnect the actuating rods from the liftgate latch remote control (see illustrations).
3   Remove the liftgate latch remote control electrical connectors (see illustration).
4   Remove the screw and nuts securing the remote control to the door (see illustration). Remove the remote control assembly through the door opening.
5   Installation is the reverse of removal.

Liftgate window latch

6   Open the liftgate and remove the door trim panel and watershield as described in Section 13.
7   Disconnect the liftgate window ajar switch, and on 2007 and earlier models, disconnect the liftgate window latch actuating rod (see illustration).
8   Mark the location of the liftgate window latch to insure correct alignment (see illustration).
9   Remove the liftgate window latch mounting bolts. Separate the assembly from the liftgate.

Liftgate latch

10   Disconnect the liftgate window ajar switch, and on 2007 and earlier models, disconnect the liftgate window latch actuating rod (see illustration 20.7).
11   Remove the liftgate latch mounting bolts and the liftgate latch (see illustration).
12   The remainder of the installation is the reverse of removal.

Liftgate release switch (2008 and later models)

13   Refer to the replacement procedure mentioned under *2013 and later models*.

2013 and later models

14   Refer to Section 13 and remove the liftgate trim panel.

Liftgate release switch

15   Disconnect the electrical connector from the liftgate switch (see illustration).
16   Remove the trim panel mounting bolts from the inside, then carefully pry-off the outside trim panel that surrounds the release switch.
17   Release the switch tabs and lift the switch out.
18   Installation is the reverse of removal. Make sure that the switch is correctly aligned.

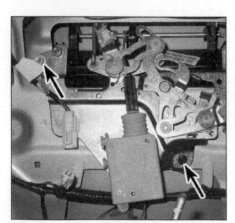

20.4 Location of the liftgate latch remote control fasteners

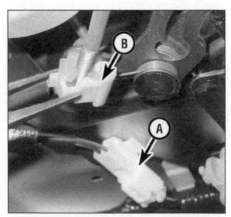

20.7 Disconnect the liftgate window ajar switch (A) and release the actuating rod from the window latch (B)

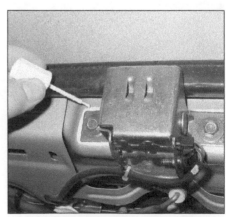

20.8 Mark the location of the liftgate window latch

20.11 Liftgate latch mounting bolts

20.15 Liftgate release switch electrical connector

21.3 Remove the finish panel from the console by carefully lifting the edges out of the locking tabs

Liftgate latch

19 Disconnect the wiring from the liftgate latch.
20 Remove the liftgate latch mounting bolts (see illustration 20.11), then remove the latch.
21 Installation is the reverse of removal.

21 Center console - removal and installation

Warning: *The models covered by this manual are equipped with a Supplemental Restraint System (SRS), more commonly known as airbags. Always disarm the airbag system before working in the vicinity of any airbag system component to avoid the possibility of accidental deployment of the airbag, which could cause personal injury (see Chapter 12).*

2004 and earlier models
Automatic transaxle models
1 Disconnect the cable from the negative battery terminal (see Chapter 5).
2 Raise the emergency brake handle.
3 Remove the floor console finish panel from the console (see illustration).
4 Remove the front bolts from the floor

console (see illustrations).
5 Remove the bolts from the rear section of the center console (see illustration).
6 Lift the center console from the vehicle (see illustration).
7 Installation is the reverse of removal.
8 Reconnect the battery. After you're done, the Powertrain Control Module (PCM) must relearn its idle and fuel trim strategy for optimum driveability and performance (see Chapter 5).

Manual transaxle models
9 Remove the manual transaxle shift knob from the shift lever.
10 Remove the floor console front finish panel.
11 Raise the parking brake handle.
12 Remove the floor console rear finish panel.
13 Remove the floor console front and rear bolts.
14 Lift the center console from the vehicle.
15 Installation is the reverse of removal.

2005 through 2007 models
16 Slide the front seats fully forward, then remove the bolts at the bottom rear of the console.

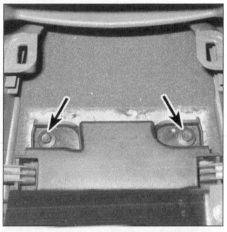

21.4a Location of the floor console front bolts - 2007 and earlier models

17 Slide the seats fully rearward, then disconnect the cable from the negative battery terminal (see Chapter 1).
18 Pull up the parking brake handle.
19 Use a plastic trim tool or a screwdriver wrapped with tape to carefully pry up the bezel from around the shift lever.

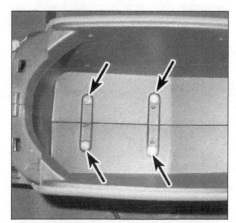

21.4b Location of the floor console interior mounting bolts - 2007 and earlier models

21.5 Location of the floor console rear bolts - one on each side

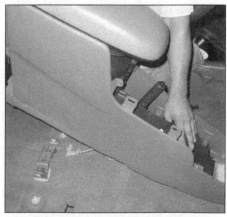

21.6 Lift the center console from the vehicle

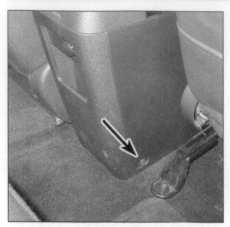

21.26 Remove these screws from the rear of the console

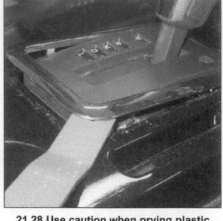

21.28 Use caution when prying plastic interior trim pieces - a plastic trim tool should be used

21.31 After removing the bezel from the shifter, pry up and remove the top of the console

20   On Mariner models, pry off the small forward top panel from the console.
21   Remove the console top panel.
22   Detach the parking brake lever boot from the top panel.
Note: *On Escape models, squeeze the front and the rear of the storage compartment to release the top panel tabs.*

23   Remove the six mounting bolts and remove the console.
24   Installation is the reverse of removal.

2008 through 2012 models

25   On manual transaxle models, set the parking brake and place the shifter in Neutral.
26   Slide the front seats fully forward, then

remove the bolts at the bottom rear of the console (see illustration).
27   Slide the seats fully rearward, then disconnect the cable from the negative battery terminal (see Chapter 1).
28   On automatic transaxle models, remove the trim ring from around the shift lever (see illustration).

21.32a Side console fastener locations

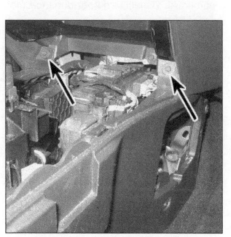

21.32b Front console fastener locations

21.35a Pry off the shifter trim bezel

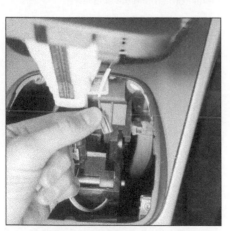

21.35b Disconnect the shift bezel electrical connector

21.35c Pry off the shift bezel surrounding trim

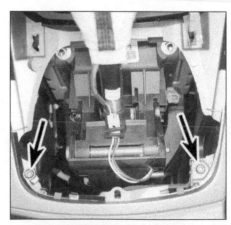

21.35d Remove the trim panel retaining screws

21.35e Pry off the right side of the panel

21.35f Then, pry off the left side and remove the entire panel

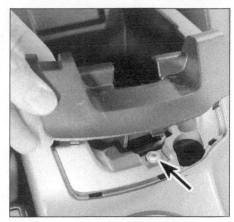

21.35g Open the center console lid, pry off the small cover underneath and remove the retaining screw

29   On manual transaxle models, detach the shift lever boot trim ring.
30   Remove the console storage bin.
31   Use a plastic trim tool or a screwdriver wrapped with tape to pry up and remove the console top panel (see illustration). Disconnect any wiring.
32   Remove the eight mounting bolts and

remove the console (see illustrations).
33   Installation is the reverse of removal.

2013 and later models
Note: *If the vehicle is equipped with power seats, only connect the negative battery cable when sliding the seat(s) forward and backward to access the center console mounting bolts. Otherwise, make sure the negative battery*

cable is disconnected during the procedure.
34   Set the parking brake and place the shifter in Neutral.
35   For the removal of the center console, follow the instructions (in sequence) provided in the photos below (see illustrations). Use a plastic trim tool when prying off the trim pieces.
36   Installation is the reverse of removal.

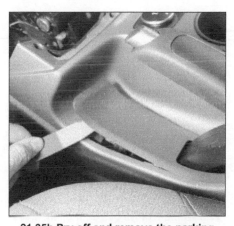

21.35h Pry off and remove the parking brake trim cover

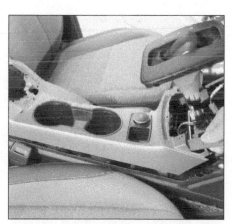

21.35i Pry and remove the center console top cover

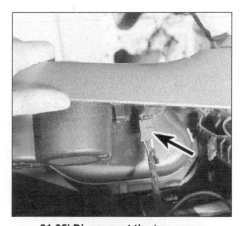

21.35j Disconnect the top cover electrical connector

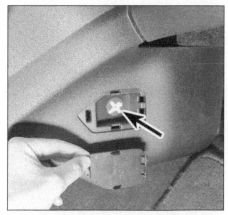

21.35k Remove the center console side panel retaining clips on each side

21.35l Remove the center console side panel screws on each side

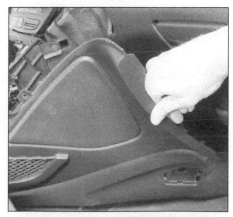

21.35m Pry along the side panels to release the pressure clips, then remove both side panels

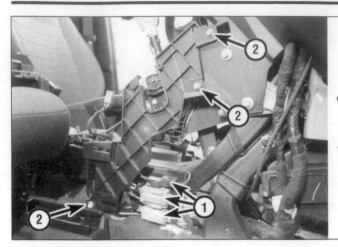

21.35n Disconnect the center console electrical connectors (1) and release the connectors from their brackets, then remove the center console front mounting bolts (2) on each side

21.35o Remove the center console rear mounting bolts on each side, then guide the center console out of the vehicle

22 Dashboard trim panels

Warning: *The models covered by this manual are equipped with a Supplemental Restraint System (SRS), more commonly known as air-bags. Always disarm the airbag system before working in the vicinity of any airbag system component to avoid the possibility of acci-dental deployment of the airbag, which could cause personal injury (see Chapter 12).*

2007 and earlier models
Instrument cluster bezel

1   Disconnect the cable from the negative battery terminal (see Chapter 5).
2   Remove the steering wheel (see Chap-ter 10).
3   Remove the screws at the top of the instrument cluster bezel (see illustration).
4   Grasp the bezel with both hands and pull straight out to disengage the bezel from

the instrument panel.
5   Installation is the reverse of the removal procedure. Make sure the clips are engaged properly before pushing the bezel firmly into place.
6   Reconnect the battery. After you're done, the Powertrain Control Module (PCM) must relearn its idle and fuel trim strategy for optimum driveability and performance (see Chapter 5).

22.3 Remove the screws from the top of the instrument cluster bezel - 2007 and earlier models

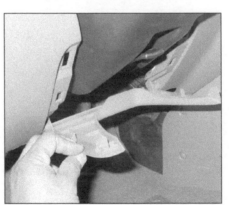

22.7 Release the upper locking tabs and swing the trim panel down - 2007 and earlier models

22.9 With the glove box door open, press in on the corner tabs and pull down to drop the door

22.10 Drop the lever, install a dowel into the lock position (A) and release the temperature control cable assembly by pulling the release tab (B) away from the selector shaft - 2007 and earlier models

22.11 Remove the climate control knobs from the control assembly

22.12 Use a panel tool to pry the center trim panel from the instrument panel - 2007 and earlier models

22.13 Disconnect the electrical connectors from the center trim panel control assembly

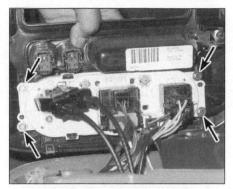

22.14 Location of the climate control assembly mounting screws - 2007 and earlier models

22.25 Carefully pry off the center instrument panel trim with a plastic trim tool

Lower trim panel

7   Release the locking tabs from the upper latches, swing the trim panel down and remove it from the instrument panel (see illustration).

Center trim panel

8   Disconnect the cable from the negative battery terminal (see Chapter 1).
9   Open the glove box door. Press the release tabs toward the interior of the glove box while pulling down on the door (see illustration).
10   Disconnect the temperature control cable from the heater/air conditioner housing (see illustration).
11   Remove the climate control knobs from the center control assembly (see illustration).
12   Use a panel tool to pry the center trim panel out slightly (see illustration).
Note: *It will be necessary to pull the center trim panel out slightly to gain access to the remaining electrical connectors.*
13   Disconnect the electrical connectors from the accessories (see illustration).
14   Remove the climate control assembly mounting screws (see illustration).
15   Remove the center trim panel from the dash area.
16   Installation is the reverse of removal. Make sure the clips are engaged properly before pushing the bezel firmly into place.
17   Reconnect the battery. After you're done, the Powertrain Control Module (PCM) must

relearn its idle and fuel trim strategy for optimum driveability and performance (see Chapter 5).

Glove box

18   Open the glove box and squeeze the sides in to allow the compartment to drop, then remove the mounting screws (see illustration 22.9).

2008 through 2012 models
Instrument cluster bezel

19   Tilt the steering column down as far as possible.
20   Use a plastic trim tool or a screwdriver wrapped with tape to carefully pry the bezel free.
21   Installation is the reverse of removal.

Center upper trim panel

22   Disconnect the cable from the negative battery terminal (see Chapter 1).
23   Open the glove compartment door, then fully release it so that it drops down.
24   Disconnect the wiring from the passenger's airbag, located at the left of the glove box opening.
25   Use a plastic trim tool or a screwdriver wrapped with tape to carefully pry the upper center instrument panel free, releasing each clip as you work your way around it (see illustration).
26   Installation is the reverse of removal.

Center lower trim panel

27   On manual transaxle models, set the

parking brake, then shift into Neutral. Detach the shift lever boot trim ring.
28   On automatic transaxle models, remove the shift lever trim ring (see Section 21).
29   Remove the storage compartment from the console.
30   Use a plastic trim tool or a screwdriver wrapped with tape to pry up and remove the top panel from the console. Disconnect any wiring.
31   Use the same tool to pry off the center lower trim panel. Pull the panel free, then disconnect the wiring (see illustration).
32   Installation is the reverse of removal.

Center middle trim panel

33   Refer to Steps 22 through 25 and remove the upper center trim panel.
34   Remove the lower center trim panel (see Steps 27 through 31).
35   Remove the four screws from the center middle trim panel.
36   Use a plastic trim tool or a screwdriver wrapped with tape to pry out the middle trim panel, then disconnect the wiring (see illustration).
37   Installation is the reverse of removal.

End trim panels

38   Use a plastic trim tool or a screwdriver wrapped with tape to pry the end panel caps free (see illustration).
39   Installation is the reverse of removal.

22.31 Pull out the panel to disconnect the wiring

22.36 Remove the four screws, then pull the center audio panel rearward to disconnect the wiring harnesses

22.38 Both instrument panel end trims are easily removed with gentle prying

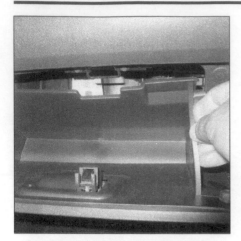

22.42 Push inward on the sides of the glove box to lower it

22.43 Remove these screws, then remove the glove box

22.46 Pry out on the instrument cluster cover and release it from the clips

Driver's knee trim panel

40   To remove the panel, simply use a plastic trim tool or a screwdriver wrapped with tape to pry the knee trim panel free from the retaining clips.

41   Installation is the reverse of removal.

Glove box

42   Open the glove box, then release the tabs to lower it fully (see illustration).

43   Remove the screws from the lower edge of the glove box and lift it out (see illustration).

44   Installation is the reverse of removal.

2013 and later models

Instrument cluster bezel

45   Position the steering wheel all the way outward and downward, then lock the steering column position lever.

46   Pry out and remove the instrument cluster cover (see illustration).

Center upper trim panels

47   Remove the center upper trim and vent panels, following the steps (in sequence) in the photos below (see illustrations).

48   Installation is the reverse of removal.

End trim panels

49   Use a plastic trim tool or a screwdriver wrapped with tape to pry the end panel caps free (see illustration).

50   Installation is the reverse of removal.

OBD port trim panel

51   Remove the OBD access port trim panel (see illustration).

52   Remove the end trim panel (see illustration 22.49).

Note: *The trim panel to the right is blocking a retaining screw, and must be pried out enough to access the screw*

53   Remove the mounting screws (see illustration) and pry off the trim panel.

54   Installation is the reverse of removal.

Driver's knee trim panel (airbag)

55   For the removal of the driver's knee airbag, refer to Chapter 12 for replacement and follow all airbag warnings closely.

Glove box

56   For the removal of the glove box, follow the photos in sequence below (see illustrations).

57   Installation is the reverse of removal.

Instrument panel upper section

58   Disconnect the cable from the negative battery terminal (see Chapter 5).

59   Remove the center console (see Section 21) and transmission selector lever assembly (see Chapter 7B).

60   Remove the defroster vent panel (see Section 24).

61   Remove the upper center speaker (see Chapter 12).

62   Remove the driver's knee bolster airbag (see Chapter 12).

63   Remove the dashboard end trim panels and the OBD port trim panel (see previous Steps).

64   Remove the instrument cluster (see Chapter 12).

65   Remove the steering column (see Chapter 10).

66   Disconnect the electrical connector and remove the GPSM bolt (the module located behind the instrument cluster).

67   Remove the HVAC control assembly (see Chapter 3), then disconnect the electrical connector from the antenna module located behind the HVAC control assembly.

68   Remove the glove box (see previous Steps).

22.47a Pry the upper trim panel off with a flat trim tool

22.47b Rotate the trim up, then disconnect the electrical connections to remove it

22.47c Remove these two fasteners . . .

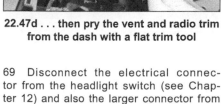

22.47d . . . then pry the vent and radio trim from the dash with a flat trim tool

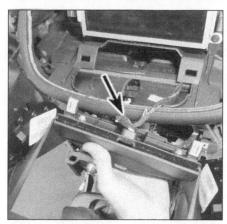

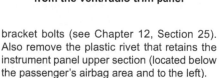

22.47e Disconnect the electrical connector from the vent/radio trim panel

22.49 Both instrument panel end trims are easily removed with gentle prying

69   Disconnect the electrical connector from the headlight switch (see Chapter 12) and also the larger connector from the left-hand side of the heater core (may be slightly obstructed by the instrument panel center brace).

70   Disconnect the passenger's side airbag electrical connectors and remove the airbag

bracket bolts (see Chapter 12, Section 25). Also remove the plastic rivet that retains the instrument panel upper section (located below the passenger's airbag area and to the left).

71   Remove the HVAC ducting (see Section 24).

72   Remove the lower radio module (FDIM) (see Chapter 12).

73   If equipped, disconnect the HVAC blend door electrical connector(s) from the center area (near the radio modules) of the upper instrument panel.

74   Remove the various mounting bolts that retain the upper section of the instrument panel. Using an assistant's help, if necessary, carefully and slowly lift off the upper section, making sure

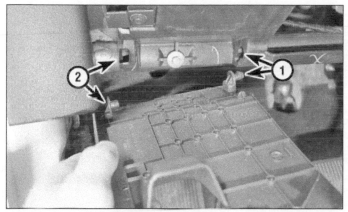

22.51 Disengage the right-side notch on the panel first (1), then slide the panel to the left and disengage the left side notch (2)

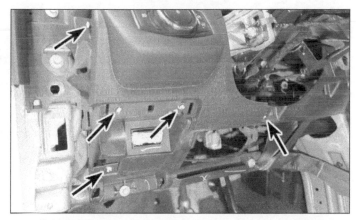

22.53 Remove the OBD port trim panel mounting screws

22.56a Remove the lower mounting bolts

22.56b From inside the glove box, remove these two screws

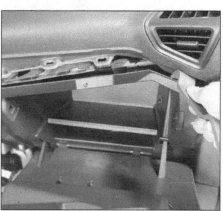

22.56c Pry the glove box free, then disconnect the electrical connector and remove the door and the glove box as one assembly

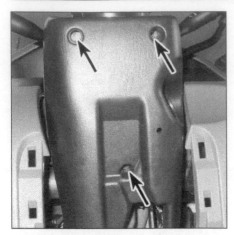

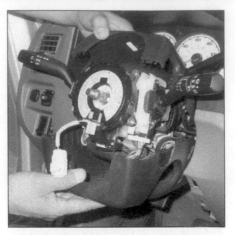

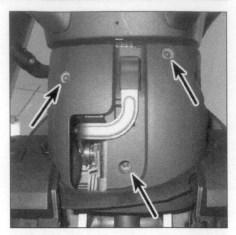

23.2a Location of the steering column cover screws - 2007 and earlier models

23.2b Separate the upper and lower steering column covers (steering wheel removed for clarity)

23.6 Lower steering column cover screws

to disconnect any remaining electrical connectors, harness clips, or fasteners. Remove the instrument panel upper section.
75   Installation is the reverse of removal.

Install all the bolts loosely by hand at first, positioning or adjusting the panel as necessary, then tighten the instrument panel upper section fasteners securely.

23   Steering column covers - removal and installation

Warning: *The models covered by this manual are equipped with a Supplemental Restraint System (SRS), more commonly known as airbags. Always disarm the airbag system before working in the vicinity of any airbag system component to avoid the possibility of accidental deployment of the airbag, which could cause personal injury (see Chapter 12).*

2007 and earlier models

1   On tilt steering columns, move the column to the lowest position. Refer to Chapter 12 and remove the ignition key lock cylinder.
2   Remove the screws, then separate the halves and remove the upper and lower steering column covers (see illustrations).
3   Installation is the reverse of the removal procedure.

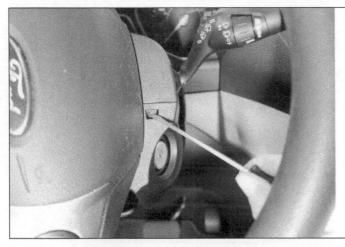

23.9a Rotate the steering wheel to a 90 degree angle (from straight), then use a small flat screwdriver to press in on the locking tabs. Repeat this process for the other tab, with the steering wheel turned in the opposite direction

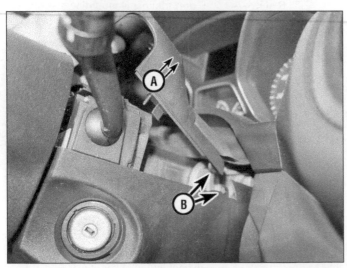

23.9b Lift up and remove the upper column cover (A) - notice that the circular clip needs to engage the hook (B) when installing

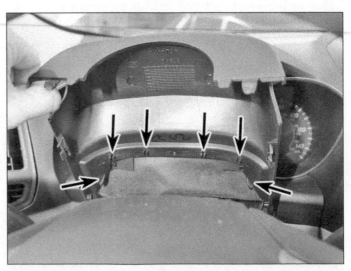

23.9c With the upper cover lifted up, disengage the leather trim clips and remove the cover

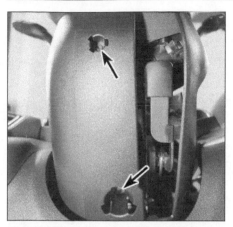

23.9d The lower cover is held in place with two screws

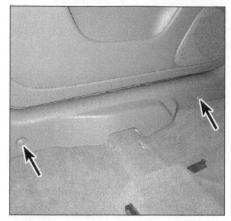

24.6 Door scuff plate retainer locations

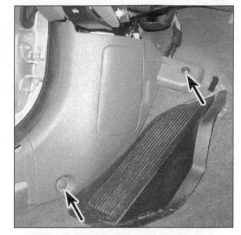

24.7 Remove the fasteners from the A-pillar lower trim panels - left side shown - 2007 and earlier models

2008 to 2012 models

4   Refer to Section 22 and remove the driver's knee trim panel.
Caution: *Be careful. It's easy to break the retaining tabs. Lift the upper cover off.*
5   Gently push in on the lower column cover, then use a plastic trim tool or a screwdriver wrapped with tape to separate the upper cover.
6   Remove the three screws from the lower cover (see illustration).
7   Release the tilt lever, then remove the lower cover.
8   Installation is the reverse of removal.

2013 and later models

9   To remove the steering column cover(s), follow the steps (in sequence) mentioned in the photos below (see illustrations).
10   Installation is the reverse of removal.

24   Instrument panel - removal and installation

Warning: *The models covered by this manual are equipped with a Supplemental Restraint*

System (SRS), more commonly known as airbags. Always disarm the airbag system before working in the vicinity of any airbag system component to avoid the possibility of accidental deployment of the airbag, which could cause personal injury (see Chapter 12).
Warning: *For models that require the a/c system to be discharged - the air conditioning system is under high pressure. DO NOT loosen any fittings or remove any components until after the system has been discharged. Air conditioning refrigerant must be properly discharged into an EPA-approved container at a dealer service department or an automotive air conditioning repair facility. Always wear eye protection when disconnecting air conditioning system fittings.*

2007 and earlier models

1   Disconnect the cable from the negative battery terminal (see Chapter 5).
2   On tilt steering columns, move the column to the lowest position. Refer to Chapter 10 and remove the steering wheel.
3   Remove the passenger's side airbag (see Chapter 12).
4   Remove the audio unit from the center of

the dashboard (see Chapter 12).
5   Remove the heater/air conditioner control assembly (see Chapter 3).
6   Remove the pin-type retainers from the front door scuff plates (see illustration), and remove the plates.
7   Remove the pin-type retainers from the A-pillar lower trim panels (see illustration), then remove the panels.
8   Disconnect the instrument panel harness connectors located by the cowl (see illustration).
9   Remove the hood latch bolts and position the assembly off to the side without disconnecting the cable (see Section 8).
10   Remove the utility compartment (see illustration).
11   Remove the steering column covers (see Section 23).
12   On manual transaxle models, disconnect the shift cable.
13   Disconnect the steering column intermediate shaft U-joint.
14   Remove the cover panel (see illustration).

24.8 Location of the instrument panel connectors

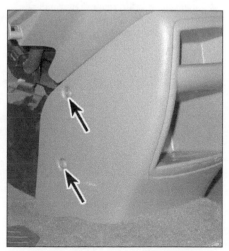

24.10 Remove the utility compartment mounting screws - two each side - 2007 and earlier models

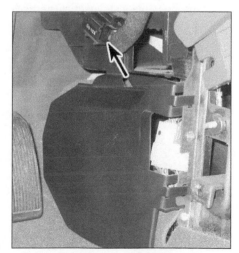

24.14 Remove the cover panel fastener

24.15 Disconnect the instrument panel connectors below the center dash - 2007 and earlier models

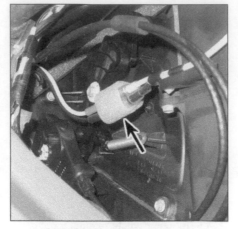

24.16 Disconnect the vacuum harness connector

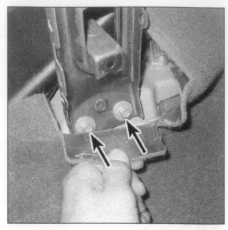

24.17 Remove the instrument panel center brace mounting bolts - 2007 and earlier models

15   Disconnect the harness connectors (see illustration).

16   Disconnect the climate control vacuum harness (see illustration).

17   Remove the instrument panel center brace (see illustration).

18   Disconnect the temperature control cable from the heater/air conditioner assembly (see Section 22).

19   Press the glove box tabs toward the interior of the glove box and pull the door down (see Section 22).

20   Disconnect the blower electrical connectors.

21   Disconnect the interior antenna cable connection (see illustration).

22   Remove the A-pillar passenger's assist handle (see illustration).

23   Remove the A-pillar moldings by prying the edges with a panel tool.

24   Remove the small cover from the center of the instrument panel (near the windshield), then unscrew the bolt.

25   Remove the instrument cluster (see Chapter 12).

26   Working in the instrument cluster cav-

ity, remove the instrument panel bolt (see illustration).

27   Remove the finish panels from each side of the instrument panel (see illustration).

28   Remove the instrument panel mounting bolts from the support structure (see illustration).

29   Remove the center floor console (see Section 21).

30   Remove the lower trim panel (see Section 22). Then detach the bolts securing the steering column and lower it away from the instrument panel.

31   A number of electrical connectors must be disconnected in order to remove the instrument panel. Most are designed so that they will only fit on the matching connector (male or female), but if there is any doubt, mark the connectors with masking tape and a marking pen before disconnecting them.

32   Once all the fasteners are removed, lift the instrument panel, then pull it away from the windshield and take it out through the driver's door opening.

Note: *This is a two-person job.*

33   Installation is the reverse of removal.

2008 through 2012 models
Removal

34   Disconnect the cable from the negative battery terminal (see Chapter 5).

35   Refer to Section 21 and remove the center console.

36   Refer to Chapter 10 and remove the steering wheel.

37   Use a plastic trim tool or a screwdriver wrapped with tape to pry off the trim panels from the windshield pillars.

38   Remove the retainers, then remove the door opening sill plates.

39   Remove the lower cowl kick panels.

40   Refer to Section 22 and remove the driver's knee trim panel and the instrument panel end trim panels.

41   Disconnect the two wiring harness connectors under the left end trim panel. Disconnect the steering module wiring harness connector.

42   Unbolt the hood and brake release handles. Move them out of the way.

43   Release the cover from the bottom of the steering column, then slide the cover up the steering column.

Note: *The manufacturer recommends that*

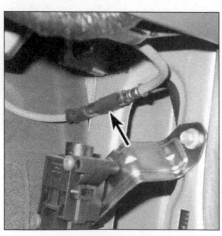

24.21 Disconnect the antenna connector

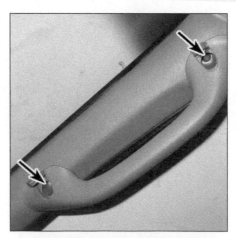

24.22 Remove the screw covers and the A-pillar assist handle mounting bolts

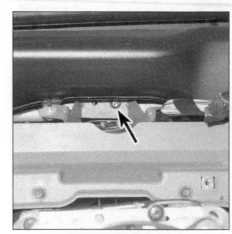

24.26 Remove this bolt from the instrument cluster cavity - 2007 and earlier models

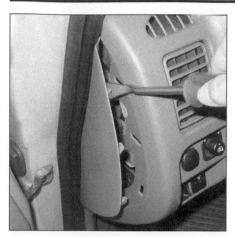

24.27 Remove the finish panels

24.28 Remove the bolts from the support structure on each side

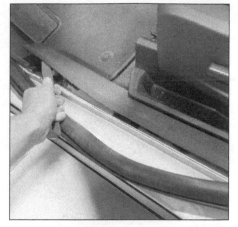

24.65a On both sides, pry up and remove the scuff plates from the bottom front door sills . . .

this pinch bolt be discarded and replaced with a new one.

44   Remove the pinch bolt from under the cover and move the intermediate steering shaft rearward to disconnect it from the coupling.

45   Remove the cover from the RCM module under the right kick panel, then disconnect the module wiring.

46   Refer to Chapter 7B and remove the shifter assembly.

47   Disconnect the wiring from the smart junction box (under the center middle instrument panel trim panel). Remove the smart junction box.

48   Disconnect the two large wiring harness connectors from the right end of the instrument panel.

49   Disconnect the antenna cable.

50   Remove the cowl grille (see Section 12).

51   Remove the wiper arm pivot shafts.

52   Remove the three instrument panel upper cowl bolts, the four center brace bolts and the four side bolts.

53   With the help of an assistant, remove the instrument panel and slide it out through the right door opening.

Installation

54   With the help of an assistant, slide the assembly in through the right door opening, loosely install the end bolts, then check carefully for correct wire routing and pinched wires.

55   Install the three upper cowl bolts, the wiper pivot shafts and arms and the cowl panel cover. Tighten the four instrument panel end bolts.

56   The remainder of installation is the reverse of removal.

2013 and later models

Removal

57   Have the air conditioning system discharged by an approved automotive repair shop (see Chapter 3).

58   Drain the cooling system (see Chapter 3).

59   Disconnect the cable from the negative battery terminal (see Chapter 1).

60   Remove the cowl paneling (see Section 12).

61   Working on the outside of the firewall, remove the evaporator flange pipe nut and disconnect the a/c tubing flange from the TXV valve. Also, disconnect the coolant hoses from the firewall (see Chapter 3).

62   Remove both front doors (see Section 14).

63   Remove the center console (see Section 21).

64   Remove the transmission selector lever assembly (see Chapter 7B).

65   Remove the door scuff plates, then remove the kick panels (see illustrations). Note that the driver's side kick panel requires the removal of the hood release lever first (see Section 8).

66   Remove the instrument panel end caps (see Section 22).

67   Remove the footrest on the driver's side, then peel back the carpeting as far as possible on both the driver's and passenger's sides.

68   Remove the glove box (see Section 22) and the insulation trim panel beneath it.

Note: *The manufacturer recommends that this pinch bolt be discarded and replaced with a new one.*

69   Remove the steering column shaft-to-steering gear pinch bolt, then separate the steering shafts (see Chapter 10).

70   Disconnect the RCM (airbag module) and the emergency brake electrical connections with the center console removed (see illustration).

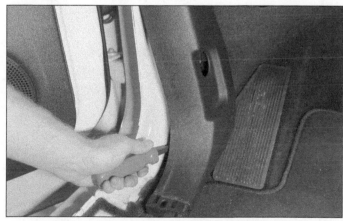

24.65b . . . then pry out and remove the driver's and passenger's side kick panels.

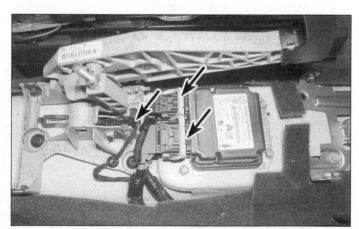

24.70 Disconnect the electrical connectors

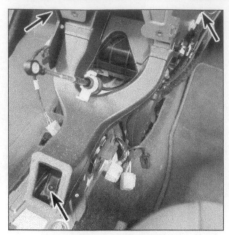

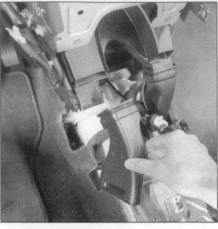

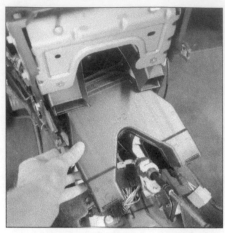

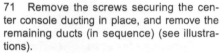

24.71a Three screws secure the ducting in place - one lower screw and two higher up on the outside edges

24.71b The side ducts are held in place by overlapping the adjacent ducting sections

24.71c The lower duct is held in the same way - work the duct free and remove it

71   Remove the screws securing the center console ducting in place, and remove the remaining ducts (in sequence) (see illustrations).
72   Disconnect the large electrical connector (see illustration).
73   Remove the retaining bolts, then unclip the ground wires attached to the floor on the driver's and passenger's sides.

74   Remove the dashboard center upper cover (see Section 22).
75   Use a plastic trim tool or a screwdriver wrapped with tape to pry off the trim panels from the windshield side trim (A pillars) (see illustrations).
76   Remove the upper defroster trim panel (see illustrations).
77   Disconnect the wiring from the smart

junction box (viewed from the glove box opening) (see illustration).
78   If applicable, disconnect the two large wiring harness connectors from the center of the instrument panel.
79   Disconnect the antenna cable from the passenger's side kick panel area.
80   Disconnect and remove the evaporator drain tube (see illustration).

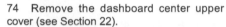

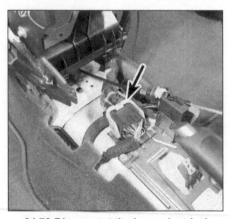

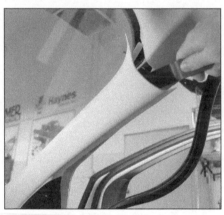

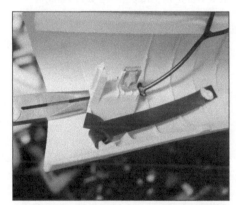

24.72 Disconnect the large electrical connector by releasing the lever and pulling it carefully apart

24.75a Pry off both A-pillar trim panels

24.75b Detach the retaining cable from the A-pillar trim panel, then disconnect the electrical connector (if equipped) and remove the panel

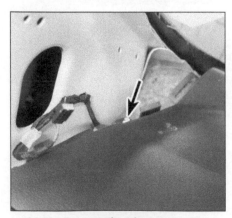

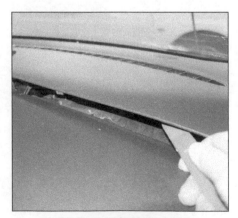

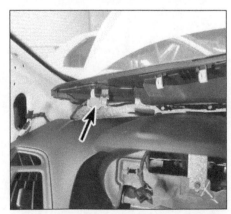

24.76a Remove the trim panel screw at each side

24.76b Pry up along the entirety of the panel

24.76c Disconnect the panel electrical connector and remove the defroster panel

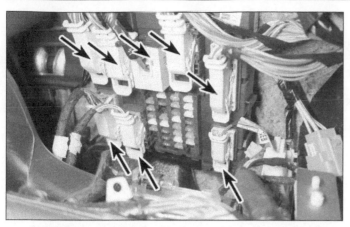

24.77 Disconnect the electrical connections from the SJB

24.80 Remove the rubber evaporator drain tube

81 Remove the instrument panel upper (driver's side) bolts (see illustration).
82 Remove the instrument panel interior crossbeam bolt at each end (see illustration).
83 Remove the instrument panel bottom-center mounting bolts (see illustration).
84 Remove the remaining bolts on both ends of the instrument panel (see illustrations).
85 With the help of an assistant, rotate the instrument panel face-downward, then

remove the instrument panel by carrying it out through the left door opening.
Note: *Start by lifting the instrument panel a few inches and check to see if you've missed anything. Rotate the instrument panel away from the firewall while observing all of the attachment points and the electrical connectors. Stop if anything is obstructing your removal and make the necessary disconnections before continuing.*

Installation

86 With the help of an assistant, slide the assembly in through the right door opening, loosely install the end bolts, then check carefully for correct wire routing and pinched wires.
87 Installation is the reverse of removal. If applicable, have the air conditioning system properly serviced back to operating condition by a certified air conditioning service technician at a repair facility.

24.81 Remove the instrument panel upper bolts (seen through windshield)

24.82 Remove the interior crossbeam bolt (hidden by the plastic cap) from each end

24.83 Instrument panel bottom-center mounting bolts

24.84a Remove the outer end bolts (passenger's side)

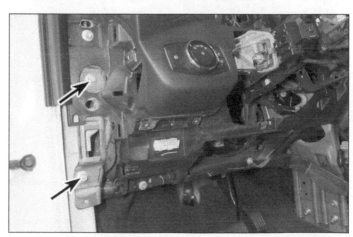

24.84b Remove the outer end bolts (driver's side)

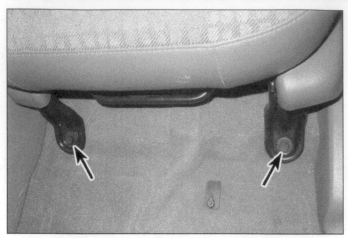

25.2a Location of the forward mounting bolts

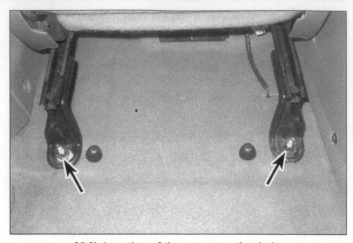

25.2b Location of the rear mounting bolts

25   Seats - removal and installation

Warning: *These vehicles are equipped with side-impact airbags located in the outboard sides of the front seat backs. Most models are also equipped with seat belt pre-tensioners, which are also explosive devices. Always disarm the airbag system before working in the vicinity of any airbag system component to avoid the possibility of accidental deployment of the airbag, which could cause personal injury (see Chapter 12).*

2004 and earlier models

Front seat

1   Pry out the plastic covers to access the seat tracks and their mounting bolts.
2   Remove the retaining bolts (see illustrations).

3   Tilt the seat upward to access the underside, then disconnect any electrical connectors (see illustration) and lift the seat from the vehicle.
4   Installation is the reverse of removal.

Rear seat

5   Fold the seat forward.
6   Remove the hinge retaining bolts and detach the seat from the vehicle (see illustration).
7   Installation is the reverse of removal.

2005 through 2007 models

Front seat

8   Slide the seat forward and back to access the bolts and their covers. Remove the covers and the bolts.
9   Tilt the seat rearward, then disconnect the wiring harnesses under the seat.

10   Carefully lift the seat out.
11   Installation is the reverse of removal.

Rear seat

Cushions

12   Fold the seat back forward.
13   Release the lock tab, then slide the cushion outward and remove it.
14   Installation is the reverse of removal.

Seat back

15   Tilt the rear seat cushions and seat back forward.
16   Detach the rear carpet from the seats.
17   Remove the nuts and bolt from the smaller seat back bracket, then tilt the seat back upright.
18   Remove the bolt from the smaller seat back bracket, then remove it.
19   Remove the nut from the larger seat back bracket, then tilt it upright.

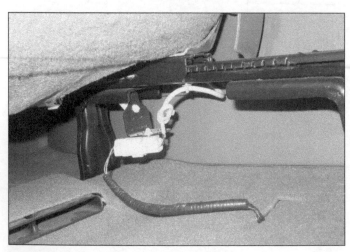

25.3 Disconnect any electrical connectors under the seat

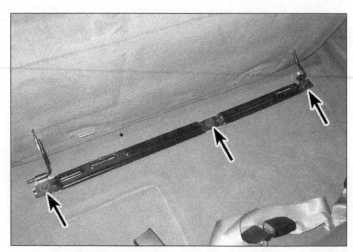

25.6 Location of the rear seat hinge bolts

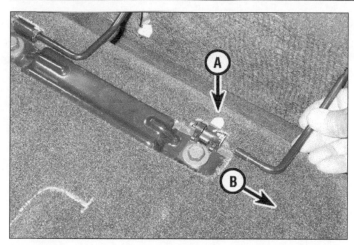

25.30 Press the tab (A), then slide the seat cushion outward (B) to remove it

25.34 Right rear seat back front mounting bolts

20   Remove the rear seat belt anchor bolt.
21   Remove the bolt from the larger seat back and remove it.
22   Installation is the reverse of removal.

2008 and later models

Front seat

23   Slide the seat to the rear, then pry open the bolt covers and remove the two front mounting bolts. Slide the seat to the front.
Note: *On some models, the right and left seat use different bolts. Don't get them mixed up.*
24   Remove the covers from the seat rear bolts, then remove the bolts.
25   Disconnect the cable from the negative battery terminal (see Chapter 5).
26   Tilt the seat for access (if necessary) and disconnect the electrical connector(s) the seat.
27   Remove the seat.
28   When installing it, tighten the seat fasteners in the following order: inner rear nut (or bolt), outer rear nut (or bolt), inner front bolt, outer front bolt.

Rear seat

Cushion (2008 through 2012 models)

29   Tilt the seat cushion forward.
30   Release the lock tab, then slide the cushion outward and remove it (see illustration).
31   Installation is the reverse of removal.

Seat back (2008 through 2012 models)

32   Tilt the seat cushions forward.
33   Free the seat belts from the elastic retainers.
34   Remove the outer seat back-to-floor bolts (see illustration).
35   Remove the center seat belt anchor bolt from under the left seat back.
36   Fold down both seat backs.
37   Unclip the carpet and move it out of the way.
38   Remove the rectangular bolt cover.

Remove the right seat outer and inner seat back-to-floor nuts (see illustration).
39   Remove the right seat back.
40   Remove the left seat back-to-floor outer and inner nuts (see illustration).
41   Remove the left seat back.
42   Install the left seat back first. The remainder of the installation is the reverse of removal.

Rear seat assembly (2013 and later models)

43   Release the seatbelt from the shoulder guide.
44   Fold the seat back downward, position the cargo floor extension rearward, then remove the trim piece - exposing the rear seat rear mounting bolts. Remove the bolts retaining the seat to be replaced.
45   Remove the rear seat front trim covers, then remove the front mounting bolts and lift out the seat as an assembly.
46   Installation is the reverse of removal.

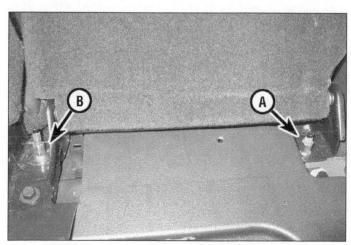

25.38 Outer (A) and inner (B) right rear seat back mounting fasteners

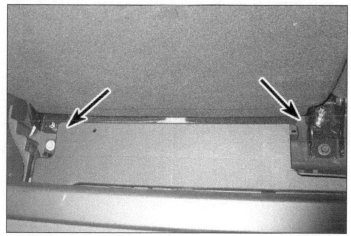

25.40 Mounting fasteners for the left section of the rear seat back

Notes

Chapter 12
Chassis electrical system

Contents

1 General Information

1 The electrical system is a 12-volt, negative ground type. Power for the lights and all electrical accessories is supplied by a lead/acid battery, which is charged by the alternator.

2 This Chapter covers repair and service procedures for the various electrical components not associated with the engine. Information on the battery, alternator and starter motor can be found in Chapter 5.

3 It should be noted that when portions of the electrical system are serviced, the negative battery cable should be disconnected from the battery to prevent electrical shorts and/or fires.

2 Electrical troubleshooting - general information

1 A typical electrical circuit consists of an electrical component, any switches, relays, motors, fuses, fusible links or circuit breakers related to that component and the wiring and connectors that link the component to both the battery and the chassis. To help you pinpoint an electrical circuit problem, wiring diagrams are included at the end of this Chapter.

2 Before tackling any troublesome electrical circuit, first study the appropriate wiring diagrams to get a complete understanding of what makes up that individual circuit. Trouble spots, for instance, can often be narrowed down by noting if other components related to the circuit are operating properly. If several components or circuits fail at one time, chances are the problem is in a fuse or ground connection, because several circuits are often routed through the same fuse and ground connections.

3 Electrical problems usually stem from simple causes, such as loose or corroded connections, a blown fuse, a melted fusible link or a failed relay. Visually inspect the condition of all fuses, wires and connections in a problem circuit before troubleshooting the circuit.

4 If test equipment and instruments are going to be utilized, use the diagrams to plan ahead of time where you will make the necessary connections in order to accurately pinpoint the trouble spot.

5   The basic tools needed for electrical troubleshooting include a circuit tester or voltmeter (a 12-volt bulb with a set of test leads can also be used), a continuity tester, which includes a bulb, battery and set of test leads, and a jumper wire, preferably with a circuit breaker incorporated, which can be used to bypass electrical components (see illustrations). Before attempting to locate a problem with test instruments, use the wiring diagram(s) to decide where to make the connections.

Voltage checks

6   Voltage checks should be performed if a circuit is not functioning properly. Connect one lead of a circuit tester to either the negative battery terminal or a known good ground. Connect the other lead to a connector in the circuit being tested, preferably nearest to the battery or fuse (see illustration). If the bulb of the tester lights, voltage is present, which means that the part of the circuit between the connector and the battery is problem free. Continue checking the rest of the circuit in the same fashion. When you reach a point at which no voltage is present, the problem lies between that point and the last test point with voltage. Most of the time the problem can be traced to a loose connection. **Note:** *Keep in mind that some circuits receive voltage only when the ignition key is in the Accessory or Run position.*

Finding a short

7   One method of finding shorts in a circuit is to remove the fuse and connect a test light or voltmeter in place of the fuse terminals. There should be no voltage present in the circuit. Move the wiring harness from side-to-side while watching the test light. If the bulb goes on, there is a short to ground somewhere in that area, probably where the insulation has rubbed through. The same test can be performed on each component in the circuit, even a switch.

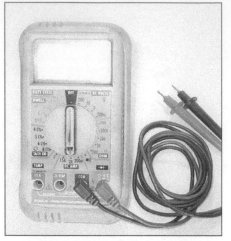

2.5a The most useful tool for electrical troubleshooting is a digital multimeter that can check volts, amps, and test continuity

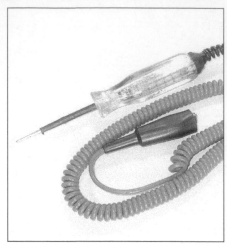

2.5b A simple test light is a very handy tool for testing voltage

Ground check

8   Perform a ground test to check whether a component is properly grounded. Disconnect the battery and connect one lead of a continuity tester or multimeter (set to the ohms scale), to a known good ground. Connect the other lead to the wire or ground connection being tested. If the resistance is low (less than 5 ohms), the ground is good. If the bulb on a self-powered test light does not go on, the ground is not good.

Continuity check

9   A continuity check is done to determine if there are any breaks in a circuit - if it is passing electricity properly. With the circuit off (no power in the circuit), a self-powered continuity tester or multimeter can be used to check the circuit. Connect the test leads to both ends of the circuit (or to the power end and a good ground), and if the test

light comes on the circuit is passing current properly (see illustration). If the resistance is low (less than 5 ohms), there is continuity; if the reading is 10,000 ohms or higher, there is a break somewhere in the circuit. The same procedure can be used to test a switch, by connecting the continuity tester to the switch terminals. With the switch turned On, the test light should come on (or low resistance should be indicated on a meter).

Finding an open circuit

10   When diagnosing for possible open circuits, it is often difficult to locate them by sight because the connectors hide oxidation or terminal misalignment. Merely wiggling a connector on a sensor or in the wiring harness may correct the open circuit condition. Remember this when an open circuit is indicated when troubleshooting a circuit. Intermit-

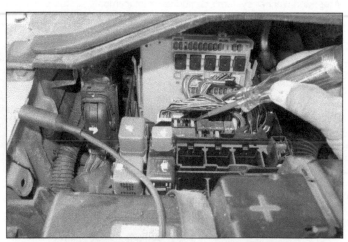

2.6 In use, a basic test light's lead is clipped to a known good ground, then the pointed probe can test connectors, wires or electrical sockets - if the bulb lights, the circuit being tested has battery voltage

2.9 With a multimeter set to the ohm scale, resistance can be checked across two terminals - when checking for continuity, a low reading indicates continuity, a high reading or infinity indicates high resistance or lack of continuity

tent problems may also be caused by oxidized or loose connections.

11   Electrical troubleshooting is simple if you keep in mind that all electrical circuits are basically electricity running from the battery, through the wires, switches, relays, fuses and fusible links to each electrical component (light bulb, motor, etc.) and to ground, from which it is passed back to the battery. Any electrical problem is an interruption in the flow of electricity to and from the battery.

Connectors

12   Most electrical connections on these vehicles are made with multiwire plastic connectors. The mating halves of many connectors are secured with locking clips molded into the plastic connector shells. The mating halves of large connectors, such as some of those under the instrument panel, are held together by a bolt through the center of the connector.

13   To separate a connector with locking clips, use a small screwdriver to pry the clips apart carefully, then separate the connector halves. Pull only on the shell, never pull on the wiring harness as you may damage the individual wires and terminals inside the connectors. Look at the connector closely before trying to separate the halves. Often the locking clips are engaged in a way that is not immediately clear. Additionally, many connectors have more than one set of clips.

14   Each pair of connector terminals has a male half and a female half. When you look at the end view of a connector in a diagram, be sure to understand whether the view shows the harness side or the component side of the connector. Connector halves are mirror images of each other, and a terminal shown on the right side end view of one half will be on the left side end view of the other half.

Electrical connectors

Most electrical connectors have a single release tab that you depress to release the connector

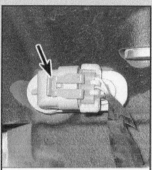

Some electrical connectors have a retaining tab which must be pried up to free the connector

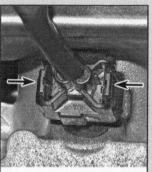

Some connectors have two release tabs that you must squeeze to release the connector

Some connectors use wire retainers that you squeeze to release the connector

Critical connectors often employ a sliding lock (1) that you must pull out before you can depress the release tab (2)

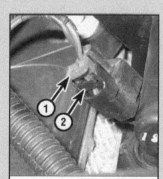

Here's another sliding-lock style connector, with the lock (1) and the release tab (2) on the side of the connector

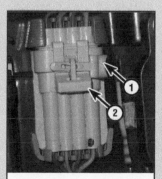

On some connectors the lock (1) must be pulled out to the side and removed before you can lift the release tab (2)

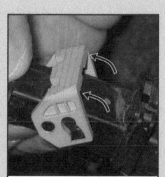

Some critical connectors, like the multi-pin connectors at the Powertrain Control Module employ pivoting locks that must be flipped open

3.1a 2012 and earlier models: The fuse/relay box is mounted on the left side of the engine compartment; it contains diodes (for the starter relay and for the air conditioning compressor clutch), fuses and relays, all of which are listed by location and function on the underside of the fuse/relay box cover

3.1b 2013 and later models: The BJB (Battery Junction Box) is mounted on the left side of the engine compartment; it contains diodes (for the starter relay and for the air conditioning compressor clutch), fuses and relays, all of which are listed by location and function on the underside of the fuse/relay box cover

3   Fuses and fusible links - general information

Fuses

1   The electrical circuits of the vehicle are protected by a combination of fuses, circuit breakers and fusible links. Fuse blocks are located under the instrument panel and in the engine compartment (see illustrations). Each of the fuses is designed to protect a specific circuit, and the various circuits are identified on the fuse panel cover. If the fuse panel cover is difficult to read, or missing, you can also refer to your owner's manual, which includes a complete guide to all fuses and relays in both fuse/relay boxes.

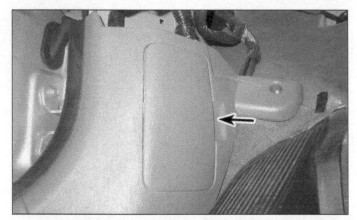

3.1c 2012 and earlier models: To open the fuse panel cover from the left kick panel (early models) or the right-side of the console (later models), pull here, then disengage the two cover locator tabs and remove the cover

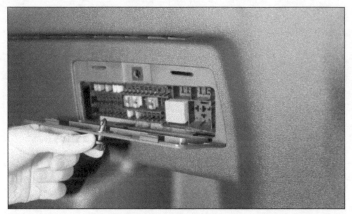

3.1d 2013 and later models: The rear fuse box is located on the right rear quarter panel area

3.1e 2012 and earlier models: The passenger compartment fuse/relay box is located either in the left kick panel as shown, or on the right-hand side of the center console

3.1f 2013 and later models: The Body Control Module (BCM) fuses are located under the passenger's side of the dash behind the glove box

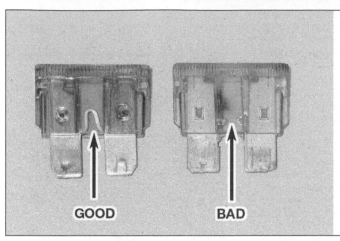

3.2 When a fuse blows, the element between the terminals melts - the fuse on the right is blown, the fuse on the left is good

2   Miniaturized fuses are employed in the fuse blocks. These compact fuses, with blade terminal design, allow fingertip removal and replacement. If an electrical component fails, always check the fuse first. The best way to check a fuse is with a test light. Check for power at the exposed terminal tips of each fuse. If power is present on one side of the fuse but not the other, the fuse is blown. A blown fuse can also be confirmed by visually inspecting it (see illustration).

3   Be sure to replace blown fuses with the correct type. Fuses of different ratings are physically interchangeable, but only fuses of the proper rating should be used. Replacing a fuse with one of a higher or lower value than specified is not recommended. Each electrical circuit needs a specific amount of protection. The amperage value of each fuse is molded into the fuse body.

4   If the replacement fuse immediately fails, don't replace it again until the cause of the problem is isolated and corrected. In most cases, this will be a short circuit in the wiring caused by a broken or deteriorated wire.

Fusible links

5   Some circuits are protected by fusible links. The links are used in circuits which are not ordinarily fused, or which carry high current, such as the starter circuit. If your vehicle uses any fusible links, they'll be located in the engine compartment fuse/relay box. Cartridge-type fusible links are similar in appearance to a large fuse. After disconnecting the cable from the negative battery terminal (see Chapter 5), simply unplug (or unbolt) and replace a fusible link with one of the same amperage.

4   Circuit breakers - general information

1   Circuit breakers protect certain circuits, such as the power windows or heated seats. Depending on the vehicle's accessories, there may be one or two circuit breakers, located in the fuse/relay box in the engine compartment.

2   Because the circuit breakers reset automatically, an electrical overload in a circuit breaker-protected system will cause the circuit to fail momentarily, then come back on. If the circuit does not come back on, check it immediately.

3   For a basic check, pull the circuit breaker up out of its socket on the fuse panel, but just far enough to probe with a voltmeter. The breaker should still contact the sockets. With the voltmeter negative lead on a good chassis ground, touch each end prong of the circuit breaker with the positive meter probe. There should be battery voltage at each end. If there is battery voltage only at one end, the circuit breaker must be replaced.

4   Some circuit breakers must be reset manually.

5   Relays - general information and testing

General information

1   Several electrical accessories in the vehicle, such as the fuel injection system, horns, starter, and fog lamps use relays to transmit the electrical signal to the component. Relays use a low-current circuit (the control circuit) to open and close a high-current circuit (the power circuit). If the relay is defective, that component will not operate properly. Most relays are mounted in the engine compartment fuse/relay box or SJB with some specialized relays located in the interior fuse boxes or BCM/fuse box (see illustrations 3.1a and 3.1f). If a faulty relay is suspected, it can be removed and tested using the procedure below or by a dealer service department or a repair shop. Defective relays must be replaced as a unit.

Testing

2   Most of the relays used in these vehicles are of a type often called "ISO" relays, which refers to the International Standards Organization. The terminals of ISO relays are numbered to indicate their usual circuit connections and functions. There are two basic layouts of terminals on the relays used in these vehicles (see illustrations).

3   Refer to the wiring diagram for the circuit to determine the proper connections for the relay you're testing. If you can't determine the correct connection from the wiring diagrams, however, you may be able to determine the test connections from the information that follows.

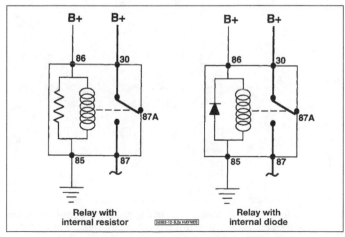

5.2a Typical ISO relay designs, terminal numbering and circuit connections

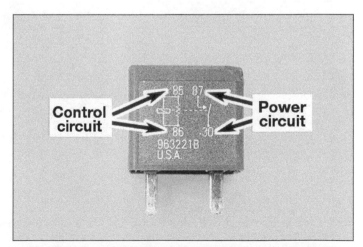

5.2b Most relays are marked on the outside to easily identify the control circuits and the power circuits - four terminal type shown

4   Two of the terminals are the relay control circuit and connect to the relay coil. The other relay terminals are the power circuit. When the relay is energized, the coil creates a magnetic field that closes the larger contacts of the power circuit to provide power to the circuit loads.

5   Terminals 85 and 86 are normally the control circuit. If the relay contains a diode, terminal 86 must be connected to battery positive (B+) voltage and terminal 85 to ground. If the relay contains a resistor, terminals 85 and 86 can be connected in either direction with respect to B+ and ground.

6   Terminal 30 is normally connected to the battery voltage (B+) source for the circuit loads. Terminal 87 is connected to the circuit leading to the component being powered. If the relay has several alternate terminals for load or ground connections, they usually are numbered 87A, 87B, 87C, and so on.

7   Use an ohmmeter to check continuity through the relay control coil.

  a) *Connect the meter according to the polarity shown in the illustration for one check; then reverse the ohmmeter leads and check continuity in the other direction.*

  b) *If the relay contains a resistor, resistance will be indicated on the meter, and should be the same value with the ohmmeter in either direction.*

  c) *If the relay contains a diode, resistance should be higher with the ohmmeter in the forward polarity direction than with the meter leads reversed.*

  d) *If the ohmmeter shows infinite resistance in both directions, replace the relay.*

8   Remove the relay from the vehicle and use the ohmmeter to check for continuity between the relay power circuit terminals. There should be no continuity between terminal 30 and 87 with the relay de-energized.

9   Connect a fused jumper wire to terminal 86 and the positive battery terminal. Connect another jumper wire between terminal 85 and ground. When the connections are made, the relay should click.

10   With the jumper wires connected, check for continuity between the power circuit terminals. Now, there should be continuity between terminals 30 and 87.

11   If the relay fails any of the above tests, replace it.

6   Turn signal and hazard flasher - check and replacement

Warning: *The models covered by this manual are equipped with a Supplemental Restraint System (SRS), more commonly known as airbags. Always disarm the airbag system before working in the vicinity of any airbag system component to avoid the possibility of accidental deployment of the airbag, which could cause personal injury (see Section 25).*
Note: *This procedure applies only to 2004 and earlier models. Later models have a Smart Junction Box that performs the same function, as well as many others. The Smart Junction Box should not be disassembled for any reason. There are no serviceable parts within the SJB. For any repairs regarding the SJB take the vehicle to the dealership repair facility or to a qualified independent repair shop.*

1   The turn signal and hazard flashers are controlled from a single electronic flasher unit which is mounted to the left of the steering column under the instrument panel (see illustration).

2   When the flasher unit is functioning properly, an audible click can be heard during its operation. If the turn signal indicator (on the instrument panel) on one side of the vehicle flashes much more rapidly than normal, a faulty turn signal bulb is indicated.

3   If both turn signals fail to blink, the problem may be due to a blown fuse, a faulty flasher unit, a broken switch or a loose or open connection. If a quick check of the fuse box indicates that the turn signal fuse has blown, check the wiring for a short before installing a new fuse.

4   To replace the flasher, remove the flasher

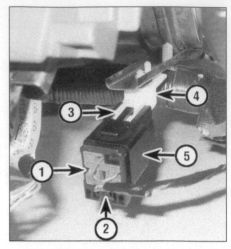

6.1 The turn signal/hazard flasher unit is located under the instrument panel, to the left of the steering column (2004 and earlier models)

1   *Electrical connector*
2   *Electrical connector release (located underneath the flasher unit, not visible in this photo)*
3   *Mounting tab (lift up with a small pocket screwdriver to release)*
4   *Plastic mounting bracket (it's not necessary to remove this part, but if you have difficulty releasing the mounting tab, you can pull the bracket down to separate it from the metal bracket above)*
5   *Turn signal/hazard flasher unit*

from its mounting bracket and then disconnect the electrical connector (see illustrations). (It's easier to detach the flasher unit first, and then disconnect the electrical connector.)

5   Make sure that the replacement unit is identical to the original. Compare the old one to the new one before installing it.

6   Installation is the reverse of removal.

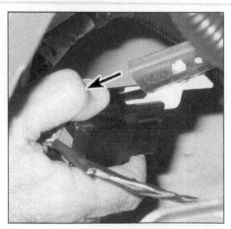

6.4a To detach the turn signal/hazard flasher unit from its mounting bracket, press down here with your thumb to release the locking tab and pull the flasher unit to the rear

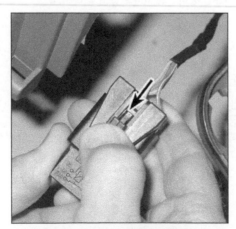

6.4b To disconnect the electrical connector from the turn signal/hazard flasher unit, depress this locking tab with your thumb . . .

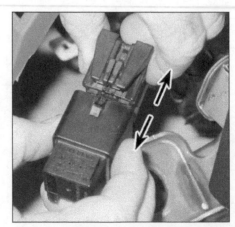

6.4c . . . and separate the connector from the turn signal/hazard flasher unit (2006 and earlier models)

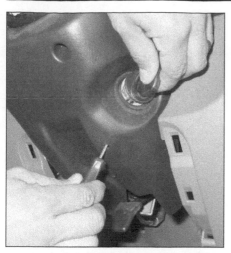

7.3a To release the key lock cylinder from the steering column assembly, insert a 1/8-inch awl or punch through this hole in the steering column cover . . .

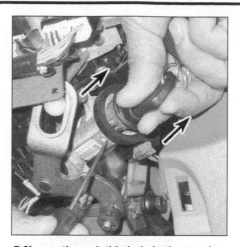

7.3b . . . through this hole in the steering column housing, then depress the release button on the lock cylinder and pull out the lock cylinder (steering column cover removed for clarity)

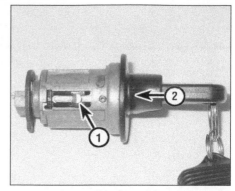

7.3c Key lock cylinder removal and installation details

1   Release button
2   When installing the key lock cylinder, align the edge of the key with the release button to put the key in the On position

7   Ignition switch, key lock cylinder, and anti-theft transceiver - removal and installation

Warning: *All models covered by this manual are equipped with a Supplemental Restraint System (SRS), more commonly known as airbags. Always disarm the airbag system before working in the vicinity of any airbag system component to avoid the possibility of accidental deployment of the airbag, which could cause personal injury (see Section 25).*
Note: *Vehicles equipped with the Passive Anti-Theft Systems (PATS), also called FordSecurilock system, the anti-theft transceiver will need to be removed before removing the lock cylinder. Follow the procedures in this section under Anti-theft transceiver removal and installation for instructions.*

Key lock cylinder equipped models

2006 and earlier models

1   Disconnect the cable from the negative battery terminal (see Chapter 5).
2   Turn the ignition key lock cylinder to the On position.
3   Insert a 1/8-inch awl or punch through the hole in the lower steering column cover. Insert it through the hole in the steering column housing, and then simultaneously depress the release button on the lock cylinder and pull the lock cylinder out of the steering column housing (see illustrations).
4   To install the lock cylinder, make sure that the ignition key is still in the On position (see illustration 7.3c). Insert the lock cylinder into the steering column housing with the wide ridge and two narrow ridges on the lock cylinder aligned with the wide groove and narrow

grooves inside the lock cylinder receptacle (see illustrations).
5   Rotate the key back to the Off position. This will allow the release button to extend itself back into the locating hole in the steering column housing.
6   Turn the lock to ensure that operation is correct in all positions.
7   Reconnect the cable to the negative battery terminal.
8   After you've reconnected the battery, the Powertrain Control Module (PCM) must relearn its idle and fuel trim strategy for optimum driveability and performance (see Chapter 1 for this procedure).

2007 and later models

9   Disconnect the cable from the negative battery terminal (see Chapter 5).
10   On 2008 and later models, refer to Chapter 11 and remove the steering column

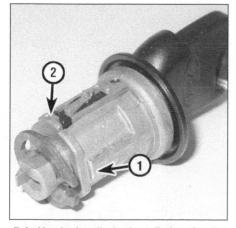

7.4a Key lock cylinder installation details

1   Narrow ridge (other narrow ridge, on other side of cylinder, not visible)
2   Wide ridge

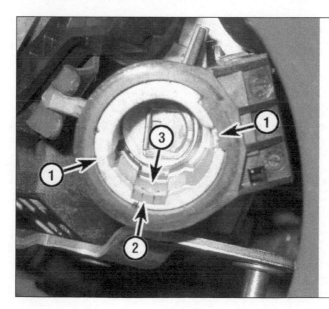

7.4b Key lock cylinder receptacle in steering column housing

1   Narrow grooves
2   Wide groove
3   Release button locating hole in steering column housing

7.17a To disconnect the electrical connector from the ignition switch . . .

7.17b . . . depress this lock and pull out the connector

7.19a To remove the ignition switch from the steering column housing, release this button on top . . .

7.19b . . . and this button on the bottom . . .

covers. Use a sharp-tipped pick to release the locking tabs on the anti-theft module and pull it off the ignition lock assembly (see Step 25).

11   Put the key into the ignition switch and turn it to the ACC position.

12   Use a thin rod such as an Allen wrench or a paper clip to depress the release pin through the access hole while pulling the key outward.

13   To install the lock cylinder, turn the key to the ACC position, then insert the lock into the housing. Align the lock pin with the access hole as the lock assembly slides in.

14   Check the lock cylinder in all positions to verify correct operation.

Ignition switch - key and tumbler type

15   Disconnect the cable from the negative battery terminal (see Chapter 5).

16   Remove the upper and lower steering column covers (see Chapter 11, Section 23)

17   Disconnect the ignition switch electrical connector (see illustrations).

18   Lower the steering column to its lowest tilt position.

19   Remove the ignition switch (see illustrations).

20   Installation is the reverse of removal.

21   After you've reconnected the battery, the Powertrain Control Module (PCM) must relearn its idle and fuel trim strategy for optimum driveability and performance (see Chapter 6 for this procedure).

22   Verify that the ignition switch operates correctly in the Lock, On, Start and ACC positions.

Anti-theft transceiver - removal and installation

23   The anti-theft transceiver is a rather delicate component and should be handled with care. The unit is used to transmit the key information to the BCM and the PCM for verify the key code is correct for the Passive Anti-Theft System (PATS), (also known as FordSecurilock system) that your car is programmed for. Each car is unique to its own key code.

24   Remove the steering column covers (see Chapter 11).

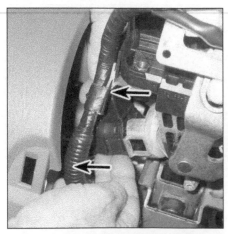

7.19c . . . then pull the ignition switch to the left to disengage it from the steering column housing

7.25 These are very small plastic connectors fastening the wires to the transceiver. Gently pry them away just enough to free the electrical connector to keep them from breaking off

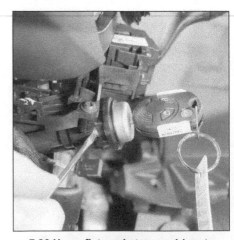

7.26 Use a flat pocket screwdriver to gently ease the connectors loose

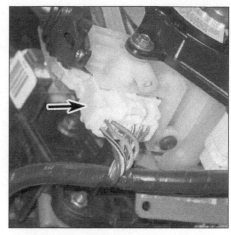

7.27a Pull the transceiver off as gently as possible

7.27b Lay the transceiver down where it will not be bent or broken

8.4 Disconnect the electrical connector from the multi-function switch

25 Disconnect the electrical connection to the transceiver (see illustration).
26 Carefully disengage the plastic clips on either side of the transceiver to disconnect it from the lock cylinder (see illustration).
27 Gently pull the transceiver off of the ignition switch housing (see illustrations).
28 Installation is the reverse of removal.

Push button start system
29 The push button starting system takes the place of the metal key type ignition system. Battery voltage is supplied to the push button switch which in turn provides a signal to the BCM, PCM and other modules that are needed for general engine and interior operations. The system also is incorporated into the anti-theft system. A programmed key (proximity key) is required to change the ignition state from Off to On.
30 The starting system is electronically controlled by the various computers and is entirely integrated into the computer systems for that particular car. Diagnosing a no start condition (no starter crank for example) can

involve extensive research by a trained technician with the proper testing equipment.
31 We recommend that in the event of a problem with the push button starting system, to take your vehicle to a qualified independent repair facility or to your local Ford dealership for evaluation and service.

8 Steering column switches - replacement

Warning: *The models covered by this manual are equipped with a Supplemental Restraint System (SRS), more commonly known as airbags. Always disarm the airbag system before working in the vicinity of any airbag system component to avoid the possibility of accidental deployment of the airbag, which could cause personal injury (see Section 25).*

2004 and earlier models
Multi-function switch
1 Disconnect the negative battery ter-

minal from the negative battery post (see Chapter 5).
2 Set parking brake with the front wheels in the straight ahead position.
3 Remove the clockspring (see Section 25).
4 Disconnect the electrical connector from the multi-function switch (see illustration).
5 Remove the multi-function switch retaining screws, then detach the switch from the steering column (see illustrations).
6 Installation is the reverse of removal.

Windshield wiper/washer switch
7 Disconnect the cable from the negative battery terminal (see Chapter 5).
8 Set parking brake with the front wheels in the straight ahead position.
9 Remove the key lock cylinder (see Section 7) and the steering column covers (see Chapter 11).
10 Disconnect the electrical connector from the windshield wiper/washer switch (see illustration).

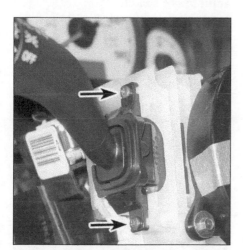

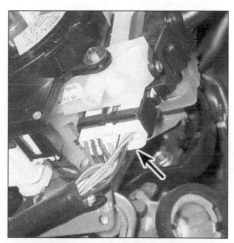

8.5a To detach the multi-function switch, remove these two screws . . .

8.5b . . . then remove the switch from the steering column (2004 and earlier models)

8.10 Disconnect the electrical connector from the windshield wiper/washer switch (2004 and earlier models)

8.11a To detach the windshield wiper/washer switch, remove these two retaining screws . . .

8.11b . . . then remove the switch from the steering column (2004 and earlier models)

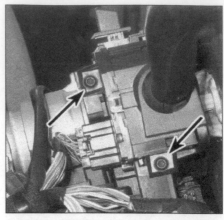

8.21 On 2005 through 2007 models the multi-function switch is one piece - remove these screws from each side of the steering column, then remove the switch

11   Remove the windshield wiper/washer switch retaining screws, then detach the switch from the steering column (see illustrations).
12   Installation is the reverse of removal.
13   After you've reconnected the battery, the Powertrain Control Module (PCM) must relearn its idle and fuel trim strategy for optimum driveability and performance (see Chapter 5, Section 1 for this procedure).

2005 through 2012 models - multi function switch

Note: *On 2008 and later models, the clockspring does not have to be removed.*
14   Disconnect the cable from the negative battery terminal (see Chapter 5).
15   Aim the front wheels straight ahead, then refer to Chapter 10 and remove the steering wheel.
16   Tilt the steering column fully down.
17   Remove the steering column covers (see Chapter 11).
18   2005 through 2007 models: If you're going to be installing the same clockspring, tape it so that it can't rotate.
19   On 2005 through 2007 models, remove the clockspring mounting screws.

20   On 2005 through 2007 models, lift the clockspring, then disconnect the wiring and remove it. (2008 and later the clockspring does not have to be removed.)
21   Remove the multi-function switch screws (see illustration).
22   Disconnect the wiring and remove the switch.
23   Installation is the reverse of removal. On 2005 through 2007 models, if the clockspring has rotated off center, hold the outer housing and gently turn the rotor clockwise until resistance is felt. Turn it counterclockwise until the yellow band shows in the window and the arrows are aligned. It's now centered and ready for installation.
24   When installing the clockspring on 2005 through 2007 models, make sure that the wide and narrow slots are correctly aligned with the two tabs.

2013 and later models - multi-function switches

Note: *Both the RH and LH switches are replaced in the same manner.*
25   Remove steering column covers (see Chapter 11).

26   Remove the two retaining screws (see illustrations).
27   Press in on the release tabs and draw the switch outward away from the column (see illustration).
28   Installation is the reverse of removal.

9   Instrument panel switches - replacement

Warning: *The models covered by this manual are equipped with a Supplemental Restraint System (SRS), more commonly known as airbags. Always disarm the airbag system before working in the vicinity of any airbag system component to avoid the possibility of accidental deployment of the airbag, which could cause personal injury (see Section 25).*

2007 and earlier models
Left/right power mirror adjustment switch
1   Disconnect the cable from the negative battery terminal (see Chapter 5).

8.26a LH switch retaining screws

8.26b RH retaining screws shown

8.27 Press the tabs in and pull outward

9.2 Carefully pry the left/right power mirror adjustment switch from the dash with a small screwdriver (the switch is secured to the dash by a small metal tab on each side)

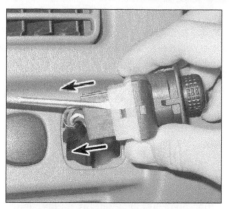

9.3 Disconnect the electrical connector from the left/right power mirror adjustment switch; if the connector is difficult to separate from the switch, use a small screwdriver to carefully pry it loose

9.7 Carefully pry the instrument panel dimmer switch from the dash

2   Carefully pry the power window switch from the instrument panel (see illustration).
3   Disconnect the electrical connector (see illustration) and remove the switch.
4   Installation is the reverse of removal.
5   After you've reconnected the battery, the Powertrain Control Module (PCM) must relearn its idle and fuel trim strategy for optimum driveability and performance (see Chapter 5, Section 1 for this procedure).

Instrument panel dimmer switch
6   Disconnect the cable from the negative battery terminal (see Chapter 5).
7   Carefully pry the switch from the instrument panel (see illustration).
8   Disconnect the electrical connector (see illustration) and remove the switch.
9   Installation is the reverse of removal.
10   After you've reconnected the battery, the Powertrain Control Module (PCM) must relearn its idle and fuel trim strategy for optimum driveability and performance (see Chapter 5, Section 1 for this procedure).

Hazard flasher switch
11   Disconnect the cable from the negative battery terminal (see Chapter 5).

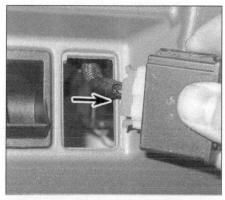

9.8 Squeeze the two tabs and disconnect the electrical connector from the switch

12   Pry the center trim panel loose from the dash (see illustration). (It's not necessary to actually remove it.)
13   Disconnect the hazard flasher electrical connector (see illustration).
14   To remove the hazard flasher switch, squeeze the two locking tangs (see illustrations) and push the switch out of the center trim panel.
15   Installation is the reverse of removal.

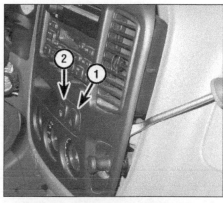

9.12 Carefully pry the center trim panel loose (it's not necessary to remove it)

1   Hazard flasher switch
2   Rear window defogger switch

16   After you've reconnected the battery, the Powertrain Control Module (PCM) must relearn its idle and fuel trim strategy for optimum driveability and performance (see Chapter 5, Section 1 for this procedure).

9.13 Disconnect the electrical connector from the hazard flasher switch

9.14a To detach the hazard flasher switch from the center trim panel, squeeze these two locking tangs . . .

9.14b . . . with a pair of needle-nose pliers and push the switch out of the trim panel; push it out from the rear side; don't try to push the switch through the trim panel from the front (center trim panel removed for clarity)

1   Hazard flasher switch
2   Rear window defogger switch
     (remove the same way)

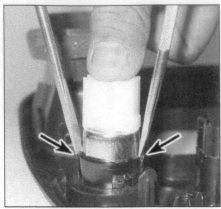

9.20a To release the cigarette lighter
from the center trim panel, pry these two
locking tangs out . . .

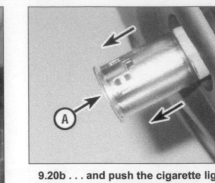

9.20b . . . and push the cigarette lighter
out of the trim panel from the back side;
when installing the cigarette lighter, make
sure that the dimpled area (A) on each
side of the lighter is aligned with a
locking tang, or the lighter won't be
securely fastened to the trim panel

9.27a 2008 and later models, the
instrument panel switches can be
removed by pushing them from the rear
as shown, or using a plastic tool to
carefully pry them out . . .

9.27b . . . on 2013 and later models, the light
switch can be removed in the same manner

Cigarette lighter

18   Disconnect the cable from the negative battery terminal (see Chapter 5).
19   Remove the center trim panel (see Chapter 11).
20   Pry open the two locking tangs and push the cigarette lighter out of the trim panel (see illustrations).
21   When installing the cigarette lighter in the center trim panel, be sure to push the lighter into the trim panel until it snaps into place, indicating that it's fully seated.
22   Installation is otherwise the reverse of removal.
23   After you've reconnected the battery, the Powertrain Control Module (PCM) must relearn its idle and fuel trim strategy for optimum driveability and performance (see Chapter 5, Section 1 for this procedure).

2008 and later models - headlight switch

Note: All other switches are integral with the component they are associated with.
24   Disconnect the cable from the negative battery terminal (see Chapter 5).

25   If you're working on a switch in one of the center instrument panel trim panels, refer to Chapter 11 and remove the panel.
26   If you're working on a switch to the left of the steering column, it's easiest to first remove the driver's knee trim panel.
27   Use a plastic trim tool or a screwdriver wrapped with tape to carefully pry the switch out of the panel or push it out from the rear (see illustrations).
28   Disconnect the wiring from the switch and remove it (see illustrations).
29   Installation is the reverse of removal.

10   Instrument cluster - removal and installation

Warning: The models covered by this manual are equipped with a Supplemental Restraint System (SRS), more commonly known as airbags. Always disarm the airbag system before working in the vicinity of any airbag system component to avoid the possibility of accidental deployment of the airbag, which could cause personal injury (see Section 25).
Caution: This procedure is intended for re-

Rear window defogger switch

17   The rear window defogger switch, which is located to the left of the hazard flasher switch, is removed and installed exactly the same way as the hazard flasher switch (see Steps 11 through 16).

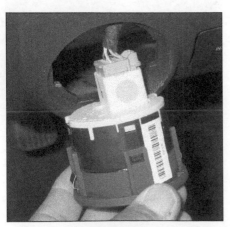

9.28a Once free of the instrument panel,
the wiring can be disconnected from
the switches . . .

9.28b . . . 2013 and
later models are
shaped slightly
different but can be
disconnected in the
same manner

10.3 To detach the instrument cluster finish panel, remove these two screws (2007 and earlier models)

10.4a To detach the instrument cluster, remove these four screws . . .

10.4b . . . pull out the cluster and disconnect the electrical connectors from the back side

moving the instrument cluster for access to other components (such as during instrument panel removal) only. If the instrument cluster requires replacement, the job will have to be performed at a dealer service department or other qualified repair shop equipped with the special scan tool that is required to upload the module configuration data and download this data into the new cluster.

Note: *If you are replacing with a new instrument cluster, all of the keys associated with the vehicle will be erased during the parameter reset procedure. The reset procedure requires two keys in order re-establish communication with the theft deterrent system. Be sure to have two usable keys before attempting this procedure.*

Note: *If you are removing the original instrument cluster for any length of time, prior to removal, take the vehicle to a service center that can download the instrument cluster data (module configuration information) and store it in their scanner for later retrieval. This information must be entered into the replacement instrument cluster and/or to re-establish communication with the original cluster.*

Caution: *It is not recommended to swap an instrument cluster with a salvage yard cluster. This can lead to severe data corruption, and*

in some cases the vehicle will not start. Each instrument cluster is specifically programmed to each individual vehicle. The steps provided are for removal and installation. All service work involving replacing, repairing, or in any way working on the internal electronics of the instrument cluster should be left to a professional with the proper equipment.

2012 and earlier models

1   Disconnect the cable from the negative battery terminal (see Chapter 5).
2   Tilt the steering wheel to its lowest position and move the transaxle shift lever out of the way.
3   On 2007 and earlier models, remove the cluster bezel screws (see illustration) and remove the cluster bezel. On 2008 and later models, use a plastic trim tool or a screwdriver wrapped with tape to carefully pry off the bezel.
4   Remove the instrument cluster retaining screws (see illustration), then pull out the cluster and disconnect the electrical connectors from the back side (see illustration).
5   Installation is the reverse of removal.
6   After you've reconnected the battery, the Powertrain Control Module (PCM) must relearn its idle and fuel trim strategy for opti-

mum driveability and performance (see Chapter 5, Section 1 for this procedure).

2013 and later models

Caution: *If you are installing a new IPC (Instrument Panel Cluster) the IPC configuration must be uploaded from the original IPC to the scan tool for later re-installation to the replacement IPC. Record the original odometer reading before removing the old cluster. Use the odometer reading in the replacement (new) IPC. If you do not have the original odometer reading you will have to provide proof of the odometer setting for the technician to install it into your replacement cluster. (In some states the invoice or work order showing the cluster has been replaced must remain with the vehicle at all times and be present at the time the vehicle is sold.)*

7   Disconnect the cable from the negative battery terminal (see Chapter 5).
8   Tilt the steering wheel to its lowest position.
9   Release the steering column upper cover (see illustration). For detail removal instructions (see Chapter 11).
10   With the upper cover removed, remove the screws securing the IPC to the instrument panel (see illustration).

10.9 Steering wheel has been removed for clarity. Work the upper cover and trim off with a flat trim tool

10.10 Remove the screws securing the IPC to the instrument panel

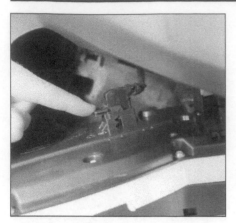

10.11 Press the tab down to release the electrical connector

11.2a Remove each windshield wiper arm retaining nut . . .

11.2b . . . mark the relationship of the wiper arm to the shaft . . .

11   Pull the IPC out of and turn it over to expose the wiring harness. Disconnect the electrical connector from the IPC (see illustration).

12   Installation is the reverse of removal.

13   Perform any programming procedures necessary for the replacement IPC (if applicable).

11   Wiper and washer system - component replacement

1   Disconnect the negative battery cable (see Chapter 5).

Windshield wiper motor

2   Remove the wiper arm retaining nuts (see illustration). Mark the positions of the wiper arms to their shafts, then remove the wiper arm (see illustrations).

3   Remove the cowl cover (see Chapter 11).

2012 and earlier models

4   Disconnect the wiper motor harness connector from the windshield wiper motor (see illustration).

5   Remove the windshield wiper motor/linkage assembly mounting bolts (see illustration).

11.2c . . . then pull the windshield wiper arm off the shaft

6   Lift the windshield wiper motor assembly from the cowl area.

7   Mark the position of the motor arm to its shaft. Remove linkage, arm and the windshield wiper motor mounting bolts (see illustration) and separate the motor from its mounting bracket.

8   Installation is the reverse of removal.

11.2d If the wiper arm is stuck to the shaft, use a small puller to remove it. Don't try to pry it off

Be sure to align the marks you made on the motor arm and the windshield wiper arms and the wiper arm shafts.

9   After you've reconnected the battery, the Powertrain Control Module (PCM) must relearn its idle and fuel trim strategy for optimum driveability and performance (see Chapter 5, Section 1 for this procedure).

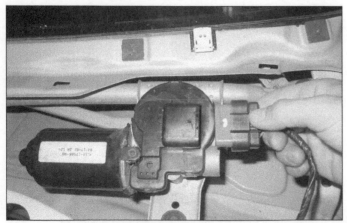

11.4 Disconnect the electrical connector from the windshield wiper motor

11.5 To detach the windshield wiper motor from the vehicle, remove these bolts

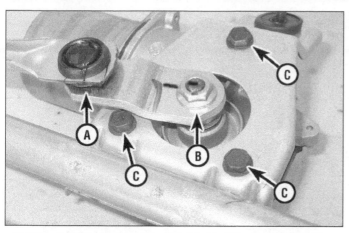

11.7 Pry the linkage from the wiper motor (A); mark the position of the arm to its shaft and remove the nut (B) and arm, then remove the motor mounting bolts (C)

11.10 Remove the mounting bolts first, then remove the motor from the mounting bracket

1   *Mounting bracket fasteners*
2   *Wiper motor-to-bracket bolts*

11.13a 2007 and earlier models: To detach the rear window wiper arm from the shaft, pull this locking tab straight up with a pair of needle-nose pliers, then pull the wiper arm straight off the shaft

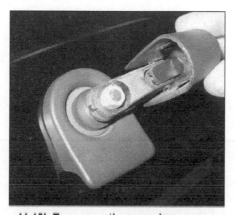

11.13b To remove the rear wiper arm on 2008 and later models, remove this nut. If the arm is stuck, wiggle it up-and-down to free it (a small puller may be required)

11.14a Remove this protective cover for the rear window wiper motor shaft nut . . .

2013 and later models

Note: *2013 and later models have an individual motor for each wiper arm.*

10   Remove the bolts securing the motor assembly in place, then lift up the motor enough to disconnect the electrical connection. Separate the motor from the mounting bracket (see illustration).

11   Remove the motor assembly.
12   Installation is the reverse of removal.

Rear window wiper motor

13   Remove the rear window wiper arm (see illustrations).
14   Remove the protective cover for the rear

window wiper motor shaft nut, then remove the nut (see illustrations).
15   Open the liftgate and remove the plastic cover for the rear window wiper motor (see illustration).
16   Disconnect the electrical connector from the rear window wiper motor (see illustrations).

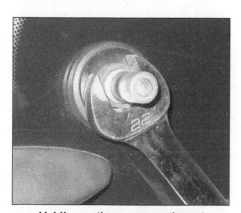

11.14b . . . then unscrew the nut

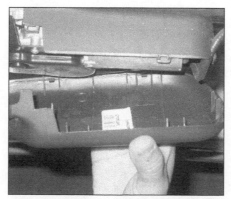

11.15 To remove the plastic cover for the wiper motor, simply pop it off

11.16a Using a panel removal tool or a screwdriver, detach the electrical connector from its bracket . . .

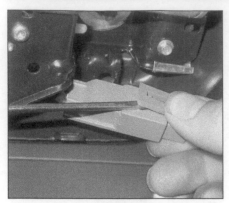

11.16b . . . then release this locking tab with a small screwdriver and unplug the connector

11.17a To detach the rear window wiper motor from the rear window, remove the mounting nut (1) and bolt (2)

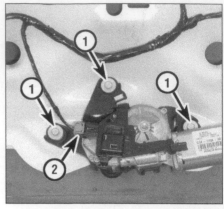

11.17b 2013 and later models are slightly different but are removed in the same manner

1   *Mounting bolts*
2   *Electrical connector*

17   Remove the mounting nut and bolt (see illustrations) and remove the rear window wiper motor.

18   Installation is the reverse of removal. After you've reconnected the battery, the Powertrain Control Module (PCM) must relearn its idle and fuel trim strategy for optimum driveability and performance (see Chapter 5, Section 1 for this procedure).

19   After you've reconnected the battery, the Powertrain Control Module (PCM) must

relearn its idle and fuel trim strategy for optimum driveability and performance (see Chapter 5, Section 1 for this procedure).

Windshield washer nozzles

Note: *The accompanying photo sequence shows windshield washer nozzle removal for 2013 and later models. Earlier models are similar.*

20   Follow the photo sequence for the nozzle removal procedure (see illustrations).

21   Installation is the reverse of removal.

Windshield washer pump motor

Note: *The front and rear washers utilize the same washer pump motor, located on the washer reservoir.*

22   Remove the passenger's side inner fender splash shield (see Chapter 11).

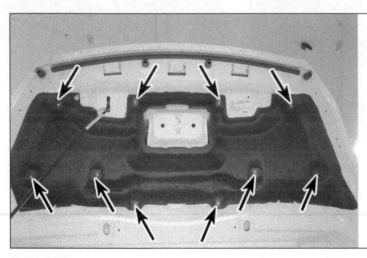

11.20a Locate the fasteners that secure the hood insulation to the hood

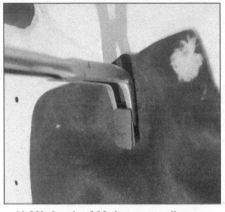

11.20b A pair of 90 degree needle-nose pliers works well to pry the clips free

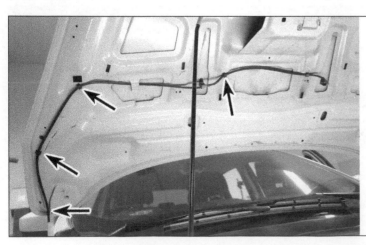

11.20c With the hood insulation removed, pry the clips free that secure the hose to the hood

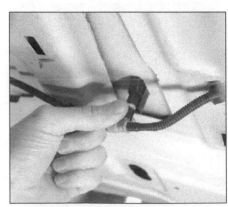

11.20d Rotate the collar to remove the hose from the nozzle

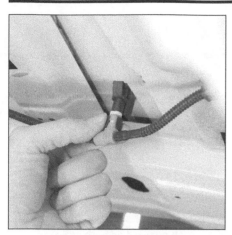

11.20e With the collar turned, pull the hose free

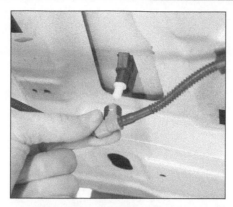

11.20f The hose should slide off. Do not force it off or start bending it from side to side. This will most likely break the plastic nozzle. Work the collar ring back and forth while pulling on the hose

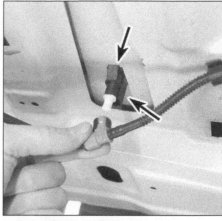

11.20g With the hose removed, press in on the two outer tabs. Push the tabs inward and press the nozzle out of the hood

23   Drain the washer bottle by either removing the washer tube, or by sucking the fluid out of the washer bottle with a syringe or siphon hose.

Note: *Place a drain pan underneath the washer bottle to catch spills.*

24   Disconnect the electrical connectors and washer hoses to the washer pump motor (see illustrations).

25   Tip the washer pump motor out at the top edge. The motor is held in place by the sides of the washer bottle by pinching the pump into place.

26   With the motor tipped outward at the top, pull upwards to remove the pump from the washer bottle.

27   Installation is the reverse of removal.

Washer fluid reservoir - removal and installation

28   Remove the passenger's side inner fender splash shield (see Chapter 11).

29   Drain the washer bottle by removing the washer motor hoses or using a syringe

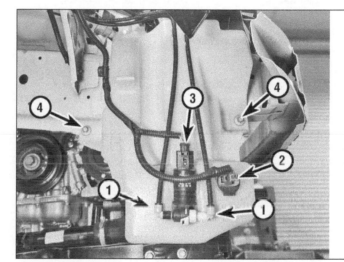

11.24a Washer fluid reservoir and pump motor details

1   *Washer hoses*
2   *Fluid level sensor electrical connector*
3   *Pump motor electrical connector*
4   *Reservoir mounting bolts*

or siphon pump.

Note: *Place a drain pan below the washer bottle to collect any spills.*

30   Disconnect the electrical and hose connections from the washer motor (see illustrations 11.24a and 11.24b).

31   Remove the plastic retainer securing the washer reservoir neck (see illustration).

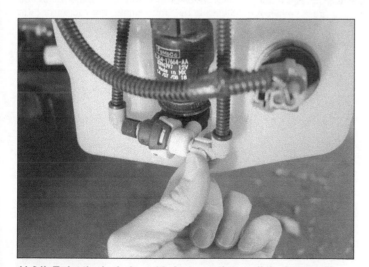

11.24b Twist the lock rings 90-degrees, then pull the hoses off

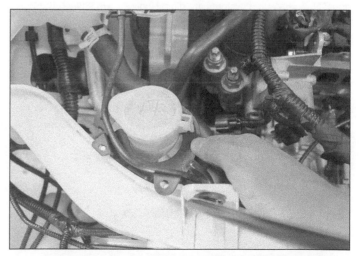

11.31 Pull the plastic retainer from the radiator support and reservoir neck

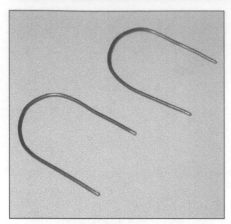

12.5a You'll need a pair of special radio removal tools (available at most auto parts stores) or make your own from a couple of pieces of coat hanger wire bent into a U-shape as shown (2007 and earlier models)

12.5b Insert the special radio removal tools into the holes at each end of the radio as shown, then push them in until you feel the internal clips release (2007 and earlier models)

12.6 Push outward on both radio removal tools and simultaneously pull out the radio

32 Disconnect the washer hoses and electrical connectors (see illustrations 11.24a and 11.24b).

33 Remove the washer reservoir mounting bolts.

34 Pull the washer bottle out through the passenger's side wheel well.

35 Installation is the reverse of removal.

---

12 Radio and speakers - removal and installation

Warning: *The models covered by this manual are equipped with a Supplemental Restraint System (SRS), more commonly known as airbags. Always disable the airbag system before working in the vicinity of any airbag system component to avoid the possibility of accidental deployment of the airbag, which could cause personal injury (see Section 25).*

1 Audio systems on later models are far more than just a receiver for AM and FM stations. Now, the audio systems are digital highways of information from navigation to where

the next gas station is located and how far it is from your exact location. New terms and abbreviations will need to be understood in order to help you distinguish which part a procedure is referring to as well as which component you need to ask for at the parts counter. Here is a list of some of these abbreviations and what they stand for.

  ACM - Audio Control Module
  FDIM - Front Display Interface Module
  FCIM - Front Control Interface Module
  GPSM - Global Positioning System Module
  APIM - Audio Protocol Interface Module

2 To receive a radio signal, several modules have to work together to get the information to the FCDIM. A signal is received by either the GPSM, the various antennas, USB attached device, CD player, or the in-vehicle warning systems. The signal then passes to the APIM which in turn sends the signals to the FCDIM unit and onto the speakers through the ACM.

3 Trouble codes are stored as U codes (communication codes) for the service technician to use as a diagnostic aid. Codes are not a solution to the problem, but are a pathway for determining the problem. Do not assume a

service code is the answer to any electrically related problem. All codes should be diagnosed with the proper testing equipment by a trained technician.

4 Disconnect the cable from the negative battery terminal (see Chapter 5).

Radio

2007 and earlier models

5 For theft protection, the radio receiver is retained in the instrument panel by special clips. Releasing these clips requires the use of a pair of special radio removal tools, available at most auto parts stores, or two short lengths of coat hanger wire bent into U-shapes (see illustration). Insert the tools into the holes at the corners of the radio until you feel the internal retaining clips release (see illustration).

6 Push outward on both removal tools and simultaneously pull the radio assembly out of the instrument panel (see illustration).

7 Disconnect the antenna and the electrical connector from the back of the radio (see illustrations) and remove the radio from the vehicle.

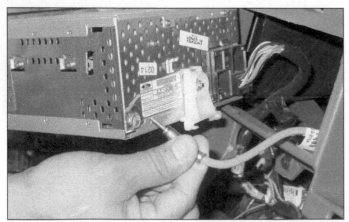

12.7a When the radio has been pulled out far enough to access the back side, disconnect the antenna cable . . .

12.7b . . . and the electrical connector from the radio

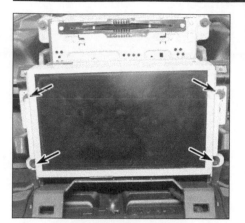

12.12 On 2013 and later models, remove these fasteners

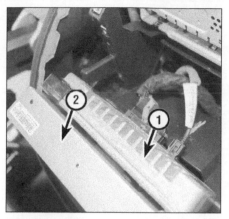

12.13 If the FDIM and the APIM are separated, the display will have to be recalibrated

| 1 | APIM | 2 | FDIM |

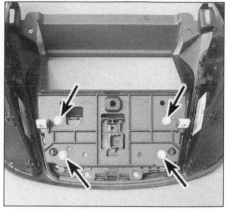

12.17 Place the center middle trim face down and remove the FCIM fasteners

8   Installation is the reverse of removal. After reconnecting the electrical connector and antenna cable, insert the radio into the dash and slide it into place until you feel and/or hear the retaining clips snapping into place.

9   After you've reconnected the battery, the Powertrain Control Module (PCM) must relearn its idle and fuel trim strategy for optimum driveability and performance (see Chapter 5, Section 1 for this procedure).

2008 and later models

Note: *If a new ACM (Audio Control Module) or FCIM (Front Control Interface Module) is being installed, download the original ACM/FCIM configuration data to a scan tool, which can be later uploaded into the replacement ACM/FCIM. This procedure needs to be performed by a dealer or independent repair facility with the proper equipment.*

10   Remove the center trim panel (see Chapter 11).

11   2008 through 2012 models - Remove the four audio unit mounting screws and separate the unit from the trim panel.

12   2013 and later models - Remove the fas-

teners securing the FDIM to the instrument panel (see illustration).

13   To remove the FDIM from the APIM, remove the fasteners at each corner and separate the units (see illustration). Then disconnect the electrical connections.

14   If equipped with a navigation system, disconnect the antenna and other wiring.

Note: *The side brackets can be removed if necessary.*

15   Installation is the reverse of removal.

FCIM replacement

16   Remove center middle instrument panel trim (see Chapter 11).

17   Turn the center middle trim over, then remove the FCIM fasteners securing it to the panel trim (see illustration).

18   Installation is the reverse of removal.

ACM

19   Remove the center middle instrument panel trim (see Chapter 11).

20   Remove the FDIM assembly.

21   Remove the two fasteners securing the ACM to the instrument panel (see illustration).

22   Pull the unit away from the instrument panel and disconnect the electrical connectors and antenna leads (see illustration).

23   Installation is the reverse of removal.

GPSM replacement

24   Remove the IPC (see Section 10).

25   The GPSM is mounted behind the IPC. Remove the single bolt attaching the GPSM to the dash support bracket.

26   Disconnect the electrical connector.

27   Installation is the reverse of removal.

Touch screen calibration

Note: *Whenever the FDIM and the APIM are separated, the screen calibration needs to be reset. A scanner can be used to reset the calibration or follow the steps in this section.*

28   Have the vehicle battery checked for the state of charge. The battery should be fully charged before attempting the calibration.

29   Turn on the radio.

30   Set the radio to an AM or FM station.

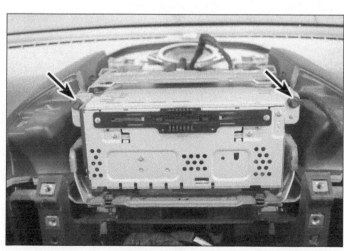

12.21 Remove the two fasteners and pull the unit from the instrument panel

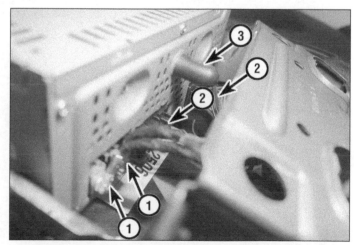

12.22 Pull the unit out far enough to disconnect the leads

1 Antenna leads
2 Electrical connectors

3 Rubber alignment post (do not remove)

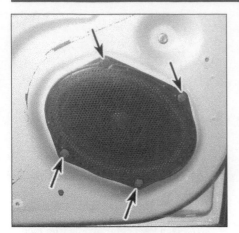

12.38a Speaker mounting screws - 2012 and earlier models

12.38b Speaker mounting screws - 2013 and later models

12.39 Depress the tab and disconnect the electrical connector

31　Vehicles with a CD player, press and Hold the eject button. Within one second, press and hold the Seek up button. If equipped, the speaker system will perform its "walk-around" check. The display will indicate that each speaker has been tested.

32　Vehicles without a CD player, use the steering wheel controls. Press and hold the Seek up button and the Tune up buttons. If equipped, the speaker system will perform its "walk-around" check. The display will indicate that each speaker has been tested.

33　Press and Hold the Seek down button to gain access to the touchscreen calibration.

34　Touch and hold the touch points as they appear on the screen.

35　Wait for the message, "New calibration settings have been measured. Press ENTER to accept the new settings" to appear on the screen.

36　Press anywhere on the screen to accept the new settings.

Speakers

Front door speakers

37　Remove the door trim panel (see Chapter 11).

38　Remove the speaker mounting screws (see illustrations) and pull the speaker out of its receptacle in the door.

39　Disconnect the electrical connector (see illustration) and remove the speaker from the vehicle.

40　Installation is the reverse of removal.

Rear door speakers

41　Remove the door trim panel (see Chapter 11).

42　Remove the speaker mounting screws (see illustration) and pull the speaker out of its receptacle in the door.

43　Disconnect the electrical connector (see illustration) and remove the speaker from the vehicle.

44　Installation is the reverse of removal.

Rear subwoofer speaker

45　Remove the rear door sill trim panel access cover and remove the retainer.

46　Remove the rear door sill trim panel.

47　Remove the liftgate sill panel.

48　Pull off the liftgate weatherstrip from the area of the quarter trim panel.

49　Remove the cargo net hold-down.

50　Use a plastic trim tool or a screwdriver

wrapped with tape to remove the rear quarter trim panel.

51　Disconnect the wiring from the speaker assembly.

52　Remove the mounting bolts and the speaker assembly.

Note: *The assembly consists of the speaker, the enclosure and the amplifier. The components can't be replaced individually.*

53　Installation is the reverse of removal.

Dash speaker

54　Remove center middle instrument cluster trim (see Chapter 11).

55　Unscrew the fasteners for the center speaker (see illustration), then lift up to disconnect the electrical connection. Remove the speaker.

56　Installation is the reverse of removal.

A pillar speaker

57　Remove the A pillar trim (see Chapter 11).

58　Release the retaining tabs and pull the speaker from its holder (see illustration).

59　Disconnect the electrical connector and remove the speaker.

60　Installation is the reverse of removal.

12.42 To detach a speaker from a rear door, remove these four screws

12.43 Before removing the speaker from the rear door, disconnect the electrical connector

12.55 Two fasteners hold the speaker in place

12.58 Carefully pry out the tabs to release the speaker from its holder

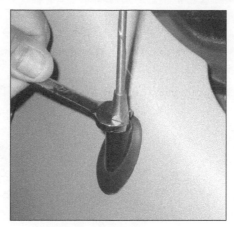

13.1 Unscrew the antenna mast with a 10 mm wrench

13.3a Detach these two cable clips . . .

13 Antenna and cable - removal and installation

Warning: *The models covered by this manual are equipped with a Supplemental Restraint System (SRS), more commonly known as airbags. Always disable the airbag system before working in the vicinity of any airbag system component to avoid the possibility of accidental deployment of the air-bag, which could cause personal injury (see Section 25).*

Antenna (and outside antenna cable)

Note: *The antenna cable actually has two sections. This procedure describes the removal procedure for the antenna mast, the antenna mounting base and the outside part of the antenna cable, which begins at the connector below the blower motor housing and extends to the antenna mounting base*

in the right fender. For the removal procedure for the rest of the antenna cable (the part from the connector to the back of the radio) see Steps 9 through 14.

1   Remove the antenna mast (see illustration) from its mounting base.

2   Remove the right scuff plate and the right lower A-pillar trim panel (see *Instrument panel - removal and installation* in Chapter 11.

3   Working under the right end of the dash, detach the two antenna cable clips and disconnect the antenna cable connector (see illustrations).

4   Loosen the right front wheel lug nuts. Raise the vehicle and place it securely on jackstands. Remove the right front wheel.

5   Remove the right inner fender splash shield (see *Front fender - removal and installation* in Chapter 11).

6   Working inside the fender area, remove the antenna cable grommet (see illustration), pull the antenna cable into the fender area, remove the antenna support bracket nut, then pull the antenna mounting base down through

its hole in the fender and remove the mounting base, support bracket and antenna cable as a single assembly

7   Remove the electrical tape securing the antenna cable to the support bracket and discard the old antenna base and cable.

8   Installation is the reverse of removal.

Antenna cable

Note: *This procedure describes the removal procedure for the rest of the antenna cable, which begins at the connector shown in illustration 13.3b and extends to the back of the radio. If you want to remove the outside part of the antenna cable, refer to Steps 1 through 8 above.*

9   Remove the radio (see Section 12).

10   Remove the right scuff plate and kick panel (see *Front fender - removal and installation* in Chapter 11), then disconnect the inside part of the antenna cable from the outside cable (see illustrations 13.3a and 13.3b).

11   Remove the glove box (see Chapter 11).

13.3b . . . and disconnect the antenna cable connector

13.6 Remove the antenna cable grommet, pull the antenna cable into the fender area, remove the antenna support bracket nut, then pull the antenna mounting base down through its hole in the fender and remove the mounting base, support bracket and antenna cable as a single assembly

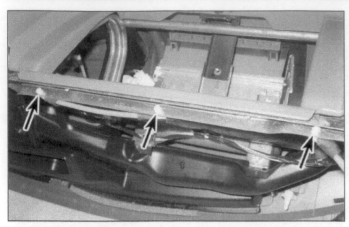

13.12 After removing the glove box, detach these three pin-type retainers from the support brace running along the lower edge of the glove box recess

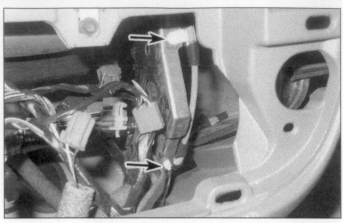

13.13 Detach these two pin-type retainers from the vertical support brace located at the right side of the radio recess

12   Detach the three pin-type retainers that secure the cable to the horizontal support brace located on the lower edge of the glove box recess (see illustration).

13   There are two more pin-type retainers (see illustration) securing the antenna cable to the vertical support brace that are located on the right side of the radio recess. Detach both of them.

14   Remove the inside antenna cable.

15   Installation is the reverse of removal.

Satellite radio antenna and cable (2008 and later models)

16   Remove the right windshield pillar trim panel.

17   Remove the right sun visor.

18   Carefully pull the front of the headliner down far enough to reach the satellite radio wiring connector, then disconnect it.

19   Loosen the antenna mounting bolt. Press the tabs to release the antenna and remove it.

20   To remove the antenna cable, remove the radio (see Section 12), then disconnect the cable from it.

21   Installation is the reverse of removal.

---

14   Rear window defogger - check and repair

---

1   The rear window defogger consists of a number of horizontal elements baked onto the glass surface.

2   Small breaks in the element can be repaired without removing the rear window.

Check

3   Turn the ignition switch and defogger system switches to the On position. Using a voltmeter, place the positive probe against the defogger grid positive terminal and the negative probe against the ground terminal. If bat-

tery voltage is not indicated, check the fuse, defogger switch and related wiring. If voltage is indicated, but all or part of the defogger doesn't heat, proceed with the following tests.

4   When measuring voltage during the next two tests, wrap a piece of aluminum foil around the tip of the voltmeter positive probe and press the foil against the heating element with your finger (see illustration). Place the negative probe on the defogger grid ground terminal.

5   Check the voltage at the center of each heating element (see illustration). If the voltage is 5- or 6-volts, the element is okay (there is no break). If the voltage is zero, the element is broken between the center of the element and the positive end. If the voltage is 10- to 12-volts the element is broken between the center of the element and ground. Check each heating element.

6   Connect the negative lead to a good body ground. The reading should stay the same. If it doesn't, the ground connection is bad.

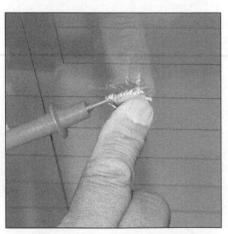

14.4 When measuring the voltage at the rear window defogger grid, wrap a piece of aluminum foil around the positive probe of a voltmeter and press the foil against the wire with your finger

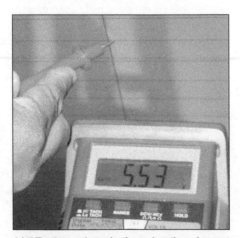

14.5 To determine whether a heating element has broken, check the voltage at the center of each element; if the voltage is 5- or 6-volts, the element is unbroken, but if the voltage is 10- or 12-volts, the element is broken between the center and the ground side; if there's no voltage, the element is broken between the center and the positive side

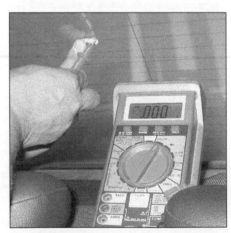

14.7 To find the break, place the voltmeter negative lead against the defogger ground terminal, place the voltmeter positive lead with the foil strip against the heating element at the positive terminal end and slide it toward the negative terminal end; the point at which the voltmeter reading changes abruptly is the point at which the element is broken

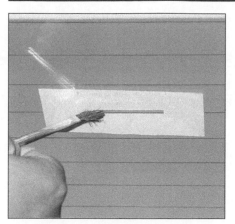

14.13 To use a defogger repair kit, apply masking tape to the inside of the window at the damaged area, then brush on the special conductive coating

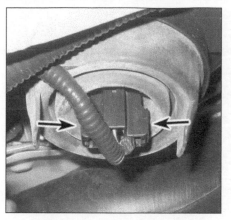

15.2a To disconnect the electrical connector from the headlight, squeeze the two locking tabs on the side . . .

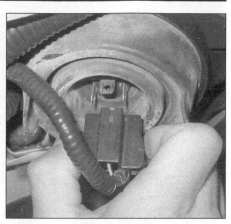

15.2b . . . and pull off the connector

7   To find the break, place the voltmeter negative probe against the defogger ground terminal. Place the voltmeter positive probe with the foil strip against the heating element at the positive terminal end and slide it toward the negative terminal end. The point at which the voltmeter deflects from several volts to zero is the point at which the heating element is broken (see illustration).

Repair

8   Repair the break in the element using a repair kit specifically recommended for this purpose, available at most auto parts stores. Included in this kit is plastic conductive epoxy.
9   Prior to repairing a break, turn off the system and allow it to cool off for a few minutes.
10   Lightly buff the element area with fine steel wool, then clean it thoroughly with rubbing alcohol.
11   Use masking tape to mask off the area being repaired.
12   Thoroughly mix the epoxy, following the instructions provided with the repair kit.

13   Apply the epoxy material to the slit in the masking tape, overlapping the undamaged area about 3/4-inch on either end (see illustration).
14   Allow the repair to cure for 24 hours before removing the tape and using the system.

15   Headlight bulb - replacement

Warning: *Halogen gas filled bulbs are under pressure and may shatter if the surface is scratched or the bulb is dropped. Wear eye protection and handle the bulbs carefully, grasping only the base whenever possible. Do not touch the surface of the bulb with your fingers because the oil from your skin could cause it to overheat and fail prematurely. If you do touch the bulb surface, clean it with rubbing alcohol.*
Warning: *High Intensity Discharge (HID) headlights: According to the manufacturer, the high voltages produced by this system*

can be fatal in the event of a shock. Also, the voltage can remain in the circuit even after the headlight switch has been turned to OFF and the ignition key has been removed. If your vehicle is equipped with HID headlights, for your safety, we don't recommend that you try to replace one of these bulbs yourself. Instead, have this service performed by a dealer service department or other qualified repair shop. Additionally, the headlight and headlight starter must be replaced as a unit.

2007 and earlier models

1   Make sure that the headlight switch and the ignition switch are both turned off, then open the hood.
2   Disconnect the electrical connector from the headlight assembly (see illustrations).
3   Remove the rubber boot from the headlight assembly (see illustration).
4   Disengage the headlight bulb retainer spring (see illustration).
5   Remove the bulb from the headlight assembly (see illustration).

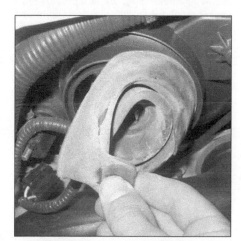

15.3 Remove the protective rubber boot from the headlight assembly (2007 and earlier models)

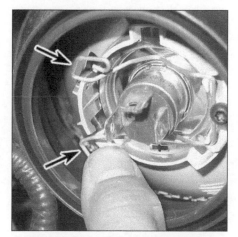

15.4 To disengage the bulb retainer spring, press each end of the spring forward and unhook it, then swing the retainer out of the way

15.5 Carefully remove the bulb from the headlight assembly

15.6 When installing the new bulb, make sure that the three metal tabs on the new bulb are aligned with the three grooves in the plastic base, and that the bulb's metal base is fully seated against the plastic base, before securing it with the spring retainer

15.7 When reconnecting the spring retainer, make sure that the retainer ends are correctly engaged with the hooks

15.13a Remove the dust covers to gain access to the bulbs

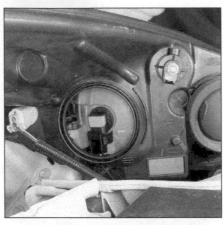

15.13b With the dust cover removed, twist the bulb counterclockwise. Then disconnect the electrical connection

15.13c Turn the bulb counterclockwise to remove it (housing removed for clarity)

6   Without touching the glass part of the new bulb with your bare fingers, insert the bulb into the headlight assembly. Make sure that the three metal tabs on the new bulb are aligned with the three grooves in the plastic base (see illustration), then push the bulb into the headlight assembly until the bulb's metal base is fully seated against the plastic base.

7   Swing the spring retainer back into place and engage the spring ends with their respective hooks on the headlight assembly (see illustration).

8   Install the protective rubber boot on the headlight assembly.

9   Plug in the electrical connector.

10   Test the headlight operation.

2008 and later models

11   Make sure that the headlight switch and the ignition switch are both turned off, then open the hood.

12   Disconnect the wiring from the headlight bulb.

13   Remove the dust cover, if applicable, then twist the bulb counterclockwise to remove it (see illustrations).

14   Installation is the reverse of removal.

15   Test the headlight operation.

16   Headlights - adjustment

Note: *The headlights must be aimed correctly. If adjusted incorrectly they could blind the driver of an oncoming vehicle and cause a serious accident or seriously reduce your ability to see the road. The headlights should*

be checked for proper aim every 12 months and any time a new headlight is installed or front end body work is performed. It should be emphasized that the following procedure is only an interim step that will provide temporary adjustment until the headlights can be adjusted by a properly equipped shop.

1   The vertical adjustment screws are

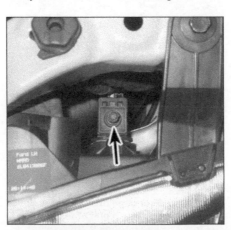

16.1a Headlight vertical adjustment screw - 2012 and earlier models

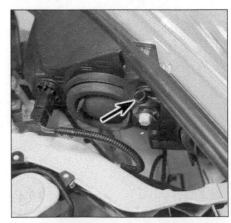

16.1b Headlight vertical adjustment screw - 2013 and later models

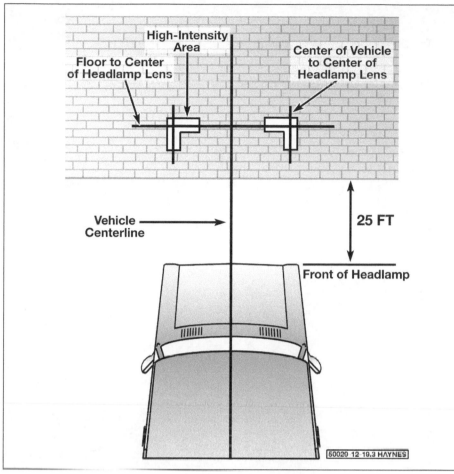

16.3 Headlight adjustment details

5   Adjustment should be made with the vehicle parked 25 feet from the wall, sitting level, the gas tank half-full and no heavy load in the vehicle.

6   Starting with the low beam adjustment, position the high intensity zone so it is two inches below the horizontal line. Adjustment is made by turning the adjusting screw clockwise to raise the beam and counterclockwise to lower the beam.

Note: *It may not be possible to position the headlight aim exactly for both high and low beams. If a compromise must be made, keep in mind that the low beams are the most used and have the greatest effect on safety.*

7   With the high beams on, the high intensity zone should be vertically centered with the exact center just below the horizontal line.

8   Have the headlights adjusted by a dealer service department or service station at the earliest opportunity.

---

17   Headlight housing - replacement

All models

1   Make sure that the headlight switch and the ignition switch are turned off.

2   Remove the front bumper cover (see Chapter 11).

3   Disconnect the electrical connector(s) from the headlight housing.

Note: *It's not necessary to remove the rubber protective boot or the bulb to remove the headlight housing. Of course, if you're planning to replace the headlight housing, and therefore need to remove the boot and bulb, it's easier to do so after you have removed the headlight housing.*

2012 and earlier models

4   Remove the two lower headlight housing mounting bolts (see illustration).

5   Remove the upper headlight housing mounting bolt and nut (see illustration).

6   Remove the headlight housing (see illustration).

7   Installation is the reverse of removal.

---

located behind each headlight (see illustrations). There are no horizontal adjustment screws.

2   There are several methods for adjusting the headlights. The simplest method requires masking tape, a blank wall and a level floor.

3   Position masking tape vertically on the wall in reference to the vehicle centerline and the centerlines of both headlights (see illustration).

4   Position a horizontal tape line in reference to the centerline of all the headlights.

Note: *It may be easier to position the tape on the wall with the vehicle parked only a few inches away.*

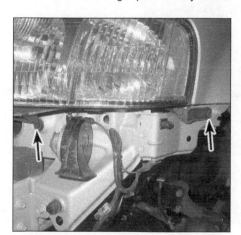

17.4 To detach the lower part of the headlight housing from the vehicle, remove these two bolts

17.5 To detach the upper part of the headlight housing from the upper radiator crossmember, remove this bolt and nut

17.6 Carefully remove the headlight housing from the vehicle

17.8 Headlight housing upper mounting screws

17.9 Headlight housing lower mounting screw

2013 and later models

8   Remove the three screws from the top side of the headlamp assembly (see illustration).

9   Remove the lower outside screw (see illustration).

10   Installation is the reverse of removal.

---

18   Horn - replacement

2012 and earlier models

Note: *These models use two horns; one mounted below each headlight.*

1   Remove the front bumper cover (see Chapter 11).

2   Disconnect the electrical connector from the horn (see illustration).

3   Remove the horn mounting nut and separate the horn from its mounting bracket.

4   Installation is the reverse of removal.

2013 and later models

Note: *These models use two horns as well, but both are mounted on a bracket forward of the right fenderwell.*

18.2 To remove the horn, disconnect the electrical connector and remove the mounting nut (left horn shown, right horn similar)

5   Loosen the right front wheel lug nuts. Raise the front of the vehicle and support it securely on jackstands, then remove the wheel.

6   Remove the inner fender splash shield

18.7 Horn mounting bracket bolt (bumper cover removed for clarity) (2013 and later models)

(see Chapter 11, Section 10).

7   Remove the bolt and detach the horn bracket and horns (see illustration), then unplug the electrical connector.

8   Installation is the reverse of removal.

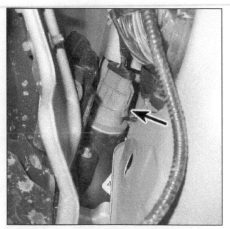

19.3a To remove the turn signal bulb holder, turn it counterclockwise 1/4-turn and pull it out of the front turn signal housing (2012 and earlier models)

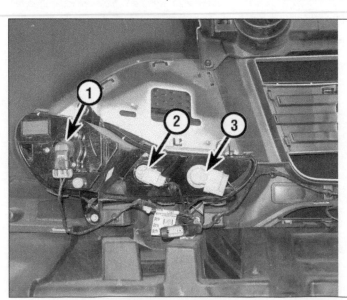

19.3b 2013 and later front lighting layout

1   *Fog lamp*
2   *Marker lamp*
3   *Turn signal and parking lamp*

19.6a To detach the high-mount brake light assembly from the roof, remove these two screws and washers; discard the old screws and washers and use new ones when you install the high-mount brake light unit

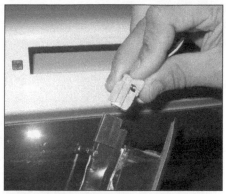

19.6b Pull out the high-mount brake light assembly and disconnect the electrical connector

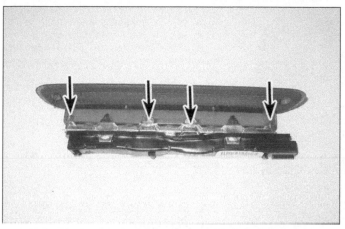

19.7a To separate the high-mount brake light bulb holder assembly from the lens, use a small screwdriver to disengage all six retaining snaps (two on other side, not visible) . . .

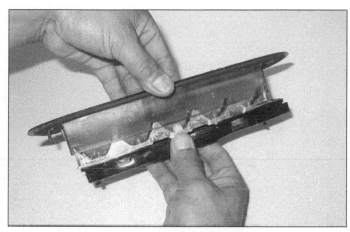

19.7b . . . then pull the bulb holder assembly out of the lens

19 Bulb replacement

Front turn signal bulbs

1   Loosen the left or right front wheel lug nuts. Raise the front of the vehicle and place it securely on jackstands. Remove the left or right front wheel.

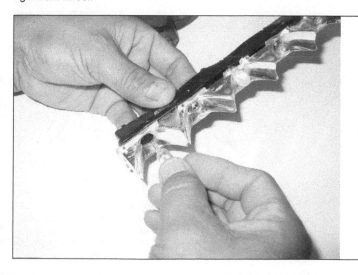

19.8 To replace a high-mount brake light bulb, simply pull it straight out of the bulb holder

2   Remove the inner fender splash shield (see *Front fender - removal and installation* in Chapter 11). Only the front section of the splash shield needs to be pried back on most models.
3   Locate the turn signal bulb holder (see illustrations). To remove the bulb holder, turn it counterclockwise and pull it out of the hous-ing. (It's not necessary to disconnect the electrical connector from the bulb holder.)
4   Remove the turn signal bulb from the bulb holder by pulling it straight out.
5   Installation is the reverse of removal.

High-mount brake light bulbs

2012 and earlier models

6   Remove the high-mount brake light assembly mounting screws (see illustration) and washers, pull the assembly out and dis-connect the electrical connector (see illustra-tion). Then place the high-mount brake light assembly on a work bench.
Note: *The manufacturer recommends that you discard the old mounting screws and washers and replace them with new ones.*
7   Separate the bulb holder assembly from the lens (see illustrations), then remove the bulb assembly from the housing.
8   To replace a bulb, simply pull it straight out of the bulb holder (see illustration).
9   To install a new bulb, push it straight into the bulb holder.
10   Installation is the reverse of removal. The manufacturer recommends that you use new mounting screws and washers.

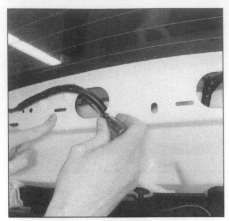

19.12a Reach in with the screwdriver to pry the retaining clips off . . .

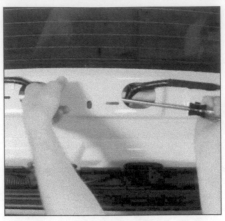

19.12b . . . some of the clips are harder to reach

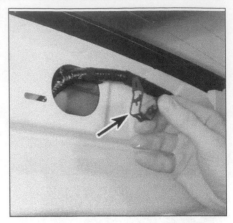

19.12c Save the clamps for reinstalling

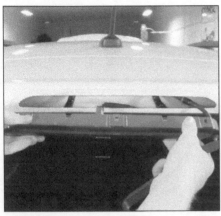

19.12d With the clips removed, pull the lamp fixture out of the liftgate

2013 and later models

Note: *On these models, the high-mount brake light uses LED bulbs, which are not replaceable individually.*

11   Remove the rear liftgate trim (see Chapter 11).

12   Use a long flat-bladed screwdriver to pry the retaining clips off (see illustrations), then

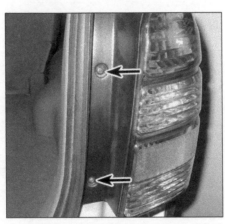

19.14a Rear taillight housing screws (2012 and earlier models)

remove the light fixture. Reuse the same clips.

13   Installation is the reverse of removal.

Brake/tail/turn/back-up light bulbs

14   Remove the two brake/tail/turn/back-up assembly mounting screws (see illustration).

To remove the assembly, pull it to the rear to disengage the two locator pins on the outer edge of the lens from their corresponding grommets in the rear quarter panel (see illustrations).

15   To replace a bulb, rotate the bulb socket counterclockwise and remove the bulb socket (see illustration).

16   To remove a bulb from its socket, pull it straight out.

17   To install a bulb in its socket, push it straight into the socket.

18   Installation is the reverse of removal.

Trunk lid (liftgate) mounted reverse lamp

19   Remove the liftgate trim panels (see Chapter 11).

20   Remove the nut securing the lens to the liftgate (see illustration).

21   Pry the lens off of the liftgate with a flat trim tool (see illustration).

22   Disconnect the bulb by turning the bulb socket counterclockwise (see illustration). Pull the bulb straight out of its socket to remove it.

23   Installation is the reverse of removal.

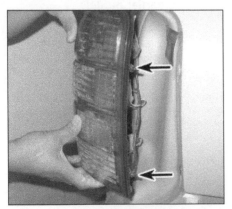

19.14b To remove the taillight housing, pop loose these two locator pins from their grommets

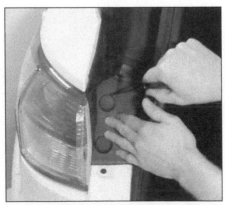

19.14c 2013 and later models have plastic caps over the mounting screws

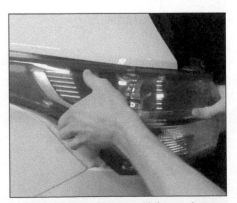

19.14d 2013 and later models pop loose similar to the earlier models. Be careful not to flex the plastic lens assembly too much from side to side. Keep the primary force straight backwards

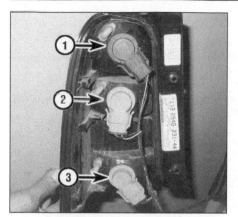

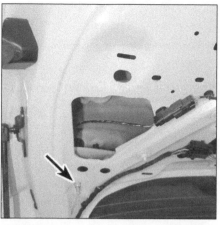

19.15 Brake/tail/turn signal/back-up light assembly (2012 and earlier models)

1   Brake/taillight bulb
2   Turn signal light bulb
3   Back-up light bulb

19.20 Liftgate-mounted light housing nut

19.21 After removing the nut, carefully pry the lens off

License plate light bulbs

24   On 2007 and earlier models, pry the light assembly from the liftgate (see illustration). On 2008 through 2012 models, slide the light housing to the right, then pull it out. On 2013 and later models, pry the small retaining tab in, then pull the light housing from the tailgate (see illustration).

25   To remove the bulb socket from the license plate light assembly, turn it counter-clockwise a 1/4-turn and pull it out (see illustrations).

26   To remove the bulb from the socket, pull it straight out.

27   Be sure the replacement bulb is fully seated into the socket, and check that it works before reinstalling.

28   Installation is the reverse of removal.

Side turn signal lamp

2012 and earlier models

29   Slide the side light housing toward the rear, then pull out on the front of the light.

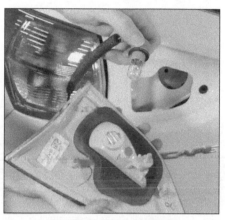

19.22 Turn the bulb holder counterclockwise to remove it from the housing

30   Disconnect the wiring and remove the light.

31   Installation is the reverse of removal.

2013 and later models

32   Raise vehicle and support the front end on jackstands. Remove the front wheels or turn the steering wheel to one side as far as

19.24a Pry the license plate light assembly from the liftgate (2007 and earlier models)

it will go to give you some room between the tire and the lamp fixture.

33   Remove the front portion of the inner fender splash shield (see Chapter 11).

34   Reach up and disconnect the bulb socket, and remove the bulb. For bulb location see front turn signal removal.

35   Installation is the reverse of removal.

19.24b On 2013 and later models, pry the lens retaining tab in with a flat screwdriver, then pull the light out of the liftgate

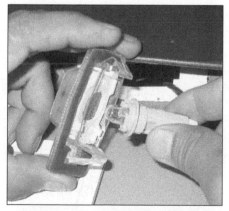

19.25a To remove the bulb socket from the license plate light assembly, turn it counterclockwise and pull it out

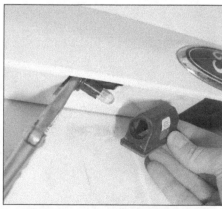

19.25b On 2013 and later models, use a pair of needle-nose pliers to gently hold the socket, then twist the housing to detach it from the socket

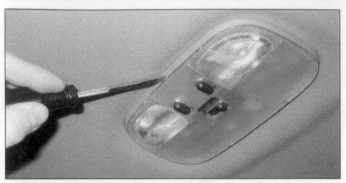

19.42 To remove the interior/map reading light lens, carefully pry it off with a small screwdriver inserted into this slot, which is provided for this purpose (prying off the lens elsewhere might damage or scratch the housing)

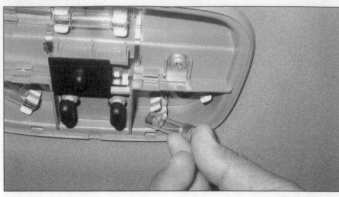

19.43 To replace a bulb in the interior/map reading light assembly, simply pull it out

Fog light bulbs

36   Raise vehicle and support the front end on jack stands. Remove the front wheels or turn the steering wheel to one side as far as it will go to give you some room between the tire and the lamp fixture.

37   Remove the fasteners from the front of the inner fender splash shield. Pull the shield back enough for access to the fog light.

38   Disconnect the wiring from the fog light.

39   2012 and earlier models - Pinch the tabs on the base of the bulb and remove it.

40   2013 and later models - Rotate the bulb socket to remove the socket from the lamp fixture, then remove the bulb. For bulb location, see illustration 19.3b.

41   Installation is the reverse of removal.

Interior/map reading light

42   Carefully pry the interior/map reading light lens loose from the light assembly (see illustration).

43   To replace one of the light bulbs in the interior/map reading light assembly, simply pull it out (see illustration).

44   Installation is the reverse of removal.

Bulb removal

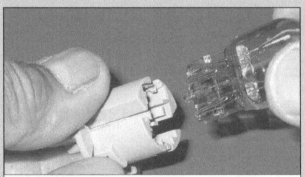

To remove many modern exterior bulbs from their holders, simply pull them out

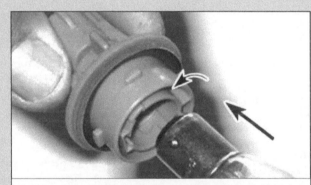

On bulbs with a cylindrical base ("bayonet" bulbs), the socket is spring-loaded; a pair of small posts on the side of the base hold the bulb in place against spring pressure. To remove this type of bulb, push it into the holder, rotate it 1/4-turn counterclockwise, then pull it out

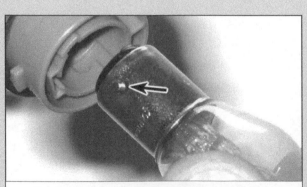

If a bayonet bulb has dual filaments, the posts are staggered, so the bulb can only be installed one way

To remove most overhead interior light bulbs, simply unclip them

20 Electric side view mirrors - description

1   Most electric rear view mirrors use two motors to move the glass; one for up and down adjustments and one for left-right adjustments.

2   The control switch has a selector portion that sends voltage to the left or right side mirror. With the ignition On but the engine Off, roll down the windows and operate the mirror control switch through all functions (left-right and up-down) for both the left and right side mirrors.

3   Listen carefully for the sound of the electric motors running in the mirrors.

4   If the motors can be heard but the mirror glass doesn't move, there's a problem with the drive mechanism inside the mirror.

5   If the mirrors do not operate and no sound comes from the mirrors, check the fuse (see Section 3).

6   If the fuse is OK, remove the mirror control switch. Have the switch continuity checked by a dealership service department or other qualified automobile repair facility.

7   If the mirror still doesn't work, remove the mirror and check the wires at the mirror for voltage.

8   If there's not voltage in each switch position, check the circuit between the mirror and control switch for opens and shorts.

9   If there's voltage, remove the mirror and test it off the vehicle with jumper wires. Replace the mirror if it fails this test.

21 Cruise control system - description

1   The cruise control system maintains vehicle speed with an electrically-operated motor located in the engine compartment, which is connected to the throttle body by a cable. The system consists of the Powertrain Control Module, the speed control actuator, the speed control cable, the speed control indicator light, the speed control actuator switches, the Brake Pedal Position (BPP) switch, the Clutch Pedal Position (CPP) switch (models with a manual transaxle) or the Transmission Range (TR) sensor (models with an automatic transaxle). The cruise control system requires diagnostic procedures that are beyond the scope of this manual. Listed below are some general procedures that may be used to locate common problems.

2   Check the fuses (see Section 2).

3   Have an assistant operate the brake lights while you check their operation (voltage from the brake light switch deactivates the cruise control).

4   If the brake lights don't come on or stay on all the time, correct the problem and retest the cruise control system.

5   Visually inspect the control cable between the cruise control motor and the throttle linkage for free movement. Replace it if necessary.

6   Test drive the vehicle to determine if the cruise control is now working. If it isn't, take it to a dealer service department or an automotive electrical specialist for further diagnosis.

22 Power window system - description and check

2004 and earlier models

1   The power window system operates electric motors, mounted in the doors, which lower and raise the windows. The system consists of the control switches, the motors, regulators, glass mechanisms and associated wiring.

2   The power windows can be lowered and raised from the master control switch by the driver or by remote switches located at the individual windows. Each window has a separate motor, which is reversible. The position of the control switch determines the polarity and therefore the direction of operation.

3   The circuit is protected by a fuse and a circuit breaker. Each motor is also equipped with an internal circuit breaker, this prevents one stuck window from disabling the whole system.

Note: *For generic power window diagnostics see the Power window system diagnostics Section in this chapter.*

2005 and later models

Note: *These models are equipped with a Smart Junction Box (SJB) (manufacturer terminology), otherwise known as a Body Control Module (BCM). Several systems are linked to this centralized control module, which allows simple and accurate troubleshooting, but only with a professional-grade scan tool. The SJB or BCM governs the door locks, the power windows, the ignition lock and security system, the interior lights, the Daytime Running Lights system, the horn, the windshield wipers, the heating/air conditioning system and the power mirrors. In the event of malfunction with this system, have the vehicle diagnosed by a dealership service department or other qualified automotive repair facility.*

4   The power window system operates electric motors, mounted in the doors, which lower and raise the windows. The system consists of the control switches, the motors, regulators, glass mechanisms, the Smart Junction Box (SJB) and associated wiring.

5   The power windows can be lowered and raised from the master control switch by the driver or by remote switches located at the individual windows. Each window has a separate motor that is reversible. The position of the control switch determines the polarity and therefore the direction of operation.

6   The circuit is protected by a fuse and a circuit breaker. Each motor is also equipped with an internal circuit breaker; this prevents one stuck window from disabling the whole system.

7   The power window system will only operate when the ignition switch is On, and for a period of time after the ignition key has been turned Off (unless one of the doors is opened). In addition, many models have a window lockout switch at the master control switch which, when activated, disables the switches at the rear windows and, sometimes, the switch at the passenger's window also. Always check these items before troubleshooting a window problem.

Power window system diagnostics for all models

Note: *These procedures are general in nature, so if you can't find the problem using them, take the vehicle to a dealer service department or other properly equipped repair facility.*

8   If the power windows won't operate, always check the fuse and circuit breaker first.

9   If only the rear windows are inoperative, or if the windows only operate from the master control switch, check the rear window lockout switch for continuity in the unlocked position. Replace it if it doesn't have continuity.

10   Check the wiring between the switches and fuse panel for continuity. Repair the wiring, if necessary.

11   If only one window is inoperative from the master control switch, try the other control switch at the window.

Note: *This doesn't apply to the driver's door window.*

12   If the same window works from one switch, but not the other, check the switch for continuity.

13   If the switch tests OK, check for a short or open in the circuit between the affected switch and the window motor.

14   If one window is inoperative from both switches, remove the switch panel from the affected door. Check for voltage at the switch and at the motor (refer to Chapter 11 for door panel removal) while the switch is operated.

15   If voltage is reaching the motor, disconnect the glass from the regulator (see Chapter 11). Move the window up and down by hand while checking for binding and damage. Also check for binding and damage to the regulator. If the regulator is not damaged and the window moves up and down smoothly, replace the motor. If there's binding or damage, lubricate, repair or replace parts, as necessary.

16   If voltage isn't reaching the motor, check the wiring in the circuit for continuity between the switches and the body control module, and between the body control module and the motors. You'll need to consult the wiring diagram at the end of this Chapter. If the circuit is equipped with a relay, check that the relay is grounded properly and receiving voltage.

17   Test the windows after you are done to confirm proper repairs.

One touch power window auto down circuit

18   The driver's window is equipped with an automatic down feature that is offered on some

models. This feature allows for a momentary contact of the driver's window switch to allow the window to reach the full open position.

19   If the auto down feature fails to operate, the most likely cause is a faulty master switch (driver's side window switch assembly). The master switch contains a sensing resistor and internal relay that determines the exact position of the window when it is fully down. The switch is not serviceable. Replace the switch and recheck the operation of the auto down feature.

23   Power door lock and keyless entry system - description, check and battery replacement

Note: *These models are equipped with a Smart Junction Box (SJB) (manufacturer terminology), otherwise known as a Body Control Module (BCM). Several systems are linked to this centralized control module, which allows simple and accurate troubleshooting, but only with a professional-grade scan tool. The SJB or BCM governs the door locks, the power windows, the ignition lock and security system, the interior lights, the Daytime Running Lights system, the horn, the windshield wipers, the heating/air conditioning system and the power mirrors. In the event of malfunction with this system, have the vehicle diagnosed by a dealership service department or other qualified automotive repair facility.*

Description and check

1   The power door lock system operates the door lock actuators mounted in each door. The system consists of the switches, actuators, Smart Junction Box (SJB) and associated wiring. Diagnosis can usually be limited to simple checks of the wiring connections and actuators for minor faults that can be easily repaired.

2   Power door lock systems are operated by bi-directional solenoids located in the doors. The lock switches have two operating positions: Lock and Unlock. These switches send a signal to the SJB, which in turn sends a signal to the door lock solenoids.

3   If you are unable to locate the trouble using the following general steps, consult your dealer service department.

4   Always check the circuit protection first. Some vehicles use a combination of circuit breakers and fuses. Refer to the wiring diagrams at the end of this Chapter.

5   Check for voltage at the switches. If no voltage is present, check the wiring between the fuse panel and the switches for shorts and opens.

6   If voltage is present, test the switch for continuity. Replace it if there's not continuity in both switch positions. To remove the switch, use a flat-bladed trim tool to pry out the door/window switch assembly.

7   If the switch has continuity, check the wiring between the switch and door lock solenoid.

8   If all but one lock solenoids operate, remove the trim panel from the affected door (see Chapter 11) and check for voltage at the solenoid while the lock switch is operated.

One of the wires should have voltage in the Lock position; the other should have voltage in the Unlock position.

9   If the inoperative solenoid is receiving voltage, replace the solenoid.

10   If the inoperative solenoid isn't receiving voltage, check for an open or short in the wire between the lock solenoid and the relay.

11   On the models covered by this manual, power door lock system communication goes through the Smart Junction Box. If the above tests do not pinpoint a problem, take the vehicle to a dealer or qualified shop with the proper scan tool to retrieve trouble codes from the SJB.

Keyless entry system

12   The keyless entry system consists of a remote control transmitter that sends a coded infrared signal to a receiver, which then operates the door lock system.

13   Replace the battery when the transmitter doesn't operate the locks at a distance of ten feet. Normal range should be about 30 feet.

Key (Ignition key/theft deterrent) programming

14   Programming a replacement key requires the use of a highly sophisticated scanning tool. The vehicle will need to be towed to a facility that can perform the key programming function, such as a dealership service center or an independent repair shop that has the needed equipment, or have a locksmith come

to the vehicle. There is no method for bypassing the theft deterrent system.

Note: *As of 2005 anyone programming keys for the purpose of starting a vehicle must also have a federal locksmith licence as well as the proper scanning equipment.*

15   For key fob battery replacement and key fob programming see Section 27 for details.

24   Daytime Running Lights (DRL) - general information

1   The Daytime Running Lights (DRL) system illuminates the headlights whenever the engine is running. The only exception is with the engine running and the parking brake engaged. Once the parking brake is released, the lights will remain on as long as the ignition switch is on, even if the parking brake is later applied.

2   The DRL system supplies reduced power to the headlights so they won't be too bright for daytime use, while prolonging headlight life.

25   Airbag system - general information and precautions

General information

1   All models are equipped with a Supplemental Restraint System (SRS), more commonly known as airbags. This system is

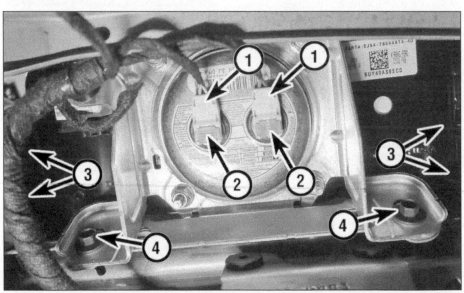

25.4 2013 and later model passenger airbag shown, looking up from the base of the instrument panel. Notice the two connectors. This is a two-stage airbag system. Each connector consists of two separate clips. A main clip (yellow) and a safety clip (orange). Removal of the connections and installation must be followed precisely

1   *This is the main connection (yellow). Do not pull this connector off until the safety (orange) clip has been lifted to the UP position first*

2   *This is the safety clip (orange). Pry this clip UP first. Lift the orange safety clip as high as it will go, then pull the main (yellow) clip off of the airbag.*

Installation of the connectors is the reverse of removal.

3   *Mounting bolts and brackets (off to the sides, not visible in frame)*

4   *Airbag module-to-instrument panel support beam bolts (must be removed for instrument panel top cover removal)*

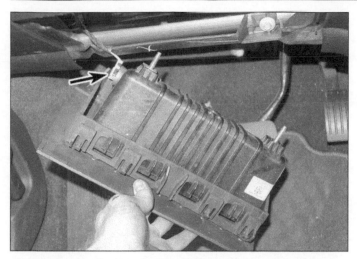

25.7 Removing the knee bolster airbag module only requires removing the mounting screws and removing the connector (shown by the arrow). Be sure to set the airbag in a safe spot with the deployment facing in a safe direction

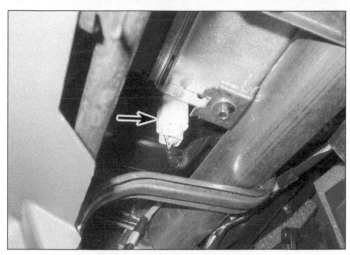

25.17 The electrical connector for the passenger's side airbag is located above the glove box recess, to the left of the airbag module (early model shown, glove box removed for clarity; for 2013 and later models, see illustration 25.4)

designed to protect the driver, and the front seat passenger, from serious injury in the event of a head-on or frontal collision and side impacts. It uses an electronic crash sensor, which is referred to by the manufacturer as the Restraints Control Module (RCM) and a side-impact sensor in each B-pillar. The airbag assemblies are mounted on the steering wheel and inside the passenger's side of the dash, and in the side of each front seat back. Later models are also equipped with a driver's knee bolster airbag and roof rail, or safety canopy airbags. Some 2001 models and all 2002 and later models are equipped with seat belt pre-tensioners. These are pyrotechnic devices controlled by the Restraints Control Module (RCM) that reduce the slack in the seat belts during an impact of sufficient force to trigger the airbags.

Airbag module

Driver's airbag

2   The airbag inflator module contains a housing incorporating the cushion (airbag) and inflator unit, mounted in the center of the steering wheel. The inflator assembly is mounted on the back of the housing over a hole through which gas is expelled, inflating the bag almost instantaneously when an electrical signal is sent from the system. A "clockspring" on the steering column under the steering wheel carries this signal to the module. This clockspring assembly can transmit an electrical signal regardless of steering wheel position. The igniter in the airbag converts the electrical signal to heat and ignites the powder, which inflates the bag.

3   For information on how to remove and install the driver's side airbag, refer to *Steering wheel - removal and installation* in Chapter 10.

Passenger's airbag

4   The airbag is mounted above the glove compartment. It consists of an inflator con-

taining an igniter, a reaction housing/airbag assembly and a trim cover (see illustration).

5   The passenger's side airbag is considerably larger than the steering wheel-mounted unit and is supported by the steel reaction housing. The trim cover has a molded seam which splits when the bag inflates.

Side impact and roof rail (or safety canopy) airbags

6   The side impact airbags are mounted on the outboard side of the front seat back frames, and the roof rail airbag assemblies run along the interior of the roof from the front A-pillar to the rear of the passenger compartment. In the event of a side impact, both airbag assemblies are activated by the sensors mounted at the base of the center pillar behind the seats.

Driver's knee bolster airbag

7   The driver's knee bolster airbag is mounted on the lower section of the instrument panel right in front of the driver's knees (see illustration). This airbag is removed for many different repair reasons and should be handled just like any other airbag. Always set the bag face up and in a safe place when it is not in the vehicle.

Restraints Control Module (RCM)

8   This unit supplies the current to the airbag system (and seat belt pre-tensioners, on models so equipped) in the event of the collision, even if battery power is cut off. It checks this system every time the vehicle is started, causing the "SRS" light to go on, then off, if the system is operating properly. If there is a fault in the system, the light will go on and stay on, or it will flash or the dash will make a beeping sound. If this happens, take the vehicle to your dealer immediately for service.

Disarming the system and other precautions

Warning: *Failure to follow these precautions could result in accidental deployment of the airbag and personal injury.*

Warning: *NEVER probe the connectors of the SRS. Do not move or remount from the original position any SRS component. Never move an SRS component with the key in the On position. Do not use any memory saver devices.*

Warning: *Whenever working in the vicinity of the steering wheel, instrument panel or any of the other SRS system components, the system must be disarmed.*

9   Point the wheels straight ahead and turn the key to the Lock position.

De-powering procedure

Note: *The airbag warning lamp illuminates when the Restraint Control Module fuse is removed and the ignition switch is On. This is normal operation and does not indicate a fault in the SRS.*

10   Turn all accessories Off.

11   Turn the ignition to Off.

12   Remove the Restraints Control Module (RCM) fuse(s) for the SRS system:

a)   *2004 and earlier models: Remove fuse F2.5 (5A) and fuse F2.7 (10A) from the Central Junction Box (CBJ) behind the left kick panel (see illustration 3.1c).*

b)   *2005 through 2007 models: Remove fuse F33 (15A) from the Smart Junction Box (SJB) at the right side of the center console.*

c)   *2008 models: Remove fuse F32 (10A) from the Smart Junction Box (SJB) at the right side of the center console.*

d)   *2009 through 2012 models: Remove fuse F31 (10A) from the Smart Junction Box (SJB) at the right side of the center console.*

e)   *2013 and later models: Remove fuse 86 (10A) from the Body Control Module (BCM) behind the glove box (see illustration 3.1f).*

25.23 To detach the passenger's side airbag from the dash, remove these four bolts, then carefully but firmly push on the airbag module to disengage the trim cover from the top of the dash

25.24 Pull the trim cover and the airbag from the dash as a single assembly. Be sure to keep your body positioned to the side of the airbag

13 Turn the ignition On and monitor the airbag warning light for a minimum of 30 seconds. If the correct RCM fuse has been removed, the light will remain lit continuously without flashing. If the airbag warning light does not remain lit continuously, STOP and remove the correct fuse and monitor the light as outlined.

14 Turn the ignition switch Off, then disconnect the negative battery cable (see Chapter 5) and wait at least two minutes before proceeding.

Re-powering procedure

Warning: *Make sure that there is no one in the vehicle and that nothing is blocking any airbag module while performing the re-powering procedure.*

a) *Turn the ignition switch from Off to On.*
b) *Reinstall the RCM fuse in the Smart Junction Box, then install the cover.*
c) *Connect the negative battery cable.*
d) *Check the system by turning the ignition switch from On to Off.*
e) *Wait 10 seconds, then turn the key back to the On position.*
f) *Visually monitor the airbag warning light. The light should remain lit for about 6 seconds, then turn off.*
g) *It may take 30 seconds or more for the RCM to complete the self tests. If any fault is present in the system, the light will either fail to light, remain lit or flash.*
h) *If the airbag warning light is inoperative and a fault exists, a chime will sound in a series of five sets of five beeps.*
i) *If there is any fault in the system, bring the vehicle to a professional automotive repair facility for proper diagnosis and repair.*

15 Whenever handling an airbag module, always keep the airbag opening (the trim side) pointed away from your body. Never place the airbag module on a bench or other surface with the airbag opening facing the surface.

Always place the airbag module in a safe location with the airbag opening facing up.

16 Never measure the resistance of any SRS component or use any electrical test equipment on any of the wiring or components. An ohmmeter has a built-in battery supply that could accidentally deploy the airbag.

17 Never use electrical welding equipment on a vehicle equipped with an airbag without first disconnecting the airbag electrical connectors, located behind a panel on the underside of the steering wheel (driver's airbag) and above the glove box (passenger's airbag) (see illustration) and under each front seat (side-impact airbags). On models with seat belt pre-tensioners, the pre-tensioner electrical connectors are also located under the seats.

18 Never dispose of a live airbag module or seat belt pre-tensioner. Return it to a dealer service department or other qualified repair shop for safe deployment and disposal.

Airbag module - removal and installation

Clockspring and driver's airbag - removal and installation

19 Refer to Chapter 10, *Steering wheel - removal and installation*, for the driver's side airbag module and clockspring removal and installation procedures.

Passenger's airbag module

20 Disarm the airbag system as described previously in this Section.

2012 and earlier models

21 Remove the glove box.
22 Disconnect the electrical connector (see illustration 25.17).
23 Remove the airbag module mounting bolts (see illustration). After removing the bolts, carefully but firmly push up on the airbag module to disengage the trim cover from the top of the dash (the airbag module and

trim cover are attached and are removed as a single assembly). Do not try to pry open the trim cover from up top, or you will damage the surface of the dash and/or the trim cover.
24 Working from above, pull the airbag trim cover and airbag module out of the dash (see illustration). Be sure to heed the precautions outlined previously in this Section.
25 Installation is the reverse of the removal procedure. Tighten the airbag module mounting bolts securely.
26 After you've reconnected the battery, the Powertrain Control Module (PCM) must relearn its idle and fuel trim strategy for optimum driveability and performance (see Chapter 5, Section 1 for this procedure).

2013 and later models

27 Remove the instrument panel upper section (see Chapter 11, Section 22). Lay it trim-side down on a clean surface.
28 Remove the airbag mounting bolts and brackets (see illustration 25.4).
29 Pry the deployment chute away from the airbag module (on the side with four retainers) to disengage the retainers, then rotate the airbag module up and out of the chute.
30 Pivot the module out of the deployment chute, disengaging the five retainers on the other side.
31 Installation is the reverse of removal, noting the following:

a) *Engage the five retaining tabs on one side of the airbag module with the five slots in the deployment chute.*
b) *Pull the other side of the deployment chute out while rotating the airbag module down into it.*
c) *Push the module down and engage the four retainers with the slots in the chute.*
d) *Install the brackets and mounting bolts, tightening them to 22 inch-lbs (2.5 Nm).*

Side-impact airbag

32 Under normal circumstances there would never be a reason to remove a seat airbag.

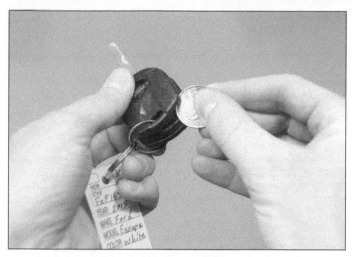

27.2a Insert the coin into the slot, then twist the coin to separate the key fob halves

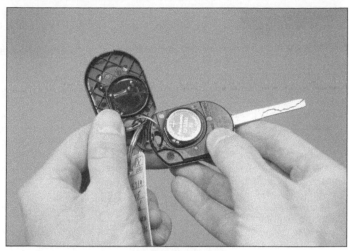

27.2b Keep the battery polarity in mind when replacing the batteries. Notice which direction the + sign is facing before snapping the halves back together

However, if it has been determined that there is a problem with the side-impact airbag module, the work must be left to a dealer service department or other qualified repair shop.

26 Back up camera - removal, installation and troubleshooting

Note: *The rear view camera or back up camera is an integral part of today's vehicles. The camera is activated with the key on and the transmission placed into reverse. A video signal is sent by way of the CAN (Computer Adaptable Network) and fed to the video screen.*

Removal and installation

1   Remove the upper liftgate molding or license plate housing (see Chapter 11).
2   Disconnect the electrical connections to the camera.
3   Remove the screw securing the camera bracket to the housing.
4   Separate the camera from the bracket.
5   Replace the camera and reinstall.
6   Installation is the reverse of removal.

Troubleshooting

No image

Turn the key on and place the vehicle into reverse. If the image does not appear, check your powers and grounds to the camera system using the wiring diagram. Check the wiring front to back and pay close attention to any areas where the wires could have rubbed or could have chaffed and separated. Repair as needed. If the image still does not appear take your vehicle to your dealership or qualified independent repair shop that is familiar with the camera systems.

Inverted image

If the image appears but is inverted (upside down) leave the key on for five minutes to allow the system to reconfigure the camera.

Poor image or wavy image

Most of the time a poor image is due to either dirt and debris on or to close to the camera lens or in some rare cases a radio frequency has interrupted the signal. Check for debris first, then check to see if anything has been added (other than factory components) to the vehicle, such as an aftermarket radio, trailer wiring, rear fog lights, etc. Disconnect any abnormal components and recheck. If it still does not appear correctly, take the vehicle to your local dealership or qualified independent repair shop that is familiar with the camera systems.

27 Key fob battery replacement and programming

Battery replacement

1   The standard replacement battery for the key fob is a CR2032. These are a flat coin type battery available anywhere batteries are sold. (You do not have to go to the dealer to purchase a replacement battery.)
Note: *Key fobs have a maximum range of 33 feet (10 m) with a fresh battery.*
2   Use a coin or a small screwdriver to separate the two halves of the key fob (see illustrations).

Programming

3   Key programming is broken into two separate segments. One segment is for the theft system and starting of the vehicle and the other segment is for the door lock and rear hatch systems.
4   The Passive Anti-Theft Systems (PATS), also known as the FordSecurilock system segment of the key fob programming is only performed by using the proper scanner and a current locksmith certification (2005 and later). Two keys have to be in your possession in order to perform this procedure. This

procedure also will need to be carried out if the keys are lost or the ignition switch lock cylinder is being replaced.
Note: *All keys should be present at the time the PATS system is being reset. All of the previous keys (not present or otherwise) are erased during the programming procedure.*
5   The door lock segment of the programming can be performed by the car owner. Follow the steps provided precisely in order to enter the programming mode of the lock system.
Note: *The door lock programming may need to be performed if the key fob battery, or in some cases the vehicle battery, has been drained or replaced.*
Warning: *The door locks must be in working order to program the door lock system.*

  a) *Unlock the doors using the door lock button (not with the key in the door).*
  b) *Turn the ignition key from Off to Run eight times in rapid succession (within 10 seconds), ending with the final turn of the key in the Run position.*
  c) *The door locks will cycle on then off, signaling that you have entered into the programming mode.*
  d) *Within 20 seconds, press any button on the key fob remote. The door locks will cycle confirming the programming has been accepted.*
  e) *Within the next 20 seconds press any key on the next key fob being programmed. The door locks will cycle confirming the programming has been accepted.*
  f) *You can program up to four remotes.*
  g) *Turn the ignition switch to Off. The door locks will cycle indicating you have ended the programming mode.*

6   Check the operation of the key fobs. If needed, perform the same procedure again, following all the steps exactly as described. If there is still a problem have your key fob checked to see if it is functioning properly.
Note: *A good method of determining if your key fob is functioning is to take your key fobs*

to any repair shop that can do tire pressure monitoring system diagnostics or a tire shop that has a TPMS tool (tire pressure monitoring system). Every TPMS tool is equipped to check RF signal strength of your key fobs. This test will not only show you the battery strength but also how strong the response signal is.

Caution: *Some of the older models have been known to go into the programming mode with as little as four key cycles from Off to On.*

Programming extra keys

Note: *You must have two programmed keys that already work with your car in order to program a third (extra key) into your vehicle.*

Caution: *The third key (extra key) must already have been cut to match your ignition and or door lock cylinder by a locksmith before attempting to program the key.*

Warning: *A maximum of eight keys can be programmed to one vehicle.*

7   Follow the steps precisely in order to complete the programming procedure. Read the entire procedure before attempting to program the extra key.

a) *Insert an existing working key into the ignition cylinder.*

b) *Turn the ignition to On then Off.*

c) *Remove the working key, then within five seconds, insert the second working key and turn it to On then Off.*

d) *Remove the second working key and insert the one programmed key into the ignition (within 10 seconds of removing the working key) and turn the key to On. Leave it in this position for one second then turn the key to Off.*

e) *Observe the security light, it should light up for three seconds to indicate the third key has entered the programming mode successfully.*

f) *Do not remove the key from the ignition. Leave the key in the ignition for approximately 10 additional minutes for the programming mode to complete.*

g) *If you have removed the key or turned the key to On without waiting the ten minutes you will have to start all over again from the first step.*

28   Wiring diagrams - general information

1   Since it isn't possible to include all wiring diagrams for every year covered by this manual, the following diagrams are those that are typical and most commonly needed.

2   Prior to troubleshooting any circuits, check the fuse and circuit breakers (if equipped) to make sure they're in good condition. Make sure the battery is properly charged and check the cable connections.

3   When checking a circuit, make sure that all connectors are clean, with no broken or loose terminals. When unplugging a connector, do not pull on the wires. Pull only on the connector housings themselves.

TRIBUTE

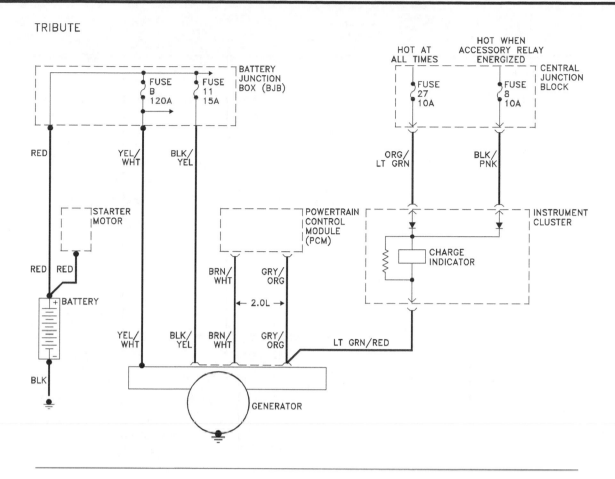

ESCAPE

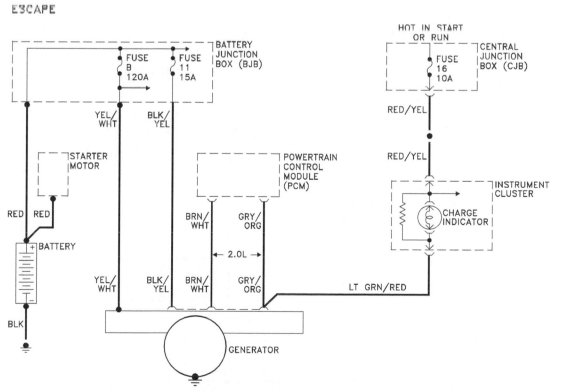

Charging system - 2006 and earlier models

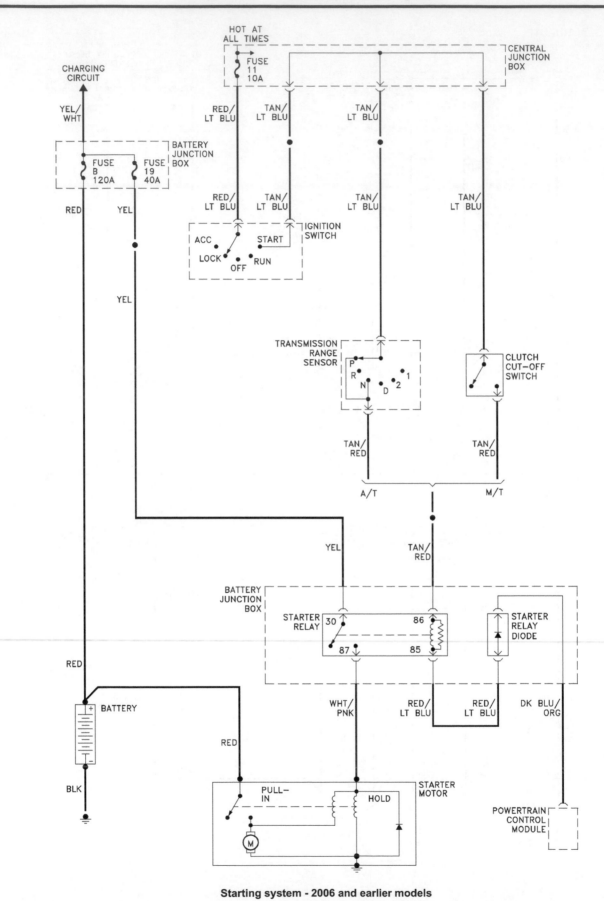

Starting system - 2006 and earlier models

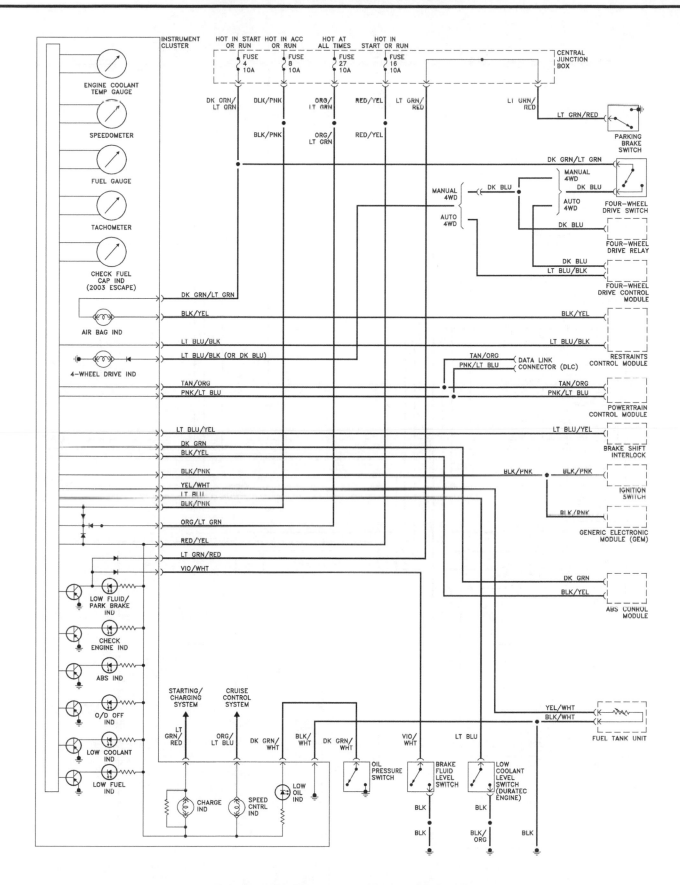

Warning lights and gauges - 2006 and earlier models

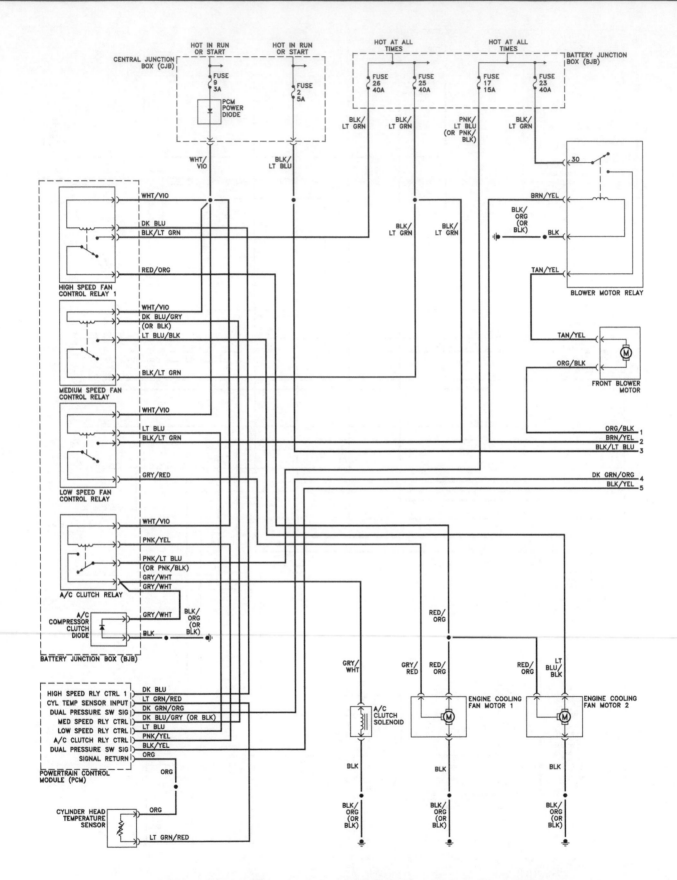

Heating and air conditioning system - 2006 and earlier four-cylinder models (1 of 2)

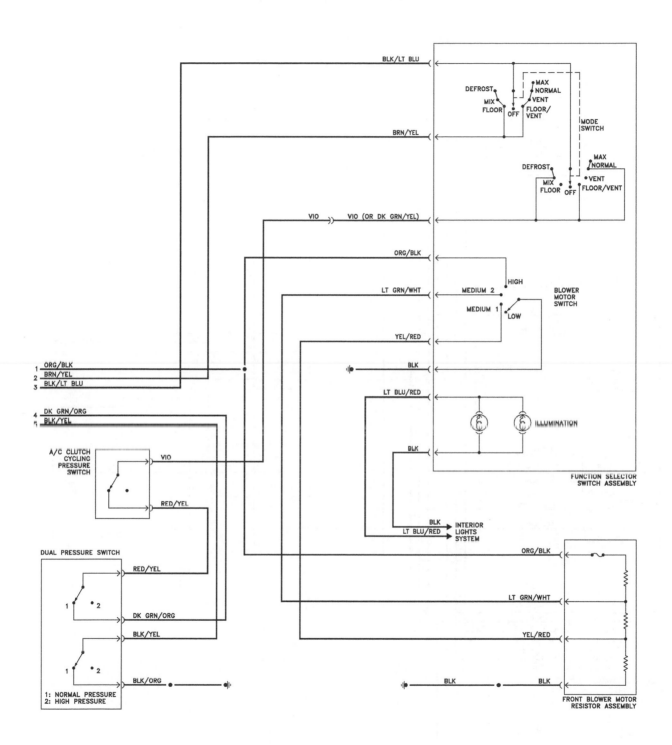

Heating and air conditioning system - 2006 and earlier four-cylinder models (2 of 2)

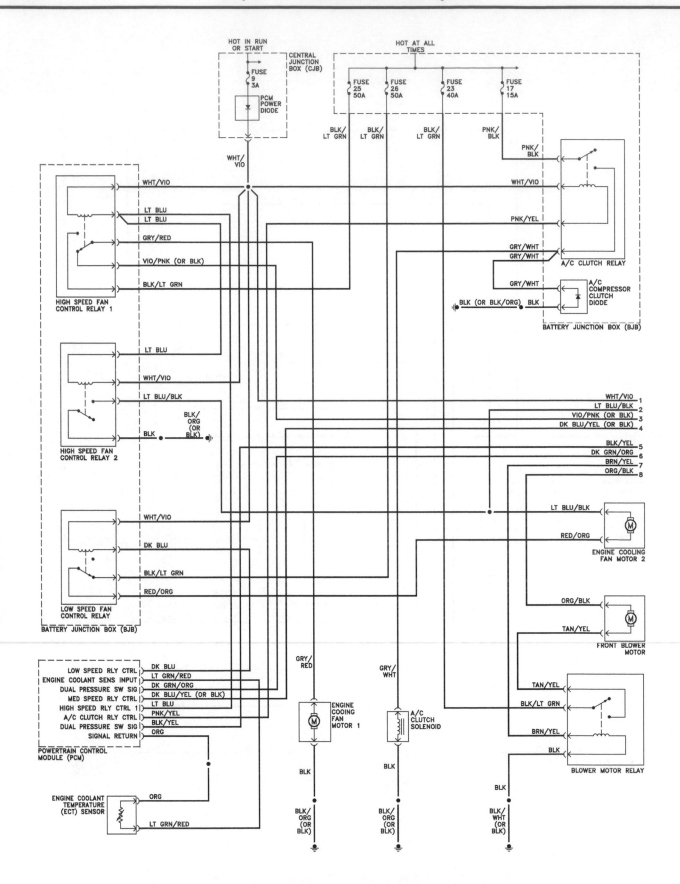

Heating and air conditioning system - 2006 and earlier V6 models (1 of 2)

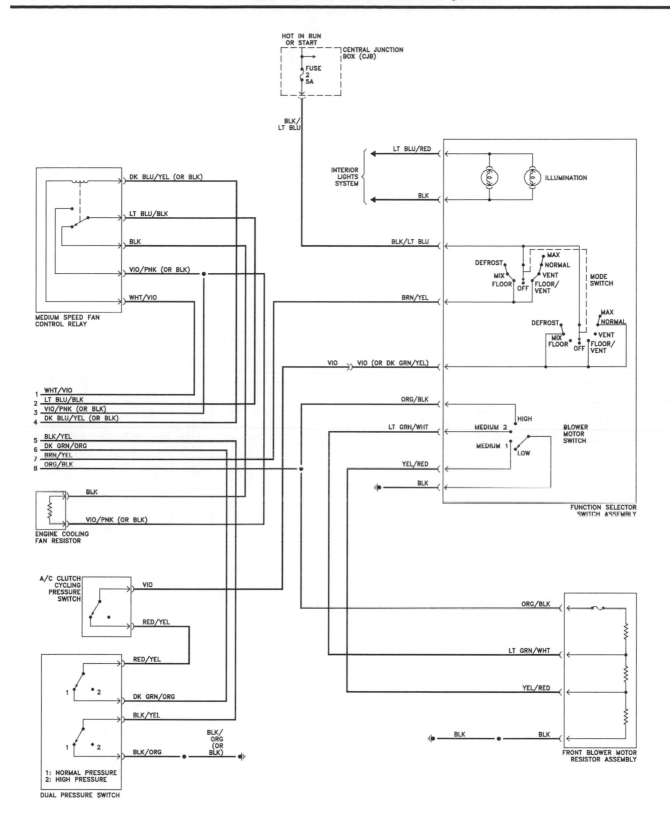

Heating and air conditioning system - 2006 and earlier V6 models (2 of 2)

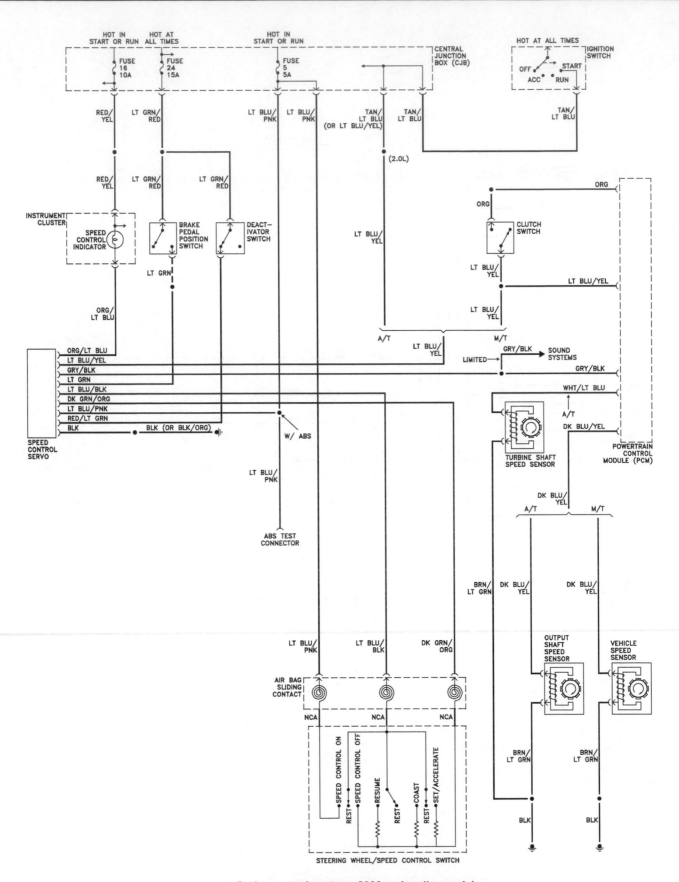

Cruise control system - 2006 and earlier models

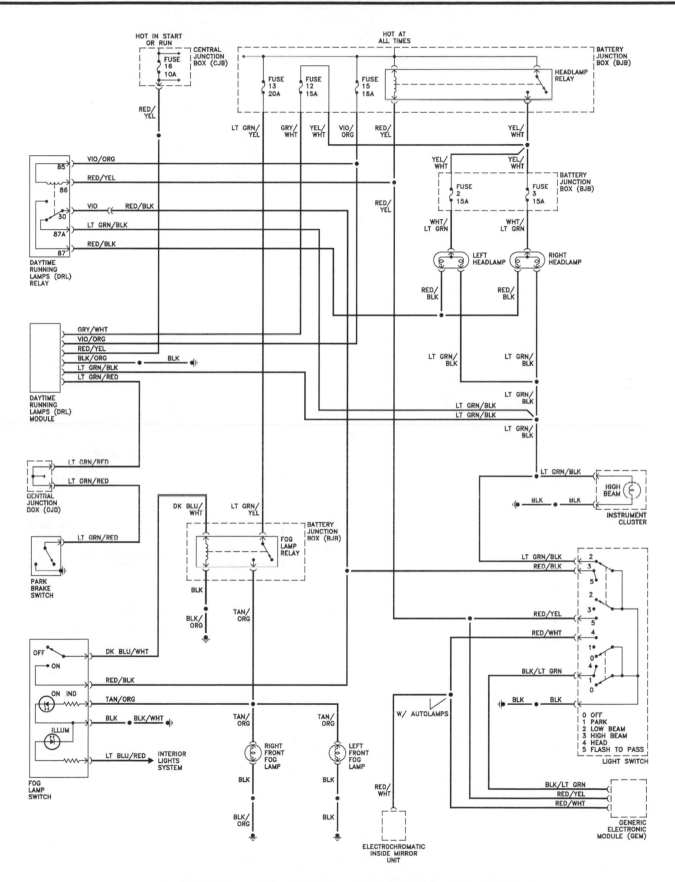

Headlight system - 2006 and earlier models (with Daytime Running Lights)

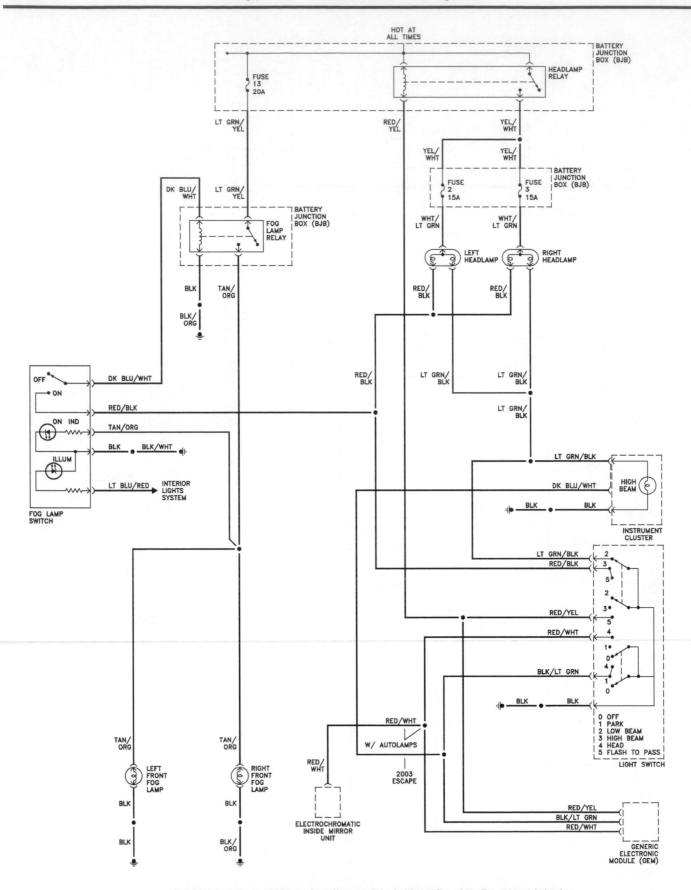

Headlight system - 2006 and earlier models (without Daytime Running Lights)

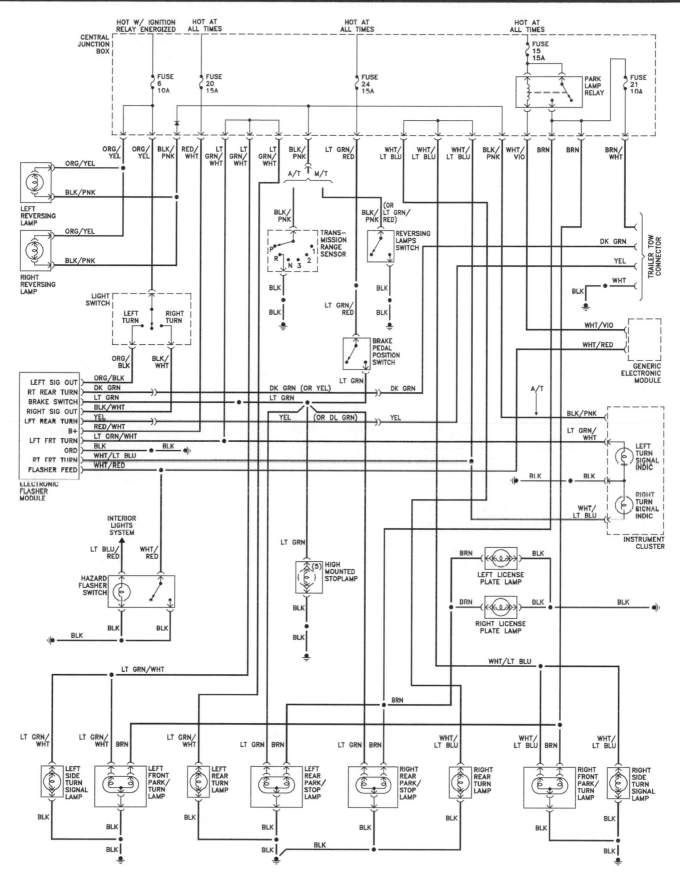

Exterior lighting system - 2006 and earlier models (except headlights)

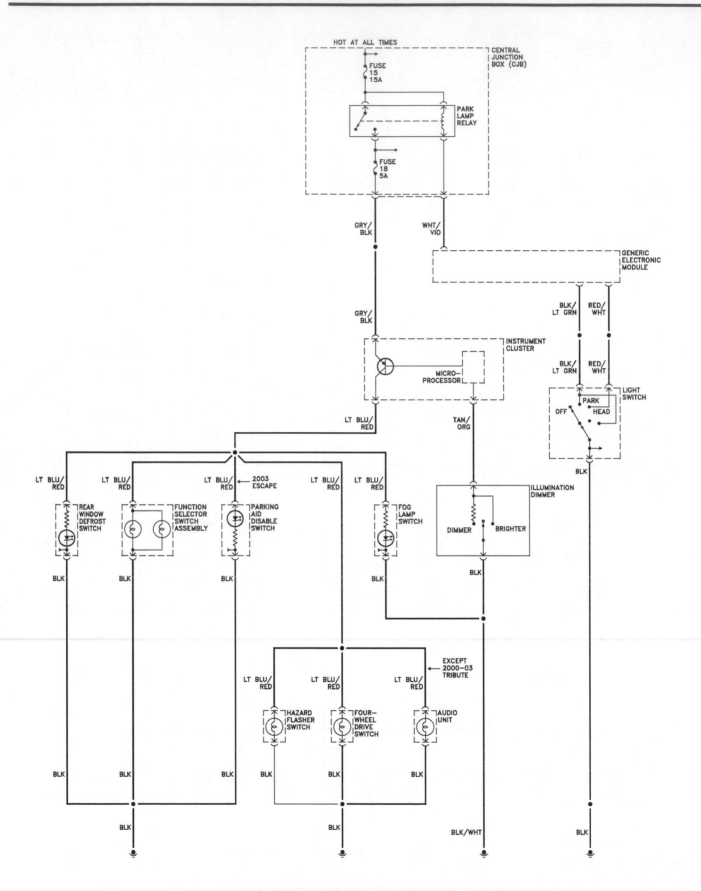

Switch illumination system - 2006 and earlier models

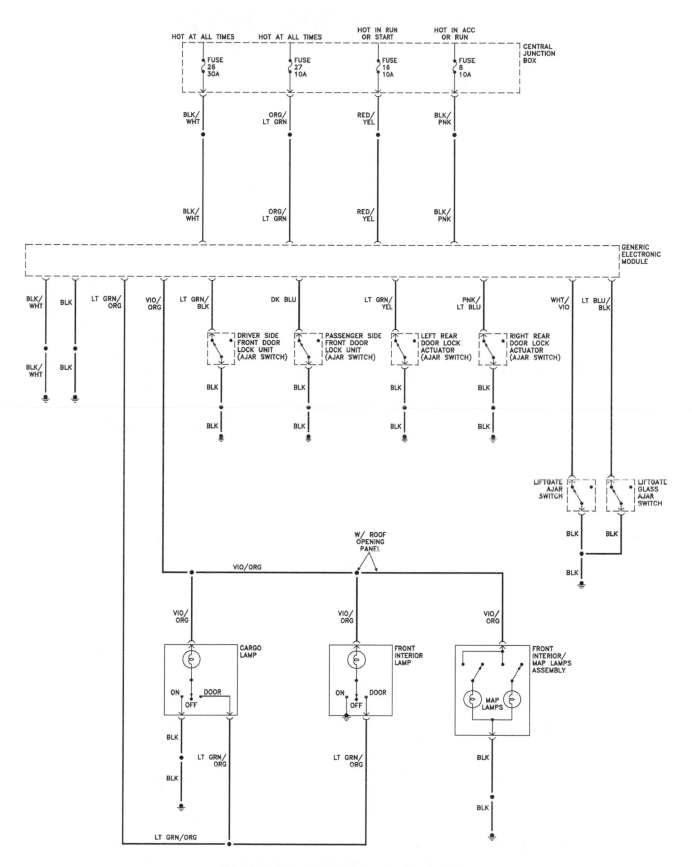

Courtesy light system - 2002 and earlier models

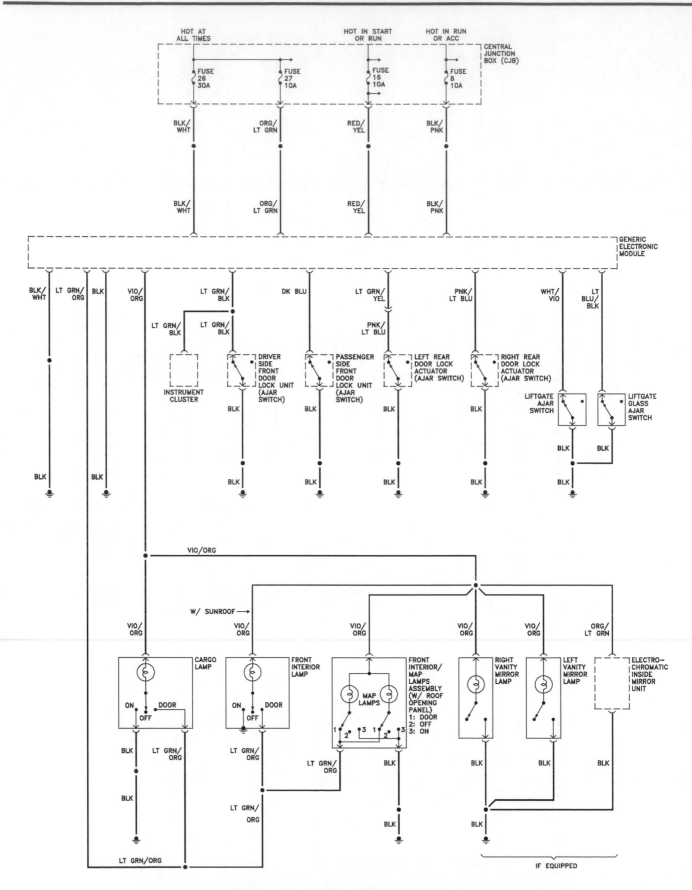

Courtesy light system - 2003 through 2006 models

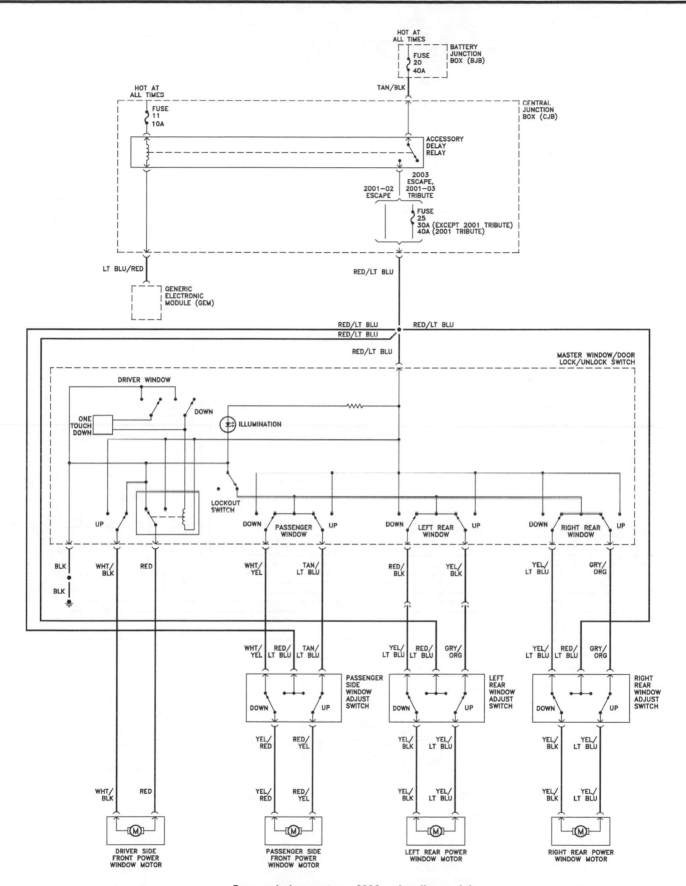

Power window system - 2006 and earlier models

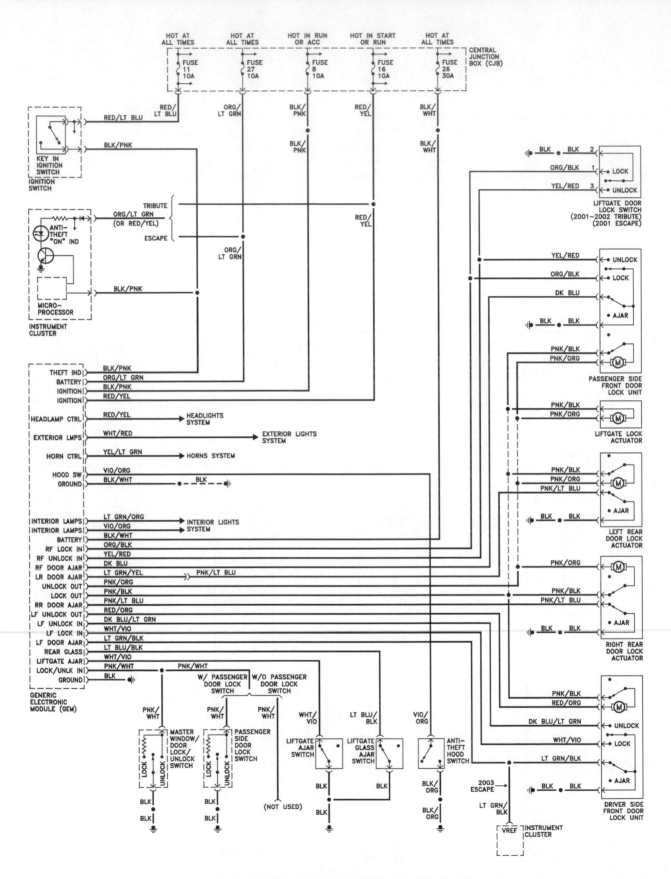

Power door lock system - 2006 and earlier models

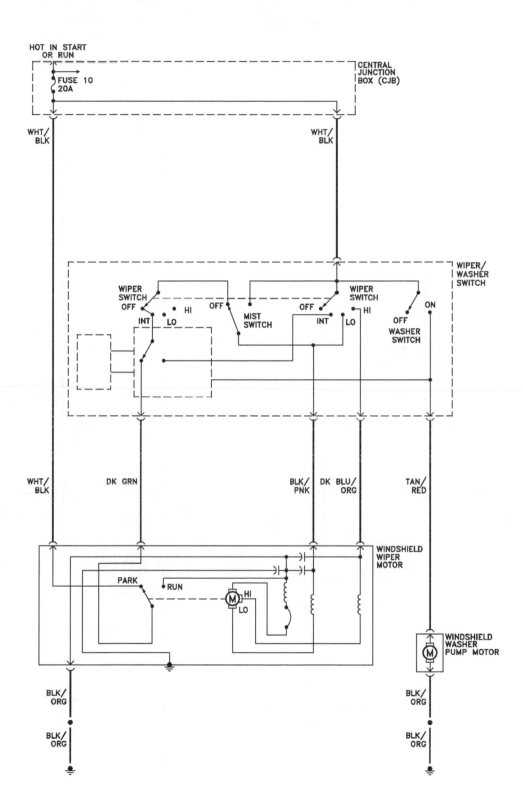

Windshield wiper/washer system - 2006 and earlier models

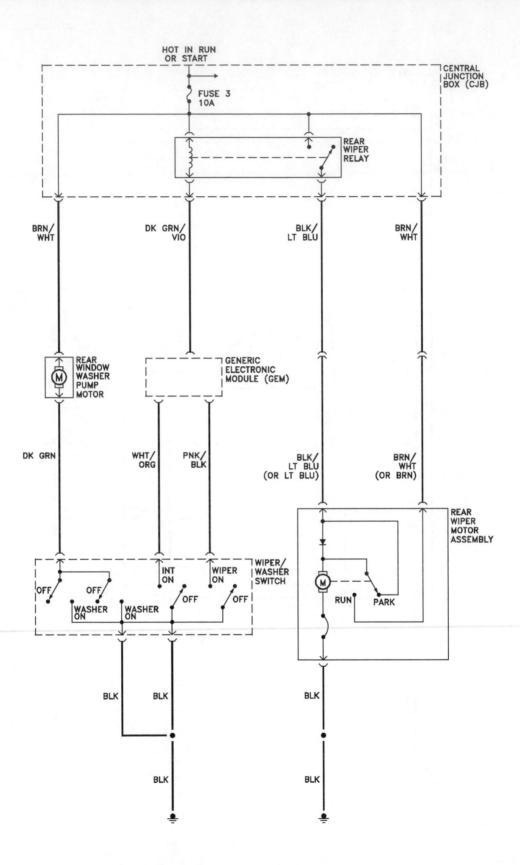

Rear wiper/washer system - 2006 and earlier models

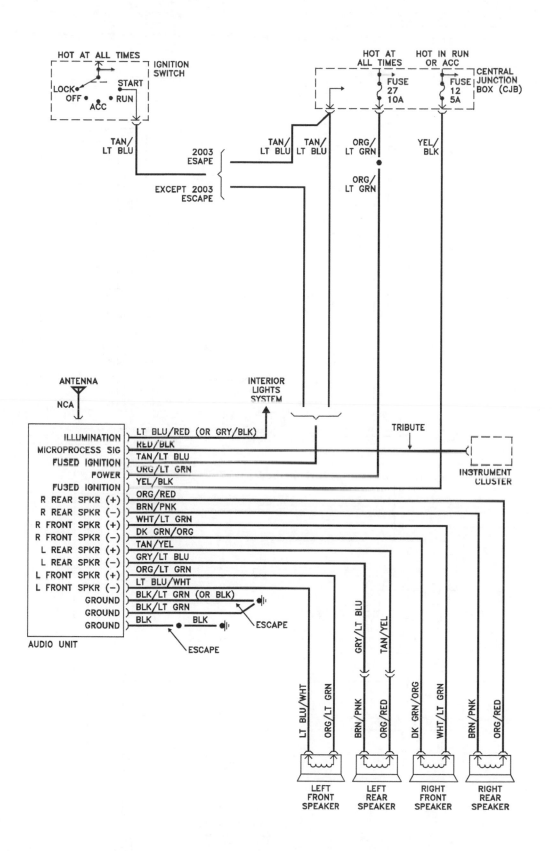

Audio system - 2006 and earlier models

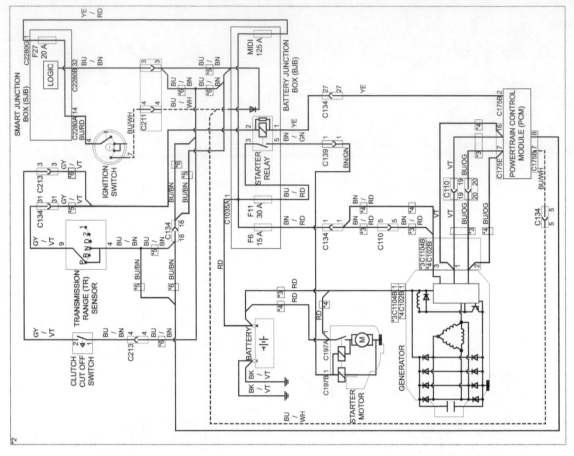

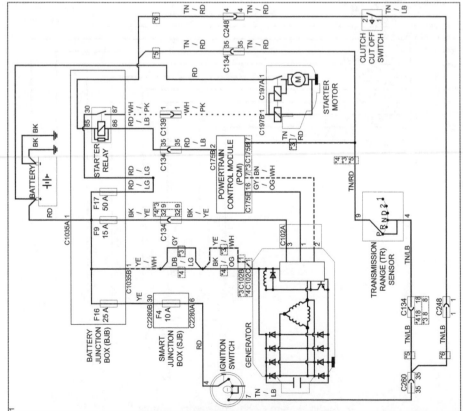

Starting and charging systems - 2007 and 2008 models

*1 For 2007
*2 For 2008
*3 Engine: 3.0 L
*4 Engine: 2.3 L
*5 Automatic transmission
*6 Manual transmission

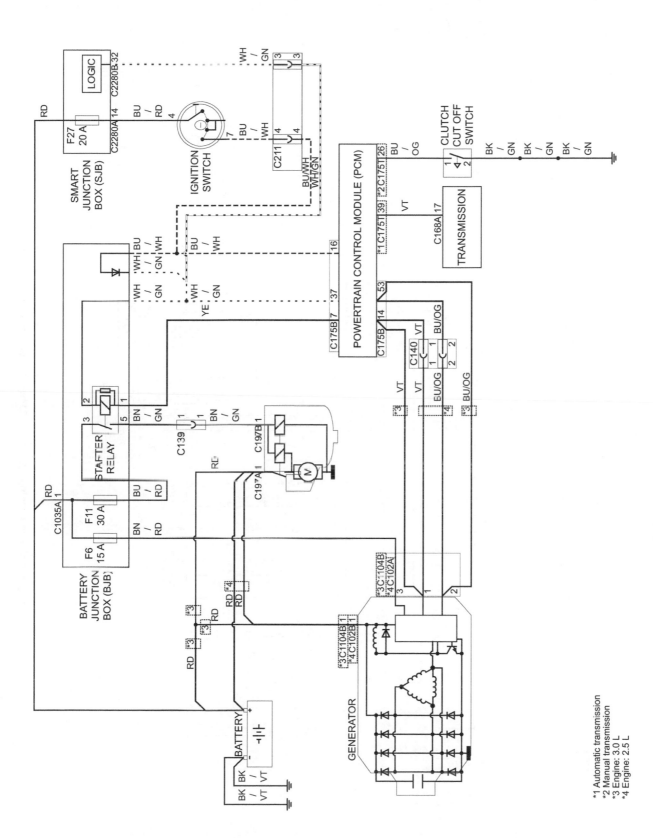

*1 Automatic transmission
*2 Manual transmission
*3 Engine: 3.0 L
*4 Engine: 2.5 L

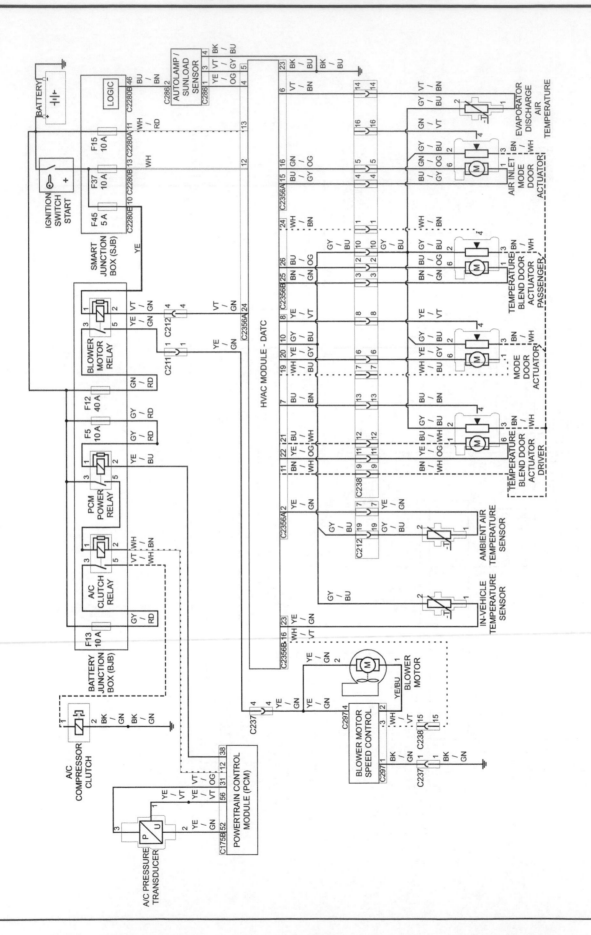

Heating and air conditioning systems (automatic) - 2008 through 2012 models

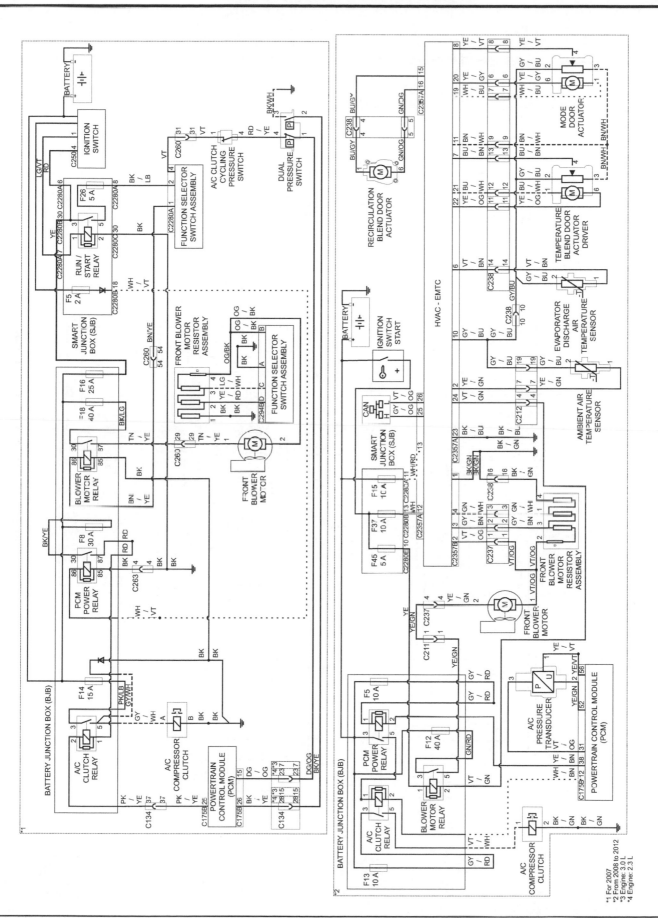

Heating and air conditioning systems (manual) - 2007 through 2012 models

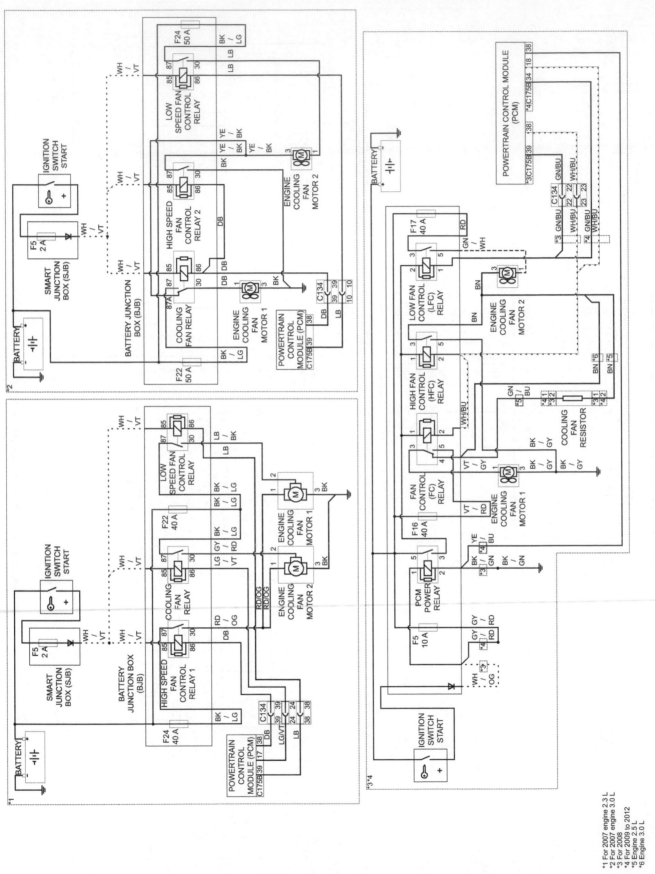

Engine cooling fan system - 2007 through 2012 models

*1 For 2007 engine 2.3 L
*2 For 2007 engine 3.0 L
*3 For 2008
*4 For 2009 to 2012
*5 Engine 2.5 L
*6 Engine 3.0 L

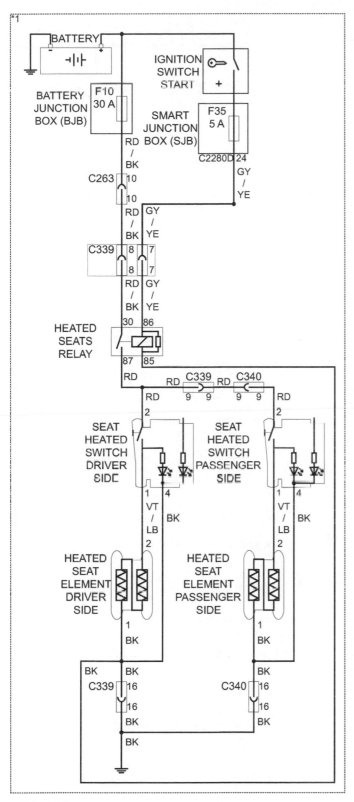
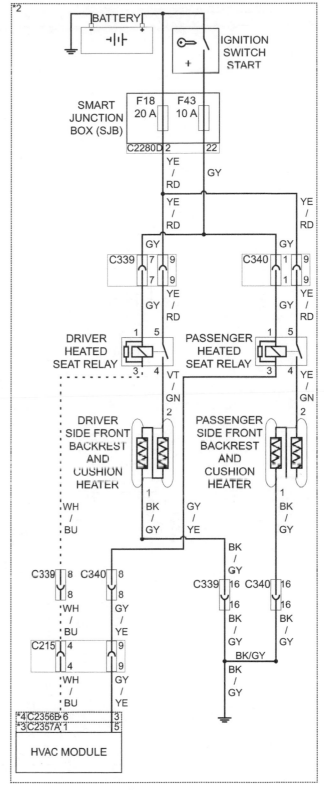

*1 For 2007
*2 From 2008 to 2012
*3 Manual air conditioning
*4 Automatic air conditioning

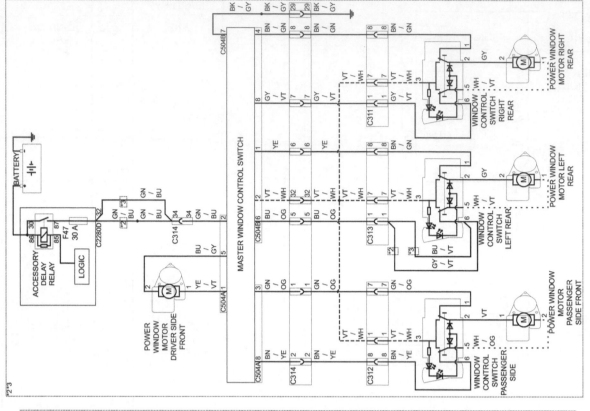

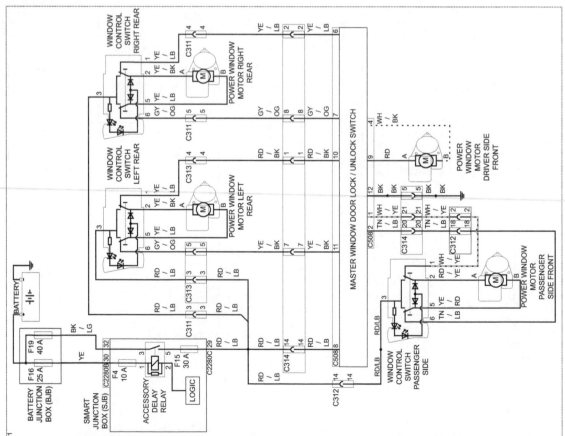

Power window system - 2007 through 2012 models

*1 For 2007
*2 For 2008
*3 From 2009 to 2012

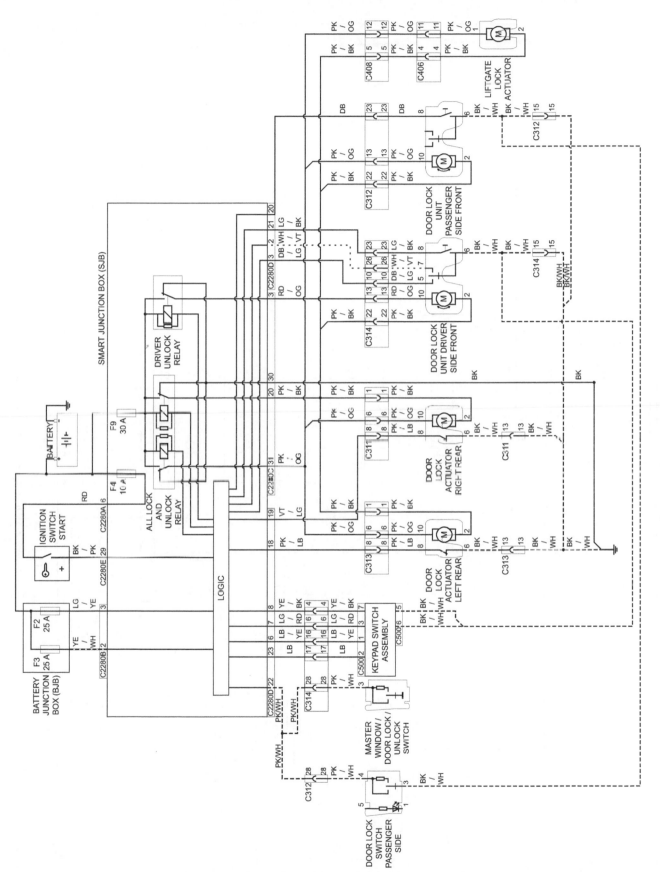

Power door lock system - 2007 models

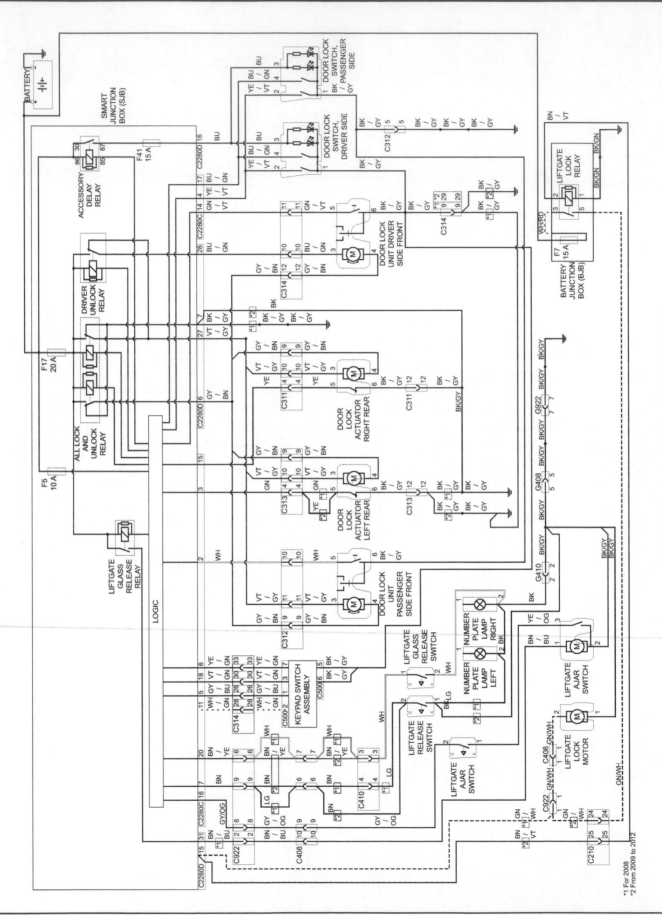

Power door lock system - 2008 through 2012 models

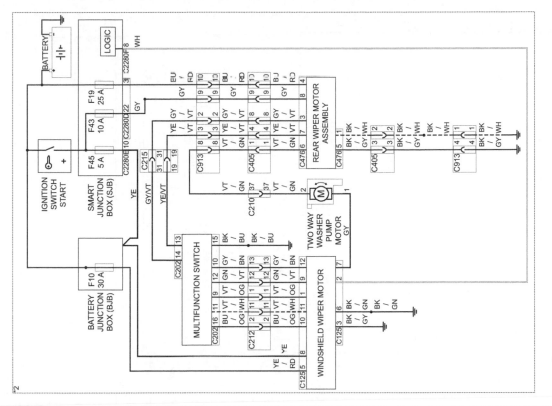

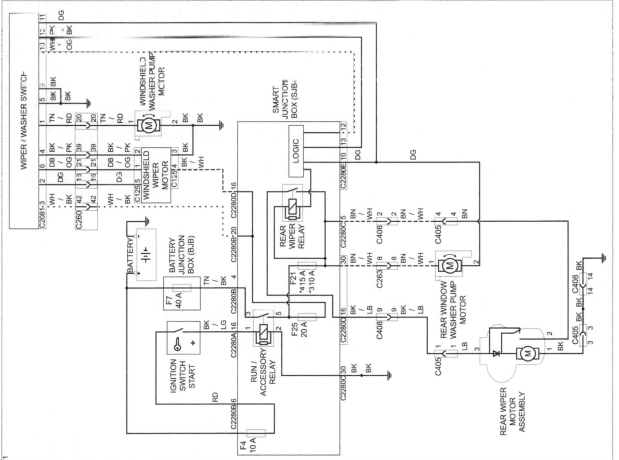

Windshield and rear window wiper/washer systems - 2007 through 2012 models

*1 For 2007
*2 From 2008 to 2012
*3 Early production
*4 Late production

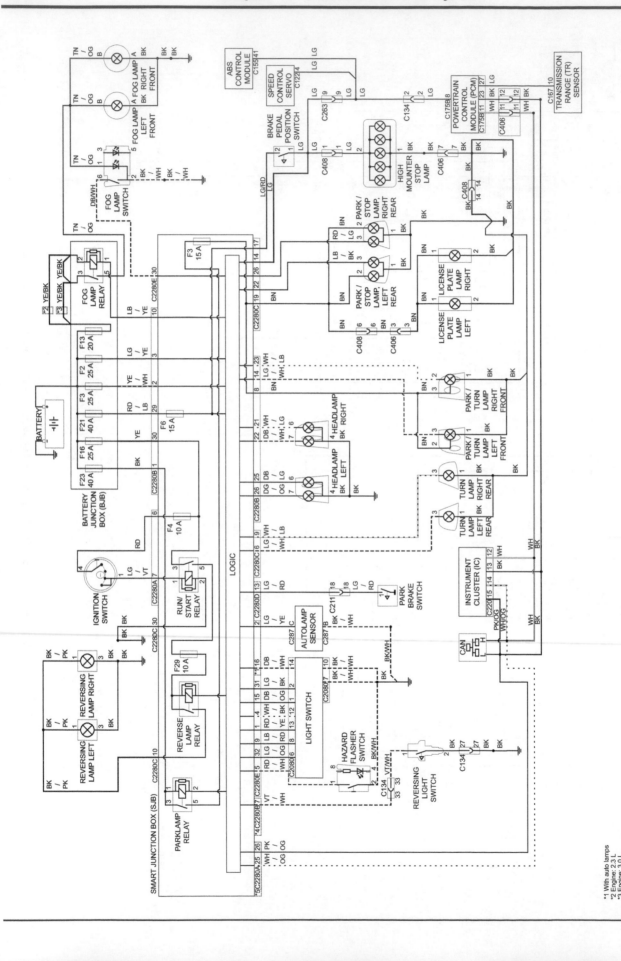

Exterior lighting system - 2007 models

*1 With auto lamps
*2 Engine: 2.3 L
*3 Engine: 3.0 L
*4 Manual transmission
*5 Automatic transmission

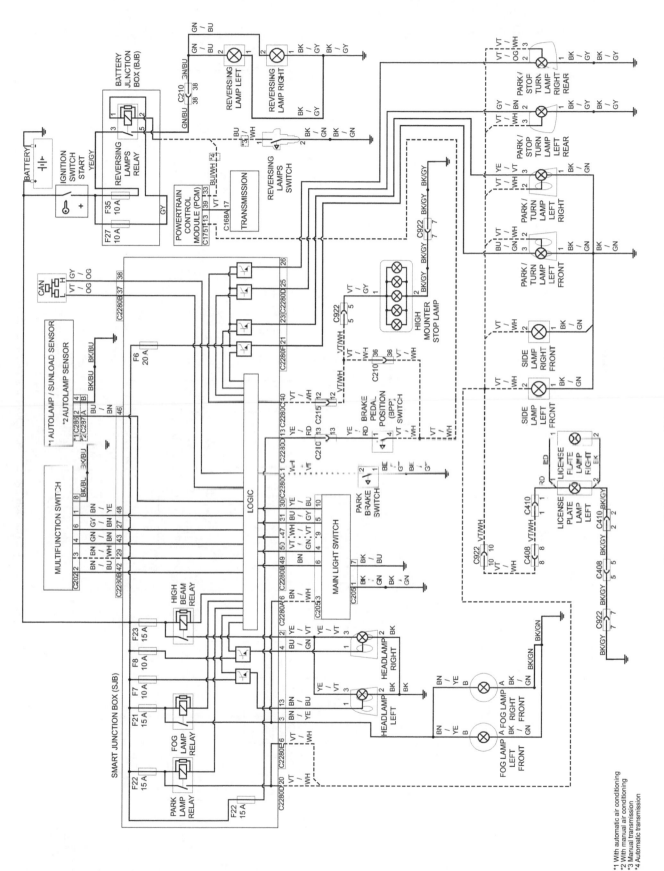

Exterior lighting system - 2008 through 2012 models

*1 With automatic air conditioning
*2 With manual air conditioning
*3 Manual transmission
*4 Automatic transmission

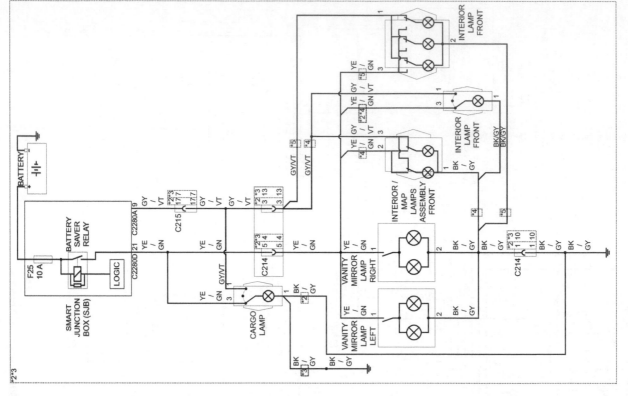

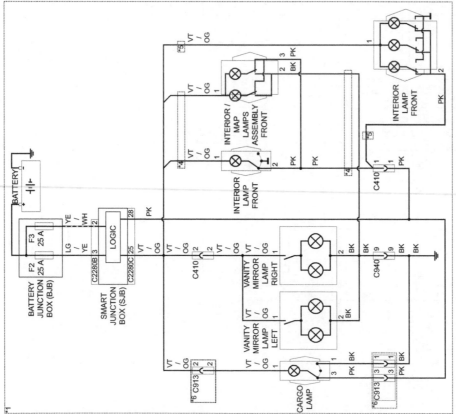

Interior lighting system - 2007 through 2012 models

*1 For 2007
*2 From 2008 to 2009
*3 From 2010 to 2012
*4 With roof opening panel
*5 Without roof opening panel
*6 Without side air curtains

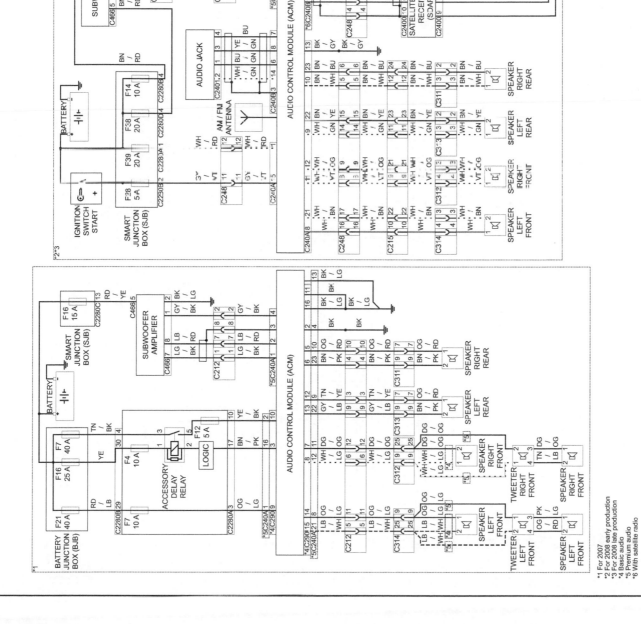

Audio system - 2007 and 2008 models

*1 For 2007
*2 For 2008 early production
*3 For 2008 late production
*4 Basic audio
*5 Premium audio
*6 With satellite radio

Audio system - 2009 through 2012 models

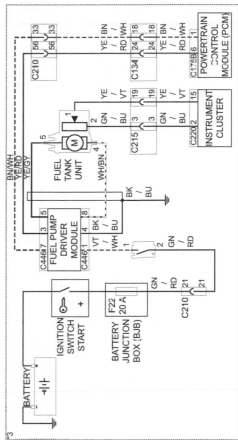

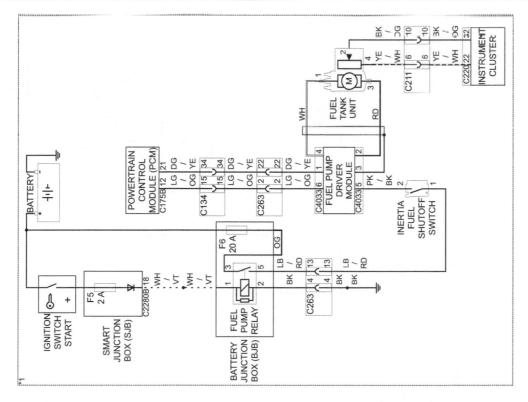

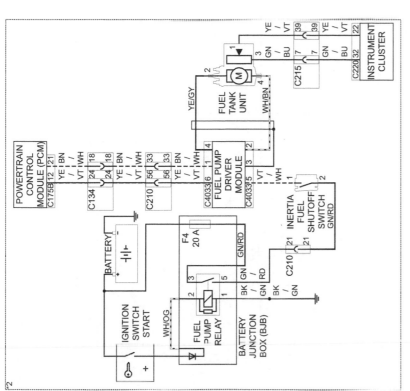

Fuel pump system - 2007 through 2012 models

*1 For 2007
*2 For 2008
*3 From 2009 to 2012

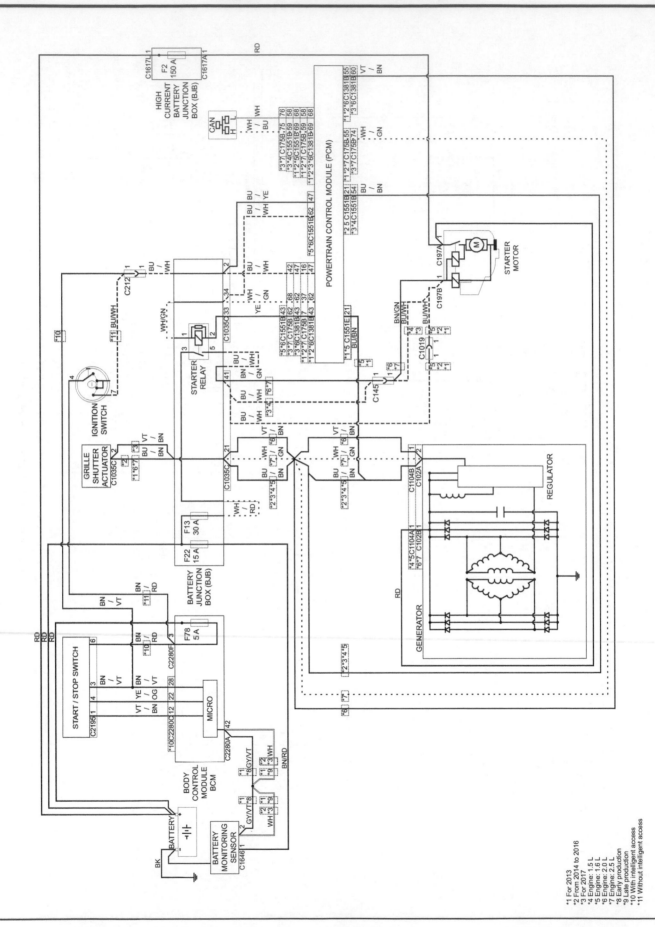

Starting and charging systems - 2013 and later models

*1 For 2013
*2 From 2014 to 2016
*3 For 2017
*4 Engine: 1.5 L
*5 Engine: 1.6 L
*6 Engine: 2.0 L
*7 Engine: 2.5 L
*8 Early production
*9 Late production
*10 With intelligent access
*11 Without intelligent access

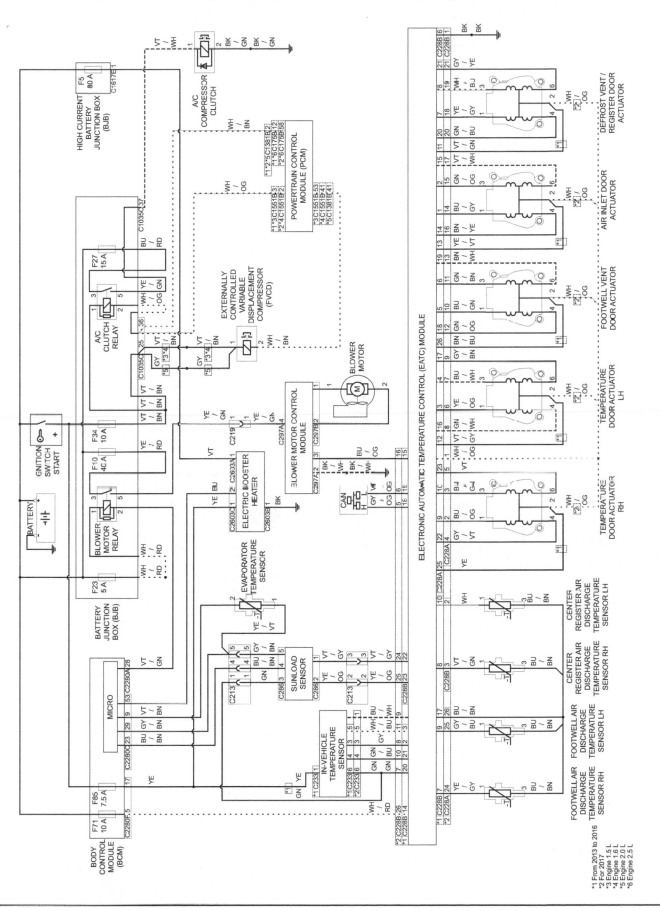

Heating and air conditioning systems (automatic) - 2013 and later models

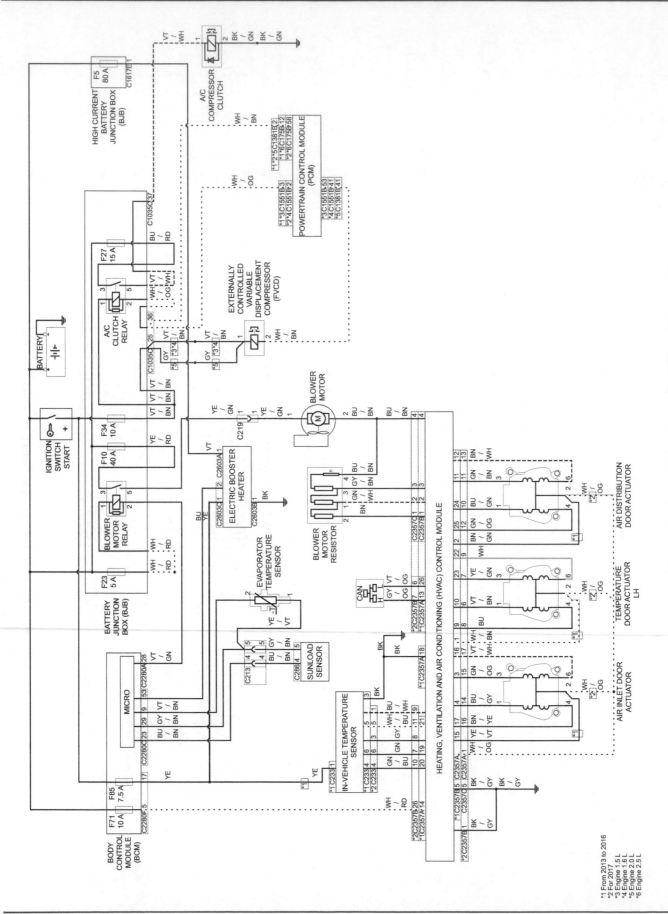

Heating and air conditioning systems - (manual) - 2013 and later models

*1 From 2013 to 2016
*2 For 2017
*3 Engine 1.5 L
*4 Engine 1.6 L
*5 Engine 2.0 L
*6 Engine 2.5 L

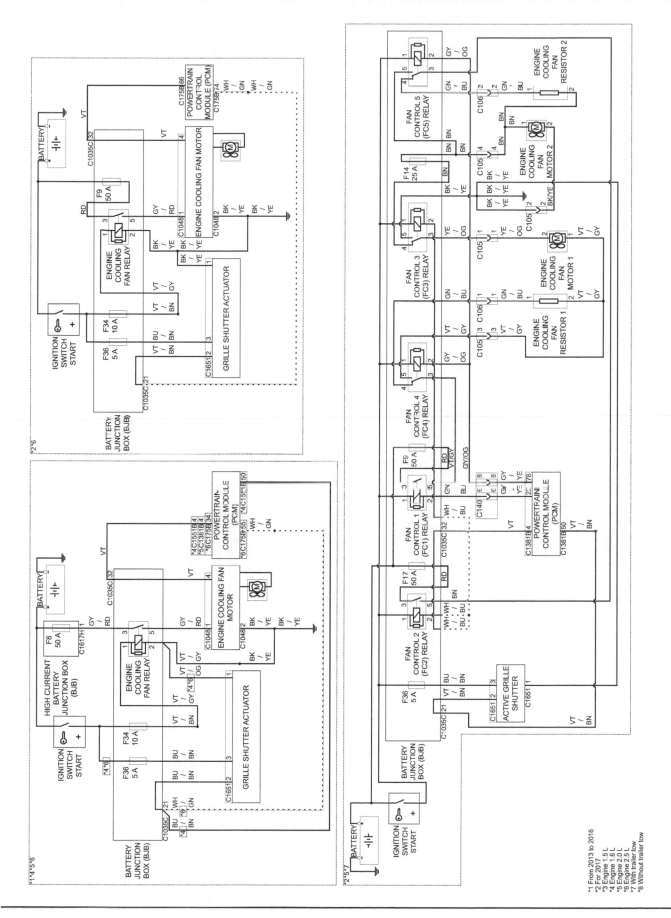

Engine cooling fan system - 2013 and later models (1 of 2)

*1 From 2013 to 2016
*2 For 2017
*3 Engine 1.5 L
*4 Engine 1.6 L
*5 Engine 2.0 L
*6 Engine 2.5 L
*7 With trailer tow
*8 Without trailer tow

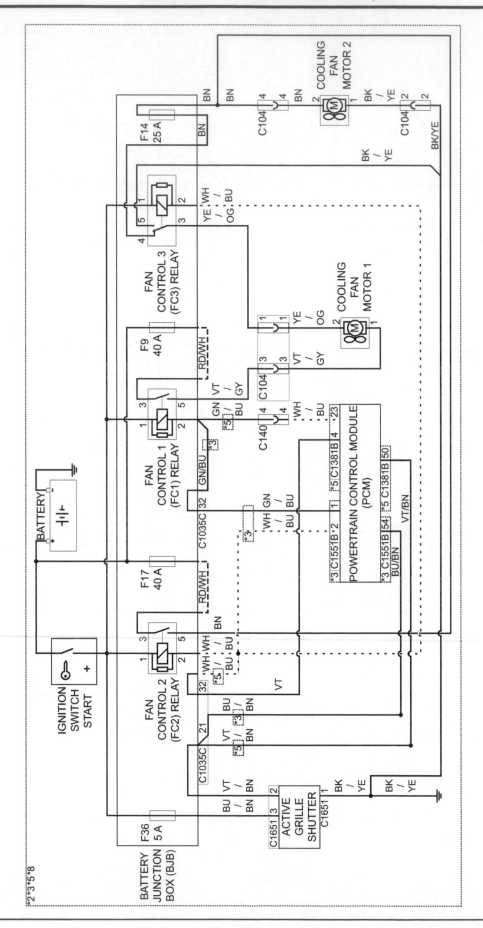

Engine cooling fan system - 2013 and later models (2 of 2)

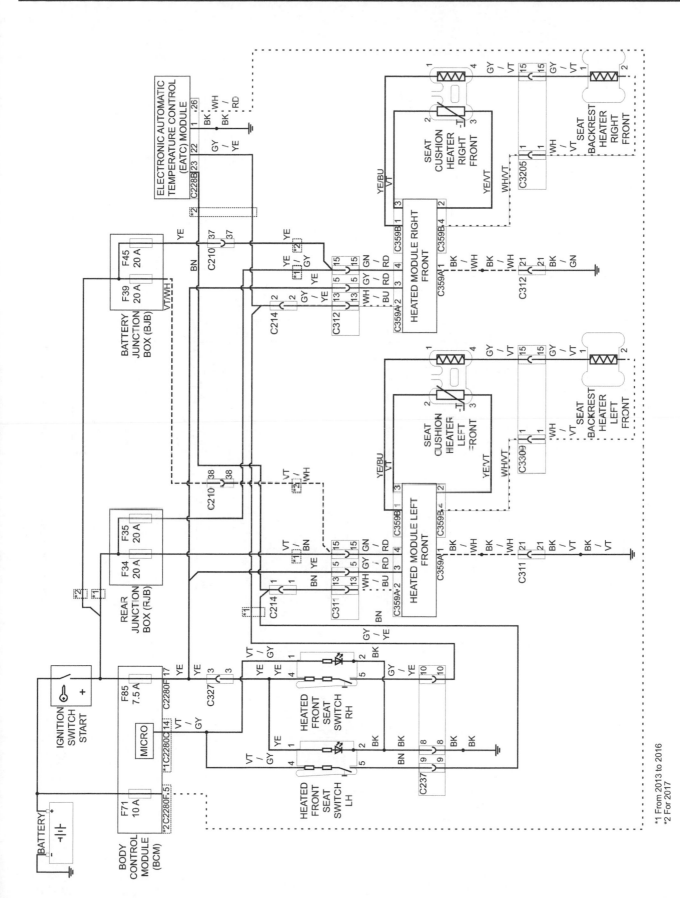

Heated seats system - 2013 and later models

*1 From 2013 to 2016
*2 For 2017

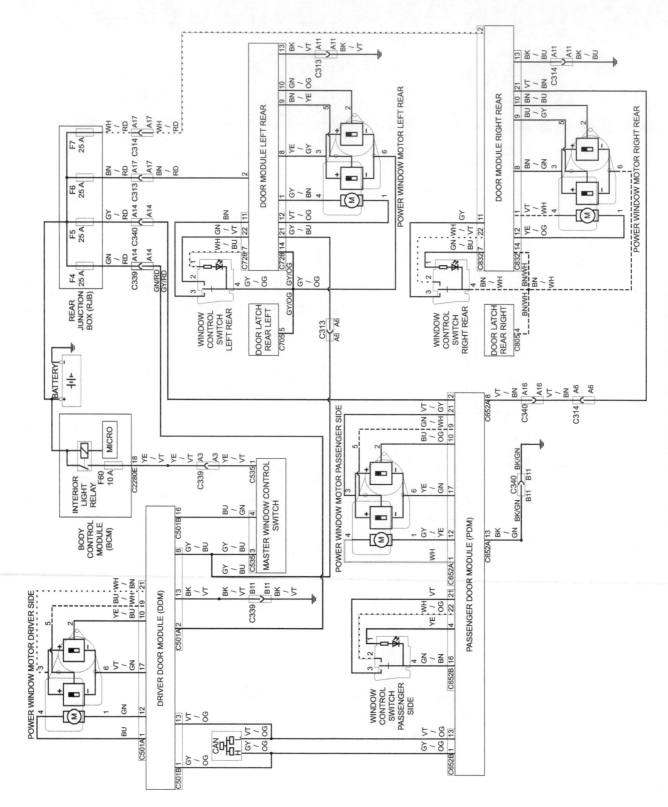

Power window system (with door module) - 2013 and later models

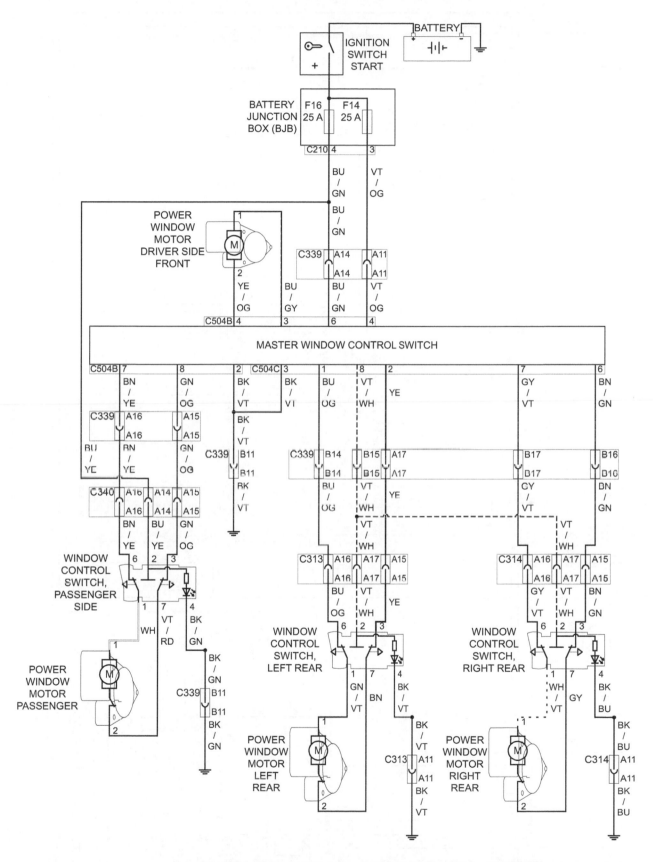

Power window system (without door module) - 2013 and later models

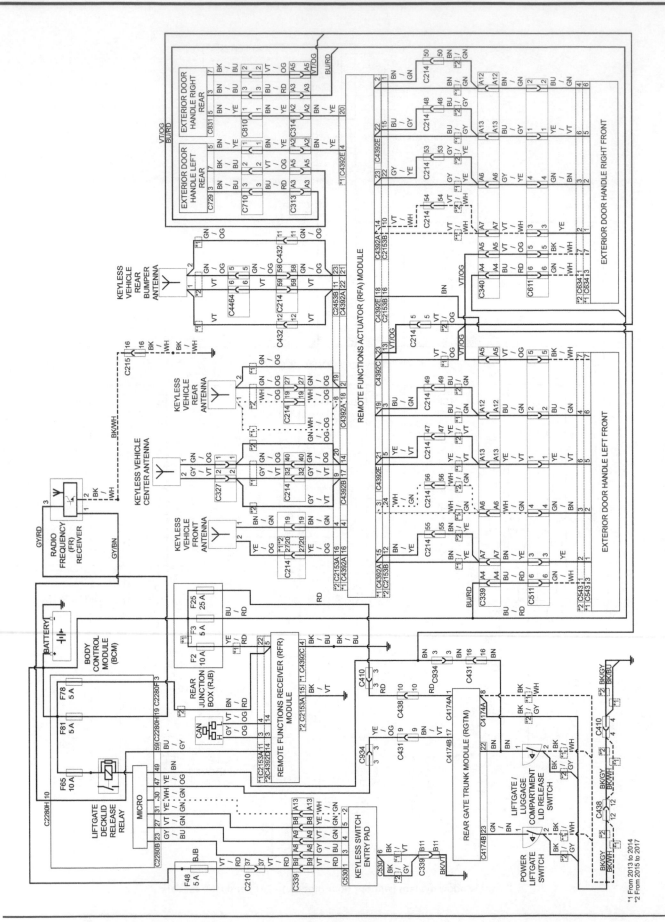

Power door lock system (keyless entry) - 2013 and later models

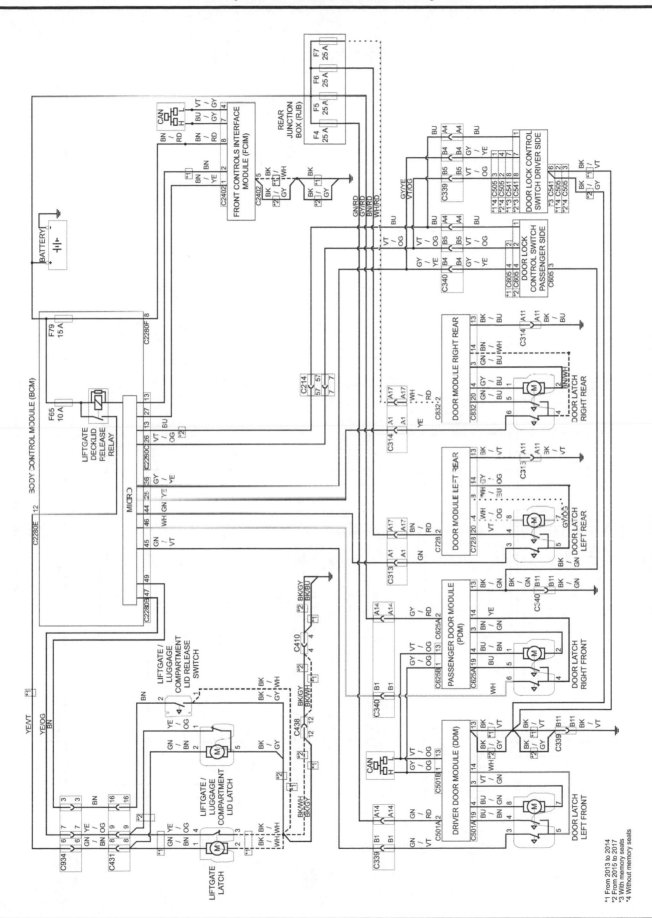

Power door lock system (with door module) - 2013 and later models

*1 From 2013 to 2014
*2 From 2015 to 2017
*3 With memory seats
*4 Without memory seats

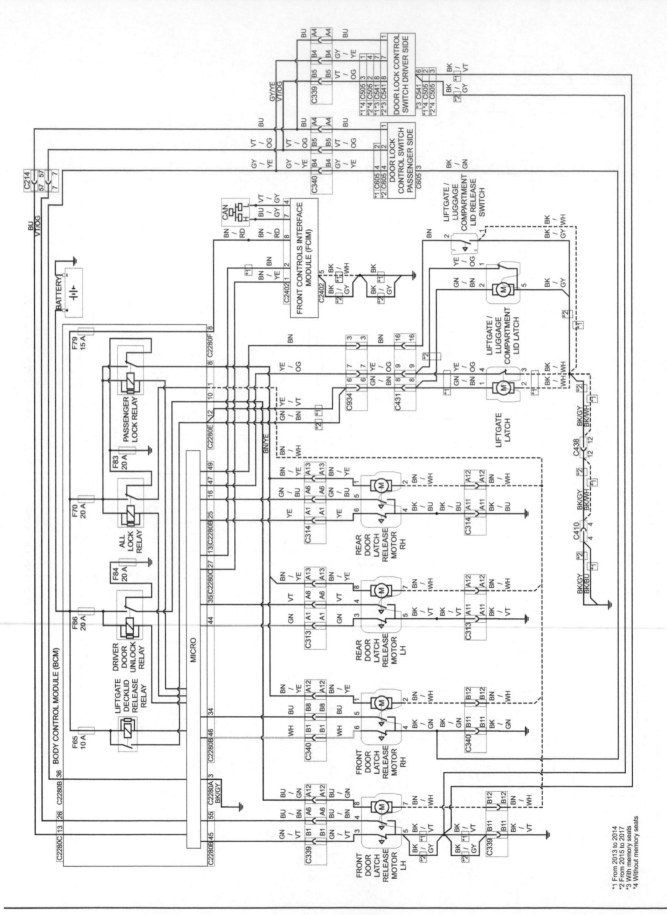

Power door lock system (without door module) - 2013 and later models

*1 From 2013 to 2014
*2 From 2015 to 2017
*3 With memory seats
*4 Without memory seats

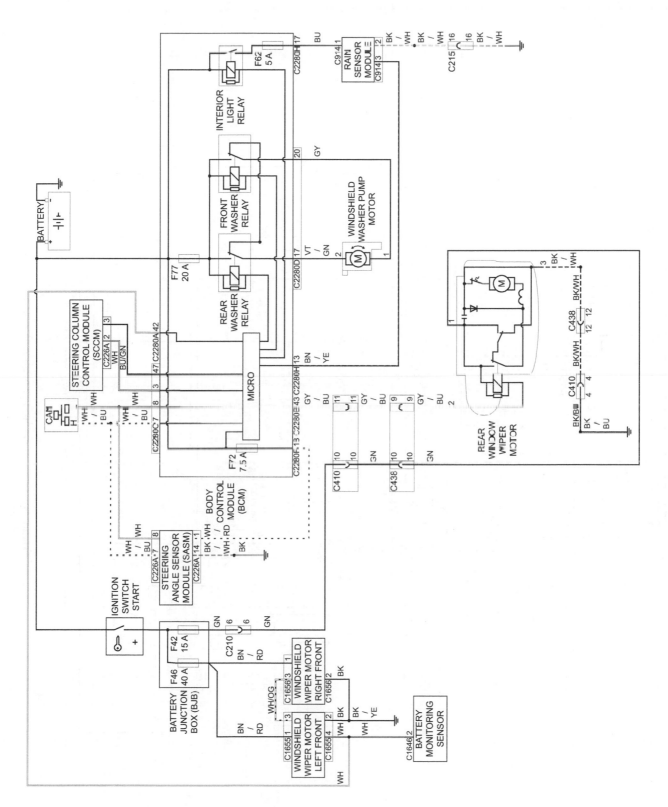

Windshield and rear window wiper/washer systems - 2013 and later models

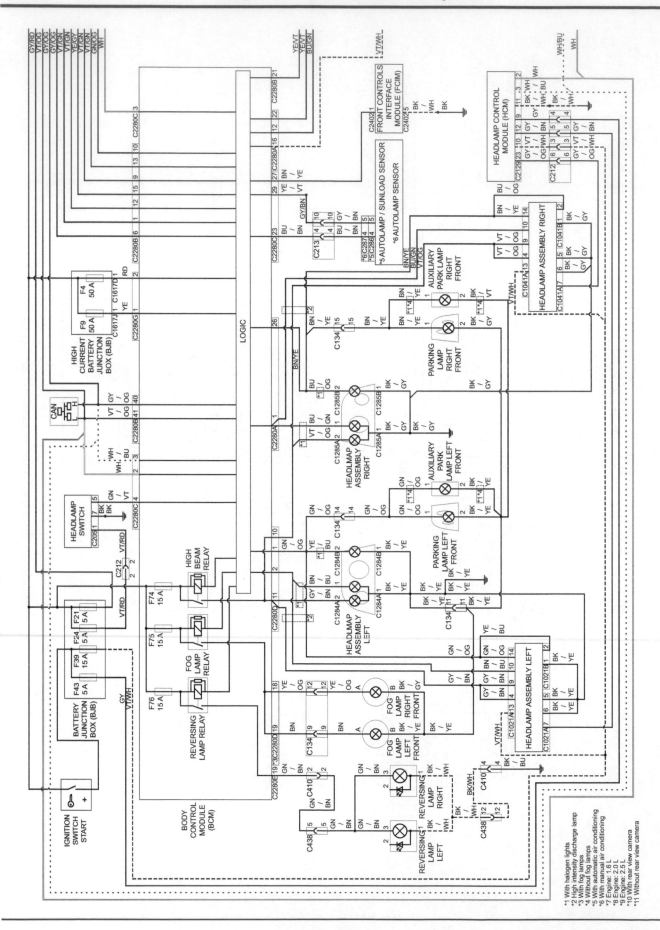

Exterior lighting system - 2013 through 2016 models (1 of 2)

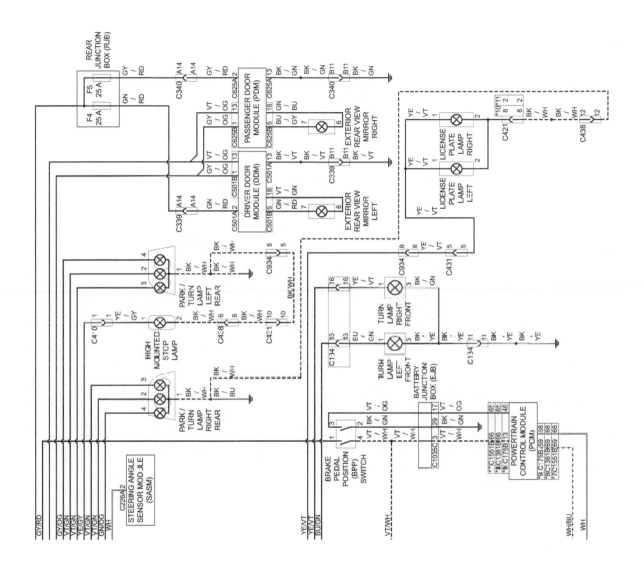

Exterior lighting system - 2013 through 2016 models (2 of 2)

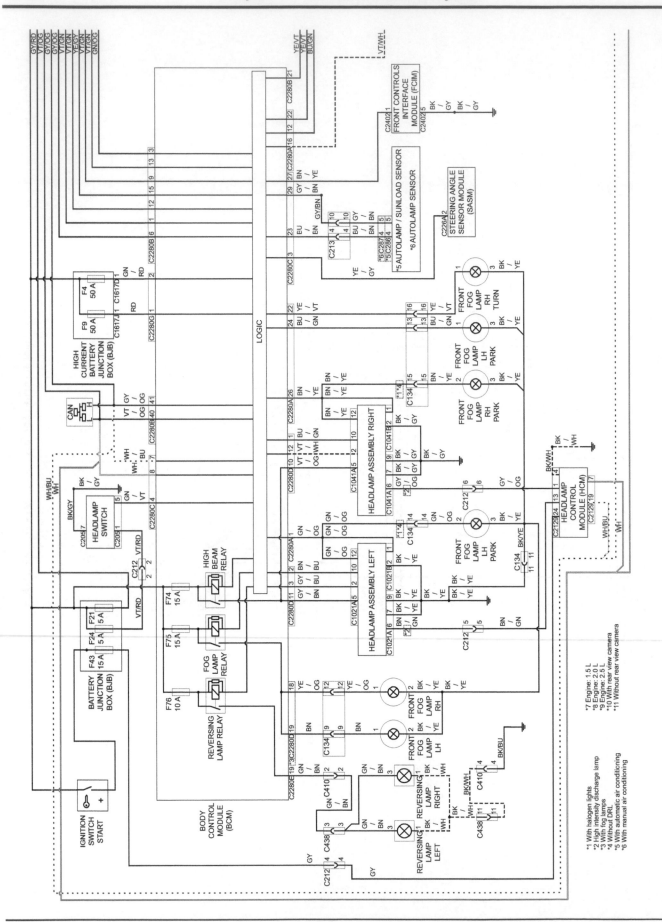

Exterior lighting system - 2017 models (1 of 2)

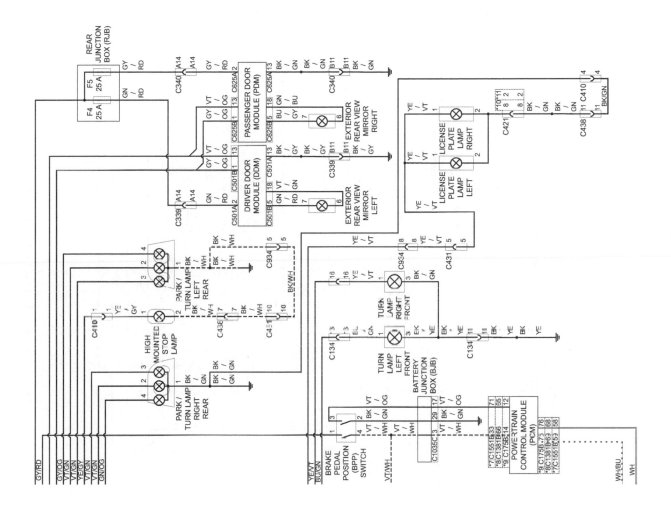

Exterior lighting system - 2017 models (2 of 2)

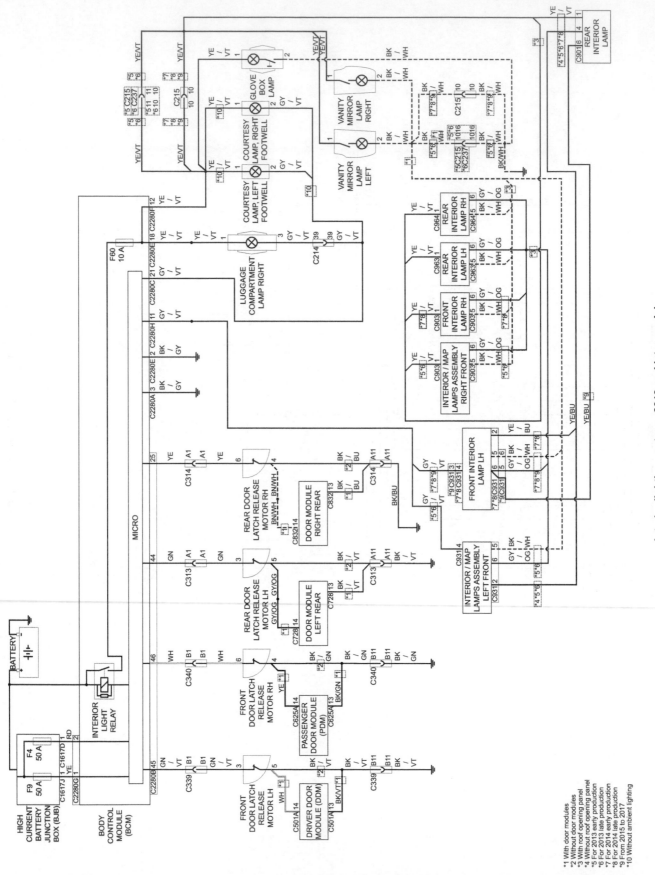

Interior lighting system - 2013 and later models

*1 With door modules
*2 Without door modules
*3 With roof opening panel
*4 Without roof opening panel
*5 For 2013 early production
*6 For 2013 late production
*7 For 2014 early production
*8 For 2014 late production
*9 From 2015 to 2017
*10 Without ambient lighting

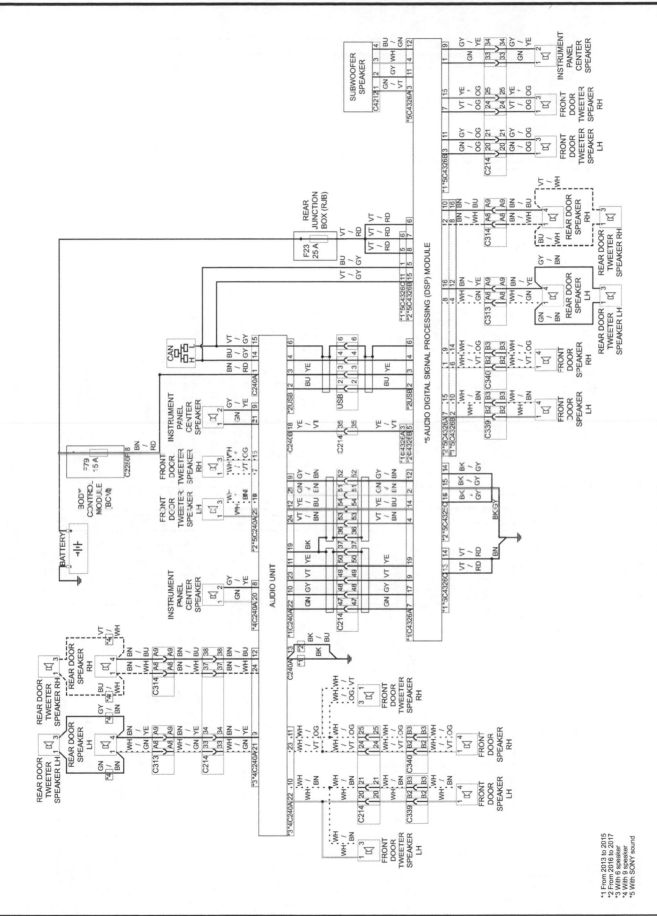

Audio system/speakers - 2013 and later models

*1 From 2013 to 2015
*2 From 2016 to 2017
*3 With 6 speaker
*4 With 9 speaker
*5 With SONY sound

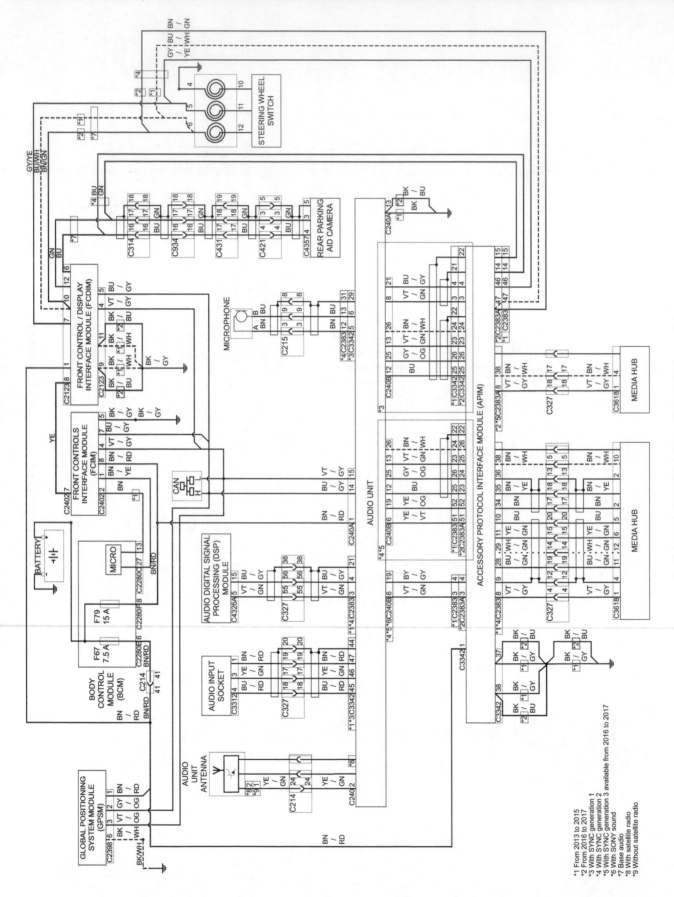

Audio system/SYNC/satellite - 2013 and later models

*1 From 2013 to 2015
*2 From 2016 to 2017
*3 With SYNC generation 1
*4 With SYNC generation 2
*5 With SYNC generation 3 available from 2016 to 2017
*6 With SONY sound
*7 Base audio
*8 With satellite radio
*9 Without satellite radio

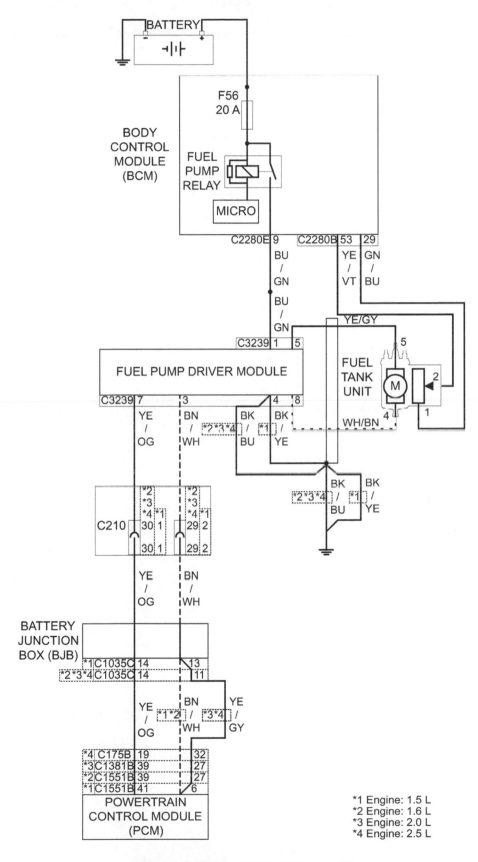

Fuel pump system - 2013 and later models

Notes

Notes

Index

Haynes Automotive Manuals

ACURA
12020 Integra '86 thru '89 & **Legend** '86 thru '90
12021 Integra '90 thru '93 & **Legend** '91 thru '95
Integra '94 thru '00 - see HONDA Civic (42025)
MDX '01 thru '07 - see HONDA Pilot (42037)
12050 Acura TL all models '99 thru '08

AMC
14020 Mid-size models '70 thru '83
14025 (Renault) Alliance & Encore '83 thru '87

AUDI
15020 4000 all models '80 thru '87
15025 5000 all models '77 thru '83
15026 5000 all models '84 thru '88
Audi A4 '96 thru '01 - see VW Passat (96023)
15030 Audi A4 '02 thru '08

AUSTIN-HEALEY
Sprite - see MG Midget (66015)

BMW
18020 3/5 Series '82 thru '92
18021 3-Series incl. Z3 models '92 thru '98
18022 3-Series incl. Z4 models '99 thru '05
18023 3-Series '06 thru '14
18025 320i all 4-cylinder models '75 thru '83
18050 1500 thru 2002 except Turbo '59 thru '77

BUICK
19010 Buick Century '97 thru '05
Century (front-wheel drive) - see GM (38005)
19020 Buick, Oldsmobile & Pontiac Full-size
(Front-wheel drive) '85 thru '05
Buick Electra, LeSabre and Park Avenue;
Oldsmobile Delta 88 Royale, Ninety Eight
and Regency; Pontiac Bonneville
19025 Buick, Oldsmobile & Pontiac Full-size
(Rear wheel drive) '70 thru '90
Buick Estate, Electra, LeSabre, Limited,
Oldsmobile Custom Cruiser, Delta 88,
Ninety-eight, Pontiac Bonneville,
Catalina, Grandville, Parisienne
19027 Buick LaCrosse '05 thru '13
Enclave - see GENERAL MOTORS (38001)
Rainier - see CHEVROLET (24072)
Regal - see GENERAL MOTORS (38010)
Riviera - see GENERAL MOTORS (38030, 38031)
Roadmaster - see CHEVROLET (24046)
Skyhawk - see GENERAL MOTORS (38015)
Skylark - see GENERAL MOTORS (38020, 38025)
Somerset - see GENERAL MOTORS (38025)

CADILLAC
21015 CTS & CTS-V '03 thru '14
21030 Cadillac Rear Wheel Drive '70 thru '93
Cimarron - see GENERAL MOTORS (38015)
DeVille - see GENERAL MOTORS (38031 & 38032)
Eldorado - see GENERAL MOTORS (38030)
Fleetwood - see GENERAL MOTORS (38031)
Seville - see GM (38030, 38031 & 38032)

CHEVROLET
10305 Chevrolet Engine Overhaul Manual
24010 Astro & GMC Safari Mini-vans '85 thru '05
24013 Aveo '04 thru '11
24015 Camaro V8 all models '70 thru '81
24016 Camaro all models '82 thru '92
24017 Camaro & Firebird '93 thru '02
Cavalier - see GENERAL MOTORS (38016)
Celebrity - see GENERAL MOTORS (38005)
24018 Camaro '10 thru '15
24020 Chevelle, Malibu & El Camino '69 thru '87
Cobalt - see GENERAL MOTORS (38017)
24024 Chevette & Pontiac T1000 '76 thru '87
Citation - see GENERAL MOTORS (38020)
24027 Colorado & GMC Canyon '04 thru '12
24032 Corsica & Beretta all models '87 thru '96
24040 Corvette all V8 models '68 thru '82
24041 Corvette all models '84 thru '96
24042 Corvette all models '97 thru '13
24044 Cruze '11 thru '19
24045 Full-size Sedans Caprice, Impala, Biscayne,
Bel Air & Wagons '69 thru '90
24046 Impala SS & Caprice and Buick Roadmaster
'91 thru '96
Impala '00 thru '05 - see LUMINA (24048)
24047 Impala & Monte Carlo all models '06 thru '11
Lumina '90 thru '94 - see GM (38010)
24048 Lumina & Monte Carlo '95 thru '05
Lumina APV - see GM (38035)
24050 Luv Pick-up all 2WD & 4WD '72 thru '82
24051 Malibu '13 thru '19
24055 Monte Carlo all models '70 thru '88
Monte Carlo '95 thru '01 - see LUMINA (24048)
24059 Nova all V8 models '69 thru '79

24060 Nova and Geo Prizm '85 thru '92
24064 Pick-ups '67 thru '87 - Chevrolet & GMC
24065 Pick-ups '88 thru '98 - Chevrolet & GMC
24066 Pick-ups '99 thru '06 - Chevrolet & GMC
24067 Chevrolet Silverado & GMC Sierra '07 thru '14
24068 Chevrolet Silverado & GMC Sierra '14 thru '19
24070 S-10 & S-15 Pick-ups '82 thru '93,
Blazer & Jimmy '83 thru '94,
24071 S-10 & Sonoma Pick-ups '94 thru '04,
including Blazer, Jimmy & Hombre
24072 Chevrolet TrailBlazer, GMC Envoy &
Oldsmobile Bravada '02 thru '09
24075 Sprint '85 thru '88 & Geo Metro '89 thru '01
24080 Vans - Chevrolet & GMC '68 thru '96
24081 Chevrolet Express & GMC Savana
Full-size Vans '96 thru '19

CHRYSLER
10310 Chrysler Engine Overhaul Manual
25015 Chrysler Cirrus, Dodge Stratus,
Plymouth Breeze '95 thru '00
25020 Full-size Front-Wheel Drive '88 thru '93
K-Cars - see DODGE Aries (30008)
Laser - see DODGE Daytona (30030)
25025 Chrysler LHS, Concorde, New Yorker,
Dodge Intrepid, Eagle Vision, '93 thru '97
25026 Chrysler LHS, Concorde, 300M,
Dodge Intrepid, '98 thru '04
25027 Chrysler 300 '05 thru '18, Dodge Charger
'06 thru '18, Magnum '05 thru '08 &
Challenger '08 thru '18
25030 Chrysler & Plymouth Mid-size
front wheel drive '82 thru '95
Rear-wheel Drive - see Dodge (30050)
25035 PT Cruiser all models '01 thru '10
25040 Chrysler Sebring '95 thru '06, Dodge Stratus
'01 thru '06 & Dodge Avenger '95 thru '00
25041 Chrysler Sebring '07 thru '10, 200 '11 thru '17
Dodge Avenger '08 thru '14

DATSUN
28005 200SX all models '80 thru '83
28012 240Z, 260Z & 280Z Coupe '70 thru '78
28014 280ZX Coupe & 2+2 '79 thru '83
300ZX - see NISSAN (72010)
28018 510 & PL521 Pick-up '68 thru '73
28020 510 all models '78 thru '81
28022 620 Series Pick-up all models '73 thru '79
720 Series Pick-up - see NISSAN (72030)

DODGE
400 & 600 - see CHRYSLER (25030)
30008 Aries & Plymouth Reliant '81 thru '89
30010 Caravan & Plymouth Voyager '84 thru '95
30011 Caravan & Plymouth Voyager '96 thru '02
30012 Challenger & Plymouth Sapporro '78 thru '83
30013 Caravan, Chrysler Voyager &
Town & Country '03 thru '07
30014 Grand Caravan &
Chrysler Town & Country '08 thru '18
30016 Colt & Plymouth Champ '78 thru '87
30020 Dakota Pick-ups all models '87 thru '96
30021 Durango '98 & '99 & Dakota '97 thru '99
30022 Durango '00 thru '03 & Dakota '00 thru '04
30023 Durango '04 thru '09 & Dakota '05 thru '11
30025 Dart, Demon, Plymouth Barracuda,
Duster & Valiant 6-cylinder models '67 thru '76
30030 Daytona & Chrysler Laser '84 thru '89
Intrepid - see CHRYSLER (25025, 25026)
30034 Neon all models '95 thru '99
30035 Omni & Plymouth Horizon '78 thru '90
30036 Dodge & Plymouth Neon '00 thru '05
30040 Pick-ups full-size models '74 thru '93
30042 Pick-ups full-size models '94 thru '08
30043 Pick-ups full-size models '09 thru '18
30045 Ram 50/D50 Pick-ups & Raider and
Plymouth Arrow Pick-ups '79 thru '93
30050 Dodge/Plymouth/Chrysler RWD '71 thru '89
30055 Shadow & Plymouth Sundance '87 thru '94
30060 Spirit & Plymouth Acclaim '89 thru '95
30065 Vans - Dodge & Plymouth '71 thru '03

EAGLE
Talon - see MITSUBISHI (68030, 68031)
Vision - see CHRYSLER (25025)

FIAT
34010 124 Sport Coupe & Spider '68 thru '78
34025 X1/9 all models '74 thru '80

FORD
10320 Ford Engine Overhaul Manual
10355 Ford Automatic Transmission Overhaul
11500 Mustang '64-1/2 thru '70 Restoration Guide
36004 Aerostar Mini-vans all models '86 thru '97
36006 Contour & Mercury Mystique '95 thru '00
36008 Courier Pick-up all models '72 thru '82

36012 Crown Victoria &
Mercury Grand Marquis '88 thru '11
36014 Edge '07 thru '19 & Lincoln MKX '07 thru '18
36016 Escort & Mercury Lynx all models '81 thru '90
36020 Escort & Mercury Tracer '91 thru '02
36022 Escape '01 thru '17, Mazda Tribute '01 thru '11,
& Mercury Mariner '05 thru '11
36024 Explorer & Mazda Navajo '91 thru '01
36025 Explorer & Mercury Mountaineer '02 thru '10
36026 Explorer '11 thru '17
36028 Fairmont & Mercury Zephyr '78 thru '83
36030 Festiva & Aspire '88 thru '97
36032 Fiesta all models '77 thru '80
36034 Focus all models '00 thru '11
36035 Focus '12 thru '14
36045 Fusion '06 thru '14 & Mercury Milan '06 thru '11
36048 Mustang V8 all models '64-1/2 thru '73
36049 Mustang II 4-cylinder, V6 & V8 models '74 thru '78
36050 Mustang & Mercury Capri '79 thru '93
36051 Mustang all models '94 thru '04
36052 Mustang '05 thru '14
36054 Pick-ups & Bronco '73 thru '79
36058 Pick-ups & Bronco '80 thru '96
36059 F-150 '97 thru '03, Expedition '97 thru '17,
F-250 '97 thru '99, F-150 Heritage '04
& Lincoln Navigator '98 thru '17
36060 Super Duty Pick-ups & Excursion '99 thru '10
36061 F-150 full-size '04 thru '14
36062 Pinto & Mercury Bobcat '75 thru '80
36063 F-150 full-size '15 thru '17
36064 Super Duty Pick-ups '11 thru '16
36066 Probe all models '89 thru '92
Probe '93 thru '97 - see MAZDA 626 (61042)
36070 Ranger & Bronco II gas models '83 thru '92
36071 Ranger '93 thru '11 & Mazda Pick-ups '94 thru '09
36074 Taurus & Mercury Sable '86 thru '95
36075 Taurus & Mercury Sable '96 thru '07
36076 Taurus '08 thru '14, Five Hundred '05 thru '07,
Mercury Montego '05 thru '07 & Sable '08 thru '09
36078 Tempo & Mercury Topaz '84 thru '94
36082 Thunderbird & Mercury Cougar '83 thru '88
36086 Thunderbird & Mercury Cougar '89 thru '97
36090 Vans all V8 Econoline models '69 thru '91
36094 Vans full size '92 thru '14
36097 Windstar '95 thru '03, Freestar & Mercury
Monterey Mini-van '04 thru '07

GENERAL MOTORS
10360 GM Automatic Transmission Overhaul
38001 GMC Acadia '07 thru '16, Buick Enclave
'08 thru '17, Saturn Outlook '07 thru '10
& Chevrolet Traverse '09 thru '17
38005 Buick Century, Chevrolet Celebrity,
Oldsmobile Cutlass Ciera & Pontiac 6000
all models '82 thru '96
38010 Buick Regal '88 thru '04, Chevrolet Lumina
'88 thru '04, Oldsmobile Cutlass Supreme
'88 thru '97 & Pontiac Grand Prix '88 thru '07
38015 Buick Skyhawk, Cadillac Cimarron,
Chevrolet Cavalier, Oldsmobile Firenza,
Pontiac J-2000 & Sunbird '82 thru '94
38016 Chevrolet Cavalier & Pontiac Sunfire '95 thru '05
38017 Chevrolet Cobalt '05 thru '10, HHR '06 thru '11,
Pontiac G5 '07 thru '09, Pursuit '05 thru '06
& Saturn ION '03 thru '07
38020 Buick Skylark, Chevrolet Citation,
Oldsmobile Omega, Pontiac Phoenix '80 thru '85
38025 Buick Skylark '86 thru '98, Somerset '85 thru '87,
Oldsmobile Achieva '92 thru '98, Calais '85 thru '91,
& Pontiac Grand Am all models '85 thru '98
38026 Chevrolet Malibu '97 thru '03, Classic '04 thru '05,
Oldsmobile Alero '99 thru '03, Cutlass '97 thru '00,
& Pontiac Grand Am '99 thru '03
38027 Chevrolet Malibu '04 thru '12, Pontiac G6
'05 thru '10 & Saturn Aura '07 thru '10
38030 Cadillac Eldorado, Seville, Oldsmobile
Toronado & Buick Riviera '71 thru '85
38031 Cadillac Eldorado, Seville, DeVille, Fleetwood,
Oldsmobile Toronado & Buick Riviera '86 thru '93
38032 Cadillac DeVille '94 thru '05, Seville '92 thru '04
& Cadillac DTS '06 thru '10
38035 Chevrolet Lumina APV, Oldsmobile Silhouette
& Pontiac Trans Sport all models '90 thru '96
38036 Chevrolet Venture '97 thru '05, Oldsmobile
Silhouette '97 thru '04, Pontiac Trans Sport
'97 thru '98 & Montana '99 thru '05
38040 Chevrolet Equinox '05 thru '17, GMC Terrain
'10 thru '17 & Pontiac Torrent '06 thru '09

GEO
Metro - see CHEVROLET Sprint (24075)
Prizm - '85 thru '92 see CHEVY (24060),
'93 thru '02 see TOYOTA Corolla (92036)
40030 Storm all models '90 thru '93
Tracker - see SUZUKI Samurai (90010)

(Continued on other side)

Haynes Automotive Manuals (continued)

NOTE: If you do not see a listing for your vehicle, please visit *haynes.com* for the latest product information and check out our **Online Manuals!**

GMC

	Acadia - see GENERAL MOTORS (38001)
	Pick-ups - see CHEVROLET (24027, 24068)
	Vans - see CHEVROLET (24081)

HONDA

42010	**Accord CVCC** all models '76 thru '83
42011	**Accord** all models '84 thru '89
42012	**Accord** all models '90 thru '93
42013	**Accord** all models '94 thru '97
42014	**Accord** all models '98 thru '02
42015	**Accord** '03 thru '12 & **Crosstour** '10 thru '14
42016	**Accord** '13 thru '17
42020	**Civic 1200** all models '73 thru '79
42021	**Civic 1300 & 1500 CVCC** '80 thru '83
42022	**Civic 1500 CVCC** all models '75 thru '79
42023	**Civic** all models '84 thru '91
42024	**Civic & del Sol** '92 thru '95
42025	**Civic** '96 thru '00, **CR-V** '97 thru '01 & **Acura Integra** '94 thru '00
42026	**Civic** '01 thru '11 & **CR-V** '02 thru '11
42027	**Civic** '12 thru '15 & **CR-V** '12 thru '16
42030	**Fit** '07 thru '13
42035	**Odyssey** all models '99 thru '10
	Passport - see ISUZU Rodeo (47017)
42037	**Honda Pilot** '03 thru '08, **Ridgeline** '06 thru '14 & **Acura MDX** '01 thru '07
42040	**Prelude CVCC** all models '79 thru '89

HYUNDAI

43010	**Elantra** all models '96 thru '19
43015	**Excel & Accent** all models '86 thru '13
43050	**Santa Fe** all models '01 thru '12
43055	**Sonata** all models '99 thru '14

INFINITI

	G35 '03 thru '08 - see NISSAN 350Z (72011)

ISUZU

	Hombre - see CHEVROLET S-10 (24071)
47017	**Rodeo** '91 thru '02, **Amigo** '89 thru '94 & '98 thru '02 & **Honda Passport** '95 thru '02
47020	**Trooper** '84 thru '91 & **Pick-up** '81 thru '93

JAGUAR

49010	**XJ6** all 6-cylinder models '68 thru '86
49011	**XJ6** all models '88 thru '94
49015	**XJ12 & XJS** all 12-cylinder models '72 thru '85

JEEP

50010	**Cherokee, Comanche & Wagoneer Limited** all models '84 thru '01
50011	**Cherokee** '14 thru '19
50020	**CJ** all models '49 thru '86
50025	**Grand Cherokee** all models '93 thru '04
50026	**Grand Cherokee** '05 thru '19 & **Dodge Durango** '11 thru '19
50029	**Grand Wagoneer & Pick-up** '72 thru '91 Grand Wagoneer '84 thru '91, Cherokee & Wagoneer '72 thru '83, Pick-up '72 thru '88
50030	**Wrangler** all models '87 thru '17
50035	**Liberty** '02 thru '12 & **Dodge Nitro** '07 thru '11
50050	**Patriot & Compass** '07 thru '17

KIA

54050	**Optima** '01 thru '10
54060	**Sedona** '02 thru '14
54070	**Sephia** '94 thru '01, **Spectra** '00 thru '09, **Sportage** '05 thru '20
54077	**Sorento** '03 thru '13

LEXUS

	ES 300/330 - see TOYOTA Camry (92007, 92008)
	ES 350 - see TOYOTA Camry (92009)
	RX 300/330/350 - see TOYOTA Highlander (92095)

LINCOLN

	MKX - see FORD (36014)
	Navigator - see FORD Pick-up (36059)
59010	**Rear-Wheel Drive Continental** '70 thru '87, **Mark Series** '70 thru '92 & **Town Car** '81 thru '10

MAZDA

61010	**GLC (rear-wheel drive)** '77 thru '83
61011	**GLC (front-wheel drive)** '81 thru '85
61012	**Mazda3** '04 thru '11
61015	**323 & Protegé** '90 thru '03
61016	**MX-5 Miata** '90 thru '14
61020	**MPV** all models '89 thru '98
	Navajo - see Ford Explorer (36024)
61030	**Pick-ups** '72 thru '93
	Pick-ups '94 thru '09 - see Ford Ranger (36071)
61035	**RX-7** all models '79 thru '85
61036	**RX-7** all models '86 thru '91
61040	**626 (rear-wheel drive)** all models '79 thru '82
61041	**626 & MX-6 (front-wheel drive)** '83 thru '92
61042	**626** '93 thru '01 & **MX-6/Ford Probe** '93 thru '02
61043	**Mazda6** '03 thru '13

MERCEDES-BENZ

63012	**123 Series Diesel** '76 thru '85
63015	**190 Series** 4-cylinder gas models '84 thru '88
63020	**230/250/280** 6-cylinder SOHC models '68 thru '72
63025	**280** 123 Series gas models '77 thru '81
63030	**350 & 450** all models '71 thru '80
63040	**C-Class:** C230/C240/C280/C320/C350 '01 thru '07

MERCURY

64200	**Villager & Nissan Quest** '93 thru '01
	All other titles, see FORD Listing.

MG

66010	**MGB** Roadster & GT Coupe '62 thru '80
66015	**MG Midget, Austin Healey Sprite** '58 thru '80

MINI

67020	**Mini** '02 thru '13

MITSUBISHI

68020	**Cordia, Tredia, Galant, Precis & Mirage** '83 thru '93
68030	**Eclipse, Eagle Talon & Plymouth Laser** '90 thru '94
68031	**Eclipse** '95 thru '05 & **Eagle Talon** '95 thru '98
68035	**Galant** '94 thru '12
68040	**Pick-up** '83 thru '96 & **Montero** '83 thru '93

NISSAN

72010	**300ZX** all models including Turbo '84 thru '89
72011	**350Z & Infiniti G35** all models '03 thru '08
72015	**Altima** all models '93 thru '06
72016	**Altima** '07 thru '12
72020	**Maxima** all models '85 thru '92
72021	**Maxima** all models '93 thru '08
72025	**Murano** '03 thru '14
72030	**Pick-ups** '80 thru '97 & **Pathfinder** '87 thru '95
72031	**Frontier** '98 thru '04, **Xterra** '00 thru '04, & **Pathfinder** '96 thru '04
72032	**Frontier & Xterra** '05 thru '14
72037	**Pathfinder** '05 thru '14
72040	**Pulsar** all models '83 thru '86
72042	**Rogue** all models '08 thru '20
72050	**Sentra** all models '82 thru '94
72051	**Sentra & 200SX** all models '95 thru '06
72060	**Stanza** all models '82 thru '90
72070	**Titan pick-ups** '04 thru '10, **Armada** '05 thru '10 & **Pathfinder Armada** '04
72080	**Versa** all models '07 thru '19

OLDSMOBILE

73015	**Cutlass** V6 & V8 gas models '74 thru '88
	For other OLDSMOBILE titles, see BUICK, CHEVROLET or GENERAL MOTORS listings.

PLYMOUTH

	For PLYMOUTH titles, see DODGE listing.

PONTIAC

79008	**Fiero** all models '84 thru '88
79018	**Firebird** V8 models except Turbo '70 thru '81
79019	**Firebird** all models '82 thru '92
79025	**G6** all models '05 thru '09
79040	**Mid-size Rear-wheel Drive** '70 thru '87
	Vibe '03 thru '10 - see TOYOTA Corolla (92037)
	For other PONTIAC titles, see BUICK, CHEVROLET or GENERAL MOTORS listings.

PORSCHE

80020	**911** Coupe & Targa models '65 thru '89
80025	**914** all 4-cylinder models '69 thru '76
80030	**924** all models including Turbo '76 thru '82
80035	**944** all models including Turbo '83 thru '89

RENAULT

	Alliance & Encore - see AMC (14025)

SAAB

84010	**900** all models including Turbo '79 thru '88

SATURN

87010	**Saturn** all S-series models '91 thru '02
	Saturn Ion '03 thru '07- see GM (38017)
	Saturn Outlook - see GM (38001)
87020	**Saturn L-series** all models '00 thru '04
87040	**Saturn VUE** '02 thru '09

SUBARU

89002	**1100, 1300, 1400 & 1600** '71 thru '79
89003	**1600 & 1800** 2WD & 4WD '80 thru '94
89080	**Impreza** '02 thru '11, **WRX** '02 thru '14, & **WRX STI** '04 thru '14
89100	**Legacy** all models '90 thru '99
89101	**Legacy & Forester** '00 thru '09
89102	**Legacy** '10 thru '16 & **Forester** '12 thru '16

SUZUKI

90010	**Samurai/Sidekick & Geo Tracker** '86 thru '01

TOYOTA

92005	**Camry** all models '83 thru '91
92006	**Camry** '92 thru '96 & **Avalon** '95 thru '96
92007	**Camry, Avalon, Solara, Lexus ES 300** '97 thru '01
92008	**Camry, Avalon, Lexus ES 300/330** '02 thru '06 & **Solara** '02 thru '08
92009	**Camry, Avalon & Lexus ES 350** '07 thru '17
92015	**Celica Rear-wheel Drive** '71 thru '85
92020	**Celica Front-wheel Drive** '86 thru '99
92025	**Celica Supra** all models '79 thru '92
92030	**Corolla** all models '75 thru '79
92032	**Corolla** all rear-wheel drive models '80 thru '87
92035	**Corolla** all front-wheel drive models '84 thru '92
92036	**Corolla & Geo/Chevrolet Prizm** '93 thru '02
92037	**Corolla** '03 thru '19, **Matrix** '03 thru '14, & **Pontiac Vibe** '03 thru '10
92040	**Corolla Tercel** all models '80 thru '82
92045	**Corona** all models '74 thru '82
92050	**Cressida** all models '78 thru '82
92055	**Land Cruiser** FJ40, 43, 45, 55 '68 thru '82
92056	**Land Cruiser** FJ60, 62, 80, FZJ80 '80 thru '96
92060	**Matrix** '03 thru '11 & **Pontiac Vibe** '03 thru '10
92065	**MR2** all models '85 thru '87
92070	**Pick-up** all models '69 thru '78
92075	**Pick-up** all models '79 thru '95
92076	**Tacoma** '95 thru '04, **4Runner** '96 thru '02 & **T100** '93 thru '08
92077	**Tacoma** all models '05 thru '18
92078	**Tundra** '00 thru '06 & **Sequoia** '01 thru '07
92079	**4Runner** all models '03 thru '09
92080	**Previa** all models '91 thru '95
92081	**Prius** all models '01 thru '12
92082	**RAV4** all models '96 thru '12
92085	**Tercel** all models '87 thru '94
92090	**Sienna** all models '98 thru '10
92095	**Highlander** '01 thru '19 & **Lexus RX330/330/350** '99 thru '19
92179	**Tundra** '07 thru '19 & **Sequoia** '08 thru '19

TRIUMPH

94007	**Spitfire** all models '62 thru '81
94010	**TR7** all models '75 thru '81

VW

96008	**Beetle & Karmann Ghia** '54 thru '79
96009	**New Beetle** '98 thru '10
96016	**Rabbit, Jetta, Scirocco & Pick-up** gas models '75 thru '92 & Convertible '80 thru '92
96017	**Golf, GTI & Jetta** '93 thru '98, **Cabrio** '95 thru '02
96018	**Golf, GTI, Jetta** '99 thru '05
96019	**Jetta, Rabbit, GLI, GTI & Golf** '05 thru '11
96020	**Rabbit, Jetta & Pick-up** diesel '77 thru '84
96021	**Jetta** '11 thru '18 & **Golf** '15 thru '19
96023	**Passat** '98 thru '05 & **Audi A4** '96 thru '01
96030	**Transporter 1600** all models '68 thru '79
96035	**Transporter 1700, 1800 & 2000** '72 thru '79
96040	**Type 3 1500 & 1600** all models '63 thru '73
96045	**Vanagon Air-Cooled** all models '80 thru '83

VOLVO

97010	**120, 130 Series & 1800 Sports** '61 thru '73
97015	**140 Series** all models '66 thru '74
97020	**240 Series** all models '76 thru '93
97040	**740 & 760 Series** all models '82 thru '88
97050	**850 Series** all models '93 thru '97

TECHBOOK MANUALS

10205	**Automotive Computer Codes**
10206	**OBD-II & Electronic Engine Management**
10210	**Automotive Emissions Control Manual**
10215	**Fuel Injection Manual** '78 thru '85
10225	**Holley Carburetor Manual**
10230	**Rochester Carburetor Manual**
10305	**Chevrolet Engine Overhaul Manual**
10320	**Ford Engine Overhaul Manual**
10330	**GM and Ford Diesel Engine Repair Manual**
10331	**Duramax Diesel Engines** '01 thru '19
10332	**Cummins Diesel Engine Performance Manual**
10333	**GM, Ford & Chrysler Engine Performance Manual**
10334	**GM Engine Performance Manual**
10340	**Small Engine Repair Manual**, 5 HP & Less
10341	**Small Engine Repair Manual**, 5.5 thru 20 HP
10345	**Suspension, Steering & Driveline Manual**
10355	**Ford Automatic Transmission Overhaul**
10360	**GM Automatic Transmission Overhaul**
10405	**Automotive Body Repair & Painting**
10410	**Automotive Brake Manual**
10411	**Automotive Anti-lock Brake (ABS) Systems**
10420	**Automotive Electrical Manual**
10425	**Automotive Heating & Air Conditioning**
10435	**Automotive Tools Manual**
10445	**Welding Manual**
10450	**ATV Basics**

Over a 100 Haynes motorcycle manuals also available

10/22